Neurofuzzy Adaptive
Modelling and Control

Prentice Hall International
Series in Systems and Control Engineering

M. J. Grimble, Series Editor

Neurofuzzy Adaptive Modelling and Control

Martin Brown
Chris Harris

Southampton University

Prentice Hall

New York   London   Toronto   Sydney   Tokyo   Singapore

First published 1994 by
Prentice Hall International (UK) Limited
Campus 400, Maylands Avenue
Hemel Hempstead
Hertfordshire, HP2 7EZ
A division of
Simon & Schuster International Group

Printed and bound in Great Britain by T.J. Press (Padstow) Ltd

Library of Congress Cataloging-in-Publication Data

Available from the publisher

British Library Cataloguing in Publication Data

A catalogue record for this book is available from
the British Library

ISBN 0-13-134453-6

1   2   3   4   5   98 97 96 95 94

To Audrey and Neville

Contents

Preface

Nothing will ever be attempted if all possible objections must be overcome[1]

The drive for autonomy and inherent in-built intelligence in manufacturing and manufactured goods, coupled with increased complexity and stringent performance requirements, necessitates more sophisticated control systems; both improved performance and flexibility. Frequently these so-called intelligent controllers have to cope with ill-defined, complex dynamics, characterised by spatial and temporal variations, yet be robust to minor faults and disturbances, and deal with nonlinear relationships over wide operating envelopes. Current developments in connectionist and linguistic based learning systems offer potential opportunities for radical innovation in control system design and management. As well as generating real advances in product quality features, it also most significantly provides the opportunity to optimally control complex, hierarchical multifacited processes in a coherent manner by integrating quantitative and qualitative aspects intelligently.

Intelligent Controllers are enhanced adaptive or self-organising controllers that automatically learn by interacting with their environment, with little *a priori* knowledge. Central to any control theory is the construction of a plant's model, and rather than being based on linear physio-mathematical models, intelligent controllers derive predictive models (at the most appropriate level of complexity) based on experiential evidence. The roots of intelligent control lie in the development of a range of anthropomorphic systems and techniques, each with varying degrees of cognitive capacity. In general, there have been two main (but apparently separate) parallel methodological developments relevant to Intelligent Control: artificial neural networks and fuzzy logic. Artificial Neural Networks (ANNs) were originally developed to emulate the human brain's neuronal-synaptic mechanisms that store, learn and retrieve information on a purely experiential basis, whereas fuzzy logic was developed to emulate human reasoning, using linguistic expressions.

Neither technique has been based on the more mathematically rigorous approaches of classical control and signal processing, and both generally lack a firm theoretical basis that establishes learning convergence and stability, and modelling

---

[1]Samuel Johnson.

and approximation capability; yet such results are essential if adaptive neurofuzzy algorithms are to be used in real world applications where safety, predictability and correctness are prime concerns.

Currently there are two schools of researchers: those who develop neurofuzzy algorithms that model the human brain's reasoning and learning capability, and those who derive neurofuzzy algorithms for specialised applications such as dynamic process modelling, pattern recognition and control. Building machines that emulate a human's capacity to reason and learn (their expertise) may be very flexible and robust, yet it is ultimately limited by our frailities (such as inconsistent reasoning, temporal instability, etc.). This book adopts the more specialised engineering approach that integrates the most relevant elements of neurofuzzy algorithms and signal processing for intelligent control to ensure guaranteed network learning, convergence stability, and conditioning, yet incorporates the more usual connectionist attributes of generalisation, abstraction, and the ability to include (and extract) knowledge about the plant in the form of a set of common sense linguistic production rules. While this currently limits us to the class of associative memory networks, it does offer the potential to interchange between neural networks (with their associated convergence and stability theory) and fuzzy algorithms (which provide a natural, transparent interpretation).

The class of associative memory networks developed in this book is based on a three layer architecture, in which the input layer provides a fixed nonlinear mapping of the sensor-based input space to a higher dimensional associative layer composed of compact support basis functions. The output of the hidden layer is linearly transformed by an adaptive weight vector (or equivalently by beliefs or confidences for a fuzzy network) and this *linear* dependence on a set of adjustable parameters means that convergence conditions can be established and the rate of convergence can be directly related to the *condition* of the network's basis functions. The size, shape and position of these basis function gives each network its own specific modelling attributes, since the learning rules are germane. Therefore it is important to study the network's internal representation as this provides information about how each network generalises (both interpolation and extrapolation), the rate of parameter convergence and the type of nonlinear functions that can be successfully modelled.

Three networks are described in detail: Albus' Cerebellar Model Articulation Controller (CMAC), the B-spline network and a class of fuzzy systems. A thorough description is provided of the Albus CMAC, which illustrates its desirable features for on-line, adaptive, nonlinear modelling and control: local learning and a computational cost that depends linearly on the dimension of the input space. The modelling capabilities of the algorithm are rigorously analysed and it is shown that they strongly depend on the generalisation parameter, and a set of *consistency equations* and *orthogonal functions* are derived which specify the functions that the network can and cannot model, respectively. The adaptive B-spline network, which embodies a piecewise polynomial representation, is also described and used for nonlinear modelling, and constructing static rule bases. Fuzzy systems are typically ill-defined, but the approach taken in this book is to use *algebraic* rather than truncation operators and smooth, continuous fuzzy sets; this makes it possible to analyse the fuzzy networks, determining their modelling abilities and proving convergence and stability results.

This book focuses on the learning, modelling and representational abilities of these networks by providing a common framework for their analysis. Their desirable features (local learning, linear adjustable weight set, fuzzy interpretation, etc.) are emphasised as meaningful comparisons between neural and fuzzy systems (and also between these "intelligent" techniques and conventional control algorithms), which can only be made when their modelling and learning abilities are thoroughly understood. Otherwise it is not possible to explain why one algorithm performs better than another and any simulation results remain problem specific. It is therefore necessary to investigate a network's modelling abilities separately from the learning rules, and it should be verified that the chosen learning rule is suitable for training a particular network. This has been the approach taken in this book for both the neural and the fuzzy systems, which allows the network specific properties to be emphasised. The fuzzy and neural systems are shown to be learning equivalent (the learning input-output behaviour of these networks are identical), and this immediately raises the question about the exact nature fuzziness. It has been claimed that fuzziness and local generalisation are synonymous. However it is our view that *only* the linguistic representation is fuzzy. Once a specific meaning has been assigned to the vague terms using context dependent information, the algorithm is no longer fuzzy; it is a deterministic nonlinear multivariate function. These localised algorithms have a fuzzy representation and a linguistic rule base can be implemented by such networks, although to claim that fuzziness and generalisation are equivalent merely confuses the issue.

Fuzzy representations of Associative Memory Networks (AMNs) can therefore be generated, and this provides a powerful tool for initialising and editing a trained network. Fuzzy linguistic variables provide a convenient method for selecting particular basis functions (rules) to verify and edit, and this is seen as their *primary* use[2]. Each of the AMNs has a *localised* internal representation and this is the main feature of these networks. Localised learning, fast initial convergence rates and fuzzy representations are all due to the network's local internal representation, and these are *all* desirable properties for on-line learning modelling and control applications. These networks can be used at various levels of abstraction in a control system hierarchy (e.g. at management/coordination, design, guidance and servo/actuator levels), although in this book we concentrate on guidance and servo control tasks.

There are many ways in which the research described in this book could be continued and three different areas have been identified. The first two research categories, network structure and learning theory (including evaluation tests) follow the philosophy of this book in analysing the modelling aspects separately from the

---

[2]They offer a sense of network transparency, using a natural knowledge representation scheme.

learning rules, whereas the third, software development, is only briefly considered here (see Appendix C). These future research areas include network structure, learning theory and software development, which are briefly introduced below.

NETWORK STRUCTURE

Some of the important theoretical contributions of this book are contained in the theorems which describe the modelling capabilities of the binary CMAC. Many pertinent results are established and the resulting consistency equations can be used to determine the type of relationships which must exist in the training data for the network to be able to reproduce a function exactly. Similarly a set of orthogonal equations is derived which illustrates the type of functions that the CMAC is unable to model in the least-squares sense. Both of these sets of equations are *necessary* and *sufficient* as they form a redundant bases for the types of functions which the CMAC can and cannot model. A lower bound for the modelling error is also described, and future work should investigate the possibilities of improving this measure and deriving an upper bound.

Practical applications generally require algorithms which can deal with high-dimensional, redundant input spaces, and although the basic AMNs are unsuitable for these modelling and control tasks, the algorithms can be extended so that the local representations can be formed in higher dimensional spaces. The basic networks can be used in a hierarchical structure, which allows the network input dependencies to be resolved at the appropriate level, and the overall architecture will be composed of a set of small networks. Another approach is simply to form a linear combination of lower dimensional submodels (see Section 8.5), which has the advantage that the training signal is *directly* available to all of the subnetworks. In either case the information encoded in the network's design simplifies the learning problem, as there are fewer parameters to update at each time instant. The alternative is to design new network architectures which use local representation in a non-lattice associative memory network. Considerable success has been achieved by combining elements of Kohonen's Self-Organising Memories with Radial Basis Function networks (see Section 3.4.3). These networks generalise efficiently in high-dimensional spaces and learn fairly quickly.

Another important topic which requires further investigation is that of network structure initialisation. Throughout this book it is assumed that the network structures are fixed (apart from the description of the ASMOD and ABBMOD algorithms in Section 8.5). For many practical applications this kind of *a priori* knowledge may be unavailable, and thus algorithms which automatically determine the number and positions of the the basis functions, the order of the basis functions, and the number of relevant inputs would be very desirable. These algorithms provide a partial answer to this problem, and these solutions could be extended to include more advanced tree building algorithms which incorporate information theory measures.

LEARNING THEORY

Future research into learning theory is divided into deriving new training algorithms and proving convergence and stability results for the instantaneous adaptive rules. This book views the neurofuzzy AMNs as simply deterministic, nonlinear functions, and although they were inspired by the information processing capabilities of humans, convergence and stability results must (and can) be established for this class of networks. A large body of classical linear adaptive modelling and control theory has been derived and this is *directly* applicable to the AMNs considered in this book. Convergence and stability results must be produced if these networks are to be used in adaptive real-world applications, and the ability to adapt on-line is one of the main motivations for the development of this class of networks. New theoretical results should be able to explain the different rates of convergence of the different networks, and this information could be used to select or design the most appropriate structure for a particular application. Similarly, stability results may provide important information about the maximum allowable modelling error and the size of the learning rate, and these bounds can also be incorporated into the network's design process. These stability theories have to be derived for specific control systems, such as the *indirect* adaptive fuzzy controller proposed in this book. Convergence results can be derived for the learning fuzzy plant model, and closed loop stability results can be proven via the small gain theorem and by Lyapunov's method.

The derivation of new learning rules which take account of the sparse network structure should also be a high priority (and the formulation of their fuzzy counterparts). The most promising directions for this research are to derive learning rules from alternative viewpoints, such as stability-based algorithms, or interpreting the weights to be the parameters of a feedback control system and designing learning laws which produce certain closed-loop behaviour. For on-line, instantaneous learning, research should be focused on higher order training rules, which use several recent input/output data pairs at each optimisation step. These minimise the disadvantages in using instantaneous learning rules but retain their computational simplicity, which is particularly significant for these neurofuzzy models.

It is only possible to assess the performance of a trained network once a suitable set of evaluation tests is available. The correlation and statistical based schemes described in Section 2.5 are valuable, and further research should be aimed at extending this work both to examine what information these tests provide and to develop new nonlinear model validation schemes.

SOFTWARE DEVELOPMENT

While this book concentrates on the theoretical algorithmic development of AMNs, a set of software network library routines is also described. The basic network structure is relatively simple and a set of interface routines has been developed which allows a novice to initialise and train an AMN. A high-level graphical user interface also allows the network to be trained, verified, and edited all within a single environment.

The importance of the development of such software libraries should not be underestimated. Many control packages are being used for simulating complex systems, and a software library which enabled these techniques to be evaluated and applied within such environments would be very valuable. The basic algorithms for these networks are fairly simple and it is not necessary (or desirable) to keep "reinventing the wheel". In order to take advantage of the particular structure of the AMNs, a set of editing and network visualisation tools should also be developed. The ability to label the multivariate basis functions using fuzzy linguistic variables provides a suitable high-level interface for editing the weights (or rule confidences). Similarly, the trained weight set gives an indication of the distribution of the training data, and which behaviours may need to be included by the designer. Such a facility proved invaluable when the B-spline docking rule base is constructed in Section 9.2.

Much of this work has already been completed by the authors for the standard B-spline and CMAC networks, although it should be extended to cope with more advanced network structures. The use of these more complex structures is necessary to cope with higher dimensional input spaces, and will probably require a distributed implementation. Therefore further work should investigate the parallelisation of these algorithms and their implementation in silicon.

ACKNOWLEDGEMENTS

Much of this book is the result of research into intelligent control carried out by members of the Advanced Systems Research Group at Southampton University, in the support of multi-institutional/industrial collaborative research projects in Intelligent Autonomous Systems. We would like to thank group colleagues who have contributed in a variety of ways: Dr Edgar An for collaborating on some of the theoretical work described in Chapters 5-7 as well as the software development, Dr David Mills for helping to write Chapter 10, Dr Chris Moore for generously providing the indirect fuzzy control example described in Chapter 11 and Dr Hong Wang, Dr Nick Bridgett and Alan Lawrence for many useful suggestions, providing figures and proofreading the manuscript. Lucas Aerospace, the CEC and British Aerospace have provided generous financial support and sponsorship for this research, and this is gratefully acknowledged.

Also, outside the Advanced Systems Research Group, we would like to thank Prof. Patrick Parks for many stimulating discussions and his permission, along with Dr-Ing J. Militzer and their publishers Elsevier Science Ltd, to publish the CMAC overlay displacement tables in Appendix B. We are also grateful to Dr Bob Sutton and the publishers Taylor and Francis (London) for permission to publish parts of previously published results. Finally, we would like to thank Adrienne Perry for vastly improving the book's grammar, but mostly for her patience, support and indulgence during its writing.

Martin Brown and Chris Harris
March, 1994

Well, the way I see it, logic is only a way of being ignorant by numbers.[1]

[1] T. Pratchett. 1992. *Small Gods*, Victor Gollancz Ltd, London.

1

An Introduction to Learning Modelling and Control

1.1 PRELIMINARIES

Increasingly, control systems are required to have high dynamical performance and robust behaviours, yet are expected to cope with more complex, uncertain and highly nonlinear dynamic processes. Along with this increased process complexity is increased abstraction and uncertainty in the models and their mathematical representation. One significant approach in dealing with major changes and uncertainty in nonlinear dynamical processes is through intelligent modelling and control. Intelligent controllers are generally self-organising or adaptive and are naturally able to cope with significant changes in the plant and its environment, while satisfying the control design requirements. As with any advanced control theory, a central issue is the representation and development of appropriate process models with known approximation errors. As processes increase in complexity, they become less amenable to direct mathematical modelling based on physical laws, since they may be:

- distributed, stochastic, nonlinear and time varying;
- subject to large unpredictable environmental disturbances; and
- have variables that are difficult to measure, have unknown causal relationships, or are too difficult or expensive to evaluate in real-time.

Yet human operators have few difficulties in dealing with the regulation of medium-to-high bandwidth, nonlinear, slowly time-varying dynamic processes, while simultaneously satisfying a complex and abstract set of static and dynamic constraints. While a human operator can cope with a large set of observations and frequently conflicting constraints subject to multiple subjective and objective performance criteria, the operator can be error prone, suffer from fatigue and non-repeatability, is subjective and can generate potentially dangerous situations in safety-critical systems. One aspect of intelligent control is to develop systems that incorporate the creative, abstract and adaptive attributes of a human, while minimising the undesirable aspects such as unpredictability, inconsistency, fatigue, subjectivity and temporal instability. Intelligent controllers should be able to perform under significant process uncertainties and incompleteness in the system and its environment, being both sufficiently robust or reconfigurable to cope automatically with system failures and sufficiently adaptive to cope with new goals or tasks or unanticipated situations.

Research into the understanding and representation of intelligence is not new and some of this considerable wealth of endeavour is relevant to the development of adaptive or self-organising algorithms which are applicable to real world problem domains, whose complexity prevents the application of algebraic model-based solutions. Research into intelligent systems assimilates and integrates concepts and methodologies from a range of disciplines including neurophysiology, artificial intelligence, optimisation and approximation theory, control theory and mathematics. This integration of research fields has led to an emergent discipline, frequently referred to as connectionism or neuroscience, that inherently incorporates distributed processing concepts organised in an intelligent manner. Connectionist or neuronial systems are, unlike conventional techniques, self-programming, appearing to be stochastic or fuzzy, heuristic and associative. These models are defined on multidimensional spaces and represent input-output mappings $f : \mathbf{x} \rightarrow \mathbf{y}$ which may be finite, infinite, discrete or continuous. An approximation to the desired mapping, f, is constructed in intelligent or learning systems either *directly* via a look-up table (that may incorporate *a priori* knowledge), specification through a functional transform, or *indirectly* by the presentation of training exemplars or from a set of inputs alone in a self-organising network. Learning systems frequently evaluate the mapping f from experiential evidence via an associative memory, although to generate a parsimonious representation that satisfies the physical constraints and conditions present in many practical processes, it is highly desirable to incorporate *all* of the available *a priori* knowledge within the learning process. Frequently, this additional knowledge is symbolic and heuristic, requiring a network representation that naturally incorporates symbolic and numeric data. Intelligence in a system refers to its ability to learn or adapt, and to modify its functional dependencies in response to new experiences or due to changes in the functional relationship. For intelligent control, it is necessary to construct networks which are able to respond appropriately to input signals not contained in the training data; this is termed *generalisation*. In particular, to minimise the network's sensitivity to input errors and to ensure that the stored behaviours are consistent, similar inputs should produce similar responses. This is called *local* generalisation and is a particularly significant property of the networks derived in this book.

General learning schemas incorporate three elements:

- an **architecture** or topology that defines the network's connectivity relationship and its representational and approximation abilities;
- **learning** or adaptation algorithms that adjust the strength of connectivity (belief or weights) and possibly the network's structure; and
- a **measure** of the network's performance.

The overall modelling abilities of a network do not generally depend on a particular learning rule, but are established theoretically by proving that an optimal system exists. Whether or not a learning rule can modify the network's connections such that it approximates this optimal system, and its ability to perform these operations in a reasonable time both depend on the network's architecture. It may be that a particular learning rule is inappropriate for training a network because either its convergence rate is too slow or it may stop at a locally suboptimal solution. In either case, this does not prevent the learning algorithm being successfully used to train other networks, or mean that the network is unable to be trained. This book focuses on learning systems for real-time adaptive modelling and control, where it is highly desirable to use networks which are *linear* in their adjustable connections, as this greatly simplifies the adaptation procedure.

Equally for intelligent modelling and control it is highly advantageous if the learning schemes generate local piecewise continuous nonlinear mappings so that the desired function can be represented over local regions by its local Jacobian. Therefore, for some operating points in the input space, the network's response can be a locally linear mapping which is characterised by the local Jacobian. The network's output is represented as a concatenation of overlapping local mappings, each of which allows local adaptation. Learning algorithms should only modify the network's response surface in a *local* manner, otherwise information which relates to totally dissimilar inputs could be overwritten. This feature can be encoded internally within the network's architecture by ensuring that information is stored locally: the local generalisation property.

These locally generalising networks are universal approximation schemes, as they can approximate any continuous nonlinear function to any desired degree of accuracy, although the local mappings are only very approximate. These concepts form the basis for the class of (lattice-based) Associative Memory Networks (AMNs) which are developed in this book. An AMN's output is formed from a weighted sum of local basis functions which are only defined over a small region of the input space, and examples of this class of networks include the Gaussian Radial Basis Function (RBF) network, the Cerebellar Model Articulation Controller (CMAC), the Basis (B)-spline network and a certain class of fuzzy logic network. For these single layer associative memory systems, learning or adaptation occurs through weight or belief adjustment rather than in structural changes. Many practical neural networks, such as Multi-Layer Perceptrons (MLPs), Functional Link Networks (FLNs) and AMNs, can be viewed as functional approximation algorithms, whereas fuzzy theory is founded in symbolic logic that utilises linguistic models stored as fuzzy production rule bases, with each rule firing in accordance with its confidence of being true. It is shown in this book that the rule confidence vectors (or belief vectors) associated with a class of fuzzy systems are equivalent to the weight set of a B-spline AMN, and similarly the rule confidences may adapt or learn as experience is acquired about the desired, partially known, functional mapping. Certain classes of neural networks and fuzzy algorithms share many common features; both have the same input space, are interpolation networks, parallel distributed associative processors, are robust with respect to both noise and disturbances through learning and both have simple physical realisations. While associative memory neural networks are easy to train, have known convergence and stability properties and are temporally stable, their internal representations are partially opaque. Conversely, the generated rule base within an adaptive fuzzy algorithm may provide some functional transparency through state interrelationships or dependencies within each rule and as such the network can be initialised and validated by the designer. When a fuzzy system utilises algebraic rather than the commonly used truncation operators together with the standard centre of gravity defuzzification algorithm, it is demonstrated in this book that there exists an invertible relationship between fuzzy algorithms and associative memory networks, with each inheriting the properties of the other.

1.2 INTELLIGENT CONTROL

Intelligent Control (IC) was originally proposed by Fu [1971], to increase the flexibility and extend the range of then current automatic control systems. The approach used techniques from the fields of artificial intelligence, operational research, and dynamical control to sense, reason, plan and act in an "intelligent" or "smart" manner [Saridis and Valavanis, 1988]. This list could now be augmented to include computer science, as advanced concepts are being employed to manage the overall system complexity and its command and control infrastructure and this relationship is illustrated in Figure 1.1. IC systems are not defined in terms of specific algorithms; they employ techniques which can sense and reason about their environment and execute commands in a *flexible* and *robust* manner [Antsaklis and Passino, 1992]. The technological drive for autonomy[1] in many complex systems has motivated research into various aspects of IC: system architectures, learning control, sensory processing and data fusion, world modelling, etc., [Albus, 1991], as shown in Figure 1.2.

Central to an overall IC systems design is the architecture which determines how complexity is managed and which modules are necessary for implementing the desired behaviour, as well as specifying the command and control infrastructure necessary to link and manage the distributed processes efficiently. Two distinct approaches have emerged in recent times, *hierarchical* and *subsumption* architectures [Brooks, 1986, Maximov and Meystel, 1992], respectively, although true system autonomy will probably only be achieved through the fusion of these two ideas [Albus *et al.*, 1990]. These two architectures are described in Section 1.2.1, as well as an approach which integrates these two philosophies and introduces various concepts that are necessary for IC system vendors to compete in a common marketplace.

The IC architectures allow the work described in this book to be put into context. This research has looked at the development and application of learning

---

[1]The ability to operate without human supervision.

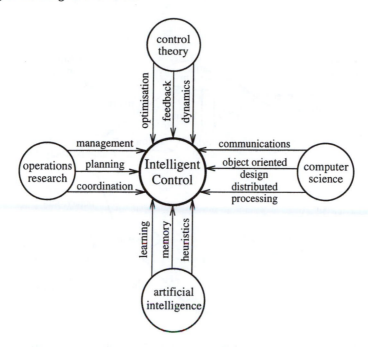

Figure 1.1 The techniques employed in intelligent controllers.

algorithms for Autonomously Guided Vehicles (AGV). AGVs are good examples of multi-level IC, since in current conventional highway systems humans control individual cars and traffic networks. Yet future systems, such as the European Prometheus programme and the USA Intelligent Vehicle Highway System (IVHS), will lead to completely automated, intelligent transportation. IC system architectures are used to decompose the complex functionality of AGVs into subtasks which have a *well-defined* interface with the remaining submodules. Specifically, the algorithms described in this book are aimed at the piloting level (low-level guidance and servo control), although it should be emphasised that these applications are merely used to demonstrate the basic concepts; they can be applied at higher levels of abstraction within the system architecture that require a greater degree of intelligence [Harris and Rayner, 1993].

1.2.1 Control System Architectures

The two important components in any system architecture are the hierarchical functional decomposition of the problem into *simpler* subtasks, and the command and control management infrastructure which sends messages both vertically and horizontally to neighbouring submodules.

Hierarchical control architectures can be used to implement systems which re-

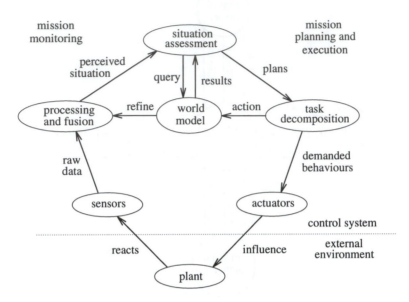

Figure 1.2   Various components necessary for intelligent behaviour, adapted from [Albus, 1991].

quire separate sensing, planning and execution phases, and to resolve complexity at various levels of understanding [Saridis, 1989]. These systems are designed according to the principle of *increasing intelligence with decreasing precision*. The highest levels in the control hierarchy use intelligent reasoning strategies to understand the processed data. As the information moves down the control structure, the information and processing algorithms become less intelligent and more precise. The exact algorithmic decomposition is problem dependent, although it generally consists of global reasoning routines, local planning algorithms and low level coordination and control techniques. Each module in the control hierarchy has just enough resources (access to data and functions) to perform its task, and knowledge is stored in a distributed manner throughout the system.

Subsumption control system architectures were originally proposed by Brooks [1986], and are based on entirely *reactive* behaviour rather than the traditional *sense, plan, action* cycle associated with hierarchical architectures. It is based on the principle that complex (and useful) behaviour can arise from the collective actions of many simple subtasks. Each processing unit implements a subtask such as *wander, explore* or *avoid objects*, etc., and each unit receives sensor signals directly and sends commands to the actuators. The overall behaviour is then a composite of the individual subtasks. The sensory processing algorithms are tightly coupled with the functions that send commands to the actuators and this is significantly different from hierarchical control systems where the sensory processing and control elements are separated. New behaviours can easily be introduced as the perception and navigation routines form part of the same task. However, this also means that the sensory processing may be redundant, with different modules performing the same data fusion techniques. Despite this minor misgiving, the basic idea performs extremely well [Brooks, 1990] and it is strongly suspected that a large part of human behaviour is entirely reactive.

Model Reference Architectures

Both of these architectures have desirable features. The hierarchical structure allows each module to be assigned a unique task and object oriented programming techniques can be used to implement such a structure efficiently, exploiting the inheritance and encapsulation features associated with this methodology. Subsumption systems use the sensory data to influence the commands issued to the actuators directly, and complex behaviours are observed from the combination of simple tasks. The combination of these two philosophies forms a very powerful system representation scheme; this is the approach of Albus who has proposed various model reference system architectures [Albus, 1991, 1992, Albus *et al.*, 1990]. As shown in Figure 1.3, the system is decomposedvertically into various levels of abstraction and reasoning, and horizontally into sensory processing, world modelling and task decomposition modules.

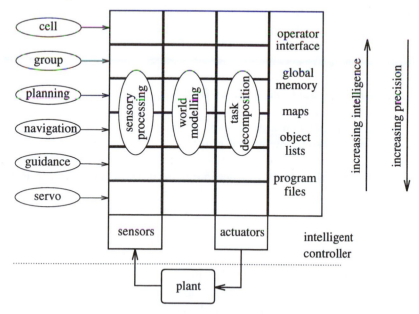

Figure 1.3 Model reference system architecture.

In order to increase both the flexibility of current autonomous systems and their rate of development, the system is designed to satisfy the following requirements:

- **extensibility**, both functional and temporal;
- a **flexible** human/computer interface;
- **real-time** operation;
- **distributed** systems which support graceful degradation; and
- **application-independent** development.

This architecture provides a convenient framework for the conceptual development of Intelligent Autonomous Systems (IAS) [Corfield *et al.*, 1991, Fraser *et al.*, 1991, Wright, 1991] and much of this book is directly applicable to this vital area of research. IASs must operate in hazardous, ill-defined, time-varying environments and complete their assigned tasks safely. They use algorithms which can learn from their interaction with the environment, resolve ambiguous and uncertain situations and operate in a fail-safe fashion. Much basic research still needs to be carried out, although a good collection of technical papers can be found in Cox and Wilfong [1990]. More recent survey articles in [Eng. App. of AI, 1991, IEEE Cont. Sys. Mag., 1993, IEEE Expert, 1991] and the proceedings of the first IFAC Intelligent Autonomous Vehicle conference [IFAC IAV, 1993] are indicative of the current state of research.

The learning algorithms necessary for modelling and controlling such a vehicle are extensively investigated in this book. While the IAS area has provided a demonstration platform, the theories, philosophy and methodologies described in this book are applicable to a wider range of complex engineering guidance, modelling, estimation and control problems.

1.3   LEARNING MODELLING AND CONTROL

The inflexibility of many industrial robots has been cited as one of the principal reasons for the reduction in the growth rate of robotic industrial applications [Albus, 1990]. It is anticipated that by introducing learning elements into the control systems, the plants will become more flexible and better able to deal with complex, real-world environments. Learning is an integral part of any IC system and exists at many levels of abstraction. At the guidance and servo levels, learning algorithms have been studied in the adaptive control field for many years [Åström and Wittenmark, 1989, Harris and Billings, 1981], and these have been complemented by research into adaptive neural and fuzzy networks [Harris *et al.*, 1993, Wang, 1994]. Higher in the IC architecture, learning can be as simple as internally updating its world model, or complex learning/exploring systems may be used [Sutton, 1990]. Each task requires an appropriate learning system, as the problem structures are generally different. Therefore learning systems should be aimed at particular problems and should always use the maximum amount of relevant *a priori* information.

1.3.1 The Need for Learning Control

The general benefits of learning have already been listed, although within the specific context of dynamic processes, learning may be required for the following reasons:

- **prior** knowledge about the plant's structure is unavailable or only partially known;
- **time-varying** plant;
- **partially known** or time-varying operational environment;
- **improve** the performance of the plant over a wide range of operating conditions;
- **increased** flexibility; and
- **reduced** design costs,

although it should always be remembered that other techniques such as gain scheduling can provide similar benefits if the problem is structured appropriately.

Learning control techniques generally use a basic model that is *inherently* non-linear. This is an important point, because it enables *global* plant models to be constructed, rather than the *locally* linear models used in adaptive control. Therefore there is no need for the continuous adaptation which is necessary to compensate for changing operating points once a satisfactory model has been learnt.

Learning and Adaptation

The benefits which are listed above can also be cited as reasons for employing adaptive modelling control schemes and therefore it is important to clarify exactly what is meant by learning and adaptation. Many attempts have been made to clarify this point and without wanting to fuel the debate even further, some personal interpretations are presented which justify the authors' decision to include the word "adaptive" rather than "learning" in the title of this book.

To adapt is to "to change a behaviour to conform to new circumstances" [Åström and Wittenmark, 1989], whereas learning generally implies a gaining or transfer of knowledge. Adaptive control techniques have been well developed over the past thirty years and many convergence (and stability) rules and theories have been developed for *linear* plants and under certain circumstances these can also be applied to various nonlinear processes as well. Learning algorithms are generally aimed at ill-defined processes and use heuristics, for instance to construct rule base systems.

One of the *main* results contained in this book is that fuzzy and certain neural systems are *equivalent* and can be trained using simple, linear rules which originally were derived in the adaptive control field. To emphasise this point, it was decided to call this book *Neurofuzzy Adaptive Modelling and Control* despite the fact that fuzzy networks learn linguistic *rules* which can be considered as a direct transfer of knowledge. This combination of simple adaptive rules which can be applied to nonlinear models coupled with a fuzzy linguistic initialisation/explanation facility generates a very powerful technique.

1.3.2 Learning Algorithms

Many different algorithms can be used within learning control schemes and their knowledge representation structures generally reflect the type of application. The three most popular types of learning algorithms are currently artificial neural networks [Miller *et al.*, 1990c, Widrow and Lehr, 1990], fuzzy logic [Harris *et al.*, 1993, Wang, 1994], and expert systems [Hunt, 1992, Quinlan, 1993]. These categories are not distinct (Figure 1.4) as there exist strong interrelationships between fuzzy and expert systems which incorporate uncertainty, as well as between fuzzy and some associative memory neural systems (see Chapter 3) and some expert systems can be implemented within a neural network architecture.

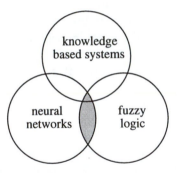

Figure 1.4   The area investigated in this book (shaded) combines techniques and concepts from fuzzy and neural algorithms and knowledge based systems.

Central to the problem of learning is the ability of the algorithm to *generalise* correctly from a limited number of training samples, which means that the algorithm must *interpolate* and locally *extrapolate* sufficiently accurately. Adaptively constructing linear models and controllers implicitly assumes a *global*, linear relationship. Whereas if a minimal amount of *a priori* information is assumed about the structure of the desired function, the algorithm has to extract the relevant knowledge from the data set and this generally requires a large number of training examples.

Most learning algorithms can be classified according to their *modelling, learning* and *validation* properties and this approach is taken in Chapter 2 where different neural networks are described and compared. The modelling capabilities of an algorithm determine the range of nonlinear functions which it can reproduce exactly and any implicit smoothness assumptions made by the network. The learning rule used does not generally affect the underlying modelling capabilities of the algorithm, although the chosen model structure influences its rate of convergence and can even determine the type of learning rule that should be used. This is also the approach used in Chapter 10, where the modelling capabilities of a set of fuzzy rules are analysed *separately* from the learning rules and it is shown that the proposed training laws are suitable for a network with that structure. Finally, any practical application of a learning algorithm requires convergence, stability and correctness tests which can verify what is being learned. If an algorithm learns, it also forgets and it must be verified that the behaviour being stored is desirable.

1.4 ARTIFICIAL NEURAL NETWORKS

Learning algorithms are required to operate in ill-defined and time-varying environments with a minimum amount of human intervention. These techniques are typically used to control plants for which a conventional mathematical analysis is not possible, and many different learning systems have been proposed for use within IC systems. Recently research interest has refocused on using biologically inspired learning algorithms and control architectures, and this has been part of a wider revival of research activity into Artificial Neural Networks (ANN). ANNs and the control community have a long history, which probably began with Wiener's book *Cybernetics* published in the late fourties [Wiener, 1948]. During the fifties and sixties, the adaptive control field grew and notable successes were achieved. The unification of several parameter estimation algorithms, coupled with the development of gradient and stability-based learning rules, provided a firm theoretical background for many of the practical applications [Åström and Wittenmark, 1989]. Several neural controllers were developed in the sixties, most notably Widrow's inverted pendulum controller in 1963 [Widrow, 1987]. More recently, the development of new network architectures and learning algorithms has provided the stimulus for control engineers to re-evaluate the potential of ANN-based controllers, or *neuro-controllers*. The new ANNs were proposed by researchers from different disciplines, such as computer science, psychology, etc., and many of the learning algorithms have their parallels in the adaptive identification and control fields.

ANN developments reflect the desire to develop alternatives to instruction-based von Neumann computing. Humans perform many tasks (speech and image recognition, complex coordination and control tasks) with relative ease that are very difficult to solve using traditional algorithmic computing techniques. The brain's architecture is vastly different from the commonly used serial computer, and neural researchers aim to endow machines with human-like information processing capabilities. Present neural-like computational architectures are based on a simplified model of the brain, with the processing tasks being *distributed* across many *simple* nodes (or neurons). The power of these algorithms comes from the collective behaviour of the simple nodes. Each ANN is completely specified (modelling and learning abilities) once the network topology, the transfer function of each node and the learning rule have been determined.

1.4.1 Neural Computing Benefits

Neural computing is *different* from conventional algorithmic computing, although the former can generally be decomposed into an algorithm and implemented on a serial machine. These apparently contradictory statements can be resolved if it is accepted that it is the *approach* which distinguishes the two techniques, rather than the final implementation. Neural networks offer solutions to problems that are very difficult to solve using traditional algorithmic decomposition techniques and the potential benefits of a neural approach are:

- **learning** from the interaction with the environment, rather than by explicit programming;
- **few** restrictions are placed on the type of functional relationship that can be learnt;
- **ability** to generalise (interpolate and extrapolate) the training information to similar situations; and
- **inherently** parallel design and the computational load can be evenly distributed across many simple processing elements. Thus the networks possess some degree of fault tolerance with respect to processor failures.

The first three properties are desirable for *any* learning algorithm, and the fourth can be used to apply these networks to larger real-time systems. If a learning algorithm possesses these properties, it can endow the control system with the following advantages [Stengel, 1992]:

- **decreasing** the required amount of human intervention;
- **increasing** the flexibility of the control system;
- **improving** the performance of the control system; and
- **reducing** the initial design time and cost.

The performance of an ANN (learning, recall, computational burden, etc.) depends on how well it satisfies the first property list, and this determines its potential for off-line design problems. For on-line adaptive modelling and control, the algorithm also needs to possess the following properties:

- **learn** significant information in a stable manner and in real-time; and
- **provable** learning convergence and stability properties.

Some of the most commonly used ANNs satisfy neither of these properties, although this means that while they can be used in neurocontrol applications, learning should only occur off-line.

1.4.2   Neural Network Terminology

The ANNs considered in this book all consist of a large number of simple processing elements called *nodes*. Signals are passed between nodes along weighted connections, where the *weights* are the network's adjustable parameters. The arrangement of the network's nodes and connections defines its architecture and there are many possible variations. One popular arrangement is shown in Figure 1.5 where the nodes are arranged into *layers* and each node in one layer has connections only

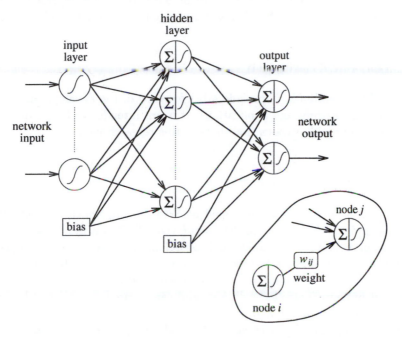

Figure 1.5   A feedforward multi-layer network where each circle corresponds to a node and each arrow represents a weighted link.

with nodes in the preceding layers. The input is presented to the first layer and this information is propagated *forwards* through the network, such that the output signal of each node never forms part of its own input. The signals at the final layer are then the network's output. *Feedback* or *recurrent* networks also have cyclic connections where the output of at least one node is propagated backwards and a modified version of this signal is used as the input to the same node. Recurrent networks have either complete or partial feedback paths and the dynamical behaviour of such networks can be extremely complex [Goles and Martinez, 1990].

Autoassociative networks attempt to reconstruct the true version of the corrupted input signal which is presented to the network. Typically this is achieved by calculating a first approximation to the desired input signal and using this as the network input at the next time instant [Hopfield, 1984]. *Heteroassociative* networks attempt to map the input to an alternative representation, and this is obviously the most general class of networks as classification problems and functional approximation tasks can be posed within a heteroassociative framework.

The learning rules used to train the networks can generally be classified as *supervised* or *unsupervised*. Supervised learning rules [Widrow and Lehr, 1990], require the desired network output to be available and they adapt the weights so that the output error is reduced. Unsupervised learning [Kohonen, 1990] is used to organise the network's structure based only on the training inputs presented to the network. This type of algorithm can be used to develop representative data set features, for vector quantisation and dimensionality reduction clustering.

1.4.3   Neurocontrol Applications

The aim of this section is to provide a brief review of the development of several neurocontrol algorithms and applications. It is not intended to be complete, as its main aim is to provide various snapshots of this research field.

Historical Perspective

The first neurocontroller was developed by Widrow and Smith in 1963 [Widrow, 1987]. A simple ADAptive LINear Element (ADALINE) was taught to reproduce a switching curve in order to stabilise and control an inverted pendulum. This ADALINE was one of the first ANNs (the Perceptron being the other [Rosenblatt, 1961]) and it has a simple architecture that has been used extensively in other ANNs. The output of neurocontroller is discrete and can be used to represent binary control actions (large negative and positive controls for instance). After Michie and Chambers [1968] and Barto *et al.* [1983] used this problem for evaluating their *reinforcement* learning control systems, the inverted pendulum problem became a standard benchtest for evaluating different learning and adaptive control schemes (see [Davison, 1990, Miller *et al.*, 1990c] where some other standard control problems are also described).

During the seventies, Albus proposed the CMAC as a tabular model of the functioning of the cerebellum and used it to control robotic manipulators. Since the early eighties, the CMAC has been used extensively to model and control highly nonlinear chemical processes [Tolle and Ersü, 1992], and in the mid-eighties it was again used in robotics applications [Miller *et al.*, 1990b]. The CMAC was originally derived from the Perceptron's architecture, but by considering the description of the ADALINE given by Widrow [1987], where a binary encoding of the input space is used, it could be considered as a modified ADALINE network which generates a pseudo-continuous output.

During the eighties, many different ANNs and IC architectures were proposed for integrating and extending these algorithms. Reinforcement learning and adaptive critic schemes have been extensively researched [Miller *et al.*, 1990c] and new ANNs such as the MLPs [Rumelhart and McClelland, 1986], RBFs [Chen and Billings, 1994], FLNs [Pao *et al.*, 1994] and B-splines [Moody, 1989] have been developed. Recurrent networks have been used in optimisation schemes [Hopfield and Tank, 1986] and for plant modelling and estimation [Williams, 1990], and some of these major developments are shown in Figure 1.6.

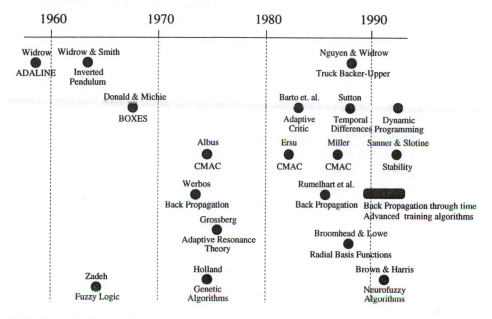

Figure 1.6   A historical perspective on neurocontrol algorithmic developments and applications.

Current Applications

ANNs have made a significant impact on the chemical industry, with applications in nonlinear process and human operator modelling, automatic plant knowledge elicitation, fault detection and monitoring, process control and optimisation and sensor validation, interpretation and fusion. For example, ANNs have been used as part of model predictive control [Saint-Donat *et al.*, 1994], for a Continuous Stirred Tank Reactor (CSTR) for pH control of sodium hydroxide, by minimising a quadratic cost function over a finite time horizon subject to constraints on the pH range. An ANN is adapted to form a dynamic model of the CSTR *off-line*, then in real-time with the ANN fixed (with variable bias), an optimiser plans a series of actions/controls over the planning horizon. The optimisation problem is one of nonlinear programming (via feasible sequential quadratic programming) that may be readily implemented on a special chip.

Fundamental to the majority of intelligent control schemes is plant identification or modelling. ANNs (just like conventional nonlinear time series methods such as NARMAX, extended Kalman filters, etc.) have been applied to dynamic process identification. In the chemical process industry, ANNs have been used in the identification of waste treatment plants, with multiple holding or settling tanks [Werbos *et al.*, 1992]. The plant is typically a 5 input, 9 output multivariable process, although it can be decomposed into three lower order sequential processing units (pretreatment, aeration and sedimentation) each represented by 3 to 3, 5 to 3 and 3 to 3 mappings. Excellent modelling results are obtained and the plant has been successfully controlled using a model predictive algorithm. Similarly, the oil refining industry, where catalytic reformers are used to rearrange hydrocarbons into higher octane aromatics for petroleum products, use ANNs as model predictors. The catalytic reforming process consists of reactors, heaters and refractionators. Feeds to the process are gasoline and naphtha whereas the product outputs are the reformate and hydrogen-rich gaseous and liquid streams. The process is a 5×2 mapping that is adequately represented by a third order nonlinear time series (typical of many adaptive control processes).

Neurofuzzy algorithms are equally finding widespread application in advanced transportation, including the European intelligent car programmes of Prometheus and Drive, and the USA IVHS programme, for collision avoidance, driver modelling, path and trajectory planning, car following, lane changing and parking. Similarly, for higher degree of freedom systems such as in the aerospace industry, neurofuzzy algorithms are being developed for full envelope nonlinear engine and flight control for advanced vehicles such as NASP X-29, avoiding the necessity of complex and expensive wind tunnel testing [White *et al.*, 1992]. Nonlinear flight control is essential for high performance, super-manoeuverable highly agile aircraft. The aircraft dynamics are highly nonlinear and multivariable due to the cross-coupling between lateral and longitudinal motion. A conventional approach to nonlinear flight (or engine) control is through nonlinear gain scheduling; indeed a neural network can achieve the same objective on-line, recognising changes in lift or load characteristics through the method of labels [Lawrence and Harris, 1992]. Conventional local linear behaviour is assumed, with local linear models being utilised to formulate the controller by classical methods, the selected operating points being based on non-dynamic aircraft data such as Mach number and air pressure. An extension of this [Wang *et al.*, 1994b] uses fixed structure ANNs, whose parameters are determined by the operating point(s). This approach allows conventional linear control design methods to be utilised within a nonlinear context. Future military and exo-atmospheric aircraft will utilise coupled mode controls, which together with enhanced operating envelopes generate highly nonlinear aircraft motions, requiring highly adaptive controls. Adaptive control laws also permit the possibility of compensation for normal aircraft deterioration, as well as dynamic reconfiguration (if there are sufficient spare degrees of freedom) to sudden faults, such as the loss of control surfaces. ANNs offer the capability of on-line system identification (and hence control) of system faults or damage.

For example, McDonnell Douglas have applied an adaptive critic ANN for evaluating the stability derivatives of an F15 aircraft, in which the induced lift and control surfaces can be evaluated to determine the aircraft's stability coefficients. Other researchers have applied ANNs to low-order stable lateral flight control with very good robust control properties. ANNs have also been benchmarked against H_∞ controllers for trajectory following for pitch rate and airspeed response for a high performance fighter aircraft. In all cases, the ANN controller exceeds existing controllers and equally significantly, McDonnell Douglas have implemented such controllers in ADA, executed in real-time and hosted in flightworthy avionics.

Neurocontrol Example

To complete this extremely brief overview of neurocontrol applications, it is worthwhile reviewing one of the most famous, which was developed in Widrow's laboratory: the truck reversing problem [Nguyen and Widrow, 1990]. The objective of this exercise is to develop a guidance level strategy which can guide the vehicle from an arbitrary initial condition to a desired vehicle state, as shown in Figure 1.7. One neural network is used to model the kinematic relationship of the truck and trailer, and a second produces the controlling steering signal which guides the vehicle safely towards the desired position. Training occurs after each manoeuvre is completed and the performance measure is based on the difference between the truck's final and desired position. This information is *back propagated* through the truck's kinematic model many times to provide the controller's training signal. Several thousand training iterations were required, although this application illustrates many of the benefits of a neural approach: the problem and solution are nonlinear and a linear approach is insufficient; the network is trained by example rather than by explicit programming and the controller is able to generalise to similar initial states.

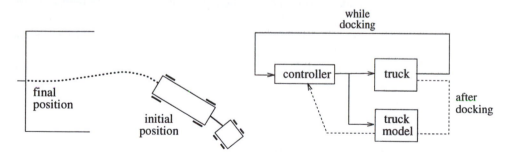

Figure 1.7 A truck reversing problem and the neurocontrol architecture used to learn an appropriate guidance strategy.

1.4.4   Further Reading

While this book provides a complete overview of the use of associative memory neural networks within IC applications, a considerable amount of research is also being performed in other research centres, and it is only due to space constraints that a more complete and balanced review cannot be given. Further information on the subject of ANNs and their control applications can be found in the publications listed below.

For an introduction to the general field of ANN research, the books by Beale and Jackson [1990], Hertz, Krogh and Palmer [1991], Pao [1989] and the introductory text [Kröse and Van der Smagt, 1993], all provide clear and concise descriptions of the most commonly used ANNs. Similarly, the two review articles by Lippmann [1987,1989], both provide excellent descriptions of the ANN within the context of pattern classification and the more recent article by Hush and Horne [1993], brings this research up to date. Also, the publication *Parallel Distributed Processing* [Rumelhart and McClelland, 1986], which provided the main stimulus to recent neural network research is still relevant and contains many useful ideas.

The IEEE Control Systems Magazine has devoted three special issues to the use of ANNs within control [1988,1990,1992]. The *Neural Networks for Control* book [Miller *et al.*, 1990c], and the more recent publications *The Handbook of Intelligent Control* edited by White and Sofge [1992], *Neural Networks for Signal Processing* [Kosko, 1992b] and *Advances in Intelligent Control* edited by Harris [1994], all provide excellent overviews of the current state of the art in the neurocontrol field. In Tolle and Ersü's book [1992], AMNs are evaluated against alternative learning algorithms and are applied to a wide range of difficult control problems. A concise survey of the neurocontrol field is provided in Hunt *et al.* [1992] and an interesting collection of papers which report on the application of ANNs to guidance and control tasks is given in AGARD [1991]. Many other neurocontrol articles can be found in the three major neural network journals: the INNS *Neural Networks* journal, the *IEEE Transactions on Neural Networks* and the MIT publication *Neural Computation* and also in most of the IEEE, INNS and NIPS neural network conferences.

Finally, any researcher investigating ANN learning theory should have a thorough understanding of the related work developed in the adaptive signal processing, filtering and control fields. This work can be found in three books by Åström and Wittenmark [1989], Goodwin and Sin [1984] and Haykin [1991].

1.5   FUZZY CONTROL SYSTEMS

Fuzzy logic was first introduced by Zadeh [1965, 1973], as a means for handling and processing vague, linguistic information. Zadeh reasoned that "conventional quantitative techniques of system analysis are intrinsically unsuited for dealing with humanistic systems", and formulated this in his *principle of incompatibility*:

> *As the complexity of a system increases, our ability to make precise and yet significant statements about its behaviour diminishes until a threshold is reached beyond which precision and significance (or relevance) become almost mutually exclusive characteristics.*

Fuzzy logic has been developed to provide *soft* information processing algorithms which can reason about and utilise *imprecise* data. It allows variables to be *partial* members of a particular set and uses generalisations of the conventional Boolean logical operators to manipulate this information. By allowing partial membership of a set, it is possible to represent the *smooth* transition from one rule to another as the input is varied smoothly, which is a very desirable property in modelling and control applications. In contrast, conventional expert systems reason using hard, *crisp* rules and are unable to represent a smooth input, output transformation.

In common with neural networks, the development of fuzzy logic and the associated fuzzy control applications have had a chequered history. The basic fuzzy theory and reasoning algorithms were developed in the late sixties, and the first control applications were investigated by Mamdani [1974]. During the seventies, Mamdani and his researchers also proposed the first self-organising fuzzy controller, [Procyk and Mamdani, 1979], although the potential of this algorithm was not fully realised in the UK. During the eighties, a small amount of research continued in the UK and the USA, but in Japan many fuzzy logic control applications were being developed. Products ranged from subway and helicopter controllers to autofocus camera mechanisms and washing machine controllers [Self, 1990], and it has also found a large product base in the automotive industry. Despite the many successful applications (one article reported a success rate of 80% for fuzzy systems compared with one of 2% for conventional rule-based systems [Schwartz, 1990]), there has been a notable lack of rigorous analysis associated with the development of the fuzzy systems, and this is possibly one reason for the lack of interest shown by the European control community during the eighties. These questions were largely ignored in the Far East where the research direction is very much applied. Whatever the merits of this approach (and the authors do believe that it is worthy of further investigation and exploitation), the European and American research effort has been belatedly increased [IEEE Fuzzy Systems, 1992, 1993].

1.5.1 Fuzzy Representations

Fuzzy information is generally represented on a computer by a set of fuzzy rules which provide relationships between vague quantities. These relationships are typically linguistic production rules of the form:

> IF   (*fuzzy logic applications are plentiful in Japan*) AND
> (*little fuzzy research takes place in the UK*)   (1.1)

THEN   (*research effort will be increased*)

Each of the *linguistic statements*, such as *fuzzy logic applications are plentiful in Japan*, is represented using fuzzy sets. A fuzzy set allows partial membership, so that it can represent the fact that fuzzy logic is popular in Japan with a degree of belief (membership) 0.75. A set of such rules forms a fuzzy *algorithm* or a rulebase which is used to store imprecise knowledge, for instance the above rule could form part of a larger fuzzy rule base which attempts to predict research funding within the UK control community. These fuzzy knowledge bases contain imprecise but *significant* information in the form of fuzzy rules. However, this imprecision is completely resolved once the fuzzy input and output sets and the knowledge manipulation routines have been defined.

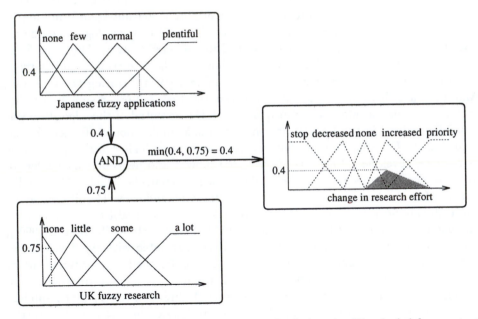

Figure 1.8   An illustration of the implementation of a fuzzy rule. The shaded fuzzy output set represents the contribution of this rule to the system output.

The fuzzy knowledge processing algorithm for the above rule is illustrated in Figure 1.8. Two precise inputs are presented to the fuzzy system: the number of applications and the current UK research funding. The degree of membership of each of the linguistic fuzzy sets is calculated (*fuzzification*), and this knowledge is combined to represent the degree of belief of the *antecedent* in the above production rule. The output of this fuzzy rule is the respective fuzzy output set which is scaled by a parameter which signifies the confidence in this rule being true, and the overall fuzzy output set is formed from the contributions of each fuzzy rule. A real-valued output is then obtained by *defuzzifying* this output set.

In practice, many fuzzy rules are used to generate the fuzzy output set and this distributed representation gives the network its ability to generalise locally (interpolate) between rules.

1.5.2 Fuzzy Control

The vast majority of fuzzy logic applications are in the modelling and control field (90% of successful applications in Japan are in control [Schwartz, 1990]). The original applications were aimed at ill-defined plants for which a complete mathematical analysis was not possible [Tong, 1977]. These included water and chemical reactor temperature regulation, pressure and mixture control, and all of these plants are highly nonlinear, noisy and subject to time delays. Not surprisingly the fuzzy controllers performed better than well-tuned PI and PID controllers. Most of these applications occurred in the UK, but during the eighties the commercial exploitation of this technique occurred in Japan [Sugeno, 1985a]. Sugeno's [1985b] fuzzy control review article in the mid-eighties cites applications for controlling cement kilns, diesel engines, traffic junctions and pump operations, amongst others. Probably the most famous fuzzy controller of this period is the subway train controller [Yasunobu and Miyamoto, 1985], which has been in constant use since 1987. The fuzzy system consists of two rule bases: the Constant Speed Controller (CSC) and the Train Automatic Stop Controller (TASC), and each rule base uses only 12 fuzzy rules which are evaluated every 100ms. A typical CSC system fuzzy control rule has the form:

IF (*the speed of the train exceeds the speed limit*)

THEN (*the maximum brake notch is selected*)

When compared with a PID controller, the fuzzy system requires less fuel, performs better and the overall ride is smoother.

More recently fuzzy rule sets have been used to control helicopters, amongst other things. The design procedure was similar to that used in the subway train controller: various control tasks were identified (hover, forward, up, etc.) and *small, separate* fuzzy controllers were designed for each subproblem. This *hierarchical* structuring of the fuzzy rule bases allows the techniques developed to be applied to high-dimensional control problems (15 in this case). The focus of many of the present day applications is self-organising or adaptive fuzzy systems. The ability to extract a set of rules which describe a complex plant or control mapping automatically, is a very desirable property for learning systems. Much of this work has been motivated by the neural network research and new algorithms are evolving which can combine the representational advantage of fuzzy systems with the learning power of neural networks (see [Kosko, 1992a, Wang and Mendel, 1992] and Chapters 10 and 11).

Adaptive fuzzy logic control is finding widespread applications in automotive products; AMNs have been used [Feldkamp and Puskorius, 1993] for on-line training of fuzzy logic controllers for car engine speed control; in this case the AMN acts as a local optimiser to initiate the fuzzy rule base. Similarly, the adaptive fuzzy logic controller has been used in the Mitsubishi Gallant 1993/1994 automotive transmission [Sakaguchi *et al.*, 1993], in which human driving characteristics are emulated by the fuzzy rule base. More interestingly, recent work on adaptive neurofuzzy control incorporates load, road and environmental conditions, as well as the *a priori* gradient curvature road conditions. The combination of active object (component) recognition and a fuzzy adaptive control system has been developed [Guo *et al.*, 1993] for polymer extrusion production for the manufacture of small aerospace tubular structures, where the control variables include temperature, torque and extrusion speed. While many of the early examples of fuzzy logic were in simple domestic products such as cameras, rice cookers and vacuum cleaners, the move is towards more complex on-line production and manufacturing processes. Two of the largest growth areas in Germany and Switzerland are the pharmaceutical and chemical industries; in North Rhine Westphalia state alone, some 88 companies are using fuzzy logic control in 265 products in 290 processes [Reusch, 1993].

Fuzzy logic controllers are also finding increasing use in aerospace with a 1/3 scale helicopter having been test flown (with a 135 fixed fuzzy set) in Japan, a 6 surface control of Rockwell's experimental advanced wing aircraft in the USA (a flexible wing structure with active leading and trailing edge control surfaces) and several attempts at fuzzy missile control, guidance and estimation. The industrial exploitation of *adaptive* neurofuzzy algorithms in safety critical areas such as advanced transportation demands provable conditions of learning convergence, process representation and network stability, and this book addresses all of these issues.

1.5.3   Further Reading

There are many introductory fuzzy papers and books being written and for further information which is outside the scope of this book, the reader should consult the following references. For a clear and concise introduction to the application of fuzzy logic controllers see Sutton and Towill [1985] where a static fuzzy rule base is used to control a simulated warship and Bernard [1988] for an application to reactor control. The two-part paper by Lee [1990] also provides an excellent source of reference. Three survey articles which describe the applications using fuzzy control in the seventies, eighties and nineties can be found in [Tong, 1977, Sugeno, 1985b, Berenji, 1992], respectively, which provide an interesting historical perspective on the development of the fuzzy control field.

An excellent collection of fuzzy papers can be found in the recent IEEE Transactions on Neural Networks special issue on fuzzy logic [1992] and in the edited book entitled *Advances in Intelligent Control* [Harris, 1994]. The first and second IEEE international conferences on fuzzy systems held in San Diego and San Francisco in March 1992 and 1993, respectively, have many interesting papers [IEEE Fuzzy Systems, 1992, 1993], and a wide range of fuzzy applications can be found in Zadeh and Yager [1992]. The authors' book [Harris *et al.*, 1993] provides an extensive treatise on self-organising fuzzy logic and is an excellent companion text, and other books [Kosko, 1992a, Pedrycz, 1993, Wang, 1994] also provide good introductions.

1.6 BOOK DESCRIPTION

This book was written so that it could be read in a modular fashion, so it is not necessary to read *every* preceding chapter. The exceptions to this rule are Chapters 3-5 which describe the AMNs within a common framework, and propose and investigate learning rules with are germane to this class of networks. The following chapters which describe the CMAC network, Chapters 6 and 7, the B-spline network and a constrained spline fitting algorithm, Chapters 8 and 9, and learning fuzzy systems, Chapters 10 and 11, can all be read independently of each other. Writing this book in this way has meant that there has been a small amount of repetition when certain aspects of the networks are described. It is hoped that this emphasises the similarities that exist between the different AMNs and that the reader does not find this too annoying. A range of possible reading orders is shown in Figure 1.9.

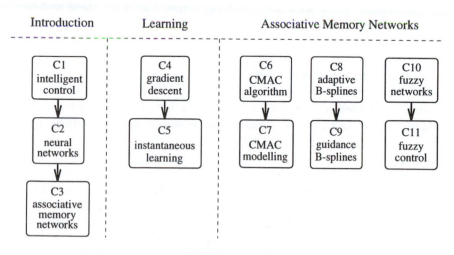

Figure 1.9 Possible reading order.

Chapter 2: Neural Networks for Modelling and Control

This chapter provides an introduction to the many different ANNs and their learning algorithms, and describes their application to IC problems. Initially, several different neuromodelling and control architectures are presented, and these systems are used to illustrate some of the desirable features which a general learning algorithm should possess: fast initial learning, guaranteed parameter convergence, local learning, network transparency, etc. Then several commonly used ANNs are presented: multi-layer perceptrons, functional link networks, radial basis functions and lattice associative memory networks. The modelling capabilities of these algorithms do *not* depend on the learning algorithm used to adapt their parameters and so the structure of these ANNs is described *separately* from the learning algorithms. Investigating the structure of these networks also enables their ability to locally generalise and extrapolate from redundant training data to be assessed and compared.

The learning algorithms play a critical role in determining the success of any application and so their importance should not be underestimated. This separation of the network's structure from its learning rule allows comparisons to be drawn with some of the system identification algorithms which have been developed and rigorously analysed. Many of these so called "neuronally inspired" learning rules are just particular applications of parameter identification algorithms and for IC systems the most efficient scheme (in terms of computational cost and parameter convergence) should always be used. Once trained, an ANN should be subject to an extensive verification stage, and the final part of this chapter describes several tests which have been developed over the past ten years for general nonlinear model evaluation. Such tests are extremely important when these systems are used in real-world applications.

Chapter 3: Associative Memory Networks

A major contribution of this book is the unification of several neural and fuzzy algorithms, and so this chapter is important because it introduces these networks within a common framework. The networks all belong to the class of AMNs, and they possess many desirable attributes for on-line, adaptive, nonlinear modelling and control applications. This description emphasises the similarities which exist between the different networks: local information storage and learning rules, three-layer network architectures, basis functions defined on a lattice, etc., as well as explicitly stating their differences. Each network has its own particular attributes: the computational cost of the CMAC is *linearly* dependent on the input space dimension, the B-spline network allows limited *a priori* knowledge to be incorporated within the network structure and automatic initialising algorithms can be developed, and fuzzy systems have a natural, linguistic, rule-based representation. This unified description illustrates how B-spline and CMAC networks can be represented as a set of fuzzy rules, and how learning and convergence results can be derived for the fuzzy networks.

Chapter 3 concludes with a discussion about the sort of learning tasks appropriate for AMNs and those which are inappropriate, as well as some possible techniques for applying these networks to high-dimensional modelling problems.

Chapter 4: Adaptive Linear Modelling

Of all the learning rules used within the neural network community, gradient descent adaptation is probably the most widely used technique. Gradient descent learning has been studied extensively as a general nonlinear optimisation technique, and this work is reviewed, using several AMNs to illustrate the major points. The rate of parameter convergence of the AMNs is shown to be directly related to the *condition* of the set of basis functions, and networks which are well-conditioned learn faster. It is also shown that network *singularity* is an important topic for AMNs, and this is illustrated by analysing the convergence of a binary CMAC which has an *infinite* number of global minima. It is important to determine the suitability of a network to be trained using a particular class of learning rules, and the work described in this chapter shows that gradient descent rules can be used to train AMNs *efficiently*.

These concepts can also be applied to MLPs in order to analyse why convergence is slow when they are trained using gradient descent rules (back propagation). The analysis assumes that for the composite learning rule to be well conditioned, each of the optimisation subproblems associated with each node must also be well conditioned. The two effects considered are that of the sigmoid's derivative on the strength of the error signal and the condition of the linear optimisation subproblem associated with each node. In both cases, the standard MLP is shown to be badly conditioned and several heuristics are described which can increase the rate of convergence.

Chapter 5: Instantaneous Learning Algorithms

Standard gradient descent rules are generally unsuitable for on-line, adaptive modelling and control algorithms because they require the complete training set to be known *a priori*. Instantaneous learning rules provide an alternative, as at each time instant the true derivative is replaced by an instantaneous estimate. This estimate depends only on the current training sample, and is suitable for performing optimisation in a *sparse* space. The weights are updated using the *principle of minimal disturbance*, the weight vector is changed by the minimal amount, such that the new information is stored. This chapter describes the basic theory behind these rules and provides new insights into how the rate of parameter convergence is related to the choice of the AMN's basis functions.

The loss in performance because of the mismatch between the true gradient descent algorithm and its instantaneous counterpart can be treated as noise which can be measured. This *gradient noise* affects both the rate of parameter convergence and the ability of the weight vector to settle down to a constant value when there exists model mismatch. The weight updates follow a "drunken path" into a domain from which they never emerge, and like true gradient descent, the amount of gradient noise depends on the condition of the basis functions. An indirect measure of gradient noise is also produced by considering the amount of *learning interference* for basis functions of different widths and shapes. Learning interference is defined as the unlearning at one point in the input space which is caused by a weight update at a neighbouring point; minimal learning interference is essential for networks which are to be used for on-line adaptive nonlinear modelling and control. This novel work provides a direct measure of the effect that different basis functions have on the network's learning ability.

Finally, a set of higher order learning algorithms are proposed and compared in an effort to overcome some of the disadvantages produced by taking instantaneous measures of the network's performance. A training store provides a library of data which can be used by more complex learning algorithms. The standard instantaneous training rules are computationally inexpensive and generally perform reasonably well, and these more advanced learning algorithms only increase the computational cost slightly, while overcoming the problems produced by correlated training data and modelling error.

Chapter 6: The CMAC Algorithm

Many of the AMNs have been considered as separate algorithms because they have been poorly described and their similarities have not been emphasised. One of the main results of the work contained in this book is that the basic algorithms *are* very similar, and in Chapter 6 a geometrical description of the CMAC algorithm is presented. This allows many of the original features (binary basis functions, overlay displacements, etc.) to be extended so that the flexibility of the basic model is increased. The basic algorithm is described as well as these improvements, and the network is applied to a nonlinear, multivariate modelling simulation. This prediction task forms a *standard* benchmark evaluation for the three networks described in the latter parts of this book (CMAC, B-splines and ASMOD), and it is used to illustrate how *a priori* knowledge can be incorporated into the B-spline network.

This CMAC description is used heavily in the next chapter, which analyses the modelling capabilities of the binary network.

Chapter 7: The Modelling Capabilities of the Binary CMAC

Chapter 7 provides a thorough and rigorous investigation of the modelling capabilities of the binary CMAC. Most of the learning algorithms considered in this book assume that the underlying network's structure is fixed, and so it is important to understand their modelling abilities. The work contained in this chapter derives a measure of the flexibility of the binary CMAC and examines its dependency on the generalisation parameter. A set of local *consistency equations* which the data must satisfy if they are to be stored exactly by the network are then derived. The equations allow the modelling abilities of various CMACs to be compared and show why the network produces locally additive models when it generalises. The set of consistency equations is then shown to be complete, in the sense that every look up which satisfies these relationships can be modelled exactly by the CMAC. Then a set of local orthogonal functions (mappings that the CMAC is unable to model in the least-squares sense) are examined. They represent the multiplicative terms that the network is unable to reproduce when it generalises and again these local relationships are shown to be complete, since any orthogonal function can be expressed as a linear combination of these local mappings. Both of these concepts (consistency equations and orthogonal functions) are then used to produce a simple and cheap lower bound for the network's modelling error. Many of the proofs in this chapter are quite involved and use the geometrical network description presented in the previous chapter extensively. This work gives a deep insight into the internal organisation of the CMAC, although it may be omitted during the first reading.

One of the main differences which exist between the AMNs is the modelling capabilities of each network. The CMAC is shown to provide an *efficient* coarse coding of the input lattice that allows local basis functions to be distributed *evenly* throughout the input space.

Chapter 8: Adaptive B-spline Networks

B-spline networks have traditionally been used for graphical applications, although their structure means that they can be considered as lattice AMNs where the weight vector is iteratively trained, using a variety of instantaneous learning rules. The B-spline network produces a piecewise polynomial representation using a well-conditioned set of local basis functions, although it is only recently that this network has been investigated in the context of learning modelling and control. This chapter provides a concise, self-contained introduction to the B-spline network and examines its modelling ability by applying it to the time series prediction problem used to evaluate the CMAC.

A set of network structuring rules are also examined and discussed. Most AMNs and fuzzy systems assume that the network's structure can be set by the designer before learning commences, as the optimisation process is then linear. However, for poorly understood problems or when the training set consists of many input variables (some of which may be redundant), it is unreasonable to expect the designer to be able to initialise a network appropriately, and an inappropriate structure produces a badly conditioned learning problem. This chapter describes two algorithms for automatically determining the B-splines's structure from a data set and compares them with other commonly used induction learning techniques. An example applies one method to the aforementioned time series prediction problem, illustrating the power of this approach.

Chapter 9: B-spline Guidance Algorithms

Two guidance level, B-spline applications are described in this chapter. Initially, a B-spline network is used to construct a static rule base which can reverse park a vehicle into a predefined slot. This application illustrates one of the main reasons for employing B-spline networks; the relationship which exists between a set of fuzzy production rules and the B-spline algorithm. A set of fuzzy production rules can be implemented using the B-spline basis functions and a B-spline network can be interpreted as a fuzzy algorithm, enabling a *linguistic* interpretation of the knowledge stored in the network to be produced. This aids the verification and validation of the rule base during the design cycle.

Next, a technique is introduced for producing constrained, desired vehicle trajectories using B-splines. A B-spline interpolant is used to fit a set of desired subgoals which specify the desired vehicle's state at specific times. The B-spline trajectory passes through these data points and provides smooth interpolation inbetween. It is also possible to constrain the desired trajectory so that it does not exceed any of the vehicle's kinematic constraints. Several algorithms have been proposed in the literature which iteratively modify the desired subgoals, such that the new desired path is physically realisable. However, this approach regards the subgoals as *soft* constraints, which should be achieved only if the *hard* kinematic constraints are not violated. The algorithm is not iterative, operates in real-time and is therefore suitable for real-world applications. The technique works well when the subgoals only violate the constraints slightly, and this assumes that the higher level task decomposition submodules have some knowledge of the vehicle's limits. This assumption is in keeping with the hierarchical control architectures which adopt the principle of increasing precision with decreasing intelligence.

Chapter 10: The Representation of Fuzzy Algorithms

The penultimate chapter analyses the properties of fuzzy systems and shows how they are also members of the class of AMNs. The modelling and learning capabilities of the fuzzy networks are investigated and a fuzzy implementation algorithm is proposed which results in a *smooth* network output. This new work provides a departure from many of the myths which currently surround this subject. This chapter investigates the fuzzy representations and the logical operators used to implement the algorithms, and it is demonstrated that *algebraic* operators generate smoother output surfaces than the more commonly used *truncation* operators. They are also consistent with the use of normalised B-spline basis functions which are proposed to represent the fuzzy linguistic variables. This approach can be used to show (trivially) that a fuzzy controller can reproduce exactly any linear controller, and to derive the relationship which exists between fuzzy systems and other AMNs. Any of the instantaneous learning rules proposed in Chapter 5 can be modified and used to train this class of fuzzy systems. This chapter also provides a thorough comparison between *continuous* and *discrete* fuzzy systems. The former is becoming increasingly popular due to the links that exist between neural and fuzzy networks, whereas fuzzy control systems have traditionally been implemented in a discrete manner.

The work addressed in this chapter is important, because it proves that there exists a *direct* link between certain fuzzy and neural systems. Therefore these systems should not be considered as separate techniques, but rather that the input/output behaviours are learning equivalent, and the networks naturally embody a *qualitative* and a *quantitative* approach to modelling and control problems.

Chapter 11: Adaptive Fuzzy Modelling and Control

This last chapter illustrates how these neurofuzzy algorithms can be applied in adaptive modelling and control schemes. To begin, several rule confidence training algorithms are described and compared, and it is shown that the adaptive fuzzy models are *learning equivalent* to the corresponding B-spline networks. However, it is argued that although the fuzzy systems can be trained directly, it is computationally more efficient to implement the fuzzy algorithms as B-spline or Gaussian RBF networks, which only produce a fuzzy rule base when explaining their learnt knowledge to a human. Several different adaptive modelling schemes are then described which can be used in *indirect* learning control systems. The majority of discrete, self-organising fuzzy controllers are *direct*; the controller is implemented using an adaptive fuzzy system and these rules are updated based on the performance of the plant. Indirect adaptive fuzzy controllers however, use an adaptive fuzzy plant model in a controller design procedure. This separation of the modelling from the controller design in the indirect approach often simplifies the theoretical analysis (convergence and stability) of the closed-loop performance, although a description is included of the more traditional indirect fuzzy control methodology and an example which illustrates its robust behaviour.

2

Neural Networks for Modelling and Control[1]

2.1 INTRODUCTION

In recent times there has been a tremendous resurgence of interest in using biologically based models and learning algorithms for adaptive modelling and control. Intelligent Control (IC) applications requires algorithms which are capable of:

- **operating** in an ill-defined, time-varying environment;
- **adapting** to changes in the plant's dynamics as well as the environmental effects;
- **learning** significant information in a stable manner; and
- **placing** few restrictions on the plant's dynamics

in order to operate autonomously in hazardous environments with the minimum amount of intervention. Human learning appears to embody elements of *all* of these properties, and currently researchers are trying to endow machines with such qualities.

The search for an algorithm which provides a universal panacea for *all* the different IC problems is a tempting but unrealistic pursuit. The algorithms that are described in this chapter and in the remainder of this book are at best *initial approximations* to a human's information processing systems (if indeed human reasoning and learning can ever be described using an algorithm) and the biological implications are *not* considered. The Artificial Neural Networks (ANNs) described in this book are useful because their modelling and learning abilities can be analysed, and this is in *direct* contrast to human behaviour. Lau and Widrow [1990] hypothesised that "it may take another 50 years before we have a solid, complete microscopic, intermediate, and macroscopic view of how the brain works. Engineers can't wait that long". Similarly, Hecht-Nielsen [1990] speculates that "the current level of understanding of brain and mind function is so primitive that it would be fair to say that not even one area of the brain or one type of mind function is yet understood at anything approaching a first-order level ... neurocomputing systems based upon these ideas probably have no close relationship whatsoever to

---

[1]An earlier version of this chapter appeared as Chapter 2 in Harris [1994].

the operation of the human brain". It is necessary to preface this introductory neural network chapter with such comments in order to emphasise that current neural network theories are far from providing a complete explanation of the operation of the human mind, and so current research can be divided into two (not necessarily distinct) categories:

- **devise** new theories about the brain's functionality;
- **application** to real-world problems.

There is constant cross-stimulation between these two research fields, but the second area should only use neural theories *if* they have something more to offer than a conventional, non-neural solution. In the past, neural algorithms have been applied to modelling and control problems without any consideration of their suitability and whether any other algorithm may be more appropriate (exactly the same comments can be made about fuzzy systems as well).

Learning algorithms have much to offer the control engineer. It is hoped that increased adaptation will result in improved system performance; increasing the quality of the solution and reducing the design and operational cost, although the current reality is far removed from this ideal. Generally the adaptive algorithms are based around *linear* plant and controller models, and a number of parameters must be chosen which determines the flexibility of the adaptation schemes. Neural networks provide *one* method with which these algorithms can be applied to nonlinear systems, although it is not the only approach and some of the "neural" learning algorithms seem primitive in comparison with the techniques developed in the adaptive control/signal processing fields. For instance, the majority of supervised learning algorithms are gradient based and it is only recently that adaptive strategies based on stability concepts have appeared [Sanner and Slotine, 1992, Tzirkel-Hancock and Fallside, 1991], mimicking the adaptive control field in the sixties, [Åström and Wittenmark, 1989, Narendra and Annaswamy, 1989].

The approach taken in this book is to evaluate the ANNs from an engineering viewpoint; the modelling capabilities are analysed *separately* from the learning algorithms. It is often claimed that the majority of ANNs are model-free estimators, but it is the authors' view that these comments are generally misinformed, as many of the neural models currently used have a *fixed* network structure and use (nonlinear) gradient descent rules to adapt the parameters. The network's structure may be very flexible due to the nonlinear adaptation, but it is *model based*, and these networks are termed *soft* or *weak* modelling schemes to distinguish them from conventional linear adaptive algorithms. Therefore it is important to examine the modelling capabilities of different ANNs, to determine what functional, representational and generalisation properties abilities they possess.

The supervised and unsupervised learning algorithms are then examined and it is shown that many of these adaptive rules can be applied to a wide range of different neural (and fuzzy, see Chapter 10) systems. The development of new learning algorithms can generally proceed *independently* of the model to which it is applied, after which the suitability of a learning algorithm for training a *particular* model should be assessed. For instance, in Section 4.5 it is argued that although nonlinear, gradient descent rules can be used to train a Multi-Layered Perceptron (MLP), the basic optimisation problem is badly conditioned and it may be more suitable to use optimisation strategies which incorporate second-order information about the performance surface [Gill *et al.*, 1981].

The final section of this chapter reviews several algorithms for assessing the trained ANN. This is a neglected topic in the neural literature, although it forms a critical part of any design methodology if these networks are to be applied in areas where *safety*, *correctness* and *certification* are prime concerns. The information provided by measuring the network's performance using the training set is described within the framework of the bias/variance dilemma, which states that the network's structure should be constrained such that it is unable to model any noise. Various forms of network testing are also proposed, based on constructing test sets, correlation-based modelling algorithms, chi-squared tests and network interrogation. The first three algorithms can be applied to nonlinear systems, and were first proposed for assessing nonlinear Volterra models, although they have been applied to ANNs as well [Billings and Chen, 1992, Billings and Voon, 1986]. The last point investigates the network's *transparency*: How easy is it to understand the knowledge stored in an ANN? Training sets rarely contain a complete description of the desired input/output relationship and once learning has ceased, it may be necessary to modify the stored information. This can only be performed if the knowledge is stored in a transparent fashion.

The class of ANNs studied in this book are Associative Memory Networks (AMNs) which are feedforward, supervised ANNs. These networks are universal approximation algorithms which can incorporate *a priori* knowledge in their structure, are suitable to be trained using instantaneous gradient descent algorithms and have a natural fuzzy interpretation, which makes the knowledge stored in the network transparent to the designer. This class of network is studied in detail from Chapter 3 onwards, although it is introduced in this chapter by considering the learning properties of an adaptive look-up table. The advantages in using these networks are that they allow conventional linear learning theory to be applied to a wide range of nonlinear modelling and control problems, and enable *a priori* functional knowledge to be incorporated into the network's structure.

2.2   NEUROMODELLING AND CONTROL ARCHITECTURES

Before the neuromodelling and control algorithms are described, it is useful to have an understanding of how the basic learning modules may be applied. These modelling and control architectures are generally network *independent*; most learning algorithms can be used, although some may be more suitable than others. The degree with which a particular learning algorithm satisfies these properties determines its suitability for on-line learning modelling control. It does not solely depend on the network's modelling capabilities or on the learning algorithm, but a combination of these two factors.

2.2.1 Representational Issues

Many neuromodelling and control algorithms are expressed in the continuous time domain, using measured variables which assess the state of the plant, the control signal and the desired plant's response, in order to predict the change in the plant's state (model) or to calculate the required change in control signal necessary to make the plant behave as required (controller). However the vast majority of neuromodelling and control applications are implemented as sampled systems, and the two sets of state equations for these two single input, single output, nonlinear systems are:

$$\dot{\mathbf{x}}(t) = \mathbf{f}(\mathbf{x}(t), u(t))$$
$$y(t) = g(\mathbf{x}(t))$$

where $\mathbf{x}(t)$ is the vector of plant states at time t, $u(t)$ is the current control signal and $y(t)$ is the observable plant output. The corresponding discrete time state equations are:

$$\mathbf{x}(t+1) = \mathbf{f}(\mathbf{x}(t), u(t))$$
$$y(t) = g(\mathbf{x}(t))$$

The majority of plant models assume that the output signal is sufficiently rich to contain information about all of the plant's states, so the above discrete time, state equation may be reformulated as:

$$y(t) = h(\mathbf{y}(t-1), \mathbf{u}(t)) \tag{2.1}$$

where $\mathbf{y}(t-1)$ is a vector of length n_y formed from the past outputs $y(t-1), \ldots, y(t-n_y)$ and $\mathbf{u}(t)$ is a vector of length n_u formed from the past and current control actions $u(t), \ldots, u(t-n_u)$.

Once a discrete or a continuous representation has been decided, the order of the plant (the delays n_y and n_u) must be determined. Over-estimating their values results in poor convergence rates and generalisation as the model is over-parameterised, although choosing too small a value means that unmodelled dynamics exist that may affect the stability of any learning control system. It has been claimed the adaptive neurofuzzy systems can be used when the order of the plant is underestimated, and although this can occur for certain controllers, neurofuzzy mappings are simply nonlinear functions and if the information which guarantees stability is not available in the input vector, the control loop can become unstable. A large body of theory has been developed for choosing these quantities when these are linear mappings, and in Section 8.5, several iterative construction algorithms which automatically determine which variables are important are presented.

NARMAX Representation

During the eighties, Billings and his collegues [1986, 1989, 1992] developed a general nonlinear modelling structure called Nonlinear AutoRegressive Moving Average models with eXogenous inputs (NARMAX), and a range of correlation and statistical tests that can assess the adequacy of the network. A NARMAX model is described by:

$$\mathbf{y}(t) = \mathbf{h}(\mathbf{y}(t-1), \ldots, \mathbf{y}(t-n_y), \mathbf{u}(t-1), \ldots, \mathbf{u}(t-n_u)) + \mathbf{e}(t) \qquad (2.2)$$

where $\mathbf{y}$, $\mathbf{u}$ and $\mathbf{e}$ are system output, input and additive disturbance vectors respectively. This is a very general relationship and many ANNs can be interpreted in this form. Volterra models and polynomials were first used for modelling nonlinear dynamic processes [Billings and Voon, 1986, Billings *et al.*, 1989], although more recently multi-layer perceptrons [Billings *et al.*, 1992] and radial basis function [Chen *et al.*, 1990] neural networks have been developed within the same modelling framework. The network performance measures can determine deficiencies in the input data as well as in the representation formed by the network (see Section 2.5.3), so that this type of theoretical framework is extremely useful for developing neuromodelling and control algorithms.

2.2.2 Modelling Strategies

There are four principal architectures which can exploit a learning modules modelling ability: as a basic plant model, an inverse plant model, a specialised inverse plant model and an operator model [Hunt *et al.*, 1992, Tolle and Ersü, 1992, Widrow and Stearns, 1985], as shown in Figure 2.1. For three of these four cases, the desired value of the network's output is *directly* available and any supervised learning rule can be used to train the weight set. The error in the specialised inverse plant modelling algorithm is formed at the *output* of the plant, whereas the network's output forms the *input* to the plant. Therefore some method is required for feeding back the plant output error, in order to train the inverse model and this is discussed later in this chapter and in Chapter 11. The success of all these schemes depends on the input signal being sufficiently exciting, in order to provide training data across the whole of the network's input space, the approximation capabilities of the network and the ability of the training rule to filter out the measurement and modelling error/noise.

Plant Modelling

A plant model may be required for a variety of reasons: to use within a larger feedback control loop which requires an estimate of the plant output, for predicting the performance of the plant when the true output is unavailable (due to time

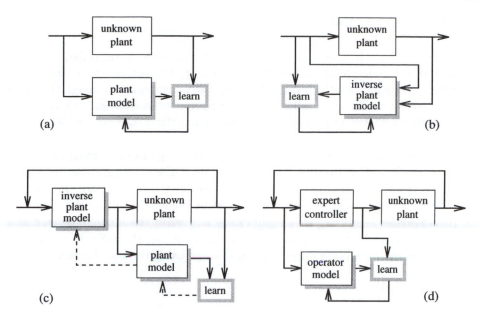

Figure 2.1 Four learning modelling architectures: (a) forward plant modelling, (b) inverse plant modelling, (c) specialised inverse plant modelling, (d) operator modelling.

delays, etc.), for fault diagnosis, or for use within an off-line controller design strategy. The basic architecture is very simple; the control signal and the plant states (suitably delayed) are sampled and this forms the network input vector. The output of the network is then calculated and compared with the measured plant output, and the difference between these two quantities is used to adjust the weight vector to reduce the network output error:

$$\epsilon_y(t) = \hat{y}(t) - y(t) \tag{2.3}$$

where $\hat{y}(t)$ is the measured plant output and $y(t)$ is the network output at time t. The learning rules are typically formulated so that they minimise the Mean Square output Error (MSE), and this cost function is given by:

$$J_y = E\left(\epsilon_y^2(t)\right) \tag{2.4}$$

where the expectation operator $E(.)$ is calculated using the network input probability density function. This is generally a discrete probability density function and the right-hand side of the above expression becomes a finite (averaged) sum of instantaneous squared output errors.

The MSE cost function gives an indication of how well the network is able to reproduce the desired function, but only if the training set *contains* a representative set of data points. When this does not occur, the model is biased towards the areas in the input space which have a greater density of training signals. Therefore care must be taken to ensure that the training data are evenly spaced in the relevant parts of the input space.

Direct Inverse Plant Modelling

The objective with inverse plant modelling is to formulate a controller, such that the overall controller/plant architecture has a *unity* transfer function. Inevitably, modelling errors perturb the transfer function away from unity, although the use of such an inverse model as a feedforward precompensator in addition to a standard, linear feedback controller generally provides good performance for a wide range of nonlinear plants.

For the inverse model to be well defined, the training examples must be unique. This is satisfied when the plant is invertible or if the training data for a non-invertible plant are contained in a restricted area of the input space, so that the plant is locally invertible. However, care must be taken when using this approach for plants whose Jacobian varies significantly and when the modelling error does not tend to zero. This is because the network minimises the MSE in the *control* space rather than the plant output space, through using the cost function:

$$J_u = E\left(\epsilon_u^2(t)\right) \tag{2.5}$$

where $\epsilon_u(t) = \hat{u}(t) - u(t)$, $u(t)$ is the control output of the network, and $\hat{u}(t)$ is the measured control signal. To a first approximation, the error in the control signal (for a single-input, single-output plant) is related to an error in the plant output by:

$$\epsilon_y(t) \approx \frac{dy(t)}{du(t)}\epsilon_u(t)$$

where $\frac{dy}{du}$ is the plant's derivative, or its *Jacobian*. Thus the two cost functions given in Expressions 2.4 and 2.5 are approximately related by:

$$J_y \approx E\left(\left(\frac{dy(t)}{du(t)}\right)^2 \epsilon_u^2(t)\right) \tag{2.6}$$

When the plant is nonlinear, the value of the Jacobian varies and different weightings are applied to the control errors. The cost functions are *not* equivalent, in the sense that one is simply a linearly scaled version of the other, and the designer should be aware that minimising one performance measure does not necessarily mean that the other is also minimised. If the output errors are uncorrelated with the plant Jacobian though, this expression simplifies to:

$$J_y \approx E\left(\left(\frac{dy(t)}{du(t)}\right)^2\right) J_u \tag{2.7}$$

and the effect of the Jacobian can be incorporated into the learning rate.

Specialised Inverse Plant Modelling

This approach again aims to provide an inverse model/plant structure which has a unity transfer function, although the method proposed is very different. A forward plant model is first constructed, and the difference between the plant's response and the desired output is used to provide an error signal, which is passed back through the forward plant model in order to adjust the inverse model's parameters [Jordan and Rumelhart, 1991, Psaltis *et al.*, 1988]. The main advantage which this approach has over the previous algorithm is that it is *goal driven*. For on-line applications, the plant output error causes the inverse model to move into previously unexplored regions of the input space, whereas the direct inverse modelling approach can only learn if the control signal is sufficiently exciting.

The learning algorithm attempts to minimise the *plant* output MSE, whereas the previous inverse modelling approach minimised the MSE *control* effort. These two quantities are approximately related by the Expressions 2.6 and 2.7, and as previously discussed if the plant is nonlinear and there exists a mismatch between the true inverse plant and the adaptive model, the direct and specialised inverse modelling approaches are *not* compatible as the optimal parameter values are generally different. As noted by Psaltis *et al.* [1987], "though there may be some benefit to performing generalised (direct) training prior to specialised training, these simulations show no clear advantage to doing so". An example is given in Brown and Harris [1993] which shows the different optimal weights, and the substantially different rates of convergence of these two modelling algorithms when the plant's gains are not close to unity and there exists modelling mismatch.

Although the inverse modelling architectures can be used to synthesise controllers, they may not be as robust as alternative learning controllers, due to the lack of feedback information [Hunt *et al.*, 1992]. This can be partially overcome by introducing on-line, inverse model adaptation, although the comments made in the previous paragraph should be taken into account.

Operator Modelling

Synthesising a controller by learning from an expert has many potential applications within the IC field [Kraiss and Küttelwesch, 1990, Shepanski and Macy, 1987, Widrow, 1987]. The learning algorithm is run in parallel with a skilled plant operator and their response forms the desired network output which is then used to train the network. This training signal typically contains a large amount of noise, due to the operator using different actions for similar inputs, and so this signal may have to be filtered [Guez and Selinsky, 1988] before the conventional network learning algorithms can be applied.

As with all modelling strategies, care must be taken to ensure that the training set contains sufficiently rich examples from the relevant operational domain, and that the network input vector contains all the information which is available to the operator. In Section 9.2, this approach is used to construct a set of fuzzy-type rules which can reverse a vehicle into a slot, and both of these points are illustrated. The supplied training data are very noisy and are distributed only in a small part of the input domain, therefore new rules had to be initialised to cope with different initial conditions. Also a new input variable had to be introduced in order to distinguish between similar situations which require very different actions. The human operator *implicitly* used this information when parking the vehicle, but it needs to be *explicitly* set for the network.

2.2.3 Supervised Control Architectures

Low-level learning algorithms need to be posed within specific modelling control and architectures, and some of the most popular are described in this section. One of the problems in formulating an on-line learning controller is that the desired control signal is rarely available, and generally only the desired plant output can be used to train the controller.

There are two distinct approaches which have been formulated in the adaptive control field: *direct* and *indirect* schemes [Åström and Wittenmark, 1989]. A direct adaptive control scheme builds an explicit model of the desired controller, whereas an indirect scheme produces a model of the plant and synthesises the control law, using a predefined optimisation/inversion calculation. For instance, the majority of self-organising fuzzy controllers have been direct, as a fuzzy rule base is used as a controller and there exists a performance index which relates errors in the plant's output to errors in the control signal in order to update the rule base. In contrast, most of the adaptive neurocontrol schemes have been indirect, as an explicit plant model is generally constructed, to be used in a predictive control algorithm for example.

Fixed Stabilising Controllers

One of the simplest *direct* learning control schemes is shown in Figure 2.2, where a fixed, stabilising, linear, feedback controller is used to train a learning network [Kraft and Campagna, 1990, Miller, 1987, Miyamoto *et al.*, 1988]. The linear controller is designed so that the closed-loop system is stable in every operating region and the control signal provides a training signal for the learning module. The performance of the closed-loop system depends on the current operating point, although the iterative training of the network gradually improves its performance (and the performance of the control signal) *on-line*. As the operating point changes, the learning controller builds up a nonlinear model of the desired control surface, such that when the plant returns to the original operating point, the learnt response has not been forgotten and it can be improved upon. This requires a learning module which is *temporally stable*; learning about one area in the input space affects

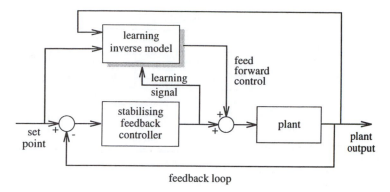

feedback loop

Figure 2.2   A direct learning controller; the fixed linear feedback controller is used to stabilise the system and to provide a training signal for the learning controller.

the knowledge stored in a different region minimally.

Despite the algorithm's simplicity, this approach has one main drawback; the design of the fixed linear controller. It has been claimed that the algorithm is robust with respect to the design of this controller, although the rate of convergence of the learning module depends on the quality of the training signal. A learning module is slow to adapt when the linear controller is performing poorly.

Predictive Learning Control Schemes

Indirect predictive learning control schemes attempt to formulate a control strategy by assessing the effect of their actions for many time-steps into the future and selecting the current "optimal" control action, which is then applied to the plant [Montague *et al.*, 1991, Saint-Donat *et al.*, 1994, Tolle and Ersü, 1992]. The architecture requires the development of a plant model, a performance function to evaluate the effect of a control action and an optimisation technique which can determine the best control action. This is illustrated in Figure 2.3, where a learning control element has also been included, so that after sufficient training, the full optimisation calculation does not need to be calculated and the computing resources can be allocated to other tasks. If the plant is time-varying, the model is generally adaptive, although the initial optimisation calculations may give very poor closed loop control if the process model is poor.

When the plant model is good and the performance function and the search strategy are appropriately chosen, this control scheme can provide excellent closed-loop control. However, the multi-step ahead optimisation calculation is generally very expensive and is only applied to systems which are not time-critical. Many simplifications of the above architecture can be performed which makes this technique more suitable for real-time control tasks, some of which are described in Tolle and Ersü [1992].

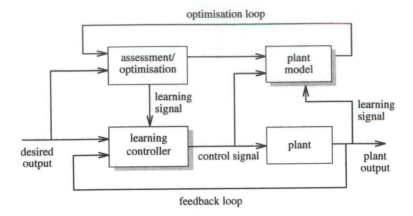

Figure 2.3 A learning predictive control architecture.

Model Reference Adaptive Control

The model reference learning control architecture has been widely used in the linear adaptive control field [Åström and Wittenmark, 1989], and is shown in Figure 2.4. The control objective is to adjust the control signal in a *stable* manner so that the

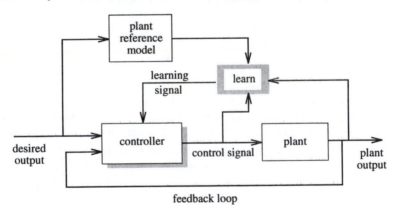

Figure 2.4 A model reference control architecture.

plant's output, $y^p(t)$, asymptotically tracks the reference model's output, $y^r(t)$, i.e.:

$$\lim_{t \to \infty} \|y^r(t) - y^p(t)\| \leq \varepsilon$$

where ε is a small positive constant [Narendra and Parthasarathy, 1990]. The performance of this algorithm depends on the choice of a *suitable* reference model and the derivation of an appropriate *learning* mechanism. Researchers in the sixties found that simple gradient-based learning rules were sometimes insufficient and there is no reason why this should not also be the case for more general nonlinear plant models and controllers.

Internal Model Control

Internal model control [Hunt and Sbarbaro-Hofer, 1991, Hunt *et al.*, 1992] uses a similar structure to the predictive learning control scheme, as shown in Figure 2.5. A (learning) module is used to model the process directly, receiving the applied control signal, rather than the reference signal which is used in the model reference adaptive control scheme. The error between the model and the measured plant output is used as a feedback signal and this is passed to the controller. The internal model controller is generally designed to be an inverse plant model (when it exists), and either of the inverse modelling schemes described in the previous section can be used to synthesise an appropriate controller.

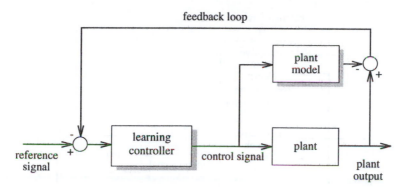

Figure 2.5 An internal model control architecture.

Many theoretical stability results about internal model control loops are available, [Hunt and Sbarbaro-Hofer, 1992, Sbarbaro-Hofer *et al.*, 1993], although they generally make assumptions the open-loop stability of the system, exact modelling and/or inverse modelling. Despite these assumptions, it is claimed that this approach extends readily to nonlinear systems and yields to robustness and stability analysis.

2.2.4 Reinforcement Learning Systems

Reinforcement learning schemes and ANNs have been very closely linked since the seminal paper by Barto *et al.* [1983]. Reinforcement control schemes are *minimally* supervised learning algorithms; the only information that is made available is whether or not a particular set of control actions has been successful. The original application attempted to balance an inverted pendulum, subject to the constraints that the platform should not move more that a certain distance from its starting point and that the inverted pendulum remained approximately upright. If either of these constraints were violated, a failure signal was sent to the learning algorithms.

From this definition it is clear that once the controller has managed to balance the inverted pendulum fairly well, very little training takes place as failures occur infrequently. The solution proposed by Barto *et al.* [1983] was to construct a control scheme which is composed of two adaptive elements; an Associative Search Element (ASE) and an Adaptive Critic Element (ACE). The ASE attempts to reproduce the optimal control signal which satisfies the given performance objectives, while the ACE attempts to monitor the performance of the controller *internally* and to provide an internal reinforcement signal which is used to train the ASE, as illustrated in Figure 2.6. The ACE is trained using the external failure/success signal. This continuous internal training of the control element has been shown to improve vastly the performance of the overall system.

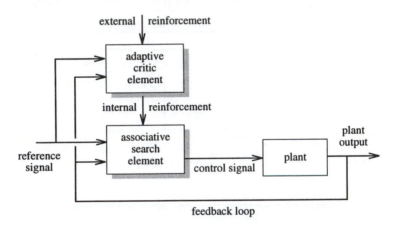

Figure 2.6 An ASE/ACE reinforcement system's architecture.

Over the past ten years, there has been a greater theoretical understanding of the overall system [Barto *et al.*, 1993, Sutton *et al.*, 1992], as well as a growing number of simulations and applications that use modified versions of this technique [Berenji and Khedkar, 1992, Millington and Baker, 1990, Porcino and Collins, 1990, Shelton and Peterson, 1992].

2.2.5   Parameterising Linear Controllers

Many different neuromodelling and control architectures have been proposed in recent years, and the previous sections have described several which are related to conventional control schemes. The novel parameter initialising neurocontrol architectures which are now described have all been developed as a result of the recent resurgence of interest in ANNs, as they attempt to exploit the ability of these adaptive systems to learn an arbitrary functional relationship. There are many reasons for utilising an intelligent *gain-scheduling* type approach: widespread industrial acceptance of linear feedback controllers, many theoretical and practical results are available about robustness and closed-loop stability and their low implementation cost. The neurocontrol schemes in this section attempt to exploit these properties and to produce algorithms which can calculate the parameters for both off-line and on-line control. Successful systems would result in reduced commissioning costs and possess the ability to adapt to time-varying process dynamics.

All of these approaches assume some previous knowledge about the plant's structure, as this simplifies the problem. In Kumar and Guez [1991], an indirect control design architecture is adopted. The plant is assumed to be a slowly varying second-order linear system and a set of *features* which describe the closed-loop response of the plant are extracted. These features could include the delay time, rise time, peak overshoot, settling time, etc., and are passed to an Adaptive Resonance Theory (ART) based classifier which predicts the parameters of the plant. This output, together with a set of desired closed-loop response characteristics, is used to produce a set of linear control gains using conventional pole placement design techniques.

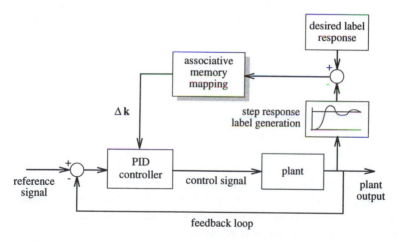

Figure 2.7 An architecture for predicting the required change in a PID controller directly based on a set of labels that describe the system's closed-loop response.

The idea of extracting the feature labels which describe the closed-loop system's response has also been independently proposed in Lawrence and Harris [1992]. The desired closed-loop response is expressed as a set of feature labels, which are then compared with the labels describing the system's closed-loop step response. This error label vector is passed to the ANN which predicts a change in the PID parameters so that the system's response will be closer to the desired one, as shown in Figure 2.7. A similar approach has also been proposed in Ruano *et al.* [1992], although this algorithm can use the step responses of both open- and closed-loop systems, and the ANN outputs the PID parameters which are optimal with respect to the Integral of Time multiplied by the Absolute Error (ITAE).

The ability to synthesise a set of PID parameters on-line, using only input/output data, has been investigated for many years [Åström and Hägglund, 1988]. The potential payback from such a system which can increase the robustness of PID controllers is large, and this research area is still in its infancy.

2.3 NEURAL NETWORK STRUCTURE

In the introduction, it was emphasised that the majority of ANNs are model based, and the structure of several such models are now described, compared with each other and contrasted with some truly *model-free* estimators. The learning capabilities of the networks are discussed in Section 2.4, as this is an important topic but is *separate* from the model description. Too often in the past Multi-Layer Perceptrons (MLPs) have been criticised for being slow to converge, when what is really meant is that MLPs *trained* using gradient descent algorithms learn slowly. The model structure *influences* the selection of the training rule, although the learning algorithm does not generally affect the flexibility of the underlying model. To begin this section, the adaptive linear combiner which forms part of most ANNs is described and its properties are discussed.

2.3.1 Linear Combiners

The Adaptive Linear Combiner (ALC) formed part of the two earliest ANNs: the ADALINE [Widrow and Lehr, 1990] and the Perceptron [Rosenblatt, 1961], and it is still used in the many of the present neural networks. The ALC simply forms a weighted sum of the inputs, and this quantity can be thresholded to produce a binary output if the network is used for pattern classification tasks. Let the p-dimensional network input vector be denoted by $\mathbf{x}$ and the weight vector by $\mathbf{w}$, the continuous ALC's output y is given by:

$$y = \sum_{i=1}^{p} a_i w_i = \mathbf{a}^T \mathbf{w} \qquad (2.8)$$

where $\mathbf{a} \equiv \mathbf{x}$, and the ALC is shown in Figure 2.8. Generally an augmented term a_0 is set equal to 1 and is known as the *bias*. The thresholded network output is:

$$y = \begin{cases} 1 & \text{if } \mathbf{a}^T \mathbf{w} > 0 \\ 0 & \text{otherwise} \end{cases} \qquad (2.9)$$

The Perceptron also has an additional layer where the network's input is preprocessed ($\mathbf{x} \rightarrow \mathbf{a}$), and a weighted combination of these modified inputs is taken.

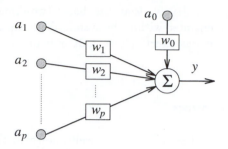

Figure 2.8 A basic adaptive linear combiner.

There are many ways in which this initial nonlinear transformation can be chosen and some of these are discussed in Chapter 3.

The weight vector, **w**, is adjusted using an error correction learning rule and training ceases when the network's overall behaviour is acceptable. The network only achieves the correct result when the set of (modified) training input examples $\{\mathbf{a}(t)\}_{t=1}^{L}$ is *linearly separable*, and in this case the learning algorithm terminates in a *finite* number of iterations. A set of training examples is linearly separable if there exists a $(p-1)$-dimensional hyperplane which can separate the training inputs in the p-dimensional input space. This hyperplane is formed from the set of weights which satisfy:

$$0 = \mathbf{a}^T \mathbf{w}$$

as this determines the classification boundary for the ALC. Figure 2.9 shows a linearly separable training set and illustrates the fact that, if there exists a single weight vector which can linearly separate the training inputs, there exists an infinite number of such hyperplanes.

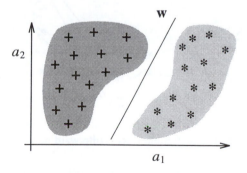

Figure 2.9 A linearly separable training set for a two example classification problem.

Most of the two-dimensional logical functions, AND, OR, NOT can be implemented in the above ALC framework. However, it was shown in Minsky and

Papert's book *Perceptrons* [1969], that the basic two-dimensional XOR logical function cannot be implemented within the standard Perceptron architecture. The result of this publication was effectively to halt much of the research into ANNs during the seventies.

2.3.2   Multi-Layer Perceptrons

Despite the limitations of the ALC, it was well known in the sixties that multi-layered networks could implement exactly the XOR and higher order logical functions, although there appeared to be no natural generalisation of the Perceptron's training algorithm and these results were of theoretical interest only.

A multi-layer ANN is a *feedforward* network where the input signal is propagated forwards through several processing layers before the network output is calculated. Each layer is composed of a number of *nodes*, and each node is (generally) composed of a simple ALC, with an appropriate transfer function which calculates the node's output from the weighted input signal. Each node has input connections with the nodes in the previous layer *only*, and the node's output is transmitted to the nodes in the next layer, as shown in Figure 2.10. Every node has an associated weight vector which *linearly* transforms its input vector.

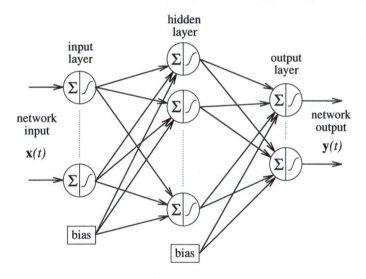

Figure 2.10 A multi-layer feedforward network.

During the recent ANN revival, a number of researchers independently derived a gradient descent algorithm suitable for training these Multi-Layer Perceptron (MLP) networks, [Le Cun, 1985, Parker, 1985, Rumelhart and McClelland, 1986, Werbos, 1974]. The transfer function in each node of these networks is a bounded, *continuously* increasing nonlinearity, rather than a binary threshold. Thus the network output is a continuous (continuously differentiable) function of every weight in the network, enabling it to be trained using gradient descent rules. The availability of such learning algorithms popularised the MLP, and at the time of writing it is probably the most widely used ANN. The model structure does not depend on the learning rule, although the rate of convergence of the learning algorithm depends on the model structure, and the quality of the final model also depends on the learning rule. For the remainder of this section, the MLP structure is discussed *without* reference to a particular the learning algorithm.

XOR solved by Multi-Layer Networks

The two-input XOR problem can be solved exactly by a three-layer MLP (one input, one hidden and one output layer) as shown in Figure 2.11. The hidden layer nonlinearly transforms the inputs into an *alternative* space in which the training

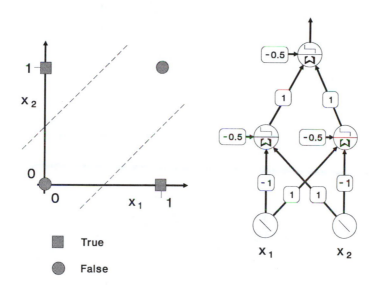

Figure 2.11 A solution for the two-input XOR logical problem using a three-layer MLP with two nodes in the hidden layer.

samples are linearly separable and a correct classification can be achieved. For the network shown, the outputs of the hidden nodes correspond to the logical functions:

(NOT x_1) AND x_2

x_1 AND (NOT x_2)

and if either of these expressions is true (logical OR in the output layer), the MLP output is also true. This construction holds for any finite dimensional logical expression, as it is possible to reduce *any* Boolean function to its equivalent *disjunctive normal form*. Thus any Boolean function can be represented by a three-layer network, where the output layer represents a multi-dimensional OR and the hidden layer nodes form multi-dimensional logical ANDs of the (possibly negated) inputs.

Functional Approximation

Using a continuous transfer function in each node means that the output is continuously dependent on the network's inputs, and there has been a lot of interest in using the MLP for functional approximation rather than classification tasks. It has been established that *any* continuous nonlinear function can be approximated to within an arbitrary accuracy by a three-layer MLP with sufficient nodes in the hidden layer [Hornik *et al.*, 1989]. Therefore the basic structure of the MLP is very flexible and can be employed in a wide variety of modelling and control tasks.

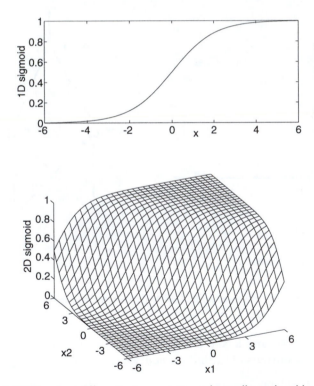

Figure 2.12 The sigmoid's output on a one- and two-dimensional input space.

It is instructive to investigate the nonlinear transformation that occurs in the hidden layer nodes, as this gives an indication of the type of problem for which the

MLP might be used successfully. Consider the commonly used *sigmoidal* transfer function:

$$f(u) = \frac{1}{1 + \exp{(-u)}} \qquad \in (0,1) \tag{2.10}$$

where $u = \mathbf{x}^T\mathbf{w}$. This function is bounded and monotonically increasing, tending to 0 as $\mathbf{x}^T\mathbf{w} \to -\infty$, and approaches 1 as the linearly combined input tends to ∞. The output of this function for a one- and two-dimensional input is shown in Figure 2.12, and is constant along the lines (three-dimensional weight space) for which:

$$w_0 + w_1 x_1 + w_2 x_2 = c$$

for some constant c, and this generalises to n-dimensional input spaces. The output of a sigmoid in the hidden layer is constant along the $(n-1)$-dimensional hyperplanes given by:

$$w_0 + \mathbf{x}^T\mathbf{w} = c$$

Thus the nodes which are composed of an ALC and a sigmoidal-type transfer function are termed *ridge functions* [Mason and Parks, 1992], as the output is *constant* along hyperplanes in their input space. If the desired function can be concisely decomposed into similar ridge functions, MLPs may be suitable models.

Sometimes in modelling and control applications the input data are *redundant* and MLPs can model this relationship by constructing hyperplanes parallel to the redundant inputs and setting the appropriate weights to zero. Thus the network model can deal efficiently with redundant data, and if a suitable network has been constructed and the input space is expanded by introducing a new, redundant input variable, no new nodes in the hidden layer need be introduced. Only a small number of weights are necessary to increase the model's size and these would be all set to zero. Thus the model's structure can incorporate redundant data *efficiently*, although it might not easily learn to recognise this redundant information.

MLPs are generally unsuitable for modelling functions which have significant local variations. The output of all of these hidden layer nodes is generally non-zero, and the resulting optimisation problem can be very complex. The theoretical modelling results guarantee that an MLP can approximate such functions arbitrarily closely, although they provide no indication about the suitability of using ridge functions as opposed to other nonlinearities in the hidden layer. Recently there has been a lot of theoretical interest in using the *Vapnik-Chervonenkis dimension* (VCdim) to investigate the complexity of MLPs [Hush and Horne, 1993], and once this number is known, it can be used to determine the amount of training data necessary for good generalisation. A realistic rule of thumb that came about from this work is that the amount of training data should be approximately ten times the VCdim, or equivalently, the number of weights in an MLP.

In conclusion, MLPs can be successfully applied to high-dimensional functional modelling and classification problems *if* the training data have redundant inputs

and the desired mapping can be approximated by a low number of ridge functions
[Wright, 1991].

2.3.3   Functional Link Networks

A functional link ANN [Pao, 1989, Pao *et al.*, 1994] has a similar structure to the
three-layer MLP, except that instead of employing ridge functions in the hidden
layer, polynomial or trigonometric terms are used and *linear* nodes are used in the
input and output layers. The use of such hidden layer transfer functions has a long
history in the nonlinear modelling community where a small number of low-order
polynomial terms or the dominant terms in a Fourier series have been used to
introduce nonlinearities into conventional linear algorithms. The ALC no longer
forms part of the nodes in the hidden layer, as shown in Figure 2.13, but is used in
the output layer. The use of such nonlinearities produces a very flexible network
[Mathews, 1991], although its usefulness for a particular application depends on
how well these nodes represent the nonlinear components of the desired function.

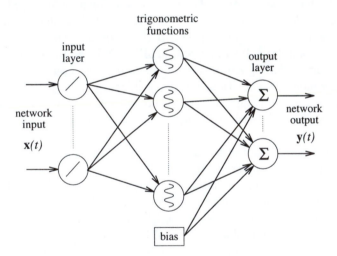

Figure 2.13 A functional link network which has trigonometric terms in the hidden layer.

These networks are universal approximation algorithms; they can approximate
a continuous nonlinear function to within an arbitrary accuracy, given a sufficient
number of nodes in the hidden layer [Cotter, 1990]. Like the ridge functions, polynomial and sigmoidal terms have a non-zero output over the whole input space,
and these are termed *globally generalising basis functions*. A successful application
of these networks requires a set of basis functions which can represent the desired
function adequately over the input domain, but not over-parameterising the network. Modifying a weight or introducing (or deleting) a new term in the hidden
layer affects the network's output globally and so it is not at all clear how the

structure should be chosen, or what type of relationship is stored in the network. There have been many *off-line* algorithms generated for polynomial and trigonometric term selection [Chen and Billings, 1994, Holden, 1994, Ivakhnenko, 1971], and the success of the network depends critically on the representations held in these hidden layer nodes.

2.3.4 Radial Basis Functions

Radial Basis Function (RBF) neural networks can be also be implemented within the standard three-layer network architecture, where the output nodes are simply ALCs, and the hidden layer nodes have a specific structure. RBF networks were first used for high-dimensional interpolation by the functional approximation community and their excellent numerical properties have been extensively investigated by Buhmann and Powell [1990] and Powell [1987]. They were first proposed within an ANN framework by Broomhead and Lowe [1988], and were used for data modelling and least-squares functional approximation. Since this paper was published, the technique has been widely adopted for off-line and on-line modelling and control tasks [Chen and Billings, 1994, Hunt and Sbarbaro-Hofer, 1992].

The (scalar) output y of an RBF network can be expressed as:

$$y = \sum_{i=1}^{p} w_i f_i \left(\|\mathbf{c}_i - \mathbf{x}\|_2 \right)$$

where w_i and $\mathbf{c}_i$ are the weight and centre, respectively, of the ith hidden layer node, and $\|.\|_2$ is the standard Euclidean norm. There are many different ways in which the univariate nonlinear functions, $f_i(.)$, can be selected and some of these are discussed in Section 3.3.4, but one important choice is the *localised* Gaussian function given by:

$$f_i(\|\mathbf{c}_i - \mathbf{x}\|_2) = \exp\left(-\frac{\sum_{j=1}^{n} (c_{ij} - x_j)^2}{2\sigma_i^2} \right) = \prod_{j=1}^{n} \exp\left(-\frac{(c_{ij} - x_j)^2}{2\sigma_i^2} \right)$$

If the training data are contained in a small region of the input space, the nodes in the hidden layer can be distributed within this region and only sparsely populate the remaining area. However, only a local model is formed and if the testing or operational data lie outside this small region, the performance of the network will be poor. Distributing the basis function centres evenly throughout the input space (all the theoretical results and some of the practical applications use this strategy) results in a more complex model, where the number of hidden layer nodes is exponentially dependent on the size of the input space, a property known as the *curse of dimensionality*. Irrelevant inputs cause the number of nodes in the hidden layer to increase dramatically with no corresponding increase in the model's flexibility.

An alternative RBF-type network that can reject irrelevant inputs was proposed by Hartman and Keeler [1991], where instead of the hidden layer nodes taking the *product* of the univariate Gaussian functions, the algebraic *sum* is used:

$$y = \sum_{i=1}^{p} \sum_{j=1}^{n} w_{ij} \exp\left(-\frac{(c_{ij} - x_j)^2}{2\sigma_{ij}^2}\right)$$

The output of the Gaussian bar network is *linearly* dependent on the nonlinear *univariate* Gaussian functions, and the network ignores irrelevant inputs by setting to zero the corresponding weights, w_{ij}. Typical two-dimensional Gaussian and Gaussian bar functions are shown in Figure 2.14. Comparing the outputs of the Gaussian and the Gaussian bar nodes shows that taking the product of the univariate Gaussian functions is similar to forming a multi-dimensional conjunction (AND) whereas summing the individual responses is reminiscent of the logical disjunction (OR). This is also very similar to Kavli's ASMOD algorithm (see Section 8.5) where a B-spline network is composed of the *sum* of several lower dimensional submodels. The Gaussian bar networks form *additive* models which cannot model any cross-product terms, although this restriction is why they are sometimes more successful than the standard algorithm and several "fuzzy" algorithms have been developed that try to produce parsimonious RBF networks with the smallest number of inputs [Tresp *et al.*, 1993].

2.3.5   Lattice Associative Memory Networks

Lattice-based Associative Memory Networks (AMNs) are the main focus of this book and the networks which are members of this class are described and compared in greater detail in Chapter 3. These networks can be mapped onto a three-layer structure with an ALC in the output node. The nodes in the hidden layer have a *localised* response and their output is non-zero in only a *small* part of the input space. In addition, the input space is normalised by an n-dimensional *lattice* and the basis functions are defined on this grid. Similar network inputs activate overlapping regions inside the network, and so these networks store information locally. The nodes in the hidden layer are termed *basis functions* and are represented by the p-dimensional vector **a**. Therefore the output of the network is given by:

$$y = \sum_{i=1}^{p} a_i(\mathbf{x})\, w_i$$

as shown in Figure 2.15.

The modelling capabilities of the network depend on the size, distribution and shape of the basis functions, and the above representation is very general. The simplest lattice AMN is probably an n-dimensional *look-up table*. Associated with each cell in the lattice is a weight and a binary basis function, where the output of the basis function is 1 if the input lies inside the cell and 0 otherwise. Thus within

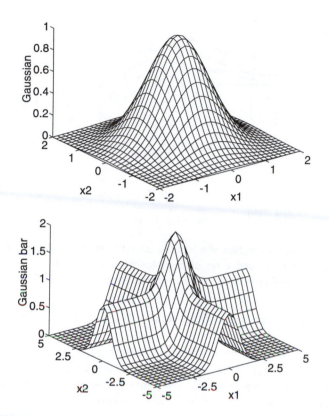

Figure 2.14 Two-dimensional Gaussian (top) and Gaussian bar (bottom) hidden layer nodes.

each lattice cell the network output is simply the corresponding weight and across the whole input domain the network's output is piecewise constant. Information about the stored functional relationship is *not* distributed to neighbouring weights and the memory requirements, p, depend exponentially on the input space dimension, but only a small fraction of the weights is involved in the network's response calculation.

Lattice AMNs partition the input space using *hard* splits (the support of each basis function is well defined), whereas Gaussian RBFs and the hierarchical networks proposed by Jacobs and Jordan [1993] provide a *soft* division of the input space. Gaussian basis functions are greater than some positive number only in a small region of the input space, hence they almost have a compact support, and the hierarchical sigmoidal decision nodes used in Jacob's tree structure (combined with a soft maximum operator) also possess this property. Truly local basis functions have the advantage that only a small region of the network contributes to the output, whereas soft split networks can potentially adapt every parameter at each time instant while still retaining a localised representation. This book concentrates on lattice AMNs, but it should be noted that these soft split networks

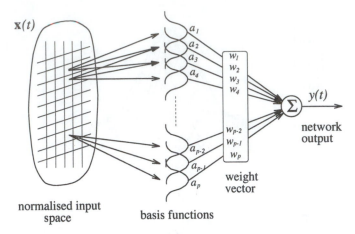

Figure 2.15   The basic architecture of a lattice associative memory network. A lattice is used to partition the input space and a set of basis functions is defined on this structure. The network's output is then formed from a linear combination of the basis functions' outputs.

provide an alternative approach which retains many of the properties of the truly local networks.

Two lattice AMNs which generalise are the Cerebellar Model Articulation Controller (CMAC) (see Chapter 6) and the B-spline network (see Chapter 8). Both networks have basis functions which are defined over more than one cell on the lattice, although in their simplest form both networks also reduce to a look-up table. The B-spline network provides *piecewise polynomial* interpolation and the local definition of the basis functions means that the basis functions can be interpreted as a set of *fuzzy* linguistic variables. The CMAC algorithm provides a coarse coding of the input lattice where the number active of basis functions does not depend on the dimension of the input space. Both networks suffer from the curse of dimensionality (this can be partially overcome using memory hashing techniques, Section 6.2.5, or by decomposing the network into submodels, Section 8.5), although very complex functional relationships can be stored, due to the local representations used in the hidden layer.

2.3.6   Model-Free Estimators

As an alternative to the model-based ANNs which have been described, a number of algorithms have been proposed which do *not* rely on any specific network structure and adapt their internal organisation in response to the training data [Atkeson, 1991, Specht, 1991]. These approaches generally store the training data (or a manageable subset) in a large memory, and use these examples to construct local models about the point of interest. These local models could simply be a low-order polynomial, or the data could be weighted using a (normalised) probability estimate.

These approaches are similar to some k-nearest neighbour algorithms, and they share the same advantages and disadvantages of this technique. For sparsely distributed data, the techniques generally produce a smooth, global response surface which closely approximates the training samples and when the data are dense, the algorithms are capable of filtering measurement noise and producing a best estimate of the true output. However, any algorithm which is based on remembering data has a high computational burden, because each training sample contributes to the network output. All of these operations can be performed in parallel, but for most practical applications the amount of training data far exceeds the number of available processors. Similarly, these algorithms *learn* by simply remembering training samples; there are no data-forgetting algorithms which could be used to model (or control) a time-varying plant.

2.3.7 Network Generalisation

All the feedforward model-based ANNs considered in this section can be represented using a three-layer network structure which has an ALC node in the output layer. The type of model and the associated generalisation characteristics are therefore *strongly* dependent on the type of nonlinearity incorporated in the hidden layer nodes. Many different sorts of nonlinearities have been described: bounded ridge functions (MLPs), trigonometric and polynomial functions (FLN), Gaussian and Gaussian bar functions (RBF) and piecewise polynomial functions (AMN). It is *not* possible to say that one type of model is always better than another; all that can be achieved is to list their desirable (and undesirable) properties.

The MLP has proved very useful for learning high-dimensional, redundant mappings, as the ridge functions partition the input space using hyperplanes. Using more than one hidden layer allows a more complex partitioning strategy to be developed, but the associated learning problem is generally much harder. The support of the hidden layer nodes (area of the input space for which the node has a non-zero output) is *global*, so adjusting any weight will affect the output of the network for *every* input. The training set should therefore contain a relatively complete coverage of the input space [Barnard and Wessels, 1992], otherwise the learnt network structure will not be of the correct form. In order to make the MLPs more *transparent* and understand the knowledge encoded in the weights, the approximation of these networks using finite polynomial and Volterra series models has been proposed. Any finite model of this form loses some information encoded in the network, where generally the higher order knowledge is lost, although it is precisely these terms which appear to give the MLP its advantages over polynomial/Volterra models. Hence the usefulness of such tests is debatable.

The success of polynomial and trigonometric functional modelling depends on the successful identification of the relevant nonlinear nodes. If any important terms are omitted, the global support of the nodes results in the model producing a globally biased solution, whereas if too many terms are included, the network tends to overfit the data, and the network output can be extremely oscillatory. However, these types of nodes have a strong theoretical background and many techniques have been developed for choosing a near optimal set of terms.

RBF networks have proved very useful for modelling highly nonlinear data, although their performance depends on the type of nonlinearity used in the hidden layer nodes as well as the distribution of their centres. It has been found that the approximation capabilities of functions with a global support appear to be slightly better than functions with a local support, although more advanced learning algorithms generally have to be used. The Gaussian bar networks, which are not strictly RBF networks, appear to be very useful when dealing with redundant data. Their support is not truly local (but more local than a sigmoid) and not truly global (but more flexible than a standard Gaussian function). The basis functions are generated by using the *addition* rather than the *product* operator to combine the univariate Gaussian functions, but only additive models are formed.

Lattice AMNs have truly local basis functions (hidden layer nodes) which provide an alternative to many of the ANN architectures currently being considered. The local support of the basis functions *forces* the network to generalise in a pre-specified manner, and therefore *a priori* knowledge about the desired function can be incorporated into the network's structure, although if the structure is inappropriate, the model will be biased (see Section 2.5.1). The networks are analogous to the piecewise polynomial data-fitting algorithms, in that the piecewise nature of the network prevents unwanted oscillatory behaviour and overfitting. However, the distribution of the basis functions on a lattice means that these networks suffer from the *curse of dimensionality*, which sometimes results in the networks being too flexible (large variance), especially in high-dimensional spaces.

Model-Based verses Model-Free Networks

Model-free algorithms provide an alternative to ANNs, and their modelling (and generalisation) performance demands that further research be performed into the relative merits of these techniques. They generate smooth, global models when the data are sparsely distributed, compared with the lattice AMNs which would produce only local approximations. Also when the training data are dense, very complex models can be produced.

Whether to use a model-based or a model-free algorithm depends on the problem and the available resources. Choosing model-based algorithms allows learning laws to be formulated for the unknown parameters and the network's response calculations do *not* depend on the number of training samples. However, the generalisation depends on the type of nonlinear hidden layer nodes and sometimes this relationship is not explicit, the exception being the lattice AMNs. The response time of a model-free algorithm generally grows linearly with the number of training samples, although no *a priori* knowledge is assumed about the form of the model and it appears to generalise sensibly.

2.4 TRAINING ALGORITHMS

The ability to *learn* complex mappings from a data set is *the* cornerstone of the recent revival of interest in ANNs. The inability of the Perceptron learning algorithm to be extended to multi-layer networks [Minsky and Papert, 1969] meant that a network could not learn to reproduce the basic XOR logical function, and so these algorithms were not considered suitable for application to more complex real-world problems. However, in the mid-eighties, when a gradient descent algorithm called Back Propagation (BP) was reinvented which enabled MLPs to learn arbitrary functional mappings, it stimulated considerable interest in learning systems.

BP is a *supervised* learning rule; the desired network output is given to the learning algorithm and the difference between this value and the actual output is used to guide the adaptive mechanisms. Many other learning algorithms have also been developed, such as *unsupervised* and *reinforcement* rules, and there are many different algorithms within these broad classifications. This section does not aim to give a complete survey of this field, but it provides a broad overview while concentrating on the rules which are applicable in the neuromodelling and control field.

2.4.1 Unsupervised Learning

Widrow termed unsupervised learning as *open-loop* adaptation, because it does not use any performance *feedback* information to update the network's parameters. This type of learning can be used in a variety of ways:

- **group** the input data into clusters, which can then be labelled in a supervised mode;
- **quantise** the continuous input space in an "optimal" manner;
- **represent** the input data in a lower dimensional space; and
- **extract** a set of features which represents the input signal.

This classification reflects the *usage* of unsupervised learning laws, as many of the above tasks are very similar. For instance, vector quantisation could be considered a subclass of feature extraction, although the latter is generally used for pattern classification, whereas the former can be used to prepare the input data for modelling algorithms.

Unsupervised learning algorithms can be used in a wide variety of networks: training the centres of an RBF network [Chen *et al.*, 1992], providing an "optimal" lattice for the AMNs, although many learning rules are closely tied to particular system architectures [Grossberg, 1988]. They have also been used for many different applications: speech recognition [Kohonen 1988, 1990], robotics modelling [Ritter *et al.*, 1989], image compression [Nasrabadi and Feng, 1988], etc. The basic unsupervised learning algorithm and the two most important unsupervised network architectures: Kohonen's Self-Organising Map and Grossberg's ART networks are described in this section.

Competitive Learning

Competitive ANNs distribute their nodes across the input space, so that they learn to represent the statistics of the input data which is presented to the network, as illustrated in Figure 2.16. The recall phase of such a network simply involves

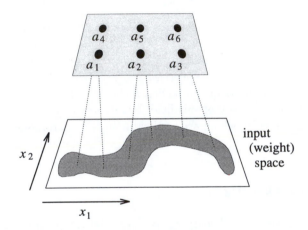

Figure 2.16   An unsupervised learning network showing how the nodes become sensitised to different parts of the input space. The input is assumed to have a uniform probability density function (shown by the shaded area) and the weight vectors (dotted lines) are distributed evenly in the relevant parts of the input space.

finding the node which best matches a particular input pattern and exporting either the index of this node or the associated classification label.

A network which is trained using the unsupervised competitive learning algorithm is composed of a group of nodes $\{a_i\}_{i=1}^{p}$, and associated with each node is a *normalised* weight vector $\mathbf{w}_i$ (each element should also be normalised so that it assumes equal importance). A normalised input, $\mathbf{x}$, is presented to the network and a match is then calculated between the input and the weight vector corresponding to each node, whose activation value is given by:

$$a_i = \sum_{j=1}^{n} w_{ij}x_j = \mathbf{x}^T\mathbf{w}_i \qquad (2.11)$$

The network then determines the node a_k such that $a_i < a_k \; \forall i \neq k$ and the output is either this index or the classification label stored at the kth node. Selecting the node with the maximal output (closest match) can also be implemented as a neural network, as described by Lippmann [1987].

The weight vector associated with the best match node is updated according to the rule:

$$\mathbf{w}_k(t) = \frac{\mathbf{w}_k(t-1) + \delta\left(\mathbf{x}(t) - \mathbf{w}_k(t-1)\right)}{\|\mathbf{w}_k(t-1) + \delta\left(\mathbf{x}(t) - \mathbf{w}_k(t-1)\right)\|} \tag{2.12}$$

and every other weight vector retains its current value.

If the input data are not normalised, the match is implemented using the Euclidean distance measure and the kth node is selected such that $\|\mathbf{x} - \mathbf{w}_k\|_2 < \|\mathbf{x} - \mathbf{w}_i\|_2 \; \forall \, i \neq k$. The learning rule is then modified to:

$$\mathbf{w}_k(t) = \mathbf{w}_k(t-1) + \delta\left(\mathbf{x}(t) - \mathbf{w}_k(t-1)\right)$$

These learning rules ($k = 1, 2, \ldots, p$) can easily be shown to form a set of p MSE gradient descent training laws, and the weight vectors converge in the limit to their optimal values with respect to the network input MSE.

The networks and the learning rules can be used as simple classification systems, where a label is associated with each node, or for vector quantisation where a value (weight) is stored at each node and the network output is simply this weight. Generalised competitive networks allow more than one node to be active at any one time (similar to k-nearest neighbour networks) and interpolate locally the information stored at each node. Initial learning is also generally faster for these networks, as more than one weight vector is updated for each input.

The Self-Organising Map

Kohonen's Self-Organising Map (KSOM) [Kohonen, 1990] is a competitive network in which the nodes are *ordered*, and its aim is to produce a low-dimensional, topology conserving representation of the input space. The nodes are regularly placed in an m-dimensional space, (generally $m < n$) and the learning algorithm attempts to formulate a mapping such that, if two inputs $\mathbf{x}(1)$ and $\mathbf{x}(2)$ are close in the input space, the two activated nodes a_i and a_k are also close in the m-dimensional user-defined space. Generally the nodes are placed on an m-dimensional *lattice* and an appropriate *kernel* function is associated with each node which tends to zero as the input moves away from the activated node. At time t, the weight vectors are updated according to the following rule:

$$\mathbf{w}_i(t) = \mathbf{w}_i(t-1) + \delta(t)h(i, k)\left(\mathbf{x}(t) - \mathbf{w}_i(t-1)\right) \tag{2.13}$$

where the kth unit is activated at time t, and $h(i, k)$ is the kernel function and $\delta(t)$ is the time-varying learning rate. If $h(i, k) = 1$ when $i = k$ and 0, otherwise the standard competitive learning algorithm is obtained. However if $h(i, k) = 1$ when $i = k$ and it tends to 0 in the locality of k, the nodes surrounding a_k receive similar training information and the weight vectors become sensitised to similar input regions. The mapping then retains the topological features of the original input space. Many different kernel functions can be used in the above algorithm: Mexican hat functions, Gaussian functions, etc., and the algorithm appears to be reasonably robust with respect to different selections.

For modelling and control applications, this algorithm has the potential to map a high-dimensional input space to a low-dimensional representation which preserves the topological ordering of the original inputs. In high-dimensional space, most of the relevant input data are contained in a much smaller subspace, and the KSOM has the ability to extract this information automatically in a computationally efficient manner [Walter and Schulten, 1993].

Adaptive Resonance Theory

Since the late sixties, Grossberg and his research group have been investigating and developing (amongst others) biologically plausible neural pattern classification architectures and the associated learning rules. The basic design question which this research addresses is development of systems that can be "designed to remain plastic, or adaptive, in response to significant events and yet remain stable in response to irrelevant events" [Carpenter and Grossberg, 1988]. This was termed the *stability-plasticity* dilemma and is a *fundamental* problem for any on-line adaptive algorithm. Grossberg proposed an Adaptive Resonance Theory (ART) for neural networks, which is designed to overcome this problem and also possesses three important properties: the input activity is normalised (similar to the competitive networks), contrast enhancement of input patterns and distinction between Short-Term Memory (STM) and Long-Term Memory (LTM) [Grossberg, 1988]. The three ART architectures proposed to date can deal with binary input signals (ART1), real-valued input signals (ART2) [Carpenter *et al.*, 1991a] and ART3 incorporates a hierarchical structure [Carpenter and Grossberg, 1990]. Other proposed systems have included a fuzzy ART [Carpenter *et al.*, 1991b], and vector associative maps for unsupervised learning and control of movement trajectories [Grossberg *et al.*, 1993].

The basic ART1 architecture is shown in Figure 2.17 with the bottom-up and top-down weight arrays ($\mathbf{W}^b$ and $\mathbf{W}^t$, respectively) playing a major part in preventing learned memories from being overwritten by new information. A bottom-up weight vector encodes the competitive element of the ART architectures, and the node with the best match is selected. The top-down weight vector associated with this node is then compared with the input pattern and if it is a good match, this categorisation is accepted, if not the node is disabled and a new best match is calculated. Once the input has been either categorised or a new node has been allocated to store this input, the LTM weights are changed.

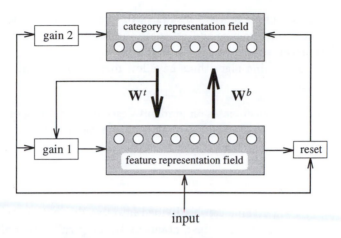

Figure 2.17 The basic ART1 architecture.

The ideas encoded in the ART architectures are *directly* relevant to on-line learning modelling and control algorithms, and any proposed system should have the ability to store new information without affecting unrelated stored data. Many learning systems do *not* possess this property, although the lattice AMNs considered in this book have this feature, as new information is stored *locally*.

2.4.2 Supervised Learning

Supervised learning rules differ from unsupervised training algorithms because the network's desired response (output) needs to be presented to the network for each input. Denoting the desired network output at time t by $\widehat{y}(t)$, the *instantaneous* performance of the network can be inferred from:

$$\epsilon_y(t) = \widehat{y}(t) - y(t) \tag{2.14}$$

and these output errors are used to update the network's weights.

Parameter convergence is the aim of this optimisation strategy, as the network is only able to *generalise* correctly if the parameters are close to their optimal values. Generalisation is *essential* if the network is trained off-line from a finite training set, and this illustrates the problem of determining *when* to stop training. If this decision is based on the size of the output error only, a low value does not always imply that the weights are correctly initialised, and this problem is discussed further in Sections 4.6 and 8.3.2.

Supervised learning is probably the most widely used training mechanism and gradient descent adaptation is probably the best known supervised learning rule. This is due to a variety of reasons:

- Widrow's original LMS learning rules are instantaneous gradient descent training algorithms.

- A large number of theoretical results have been derived, which can establish parameter convergence and estimate the rate of convergence for certain network structures.
- The resulting learning algorithms have low memory requirements and a low computational cost.

Therefore this section concentrates on gradient descent learning algorithms (batch and instantaneous), although other training rules are discussed.

Batch Learning

Supervised learning algorithms can be divided into two distinct approaches: *batch* and *instantaneous* training rules. Instantaneous training rules use only the information provided by a single training example $\{\mathbf{x}(t), \hat{y}(t)\}$, when the weight vector is updated, whereas batch learning laws generally use *all* the training data to adapt the weights. Under certain conditions, the instantaneous training rules are approximations of their batch counterparts, and so the latter are described first.

For a training set given by $\{\mathbf{x}(t), \hat{y}(t)\}_{t=1}^{L}$, a cost function which measures the current performance of the network needs to be specified. In most applications this is the MSE defined by:

$$J = E\left(\epsilon_y^2(t)\right) = \frac{1}{L}\sum_{t=1}^{L}\epsilon_y^2(t) \tag{2.15}$$

Other cost functions may improve the network's performance, although using the MSE produces computationally efficient learning algorithms and good final models.

Gradient descent learning algorithms adapt the weight vector in the direction of the *negative* gradient of the performance function:

$$\Delta\mathbf{w} = -\frac{\delta}{2}\frac{\partial J}{\partial\mathbf{w}} \tag{2.16}$$

where δ is the *learning rate* which determines the stability and the rate of convergence of the learning algorithm and $\Delta\mathbf{w}$ is the weight vector update. Applying the chain rule to the MSE cost function gives:

$$\Delta\mathbf{w} = -\frac{\delta}{2}\frac{\partial y}{\partial\mathbf{w}}\frac{\partial J}{\partial y} = \delta E\left(\frac{\partial y(t)}{\partial\mathbf{w}}\epsilon_y(t)\right) \tag{2.17}$$

Therefore the gradient update contains information about the network Jacobian $(\partial y/\partial\mathbf{w})$, and the performance of the network $\partial J/\partial y$. For the MSE cost function, the latter quantity is simply the output error, and the *efficiency* of this algorithm will depend on the structure of the network (information held in the Jacobian). If the network is linearly dependent on the weight vector $(y(t) = \mathbf{a}^T(t)\,\mathbf{w})$, the above expression simplifies to:

$$\Delta \mathbf{w} = \delta E \left(\epsilon_y(t) \mathbf{a}(t) \right)$$

and the computational simplicity of this rule is obvious.

For linear networks, it is shown in Chapter 4 that the weights trained using this learning algorithm converge to their optimal values, and the rate of convergence can be estimated. The gradient descent learning rules can also be extended for nonlinear optimisation (to train MLPs for instance, see Section 4.5), although the theoretical convergence results do not generally apply. This is illustrated in Figure 2.18, where it is shown how gradient descent rules can become trapped by local minima and plateau areas. Even the rate of convergence of linear networks depends on the *condition* of the model and slow convergence occurs when the model is poorly conditioned.

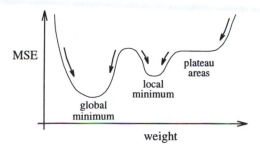

Figure 2.18   The nonlinear relationship between a weight and the mean square output error. Gradient descent training algorithms can become trapped by local minima and plateau areas.

This discussion has considered *first-order* gradient descent rules which use only the first-derivative information. Second and higher-order training algorithms can be derived which make use of second (and higher) -order derivatives [Widrow and Stearns, 1985]. Probably the best known of these is Newton's method which uses second-order curvature information and this can be expressed as:

$$\Delta \mathbf{w} = -\mathbf{D} \frac{\partial J}{\partial \mathbf{w}}$$

where $\mathbf{D}$ is (an approximation to) the inverted second-derivative matrix (the Hessian) [Gill *et al.*, 1981, Mills, 1992]. Equation 2.16 is therefore an approximation to this more complex learning rule, and they are equivalent when the cost function has constant second derivatives and there is no cross-coupling with respect to each weight, i.e. $\mathbf{D} = \mathrm{diag}\{\delta, \delta, \ldots, \delta\}$.

Instantaneous Learning

Instantaneous learning rules generally use only a single piece of training information when the weights are updated. This is closer to human learning processes, which do not use all the available information. However, this fact should not be used to justify the use of these adaptation rules in learning control applications, but should inspire their development and their performance should be *rigorously* assessed. Batch learning rules have the disadvantage that all the training information is required before the weights are updated and, although recursive algorithms are available, the large and complex structure of the networks limits the feasibility of these techniques.

Instantaneous learning algorithms generally attempt to minimise an instantaneous measure of the network's performance, and for the MSE cost function this is given by:

$$J(t) = \epsilon_y^2(t) \tag{2.18}$$

Using a gradient descent rule, the weight update is:

$$\Delta \mathbf{w} = -\frac{\delta}{2} \frac{\partial J(t)}{\partial \mathbf{w}} = \delta \epsilon_y(t) \frac{\partial y(t)}{\partial \mathbf{w}} \tag{2.19}$$

which is simply an (unbiased) estimate of the derivative of the true cost function. When the network is linear this reduces to:

$$\Delta \mathbf{w} = \delta \epsilon_y(t) \mathbf{a}(t)$$

This rule can be interpreted as updating the weights in proportion to the amount that they contributed to the output. If the vector **a** is *sparse* (many zero elements) and the non-zero elements are approximately of the same magnitude, the learning algorithm works well, as is shown in the following example and in Chapter 5.

In Section 5.3.1, it is shown that if no mismatch (modelling error or measurement noise) exists, then a weight vector trained using the above rule converges to a value that can exactly reproduce the training data. However, when modelling error exists, the noise associated with the gradient estimate means that the weight vector no longer converges to an optimal value, but lies in a domain which surrounds this quantity. The size and shape of this *minimal capture zone* (see Section 5.4.3) depends on the network's structure and the distribution of the training data, and this can be regarded as part of the cost of using an instantaneous estimate rather than the true gradient. If the learning rate is allowed to decay to zero rather than remaining constant, the gradient noise is filtered out and the weights converge to their optimal values [Luo, 1991].

These learning rules can also be used for networks which depend nonlinearly on their weights, and the original BP algorithm for training MLPs was an instantaneous learning rule [Rumelhart and McClelland, 1986]. Also higher-order learning rules, which use the past L training samples, can be derived and these generally reduce the size of the minimal capture zones and increase the rate of convergence, at a cost of increasing the computational complexity.

Example: Look-Up Table

A look-up table has an extremely simple structure, although it can provide considerable insight into the suitability of training a network, using an instantaneous learning law. It has an n-dimensional lattice defined on the input space and the output of the network is simply the ith weight, if the input lies in the ith lattice cell. The transformed vector $\mathbf{a}$ has $(p-1)$ zero elements and the ith element is unity, so it is extremely sparse.

The weight vector is updated using an instantaneous *stochastic* learning rule:

$$\Delta w_j(t-1) = \delta_j(t)\,\epsilon_y(t)\,a_j(t)$$

where $\delta_j(t)$ is the learning rate associated with the jth node at time t. Only the ith weight is updated at time t, so the learning algorithm is computationally efficient, and the time-varying learning rate is given by:

$$\delta_i(t) = \frac{1}{\sum_{k=1}^{t} a_i(k)}$$

therefore the *a posteriori* weight value at time t is:

$$
\begin{aligned}
w_i(t) &= w_i(t-1) + \frac{1}{\sum_{k=1}^{t} a_i(k)}(\widehat{y}(t) - w_i(t-1))a_i(t) \\
&= \frac{\widehat{y}(t) + w_i(t-1)\sum_{k=1}^{t-1} a_i(k)}{\sum_{k=1}^{t} a_i(k)} \\
&= \frac{\sum_{k=1}^{t} \widehat{y}(k)a_i(k)}{\sum_{k=1}^{t} a_i(k)}
\end{aligned}
$$

This last quantity is the optimal estimate of the weight (at time t), and so the instantaneous and batch learning laws are equivalent for a look-up table.

The instantaneous learning algorithm works well because the transformed input vector $\mathbf{a}$ is sparse and all the non-zero elements are equal in value. If a network is to be successfully trained using instantaneous learning rules, it should partially satisfy these two conditions, which is true for the lattice AMNs analysed in this book.

Alternative Strategies

This section has concentrated mainly on gradient descent learning rules because of their popularity and the theoretical results which can be derived. However, they are not suitable for every optimisation problem and for every network structure. Two other supervised learning algorithms which have been used for training ANNs are *stochastic* and *genetic* learning rules.

Stochastic training algorithms introduce a random element into the search for the optimal weight vector. They can be used when the network's output is *not* a continuously differentiable function with respect to weight vector. In the most general form, a random search can be performed in weight space and the weight vector which generates the lowest cost is selected, or else a stochastic element can be introduced into other learning rules (such as gradient descent) in order to reduce the chance of becoming stuck in local minima. This technique should only be used if the cost function is highly complex as they use little or no information about its shape, and gradient descent rules generally perform better [Monzingo and Miller, 1980].

Genetic algorithms have gained in popularity over the past few years, although the seminal text on the subject was written nearly twenty years ago [Holland, 1975]. The weights are represented using binary strings and the strings are concatenated to form one large string. A *population* is then created from several large, different strings and these are combined using various genetically inspired techniques such as cross-over (cutting strings and swapping the respective components) and mutation (randomly changing bits in the string). The fitness of the each new string is evaluated and a new population is created which consists of some of the new and some of the old strings, and the fittest strings have a highest probability of being members of the new population. The technique combines elements of directed (using cross-over) and random (mutation) search algorithms and has been shown to perform well in a wide variety of off-line tasks [McGregor *et al.*, 1992, Renders and Hanus, 1992], which involve highly complex optimisation calculations.

2.4.3 Reinforcement Learning

Reinforcement learning algorithms use a reduced form of performance feedback information in their updating rules. This performance information is generally binary and denotes whether the sequence of control actions has been successful. The seminal paper on this subject was written by Barto *et al.* [1983] where an inverted pendulum was trained using a reinforcement learning algorithm. The learning ANNs which implement the ASE and ACE blocks, described in Section 2.2.4, are generally similar to look-up tables, although recently more complex input space quantisation strategies have been proposed [Zhang and Grant, 1992], and more flexible networks have been applied [Anderson, 1989]. The reduced performance feedback information means that the networks which initially learn quickly perform well and this was confirmed when the CMAC and fuzzy networks were applied to the pole-balancing reinforcement learning task [Berenji and Khedkar, 1992, Lin and Kim, 1991].

These reinforcement learning rules were motivated by the adaptive automata algorithms [Barto and Anandan, 1985, Narendra and Thathachar, 1974], where they are used to update the probability vectors associated with each possible input state. A good description of the weight update equations is provided by Barto *et al.* [1983] and Millington and Baker [1990]. The ACE reinforcement weights are updated in proportion to the weighted sum of the local reinforcement signals and the ASE weights are updated in proportion to the internal reinforcement signal at time t, multiplied by a term which measures the eligibility of that particular state. The continuous reward/punish signal from the ACE means that the ASE is able to learn, even when the system does not fail, and the performance of this algorithm over a system which only uses the external reinforcement signal has been demonstrated many times.

2.5   VALIDATION OF A NEURAL MODEL

The most important part of any learning systems design procedure is the verification and validation phase. During these tests, the designer should be able to assess how well the network has learnt the training data and how successfully it is able to generalise (interpolate and extrapolate) to unforeseen cases. Most learning algorithms can successfully learn a set of training examples given a sufficiently flexible model structure or an appropriate learning algorithm, although the question of whether they possess the ability to generalise *correctly* (or sufficiently accurately) is still unresolved [Barnard and Wessels, 1992]. In Lau and Widrow [1990], it is said that "it is necessary to develop quantitative techniques to evaluate neural network's performance with real-world data ... Rigorous mathematical foundations must be developed to determine the characteristics of the training set and the network's ability to generalise from the training data". For safety and certification reasons the performance of the trained network must be completely understood; it cannot be simply regarded as a black box about which nothing can be proven. It has been argued that most of the ANNs are model-based algorithms and so tests are required to determine if the structure is sufficiently flexible and to establish if the network's input representation contains enough information. From a learning viewpoint, it is desirable to have the *smallest* acceptable network and input vector, as this forces the network to generalise sensibly.

There are many different ways in which the performance of a trained network can be assessed, the simplest (and probably the most biased and abused [Weiss and Kulikowski, 1991]) method is by simply assessing the network's performance in reproducing the training data. A better approach is to split the available data into a training and a test set, and to use the testing data to evaluate the final model. There are many different ways in which the available data can be divided, and some of these are reviewed in Section 2.5.2. This test only provides very general performance information; if the model is poor, it does not give any reasons about why this occurs. During the eighties, a number of correlation and statistical based tests were devised for validating the NARMAX models developed by Billings and his collegues [1983, 1986, 1989, 1992]. These tests are equally applicable to neurofuzzy modelling and control algorithms, and the model validation tests are reviewed in Section 2.5.3. The topic of network *transparency* is then examined, as this is one of the most desirable features of the networks discussed in this book. Network transparency refers to the manner in which information is stored, and a network is said to be transparent to the designer if the relationship between a weight and the network surface is *easily* understood, otherwise the network is said to be *opaque*. Networks with this property are described in Section 2.5.4, and the validity of assessing opaque networks by mapping them onto a transparent model is discussed.

All of these evaluation methods provide information about the generalisation ability of the trained model, and it is anticipated that additional performance tests will evolve in the future.

2.5.1   Bias Variance Dilemma

Artificial neural networks have been called *model-free* estimators, as they are flexible enough to approximate *any* smooth nonlinear function to a prespecified degree of accuracy, given sufficient resources. This is a *necessary* theoretical result, which shows that there may be a large number of modelling problems which neural algorithms could be applied to, although it does not guarantee success. Irrespective of the learning algorithm used to train the parameters, the model's structure should be appropriate. The model should be flexible enough to learn the desired mapping described by the training data, but it should not be *over-parameterised*, as this causes the model to fit the noise which is inherent in most data sets. The problem of estimating how flexible a model should be (how many nodes in each layer, number of layers, order of the splines, etc.) is well developed in the statistical modelling community where it is termed the bias/variance dilemma [Geman *et al.*, 1992]. The MSE performance measure for a particular input can be decomposed into two components which reflect the *bias* of the network error (where the average is taken over all possible training sets), and the *variance* of the estimates of the trained networks. Following the notation developed by Geman *et al.* [1992], let D denote a training set, $y(\mathbf{x}, D)$ be the output of a network trained using the data contained in D and $E_D(.)$ denote the expectation operator taken over all possible training sets, then the output error is given by:

$$\epsilon_y(\mathbf{x}) = \widehat{y}(\mathbf{x}) - y(\mathbf{x}, D)$$

The suitability or effectiveness of this network as a predictor of $\widehat{y}(\mathbf{x})$ can be measured by calculating the MSE for all possible training sets D, giving:

$$\begin{aligned} E_D\left(\epsilon_y^2(\mathbf{x})\right) &= E_D\left((\widehat{y}(\mathbf{x}) - y(\mathbf{x}, D))^2\right) \\ &= (\widehat{y}(\mathbf{x}) - E_D(y(\mathbf{x}, D)))^2 + E_D\left((\widehat{y}(\mathbf{x}) - y(\mathbf{x}, D))^2\right) \end{aligned} \qquad (2.20)$$

The first term on the right-hand side is called the *bias*, as it measures the *average* modelling error, whereas the second estimates the *variance* of the network approximations. Even if an "average" network is able to interpolate the data and make the bias zero, a high variance still causes poor performance. A large variance occurs when the performance of the network is very sensitive to the training data (if the network is too flexible) which results in a poor MSE performance.

In general, a network should be flexible enough to ensure that the modelling error (bias) is small, although the model should not be over-parameterised, as this causes its performance to strongly depend on a *particular* training set (high variance).

2.5.2 Output Error Tests

Many of the network training rules are based on minimising an MSE cost function and therefore it is natural that the same test should be applied first when the performance of the network is evaluated. The Root Mean Square (RMS) output error is generally a *negative* test; it only gives information about the training being inadequate (a high RMS), and little can be deduced from a low RMS value. This can be easily illustrated by considering a nearest neighbour data storage algorithm which simply remembers the training data. After one training cycle, the network's recall is perfect and the RMS is zero, but this testing procedure gives no information about its ability to generalise. Alternatively, the reduction in the RMS after one learning cycle for an MLP trained using a gradient descent algorithm is small and the only information provided is that the learning procedure should not be stopped.

It is important to obtain a good estimate of the true performance of the ANN, as this could be used to determine if the structure is flexible enough for a particular data set. If the modelling capabilities of the network were increased and this resulted in a lower estimate of the true RMS value, it would indicate that the original network structure was insufficient. Similarly, if the true RMS value was higher for a more complex model, this would indicate that the ANN was starting to model the disturbances present in the data and a simpler network structure would be preferred, as this forces the network to generalise. The RMS test provides no estimate of the variance of the network, and this is illustrated in Figure 2.19.

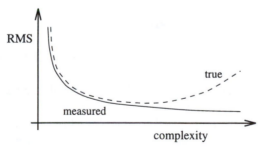

Figure 2.19 An illustration of the trade-off between network complexity and the true (dashed line) and measured (solid line) RMS.

Despite the fact that this performance measure can easily be abused, several statistical and information theory-based validation tests, see Section 8.5.4, have been developed which attempt to reproduce the true RMS, by combining the measured RMS with terms that estimate the network's complexity. These measures weigh the RMS against the number of parameters in the network and the amount of training data, producing parsimonious networks that are able to model the data adequately.

Test Set Construction

Any data set presented to a learning algorithm must be split into a part which is used to train the network, and another used for evaluating its performance by testing its ability to generalise correctly. For off-line design, where a fixed training set is presented to the network, there is generally a problem due to lack of data and so an obvious problem is to determine how much should be used for training and how much should be used for testing. One heuristic which works well in practice is to use two-thirds for training and one-third for testing. This approach is reasonable because it is not possible for a learning algorithm to perform well if there are insufficient training data, but the performance cannot be assessed accurately if the test set is too small. Random subsampling can be used to obtain a better estimate of the true RMS, and this proceeds by randomly generating different test and training sets, training a network on each of these different test sets and averaging the RMS values after each training session. These measures produce estimates for both the network's bias and its variance which can be then used to assess the model's true performance. A special case of this technique is the *leave-one-out* strategy where only one piece of data is used for testing and the remainder is used for training. The learning and testing procedures are carried out enough times so that each data pair is used as the test case just once, and the RMS value (total variance) can be calculated by averaging over each RMS estimate [Weiss and Kulikowski, 1991].

2.5.3   Correlation and Statistical Tests

The validation of a trained ANN using a test set is the first of many testing procedures which should be applied to the network, and for the nonlinear NARMAX models (see Section 2.2.1) many other evaluation algorithms have been developed. These tests can highlight deficiencies occurring in the input data and in the network's structure, so they can be used to assess the order of the nonlinear models, measurement errors that may have occurred when the data were collected and to produce parsimonious models which generally perform well.

Correlation Tests

Correlation-based evaluation tests for NARMAX models were developed after it became apparent that the linear covariance tests [Ljung and Soderstrom, 1983] were not sufficient when they are applied to nonlinear systems [Billings and Voon, 1986]. A further set of correlation-based tests was developed and any well-trained ANN model should satisfy the following five conditions:

$$
\begin{aligned}
\phi_{\epsilon\epsilon}(\tau) &= E\left(\epsilon(t-\tau)\epsilon(t)\right) = \delta(\tau) \\
\phi_{u\epsilon}(\tau) &= E\left(u(t-\tau)\epsilon(t)\right) = 0 \qquad \forall \tau \\
\phi_{u^{2\prime}\epsilon}(\tau) &= E\left(\left(u^2(t-\tau) - E\left(u(t)\right)^2\right)\epsilon(t)\right) = 0 \qquad \forall \tau & (2.21) \\
\phi_{u^{2\prime}\epsilon^2}(\tau) &= E\left(\left(u^2(t-\tau) - E\left(u(t)\right)^2\right)\epsilon^2(t)\right) = 0 \qquad \forall \tau & (2.22) \\
\phi_{\epsilon(\epsilon u)}(\tau) &= E\left(\epsilon(t)\epsilon(t-1-\tau)u(t-1-\tau)\right) = 0 \qquad \tau \geq 0 & (2.23)
\end{aligned}
$$

where $\epsilon(t) = \epsilon_y(t)$ is the current network output error. To generate meaningful results, *normalised* correlation functions are generally calculated and these are defined by:

$$
\hat{\phi}_{\psi_1\psi_2}(\tau) = \frac{\sum_{t=1}^{L-\tau} \psi_i(t)\psi_2(t+\tau)}{\left(\sum_{t=1}^{L} \psi_1^2(t) \sum_{t=1}^{L} \psi_2^2(t)\right)^{0.5}} \tag{2.24}
$$

The normalised correlation functions lie in the interval $[-1, 1]$, and a 95% confidence band $(= \pm 1.96/\sqrt{L})$ indicates if the calculated correlations are significant.

It is not known whether or not these tests are sufficient, but they form a powerful set of evaluation methods for process modelling, and have proved very useful for a wide variety of different ANNs. They provide information about the structure of the network, as well as determining if the input vector is sufficiently rich.

Chi-Squared Tests

Another evaluation algorithm which has been used for nonlinear model validation is the chi-squared statistical test [Bohlin, 1978, Leontaritis and, Billings, 1987], which indicates whether or not a model is biased. This performance measure is calculated by defining an η-dimensional vector $\Omega(t)$ as:

$$
\Omega(t) = [\omega(t), \omega(t-1), \ldots, \omega(t-\eta+1)]^T
$$

where $\omega(t)$ is a function of the previous system inputs, outputs and prediction errors. The chi-squared statistic is then given by:

$$
\zeta = L\mu^T \left(\Gamma^T\Gamma\right)^{-1} \mu \tag{2.25}
$$

where

$$
\begin{aligned}
\mu &= L^{-1} \sum_{t=1}^{L} \frac{\Omega(t)\epsilon_y(t)}{\sigma_\epsilon} \\
\Gamma^T\Gamma &= L^{-1} \sum_{t=1}^{L} \Omega(t)\Omega^T(t)
\end{aligned}
$$

where σ_ϵ^2 is the variance of $\epsilon_y(t)$. The model is generally regarded as adequate if for several different vectors $\Omega(t)$, the corresponding values of ζ lie within a 95% acceptance region:

$$\zeta < k_\alpha(s)$$

where $k_\alpha(s)$ is the critical value of the chi-square distribution with s degrees of freedom and a level of significance $\alpha = 0.05$.

2.5.4   Network Transparency

In validating a trained ANN, it would be desirable to have an understanding of the influence of each weight on the network output. This enables the designer to search for regions where the network has been insufficiently trained and aid the evaluation process as the magnitude of the parameters would be directly related to the size of the network output. However, many ANNs are *opaque*; the knowledge which is stored in the network is distributed across many parameters in a complex manner which cannot easily be understood by the designer. Some attempts have been made to interpret MLPs (for example) using a truncated Taylor series or finite Volterra model [Soloway and Bialasiewicz, 1992], although neglecting the higher order terms of a sigmoid expansion loses the fine detail, and it is conjectured that these higher order terms provide the representational advantage of globally generalising nodes for *smooth* functional approximation tasks.

One notable exception to this problem is the B-spline lattice AMN, and to a lesser extent the higher order CMAC networks. Their internal structure can be interpreted as a set of fuzzy production rules (see Section 9.2.1 and Chapter 10) such as:

> IF (x_1 *is positive small* AND x_2 *is almost zero*)
>
> THEN (y *is negative small*)

This provides the AMNs with some degree of transparency, although it is generally only true for small-dimensional spaces, as the number of rules is exponentially dependent on the number of inputs. More complex input space quantisation strategies and network architectures [Kavli, 1994, Zhang and Grant, 1992] can make these techniques applicable to higher dimensional modelling problems although, unless the redundancy is explicitly represented, the fuzzy rule base may be overly complex.

2.6   DISCUSSION

The majority of the currently used ANNs are model based. The structure of the network remains fixed during training, and only the weight vector is adapted. In this context, these algorithms should be subject to a rigorous analysis in order to determine their advantages and disadvantages when compared with other learning model based algorithms. A learning algorithm can be classified according to its basic modelling flexibility, its learning capabilities and the model transparency (the ease with which the final model can be understood by the designer). This chapter has reflected that philosophy. Human learning provided the inspiration for many of the algorithms described in this section, but these techniques must be properly understood if they are to be applied successfully. The argument that, because these algorithms are based on human thinking and human thinking cannot be analysed, *therefore* results cannot be proved about these techniques is incorrect. Many of the so-called neural learning algorithms have their counterparts in the adaptive control and signal processing literature: LMS $\equiv$ MIT rule, and NLMS $\equiv$ Kaczmarz's algorithm (Åström and Wittenmark [1989], Kaczmarz [1937] and Chapter 5), and many theoretical results are now known about their performance. The same amount of rigour should be applied to the neuromodelling and control research, otherwise results will be *strictly* problem dependent.

The biological motivations and implications of these algorithms have been deliberately understated in this chapter and in the remainder of the book. Biological research should stimulate and motivate this research, but it does not have to guide the work blindly. In the introduction it was stated that it may be another fifty years before a first-order understanding of the brain's functionality is achieved, yet control engineers want these results *today*, with provable convergence and stability properties.

The modelling capabilities of an ANN should be flexible enough to achieve the desired objectives, but if a model is too flexible, it can overfit the data and the generalisation is poor. Learning from a data set can be posed as a functional approximation task, as it is not enough to simply learn to reproduce the training data, the underlying structure of the function must also be learnt from the samples, otherwise the network is not able to generalise correctly to neighbouring inputs. The adaptive ANN should also be well conditioned, as generalisation *requires* both a good model structure and parameter convergence. If the network is badly conditioned, the designer may be fooled into thinking that it is performing well when the prediction errors are small. However, parameter convergence in a badly conditioned network (of the correct structure) occurs *slowly*, and prematurely stopping the learning procedure means that the network is unable to generalise appropriately.

One of the original objectives in the ANN field was to develop algorithms which were capable of learning information *on-line* [Narendra and Mukhopadhyay, 1992, Narendra and Parthasarathy, 1990]. Many neuromodelling and control applications have tended to ignore this and the ANNs are used solely for their non-linear approximation ability. To exploit the full potential of this area, *on-line* adaptation is *essential* in order to cope with time-varying plants operating in dynamic environments. Many of the learning rules currently developed are based on gradient descent, as originally occurred in the adaptive control field. However, it was found that under certain conditions (for instance, if the learning rates were set too high), the resulting closed-loop systems could become unstable. This led to stability-based learning laws being developed during the sixties and it is only recently that similar adaptive rules have been proposed for neural networks [Sanner and Slotine, 1992, Tzirkel-Hancock and Fallside, 1991]. The present stability theories are directly applicable to some ANNs, although it would be hoped that new results and algorithms could be developed which exploit network specific properties.

Future research must be aimed at understanding the modelling properties of the various ANNs, and proposing new architectures to improve their suitability for modelling and control applications. The inspiration for such improvements could be biologically based or come from the related fields of adaptive control and signal processing. Possibly, a more important topic is to understand and derive new learning laws which are applicable to various ANNs. The ability to prove convergence and stability is *crucial* if these learning algorithms are to be applied for on-line adaptive control, and for certain classes of networks which are *linear* in their weight vector (RBFs, lattice AMNs), the theories developed for adaptive linear systems can be modified and applied to these networks [Narendra and Annaswamy, 1989]. These algorithms should be compared with other nonlinear adaptive control strategies and the advantages in using these networks must be clearly quantified. This can only be achieved when theoretical results are available to compare the functional approximation and learning convergence of the different algorithms. Finally, the problems associated with applying these techniques to real-world tasks should not be underestimated. Algorithms need to be developed and improved for identifying the relevant network inputs, monitoring the adaptive networks and verifying and validating the knowledge stored in the network. For most practical applications these are non-trivial problems, but are essential for a successful implementation.

3

Associative Memory Networks

3.1 INTRODUCTION

The remainder of this book concentrates on a class of Artificial Neural Networks (ANNs), called Associative Memory Networks (AMNs), and investigates their suitability for both *on-line* and *off-line* nonlinear adaptive modelling and control. In order to use learning algorithms within adaptive control loops, their behaviour must be both desirable and predictable. Desirable properties include fast initial learning and long-term convergence. The behaviour of the plant must also be predictable in order to derive convergence and stability results which are *essential* when these adaptive algorithms form part of practical systems that require correctness, provability and certification. This chapter introduces the basic structure and describes the common features of, and the differences between, each AMN before the *commonly* utilised learning rules are derived and analysed in the following two chapters. These networks have specific properties, which allow them to be trained efficiently using *instantaneous linear* learning rules, and these are highlighted in this chapter. It is important to have a thorough understanding of the modelling ability of each network, as this influences their learning convergence and stability properties.

The class of AMNs has as members the Cerebellar Model Articulation Controller (CMAC), B-spline and fuzzy networks, Kanerva's Sparse Distributed Memory Model (KSDMM) and Radial Basis Functions (RBF). These networks store information *locally*, and it is this feature which distinguishes them from other neural networks, such as MLPs. This sparse internal representation also makes them suitable for instantaneous linear adaptation, and although many ANNs can be trained using such learning algorithms, learning may proceed very slowly when the underlying model structure is unsuitable. Their internal representation also makes AMNs more *transparent* than conventional ANNs, and this is especially true for fuzzy and B-spline algorithms which can be described using a set of fuzzy linguistic production rules. The link between fuzzy networks and neural networks is important, and AMNs provide a common framework with which to study both areas. This is useful, because it allows new learning rules to be developed for training fuzzy rules, and the establishment of the conditions under which convergence and stability results can be proved. It also allows the AMNs to be interpreted as a set of fuzzy rules, which can aid the initialisation and verification phases in a network design/test cycle.

The basic structure of these AMNs are now described, first at a high level, and then the features of the lattice networks (CMAC, B-spline and fuzzy) are presented in greater detail. The remainder of this book concentrates on these three networks, although the KSDMM and RBF networks provide alternative structures which are considered in this chapter. Generally, the differences between these algorithms are minor, but they give a network its specific features. This chapter concludes by reviewing the applicability of these networks to on-line modelling and control tasks. The structure of these networks has influenced their software implementation and, as described in Appendix C, it allows a *common* software data structure to be used for implementing the B-spline, fuzzy and CMAC algorithms. The data structure is constructed in a modular fashion, which allows the computational load to be evenly distributed across multiple processors.

3.2   A COMMON DESCRIPTION

The common high-level structural features of the AMNs are now described. These networks have much in common with the classical Perceptron, which was invented over thirty years ago by Rosenblatt. The Perceptron network is described first because it provides an informative historical perspective on the development of these AMNs (especially the CMAC).

3.2.1   Rosenblatt's Perceptron

AMNs have many structural similarities with the Perceptron network proposed by Rosenblatt [1961]. The classical Perceptron is an adaptive pattern classification network which uses a linear discriminant function in order to distinguish between different classes, and its three-layer network architecture is shown in Figure 3.1. Its input is an analogue-valued vector, and the nodes in the first layer are termed *sensory cells*. This input vector is typically denoted by s, but to remain consistent with the notation developed in this book, it is represented by x. The sensory cells are linked to the *association cells* in the hidden layer using a *sparse* set of random connections, and the weight on each connection is randomly set. If the summed input to an association cell exceeds 0, the output of this node is 1; otherwise it is 0. Remaining consistent with the notation developed for the Perceptron, and subsequently used in the CMAC, the output of the association cells is denoted by the vector a. The association layer is fully connected to the output node by a set of adaptive weights, w, and the output node, or response cell, sums the weighted

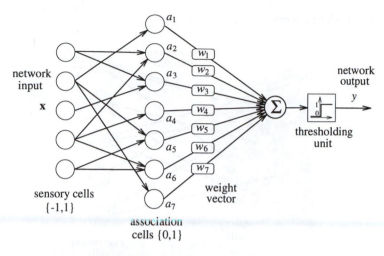

network
input

x

network
output

y

thresholding
unit

a_1
a_2
a_3
a_4
a_5
a_6
a_7

w_1
w_2
w_3
w_4
w_5
w_6
w_7

sensory cells
{-1,1}

weight
vector

association
cells {0,1}

Figure 3.1 Rosenblatt's Perceptron.

contribution from each association cell and this gives a network output, y, of 1, when this quantity is greater than 0, and 0 otherwise. This binary output represents the network classification for a two decision class problem. If there were more than two decision classes, the output layer would be composed of more than one response cell, and the adjustable weight vector would become a matrix (one vector for each response cell). When a Perceptron has m response cells, its behaviour (training and recall) is identical to m single-output networks which have the same structure, although the form of each output mapping is generally different. Therefore, it is often better (and no worse) to use multiple-input, single-output networks, and this is the approach described in this book.

Generalisation in the Perceptron

Structurally the network can be broken down into two mappings. The first map has the sensory cells as its inputs and the association cells as outputs. This $\mathbf{x} \rightarrow \mathbf{a}$ mapping is *fixed* and *nonlinear*. The second map has as its inputs the outputs of the association cells, and the output is the classification of the response cells. Here the $\mathbf{a} \rightarrow y$ mapping is *adaptive* and the input to the response cell is a *linear* sum of the inputs.

Generalisation in the Perceptron depends on the random connections between the sensory and the association cells. When the sets of non-zero association cells are *different* for two separate inputs, the output depends on different parts of the weight vector. The learning rule (which is discussed in the next section) only updates those weights which contributed to the output. Therefore, when the network is updated for one input, its ability to recall the second pattern is not affected. This *sparse* coding of the input space dramatically reduces the amount of *learning interference* in the network.

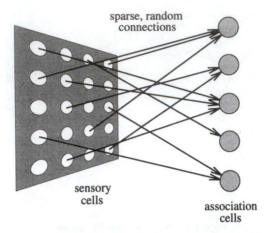

Figure 3.2   Generalisation in the Perceptron network. The random connections between the sensory and association cells means that topological structure of the input space is not preserved.

For modelling and control purposes, the desired function is generally smooth; similar network inputs are mapped to similar network outputs. This immediately imposes a topological structure on the input space. The mapping between the sensory cells and the association cells in the Perceptron is *random*, which may be useful for simulating visual phenomena, but it does not preserve the topological features of the input space. Hence, the Perceptron is not directly useful for on-line modelling and control applications. This is illustrated in Figure 3.2, where topologically different inputs are connected to the same association cell. It was this property, that the sensory/association cell mapping should conserve the topology of the input space, which influenced Albus' development of the CMAC network [Albus, 1975a].

Perceptron Training Algorithm

The Perceptron's weight vector can be trained using a simple learning rule which adapts the weight vector only when the network incorrectly classifies an input. This learning rule is guaranteed to converge when the patterns presented to the network are linearly separable [Fukunaga, 1990], otherwise the weight vector does not converge to a fixed value.

    The Perceptron learning rule for the two class problem can be expressed as follows: define **b** to be the modified output of the association cells in the hidden layer, where $b_i = -a_i$ or a_i (the outputs of the association cells), depending on whether **x** lies in the first or second class, respectively. The weight vector is then updated according to:

$$\mathbf{w}(t) = \begin{cases} \mathbf{w}(t-1) & \text{if } \mathbf{w}^T(t-1)\mathbf{b}(t) > 0 \\ \mathbf{w}(t-1) + \delta'\mathbf{b}(t) & \text{otherwise} \end{cases} \qquad (3.1)$$

where δ' is the learning rate. When the first condition is satisfied, the input has been correctly classified and the weight vector does not need to be updated. If however the input has been incorrectly classified (second condition), the weight vector should be modified so as to increase the (unthresholded) *a posteriori* network output. This is achieved, since an incorrect classification means that the unthresholded *a posteriori* network output is given by:

$$\mathbf{w}^T(t)\mathbf{b}(t) = \mathbf{w}^1(t-1)\mathbf{b}(t) + \delta'\|\mathbf{b}(t)\|_2^2 > \mathbf{w}^T(t-1)\mathbf{h}(t) \qquad \text{for } \delta' > 0$$

The learning rate δ' still needs to be selected, and three common methods for choosing it are now given. The original fixed increment rule sets δ' equal to an arbitrary positive constant, δ, and when there exists a weight vector which can correctly classify all of the training data, this learning algorithm converges in a finite number of iterations. The *error correction* rule selects δ' such that $\mathbf{w}^T(t)\mathbf{b}(t) > 0$. This occurs when the learning rate satisfies:

$$\delta' > -\frac{\mathbf{w}^T(t-1)\mathbf{b}(t)}{\|\mathbf{b}(t)\|_2^2}$$

The final method for selecting δ' is using a *gradient correction* rule, that is:

$$\delta' = \delta\left(\widehat{y}(t) - \mathbf{w}^T(t-1)\mathbf{b}(t)\right)$$

where $\widehat{y}(t)$ is the desired class of the input $\mathbf{x}(t)$. These last two methods for choosing the learning rates produce training rules which are *very* similar to the instantaneous normalised least-mean square and the least-mean square learning algorithms, respectively (see Chapter 5), that are used to train various AMNs which produce a *continuous* output.

3.2.2 Associative Memory Networks

AMNs are used generally for functional approximation tasks, although they share many common features with the pattern classification Perceptron network. All of these networks can be represented as a three-layer system, as shown in Figure 3.3. They are *model based*, having a fixed structure and a set of adaptive parameters which are iteratively trained in order to achieve a desired behaviour. However, the basic structure of the network is quite flexible, and so such schemes are termed *soft* or *weak* modelling algorithms.

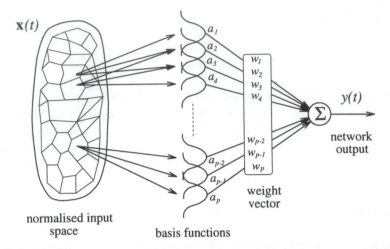

Figure 3.3   A schematic associative memory network. The association cells in the hidden layer are represented by local basis functions and this produces a topology conserving normalisation of the input space.

Structure

In an AMN, the Perceptron's sensory cells are replaced with a layer which *normalises* the original input space. This normalisation can take many forms: it could map the real input to a binary input vector and thus emulate the sensory cells in the Perceptron, it could define an n-dimensional lattice on the input space or a more complex partitioning strategy, or it could simply pass on the inputs without modification. The *basis functions*, which represent the association cells, are defined on this normalised input space. Their size, shape and overlap determine how the network generalises in n-dimensional space and also its complexity. There are many different ways in which the input space can be normalised, and also many different forms of basis functions, and these differences distinguish the various members of the class of AMNs. The outputs of the basis functions, a_i, generally lie in the interval $[0, 1]$, and when the output is zero, the associated basis function and the corresponding weight do not contribute to the network's output. In common with the Perceptron network, there is no bias term included in the hidden layer.

For some AMNs, this topology conserving input layer transformation must be specified by the designer before learning commences and this enables *a priori* model knowledge to be encoded in the network design. Other networks parameterise the basis functions and adapt these variables during learning and both types of approaches have advantages and disadvantages. The networks which adapt the basis functions are certainly the most *intelligent* ones, although when the data are biased towards one region of the input space during training, the representations formed in the remaining areas may not be sufficient. Hardwiring this information into the network at the design stage overcomes this problem, often at the cost of increased memory requirements and greater *a priori* understanding of the process. This *biases* the network (see Section 2.5.1) and reduces the variance of the parameter estimates.

The location of the basis functions in the input space can have a significant effect on the efficiency of the algorithm when implemented on a serial machine. If the basis functions are distributed randomly in the input domain (or the positions are irregular), the output of every basis function in the hidden layer needs to be calculated. However, when the basis functions are positioned in a regular fashion across the input space, the basis functions respond (non-zero output) in only a small region, a predefined algorithm can be used to calculate the addresses of these relevant basis functions. Thus an extensive search and comparison process can be avoided.

The output of the AMN, y, is formed from a linear combination of the outputs of the basis functions. It is, in general, *continuous*; the linearly combined basis functions are not thresholded (unlike the Perceptron output). The linear coefficients are the adjustable weights, w_i, and because the output is linearly dependent on the weight set, learning is simply a *linear* optimisation problem. Therefore:

$$y = \sum_{i=1}^{p} a_i w_i = \mathbf{a}^T \mathbf{w}$$

where the outputs of the basis functions depend *nonlinearly* on the inputs, and at time t this can be written as:

$$y(t) = \mathbf{a}^T(t)\,\mathbf{w}(t-1) = \mathbf{a}^T(\mathbf{x}(t))\,\mathbf{w}(t-1)$$

The first expression for $y(t)$ is generally used in this book, although the second is adopted whenever it needs to be shown explicitly that the basis function's output, a_i, depends on the network's input, $\mathbf{x}$.

Learning Rules

The instantaneous learning rules that are used to train the AMNs are derived and rigorously analysed in Chapter 5, although a basic description is given below because it highlights the type of networks with which they should be used. These training algorithms update the weight vector after the presentation of each data pair $\{\mathbf{x}(t), \hat{y}(t)\}$, where $\hat{y}(t)$ is the desired output (response) of the network. The weight vector is then adapted using one of the following rules:

$$\mathbf{w}(t) = \mathbf{w}(t-1) + \delta(\hat{y}(t) - y(t))\mathbf{a}(t) \tag{3.2}$$

$$\mathbf{w}(t) = \mathbf{w}(t-1) + \delta\left(\frac{\hat{y}(t) - y(t)}{\|\mathbf{a}(t)\|_2^2}\right)\mathbf{a}(t) \tag{3.3}$$

where δ is the learning rate. These are termed instantaneous *gradient descent* and *error correction* learning rules, respectively. Both of the weight updates have the same basic form: *scalar* multiplied by the *transformed input vector* **a**, where the scalar depends on the network output error. They update the weight vector parallel to the transformed input vector (the search direction) and the step length is given by the value of the scalar. The error correction learning rule *normalises* the step length by the size of the transformed input vector and is sometimes termed a normalised gradient descent rule (the reason for this is explained in Section 5.2.2). Like the Perceptron training rules, when the output error is zero the weights are not adapted, whereas if it is non-zero the weights are adjusted such that the instantaneous *a posteriori* network output error is reduced.

For the Perceptron network, $\mathbf{a}(t)$ is sparse; only a small number of coefficients are non-zero. If the basis functions in the AMN have a non-zero output in only a small part of the input space, the transformed input vector is also sparse and a small number of weights contribute to the network's output. Hence only these parameters should be updated when the network is trained, using an instantaneous learning rule. The nonlinear transformation is topology conserving and so similar network inputs map to similar sets of non-zero basis functions. Dissimilar inputs are mapped to completely different sets of basis functions and so knowledge is stored *locally*. This is termed *local generalisation* since learning in one area of the input space does not interfere with the knowledge pertaining to a different region, and the network distributes the information locally.

Terminology

Before the individual AMNs are described in more detail, the terminology which is used to describe networks' features is explained. The *support* of a basis function is the domain in the input space for which the basis function's output is non-zero, and this is often referred to as a *receptive field* in the neural network literature. A basis function has a *bounded support* when its support is smaller than the network's domain. This also referred to as a *compact support*. The basis functions of a *lattice* AMN are defined on an n-dimensional lattice in the input space which provides the normalisation of the input space. Since a large part of this book is concerned with lattice AMNs, the extra structure imposed on these networks will now be described.

3.2.3 Lattice-Based Associative Memory Networks

The input space of this type of AMN is normalised by a lattice on which the basis functions are defined. These networks have many desirable features: the address of the non-zero basis functions can be explicitly calculated (no search procedure is required), the transformed input vector is generally sparse which means that knowledge is stored and adapted locally and these networks are generally *transparent*, in that the weights can be interpreted as vague or fuzzy rules. However, the memory requirements for these networks are exponentially dependent on the dimension of the input space, and can therefore only be used in applications which have small- or medium-sized input spaces, such as dynamic modelling and control.

The Input Lattice

Partitioning the input space using a lattice is a very simple strategy, and this has both advantages and disadvantages. The advantages of this approach include:

- **ease of construction**, as partitioning each input axis implicitly generates the n-dimensional lattice;
- **prior network structuring**, as knowledge about the variation of the desired function with respect to a particular input can be easily incorporated into the model when the lattice is specified; and
- **computational simplicity**, as the procedure to find which cell an input lies in can be formulated as n univariate search algorithms.

However, there are some potential disadvantages with this technique:

- **curse of dimensionality**, which results in the network's memory requirements being exponentially dependent on the input space dimension; and
- **inappropriate generalisation**, where the representation formed in one part of the input space may not be appropriate in another.

The latter point can be illustrated by considering the two-dimensional lattice shown in Figure 3.4. A detailed representation may be required when both x_1 and x_2 are small and a lattice produces a fine representation in this region. However, when

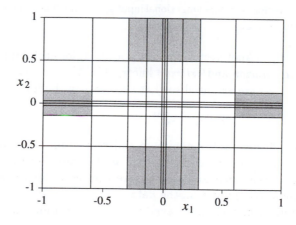

Figure 3.4   Inappropriate generalisation when a lattice is used to partition the input space. An overly complex model may be produced in the shaded areas.

either input is large, a coarse model may be sufficient and a lattice network would be over-parameterised.

These limitations should be noted, although it is possible to overcome them partially by using advanced domain decomposition algorithms based on k-trees, and some of these techniques are briefly described in Section 8.5.

Knot Placement

In order to define a lattice on the input space, a set of n knot vectors must be given, one knot vector for each input axis. These knot values give the positions of the $(n-1)$-dimensional hyperplanes which are parallel to the other $(n-1)$ axes, and the set of all the hyperplanes generates the lattice in the input space. There are usually a different number of knots on each axis and they are generally placed at different positions. Therefore the designer is able to incorporate *a priori* knowledge into the network's design when the position of these knots is specified. Figure 3.5 shows a nonlinear knot placement strategy on a one-dimensional input space.

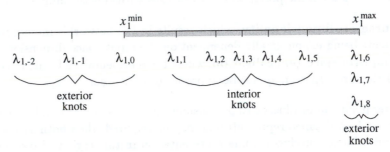

Figure 3.5   A knot vector on a one-dimensional input space which uses a nonlinear placement strategy, $r_1 = 5$ and $k_1 = 3$.

More specifically, a knot vector must be specified, for each input axis, which is composed of both *interior* and *exterior* knots. The $\sum_i r_i$ interior knots $\lambda_{i,j}$, ($j = 1, \dots, r_i$, $i = 1, \dots, n$) are arranged so that they satisfy the following relationship:

$$x_i^{\min} < \lambda_{i,1} \le \lambda_{i,2} \le \cdots \le \lambda_{i,r_i} < x_i^{\max} \tag{3.4}$$

where $x_i^{\min}$ and $x_i^{\max}$ denote the minimum and maximum values of the ith input, respectively. When two or more knots occur at the same position, this is termed a *multiple* or *coincident* knot and this feature can be useful when attempting to model underlying discontinuities in the data.

At the extremities of each input axis, a set of k_i exterior knots must be given which satisfy:

$$\lambda_{i,-(k_i-1)} \le \cdots \le \lambda_{i,0} = x_i^{\min}, \qquad x_i^{\max} = \lambda_{i,r_i+1} \le \cdots \le \lambda_{i,r_i+k_i} \tag{3.5}$$

These knots are necessary for generating the basis functions of width k_i which are close to the boundary of the lattice. The network's input space is the domain $[x_1^{\min}, x_1^{\max}] \times \cdots \times [x_n^{\min}, x_n^{\max}]$, so the exterior knots are only used for defining these basis functions and the extrema of the input lattice.

The jth univariate interval on the ith axis is denoted by $I_{i,j}$ and is defined to be:

$$I_{i,j} = \begin{cases} [\lambda_{i,j-1}, \lambda_{i,j}) & \text{for } j = 1, \ldots, r_i \\ [\lambda_{i,j-1}, \lambda_{i,j}] & \text{if } j = r_i + 1 \end{cases} \tag{3.6}$$

The (r_i+1) intervals, $I_{i,j}$, partition the ith axis, and when two or more of the knots are coincident, one of more of these intervals is *empty*.

There are two common methods for choosing the exterior knots; setting them all equal to the appropriate endpoint or placing them at equispaced intervals, and both of these strategies are illustrated in Figures 3.5 and 3.6.

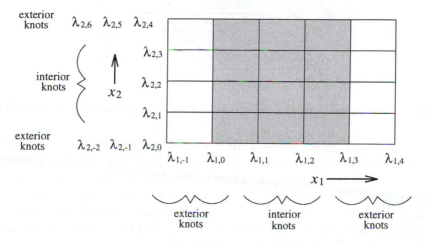

Figure 3.6 A two-dimensional lattice where the shaded area represents the input space of size (3×4). Equispaced exterior knots are used on the first axis, and coincident exterior knots on the second. The univariate basis functions' supports are two and three intervals on the first and second axes, respectively, so two or three exterior knots must be defined on the appropriate univariate input.

On each axis there are $(r_i + 1)$ intervals (possibly empty if the knots are coincident) which means that there are $p' = \prod_{i=1}^{n}(r_i + 1)$ n-dimensional *cells* in the lattice.

The Basis Functions

The output of the hidden layer in a lattice AMN is determined by a set of p basis functions which are defined on the n-dimensional lattice. The shape, size and distribution of the basis functions are characterised by the type of lattice AMN being used, and the supports of the lattice basis functions are n-dimensional rectangles which are generally bounded, as illustrated in Figure 3.7. For any input

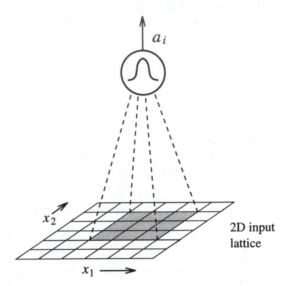

Figure 3.7   The support (shaded area) of a two-dimensional multivariate lattice basis function.

to a CMAC or a B-spline lattice AMN (and certain types of fuzzy networks), the number of non-zero basis functions is a *constant*, and is called the *generalisation parameter*, which is denoted by ρ. These lattice AMNs have a particular network structure which allows an efficient implementation in software (Appendix C) and hardware, and also provides an insight into how the networks *generalise* in n-dimensional space. This is used extensively in Chapter 7, when the modelling capabilities of the binary CMAC are investigated.

For each input, ρ basis functions contribute to the output and so they can be organised into ρ sets, where one and only one basis function in each set is non-zero. The union of the supports in each set forms a complete and non-overlapping n-dimensional *overlay*. Each overlay is complete and the supports are non-overlapping, because one and only one basis function from each set contributes to the output and it is just large enough to cover the lattice. There are ρ sets and so the lattice is covered by ρ overlays. The overlays for a two-dimensional space with $\rho = 3$ are illustrated in Figure 3.8, where the size and shape of the basis functions' supports are contained in the ρ overlays.

This decomposition of the basis functions into ρ sets of overlays illustrates how the total number of basis functions, p, is *exponentially* dependent on the input space dimension when all of the univariate supports are bounded. The total number of basis functions is greater than the number of multivariate supports on each overlay which in turn is equal to the *product* of the number of univariate basis functions

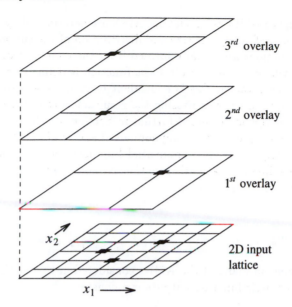

Figure 3.8 A two-dimensional lattice AMN with $\rho = 3$. Each overlay is composed of the basis functions' supports, and each network input lies in the support of exactly three basis functions.

on each axis. The univariate basis functions have a bounded support and so there exist two (or more) basis functions defined on each axis, and the number of basis functions on each overlay is bounded below by 2^n. Therefore, these networks suffer from the *curse of dimensionality* [Bellman, 1961] and are only generally applicable to small- or medium-dimensional input spaces.

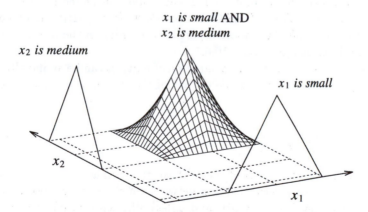

Figure 3.9 A two-dimensional basis function formed from two combined (using the *product* operator) univariate basis functions. It also represents the fuzzy intersection (AND) of two univariate membership functions.

The modelling capabilities of a lattice AMN are determined by the shape of the basis functions, as well as by the size and distribution of the supports. These multivariate basis functions are usually formed from the combination of n univariate basis functions, where one univariate basis function is defined on each input axis. Under certain circumstances this is equivalent to defining a *distance measure* on the normalised input space and passing the resultant output through a univariate basis function and this is discussed further in Section 6.4. These univariate basis functions are generally piecewise polynomials [Lane *et al.*, 1992], although other shapes based on trigonometric, exponential functions, etc., can also be used. In Figure 3.9, a two-dimensional basis function is shown, which is formed from two univariate, triangular basis functions.

Partitions of Unity

The basis functions in many of the AMNs considered in this book are *normalised*; the sum of all the basis function outputs is a constant:

$$\sum_{i=1}^{\rho} a_{ad(i)}(\mathbf{x}(t)) = \text{constant} \tag{3.7}$$

where $ad(i)$ is the address of the non-zero basis function on the ith overlay. If a network possesses this property, it is said to form a *partition of unity* or possess a *constant field strength* [Werntges, 1993]. This is a desirable property because any variation in the network's surface is solely due to the weights in the network. Otherwise the network's response may depend on the size of the field strength and vary according to the position of the input. Generally, as the dimension of the input space increases, the variance of the field strength also increases, unless the network forms a partition of unity. Of the networks considered in this book, both the B-spline and the binary CMAC lattice AMNs form partitions of unity, and the fuzzy systems *implicitly* impose a partition of unity on the input membership functions, as described in Section 10.6.1.

When a network does not form a partition of unity, it can be artificially endowed with this property by redefining the output of the ith non-zero basis function, for $i = 1, \ldots, \rho$, as:

$$a_{ad(i)}(\mathbf{x}(t)) = \frac{\tilde{a}_{ad(i)}(\mathbf{x}(t))}{\sum_{j=1}^{\rho} \tilde{a}_{ad(j)}(\mathbf{x}(t))} \tag{3.8}$$

where $\tilde{a}_k(.)$ is the "true" output of the kth basis function and $a_k(.)$ is its *normalised* version. It can be easily verified that a network which uses these normalised basis functions forms a partition of unity, and unless otherwise stated these normalised basis functions are used throughout this book in the CMAC, B-spline and fuzzy networks. In addition, the weights are sometimes initialised to a NULL value and hence do not contribute to the output until its value is set. When this situation occurs, there are (at least) two different ways in which the information can be treated. The NULL value can simply be assumed to be zero, and the basis functions are normalised as previously described. Another possible solution is not to include the basis function outputs with NULL weights in the normalising sum of Equation 3.8 and a flag is set to indicate that they should be appropriately initialised. These basis functions and their NULL weights are not included in the output calculation, and the weights are initialised to the desired network output [Tolle and Ersü, 1992], as described in Appendix A. In either case the resulting network always forms a partition of unity.

The importance of the network forming a partition of unity is demonstrated in Section 6.4.3, where it is shown that the output of a set of normalised basis functions is generally much smoother than that of the corresponding unnormalised network. This is a desirable property for modelling and control applications, and for this reason all of the lattice AMNs considered in the remainder of this book possess this feature.

3.3   FIVE ASSOCIATIVE MEMORY NETWORKS

Until now the common features of the different AMNs have been emphasised. In this section, however, the differences and the various properties of five common AMNs are discussed. These five networks described are the CMAC, B-spline and fuzzy networks which are lattice AMNs, the Radial Basis Function network and Kanerva's Sparse Distributed Memory Model.

3.3.1   The CMAC Network

The CMAC was first proposed by J.S. Albus [1975a, b]. It is a lattice AMN and has been used as a learning algorithm for modelling and controlling high-dimensional, nonlinear plants such as robotic manipulators. Its basic structure is very similar to Rosenblatt's Perceptron and the original notation used for the two networks is the same. In its simplest form, the CMAC is a *look-up table* where the basis functions generalise locally.

The CMAC has been used as a basic learning module in a variety of applications. Ersü and his co-workers have used it for the adaptive modelling and control of a wide variety of different plants: biotechnical processes, robotic manipulators, vehicle guidance systems [Tolle and Ersü, 1992, Tolle *et al.*, 1994], etc., utilising a variety of different control architectures. More recently, Miller's group in New Hampshire [Miller *et al.*, 1990a] have been using the CMAC for real-time adaptive control of multi-degree of freedom robotic manipulators and for adaptive pattern recognition and signal processing problems. The CMAC has been applied to driver modelling and vision recognition, and several research groups have looked at the algorithm from a "fuzzy" viewpoint [Ozawa and Hayashi, 1992], where fuzzy-type membership functions are used to implement the univariate basis functions. Many other different schemes have been proposed to improve the basic algorithm by An *et al.* [1991], Lane *et al.* [1992] and Parks and Militzer [1992], and these are discussed in Chapter 6.

Topology Conserving Mapping

The main difference between the CMAC and the Perceptron is the way in which the fixed nonlinear mapping is generated. For on-line modelling and control, this mapping must be topology conserving in order to learn information locally, otherwise the algorithm's behaviour can be unpredictable. Albus devised a clever encoding scheme which ensures that the user-defined generalisation parameter ρ does not depend on the input space dimension. In this scheme, the supports are distributed on the lattice, such that the projection of the supports onto each of the axes is *uniform*. This is called the *uniform projection principle*, and algorithms which satisfy this rule have their basis functions distributed *evenly* on the input lattice. A network satisfies the uniform projection principle if the following condition holds:

> *As the input moves along the lattice one cell parallel to an input axis, the number of basis functions dropped from, and introduced to, the output calculation must be a constant and not depend on the position of the input.*

This constant may be different for each axis, although it is equal to one for each input axis of the CMAC network.

When the generalisation parameter is specified, it also determines the *size* of the supports. Each basis function has a support which is ρ units wide in each dimension, and hence the supports are hypercubes of volume ρ^n defined on the lattice. These ρ overlays are then distributed such that the uniform projection principle holds. From this it can easily be seen that as ρ increases, the basis functions cover the lattice in a *sparser* fashion. The original Albus distribution places the first corner of each of overlay along the lattice diagonal, such that the first interior lattice point of the ith overlay occurs at the point $(i, i, \ldots, i)^T$, for $i = 1, \ldots, \rho$. This distribution obviously satisfies the uniform projection principle, because the set $\{1, 2, \ldots, \rho\}$ is generated by the projection of all of the supports onto each axis, and this is illustrated in Figure 3.10 for a two-dimensional CMAC with $\rho = 3$. As the input moves one cell parallel to any axis, a hyperplane (line) on one and only one overlay is crossed and the weight set shares $(\rho - 1)$ weights in common with the previous weight set. Therefore, similar inputs map to similar sets of non-zero basis functions, and the nonlinear transformation preserves the topology of the input space.

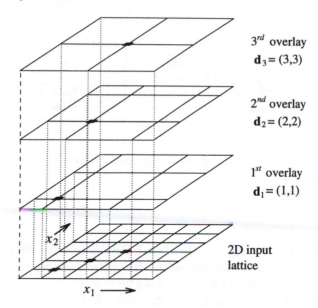

Figure 3.10 The uniform projection principle illustrated for a two-dimensional CMAC with $\rho = 3$. Each line of the input lattice corresponds to a line lying in one and only one of the overlays.

Basis Function Shape

The size and the distribution of the basis functions are determined when the designer specifies the generalisation parameter ρ. There are alternative overlay distribution strategies [Parks and Militzer, 1991a], and these are discussed in Section 6.2.3. All that remains to be specified is the shape of the basis functions.

The original CMAC had *binary* basis functions, the output of which is either on (a positive number) or off (identically zero). Within each n-dimensional cell, the network's output is constant, so the basic algorithm is reminiscent of a look-up table where the basis functions generalise locally. For this type of basis function, most of the computations can be carried out using integer operations only, so the basic algorithm is very efficient, as shown in Section 6.5.2.

Some implementations require a continuous network output (at the very least) and the only way to achieve this is to use *higher order* basis functions. The network's output is formed from a linear combination of the basis functions, with the smoothness of the basis functions being reflected in the output of the network. Binary and order 2 basis functions are shown in Figure 3.11, with the latter generating a *continuous* network output.

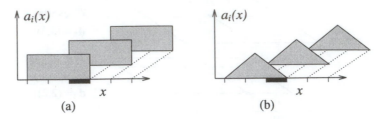

Figure 3.11   Active binary (a) and order 2 (b) univariate basis functions for an input lying in the shaded interval, $\rho = 3$.

CMAC Summary

The basic CMAC network is a distributed look-up table which generalises locally. Its internal structure is determined by the generalisation parameter which not only denotes the number of non-zero basis functions for each input, but also specifies the size of the hypercubic supports. The CMAC implicitly assumes that the basis functions have the same sized support on each axis, and then distributes the basis functions *sparsely* on the lattice. As the generalisation parameter is increased the supports become larger, and the basis functions are more sparsely distributed on the lattice. In addition, learning becomes less local and, while the initial rate of convergence may be quicker, the long-term convergence is generally slower. The modelling capabilities, the computational cost and the rate of learning of the CMAC therefore depend on the generalisation parameter (for greater detail see Chapters 6 and 7). Higher order basis functions have been proposed as a method for generating a continuous CMAC output, and this requires a set of $(\rho - 1)$ exterior knots to be specified at the extremities of each univariate axis for the basis functions to be well defined.

When the parameters of the CMAC are set sensibly, the network generally produces good results. This is especially true when the improved overlay distribution strategy, as described in Appendix B, is employed.

3.3.2   B-spline Network

B-splines are more commonly found in graphical applications [Cox, 1990]. Their use as basic surface fitting elements is commonplace, and it is slightly surprising that they have rarely been proposed for use as adaptive functional approximation algorithms. When B-splines are used to model data sets it is possible to modify the data locally, and the corresponding change in the network's response is also *local*. This is one of the main features that makes B-splines attractive for on-line adaptive modelling and control; learning about one part of the input space *minimally* affects the response of the model for different inputs.

B-splines have been used in several different areas of robotics research. Their first, and most obvious, application was in trajectory generation where a smooth spline is fitted to a set of discrete data points which describes the desired behaviour of the robot. These splines make it possible to modify the trajectory locally in response to unforeseen circumstances due to their localised nature, as is shown in Chapter 9. However, the main focus of this book is adaptive modelling and control, and the B-spline network was proposed for this purpose by Moody [1989], where the algorithm was benchmarked against a standard chaotic time series prediction problem. Applications are varied: autonomous vehicle parking algorithms, non-linear robotic compensation, robotic actuator modelling, and in Section 11.4 an algorithm is given for synthesising a controller based on a B-spline or fuzzy plant model.

The most important feature of the B-spline algorithm is the *smooth* network output which is due to the shape of the basis functions; the distribution of the basis functions is automatically determined by the algorithm. For this reason, the shape of the basis functions will now be described.

B-spline Basis Functions

The design parameters of this network are the orders of the splines. These *explicitly* determine the shape of the basis function and *implicitly* set the size of its support and the generalisation parameter. The univariate B-spline basis functions of order k have a support which is k intervals wide, and so choosing the order of the polynomial approximation also specifies the size of the support of the univariate basis function. Hence, each input is mapped to k non-zero basis functions and so the generalisation parameter is also k. A set of univariate B-spline basis functions is shown in Figure 3.12, and there exists a simple and stable recurrence relationship for evaluating the output of the basis functions, which is described in Section 8.2.2. The univariate basis functions are piecewise polynomials of order k, so that the output of the B-spline network is also a piecewise polynomial of order k. Within each interval the basis functions are polynomials of order k, and at the knots its first $(k - 2)$ derivatives are continuous. Therefore, the B-spline network is simply a piecewise polynomial approximation scheme, which is also a member of the class of lattice AMNs.

Dilated B-splines have been proposed as a method for partially relaxing the relationship which exists between the order of the spline, the size of the support and the generalisation parameter (see Section 8.2.5). This technique is similar to ordinary B-splines, in that the basis functions are simply piecewise polynomials, although they allow the size of the support (and hence the generalisation parameter) to be an integer multiple of the order of the basis functions. For example, when quadratic B-splines are used ($k = 3$) the width of the support can be set to be $3i$, for $i = 1, 2, \ldots$.

Multivariate basis functions are formed by taking the *tensor product* of the uni-

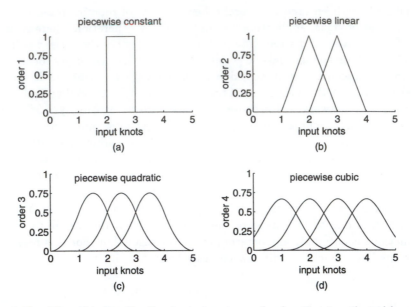

Figure 3.12   The univariate B-spline basis functions, of orders $1-4$, activated by an input lying in the third interval $(=[\lambda_2, \lambda_3))$.

variate basis functions. Thus each multivariate basis function is formed from the product of n univariate basis functions, one from each axis and every possible combination of univariate basis functions is taken (see Figure 3.9). The network output is linearly dependent on these multivariate basis functions, and because the univariate basis functions are combined using the *product* operator, their smoothness is reflected in the output of the network. Denoting the order of the univariate basis functions on the ith input by k_i, the support of each multivariate basis function is a hyperrectangle of size $(k_1 \times k_2 \times \cdots \times k_n)$ defined on the n-dimensional lattice. The generalisation parameter is also uniquely determined by the set $\{k_i\}_{i=1}^n$, and is given by $\rho = \prod_{i=1}^n k_i$. This is exponentially dependent on the n and as the dimension of the input space grows, the computational cost of implementing this algorithm increases exponentially in n.

Placement of the Basis Functions

For the multivariate B-spline network the generalisation parameter (which denotes the number of overlays) is given by $\rho = \prod_{i=1}^n k_i$, and the volume of the support of each basis function is also $\prod_{i=1}^n k_i$. Thus the only way to distribute the ρ overlays within the first hyperrectangle is to place an overlay on *every* available interior lattice point. This is in direct contrast to the CMAC network which *sparsely* covers the interior lattice points, as illustrated in Figure 3.13.

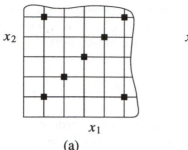
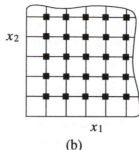

(a)   (b)

Figure 3.13   For basis functions whose support is of size (4×4), the CMAC forms 4 overlays and sparsely populates the lattice (a), whereas the B-spline algorithm generates 16 overlays (b) and *completely* covers the lattice.

It would be incorrect to say that either the CMAC or the B-spline overlay displacement strategy is better; all that can be said is that they are different. The B-spline network allows the order of the splines on each axis to be different, therefore *a priori* knowledge to be incorporated into the network design. New knots can also be introduced into the B-spline network, and the old network can be reproduced *exactly* by the new network [Kavli, 1994]. However, the computational cost is exponentially dependent on n, whereas the CMAC provides a sparser coding of the input space, and its computational cost is only linearly dependent on n. Also, new knots cannot be introduced into the CMAC without changing the modelling capabilities of the network and the generalisation is always the same on each axis.

3.3.3   Fuzzy Systems

It might be a little surprising to see fuzzy systems included in a section describing AMNs. However, on a high level the basic information processing principles are the same, and in Chapter 10 it is shown that under certain technical conditions the low-level algorithm is *also* identical. The parallel drawn between fuzzy networks and AMNs is important, because it allows the learning algorithms and convergence proofs (of AMNs) to be employed in fuzzy networks and equally it enables the information stored in an AMN to be interpreted as a set of fuzzy production rules which provide network *transparency*. It also highlights the two main areas in which research should be concentrated: data representation and knowledge processing.

Research into fuzzy information processing started in the mid-sixties when Zadeh [1965, 1973] proposed the investigation of systems from a fuzzy viewpoint. Fuzzy modelling and control algorithms were originally developed in the UK in the seventies [Procyk and Mamdani, 1979], and during the eighties and nineties fuzzy theory has been extensively applied to a wide range of consumer products in Japan [Berenji, 1992]. Among the most notable of these products have been an automatic train controller and a helicopter controller. In recent years, the link between symbolic processing (fuzzy) and numerical processing (neural) has been investigated, and this has resulted in architectures which integrate the representational ability of fuzzy systems with the learning ability of neural networks.

Fuzzy Knowledge Representation

A fuzzy system generally stores its knowledge in the form of a fuzzy algorithm, which is composed of a set of fuzzy production rules relating the network's input to its output:

$$
\begin{aligned}
r_{11,1}: \qquad & \text{IF } (x \text{ is negative small AND } \dot{x} \text{ is positive medium}) \\
& \text{THEN } (y \text{ is negative large}) \qquad\qquad (c_{11,1}) \\
r_{25,4}: \quad \text{OR} \quad & \text{IF } (x \text{ is almost zero AND } \dot{x} \text{ is almost zero}) \\
& \text{THEN } (y \text{ is almost zero}) \qquad\qquad (c_{25,4}) \qquad\qquad (3.9) \\
r_{37,5}: \quad \text{OR} \quad & \text{IF } (x \text{ is positive medium AND } \dot{x} \text{ is negative medium}) \\
& \text{THEN } (y \text{ is positive small}) \qquad\qquad (c_{37,5})
\end{aligned}
$$

The confidence, $c_{i,j}$, in the ij^{th} fuzzy rule ($r_{i,j}$) being true is stored, and the set of all the rule confidences defines the plant input/output mapping, along with the fuzzy linguistic descriptions and the logical operators.

To implement the above rule set, a description of each of the univariate fuzzy linguistic statements (e.g. *x is positive small*) needs to be given and the operators which are used to implement the underlying fuzzy logic (AND, OR, IF() THEN(), etc.) must be specified. There is no single correct implementation method, and many different methods have been proposed [Harris *et al.*, 1993]. Originally, truncation operators, *max* and *min*, were used to implement the fuzzy operators and the fuzzy linguistic variables were generally chosen to be triangular functions. Recent research into the similarities between fuzzy and neural systems has seen algebraic operators being proposed to implement the underlying fuzzy logic and alternative basis function shapes (Gaussian, B-splines, etc.) being used to represent the linguistic variables [Brown and Harris, 1991, Bruske, 1993, Pacini and Kosko, 1992, Wang, 1994, Wang and Mendel, 1992] and Chapter 10.

The fuzzy rules can be trained by adapting the rule confidences which changes the strength with which a rule fires. These rules can also be initialised using an expert's *a priori* knowledge, and in practice both of these network design methodologies are used to construct a fuzzy rule base.

The information flow through a fuzzy system is shown in Figure 3.14. A *crisp* input (a single value rather than a fuzzy or possibility distribution) is presented to the network, and the memberships of the multivariate fuzzy input linguistic variables (represented by multivariate fuzzy sets) are calculated. The confidence in each of the fuzzy output linguistic variables are then determined, and the network output is obtained by *defuzzifying* this information. It can be shown that, when certain assumptions are made (see Chapter 10), the output is linearly dependent on

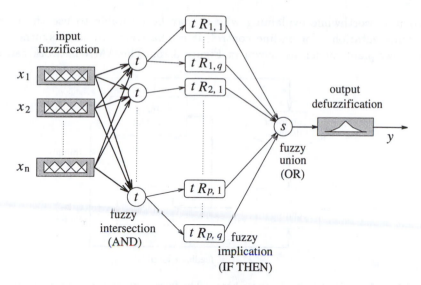

Figure 3.14 Information flow through a continuous fuzzy system. There are p multivariate fuzzy input sets and q univariate fuzzy output sets.

the multivariate fuzzy input basis functions and so there is a direct correspondence with the CMAC and B-spline lattice AMNs, and Gaussian RBF networks.

Fuzzy AMN Implementation

It has been stated that the output of a set of fuzzy rules is equivalent to the output of an AMN with appropriately defined univariate basis functions. In this case, a fuzzy network can be encoded directly as an AMN, or else the knowledge can be encoded explicitly as a set of rule confidences and a fuzzy inferencing algorithm applied to the network. The shape of the univariate fuzzy input sets defines an n-dimensional lattice, which specifies the centres of the basis functions (Gaussian fuzzy sets) or the knots (B-spline fuzzy sets). Then multivariate fuzzy membership functions are generated by finding the membership of the corresponding n univariate basis functions and combining this information, using an appropriate conjunction operator (such as *min* or *product*). The network's output is then simply a linear combination of the basis functions, although when the complete fuzzy network is implemented, a rule confidence matrix is stored instead of a weight vector and a full fuzzy inferencing calculation is performed. The result is the same in either case, and this is formally proved in Chapter 10.

This fuzzy implementation of a set of vague, linguistic rules means that the fuzzy information is encoded in a redundant fashion. Instead of storing several rule confidences which relate to the same multivariate fuzzy input set, a *single* weight can be stored. This reduces the memory requirements and the computational cost, and so it is worthwhile explaining why it may be desirable to use the original fuzzy representation. For on-line control, it is relatively simple to synthesise an AMN/fuzzy plant model, as shown in Figure 3.15. The AMN network can only

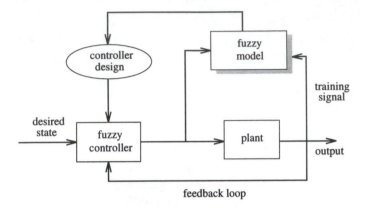

Figure 3.15   An indirect fuzzy control scheme. The fuzzy system is used to model the plant which is then linguistically inverted to synthesise a control signal.

process information in a feedforward fashion, whereas a fuzzy system makes no distinction between inputs and outputs, as knowledge is stored in terms of rule confidences. Therefore, the plant model can then be linguistically inverted such that, given both the present and the desired plant state, the network's output is the control signal.

Fuzzy Summary

The fuzzy networks considered in this book are members of the class of AMNs. The size and shape of the univariate fuzzy input sets which implement the linguistic statements *indirectly* generate an n-dimensional lattice, on which the multivariate fuzzy basis functions are defined. Fuzzy information processing is sometimes identical to multiplying each of these basis functions by a weight, which results in a simplified fuzzy output calculation. This theory assumes that algebraic operations are used to implement the underlying fuzzy logic, but does not hold for an arbitrary set of operators. However, a smooth network output is generated when algebraic operators are used, and their selection can be justified on this point alone. The fact that the fuzzy networks can be interpreted as a linear combination of multivariate fuzzy input sets means that alternative lattice sparse coding strategies can be investigated within a fuzzy framework. Thus a continuous output CMAC can be interpreted as a *sparse* set of fuzzy rules, and this network has the advantage that the number of rules contributing to the output does *not* depend on the dimension of the input space. In similar independent work [Wang and Mendel, 1992], Gaussian basis functions have been proposed to model the fuzzy linguistic variables, and the rules are selected using an orthogonal least squares learning algorithm [Chen *et al.*, 1990]. This is *only* possible because the network is *linearly* dependent on the adjustable parameters.

A control strategy based on the linguistic inversion of a fuzzy plant model has also been proposed. This approach has many desirable properties; the training signal (for the fuzzy plant model) is always directly available, no retraining is necessary if a different desired behaviour is specified, and convergence of the plant model to an output MSE can be proved. This is described and discussed in greater detail in Chapter 11.

3.3.4   Radial Basis Function Network

Radial Basis Function (RBF) networks were first used within the neural network community in 1988 [Broomhead and Lowe, 1988]. They were originally proposed as a technique for modelling data in a high-dimensional space [Powell, 1987], and it was only natural that researchers started to look at this algorithm during the recent resurgence of interest in nonlinear modelling using neural networks. RBF networks are AMNs, where the basis functions may be placed on a lattice or at arbitrary locations in the n-dimensional input space. The former is generally assumed when the interpolation ability of these networks is being analysed theoretically, although the strength of the algorithm (for practical applications) is that basis functions can be placed *anywhere* in the input space. This can result in RBF networks partially overcoming the curse of dimensionality associated with lattice AMNs.

RBF networks have been widely used for data modelling and control tasks. In a series of papers by Chen *et al.* [1990, 1992, 1994], a variety of different learning algorithms have been used to train the RBFs for a wide range of different modelling problems. Off-line orthogonal least-squares algorithms have been developed for selecting relevant basis functions, and on-line clustering methods have been used to modify the centres of these basis functions. The weights can be found by direct matrix inversion or by using a recursive least-squares algorithm. These algorithms have been used to model real and simulated nonlinear time series data collected from water pumps, diesel engines, unemployment figures, etc. An interesting approach was proposed by Lowe [1991], where an RBF network is trained using plant data. A control signal is then calculated by back propagating the plant output error through the network, and so an explicit inverse plant model does not need to be calculated [Kindermann and Linden, 1990]. These networks have been used in a wide range of control schemes from direct inverse modelling to nonlinear internal model control and predictive control [Hunt and Sbarbaro-Hofer, 1991, Sbarbaro-Hofer *et al.*, 1992, 1993]. They have also been used to compensate for the plant's nonlinearities [Sanner and Slotine, 1992, Tzirkel-Hancock and Fallside, 1991] where a sliding mode controller is used to stabilise the system when the input lies outside the network's domain. These hybrid control schemes, employing techniques from both conventional nonlinear control design and neurocontrol, seem to hold the greatest promise for real applications in the near future. Outside the control field, RBF networks have been used in speech and handwriting recognition systems and in a variety of other tasks, all of which deal with high-dimensional, highly nonlinear data [Prager and Fallside, 1989].

Radial Basis Functions

Radial basis functions have a special structure; they can be expressed as:

$$a_i(\mathbf{x}) = f(\|\mathbf{c}_i - \mathbf{x}\|_2) \qquad\qquad (3.10)$$

where $\mathbf{c}_i$ is the n-dimensional vector denoting the *centre* of the ith basis function, $\|.\|_2$ is the common Euclidean norm, and $f(.)$ is a univariate function from $\Re^+$ to $\Re$. The network output is then formed from a linear combination of these basis functions. The modelling capabilities of this network are determined by the position of the centres of the basis functions and the shape of the function, $f(.)$. Different centre placement strategies are discussed in the following section, but first the possible shapes for $f(.)$ are considered.

Let $r = \|\mathbf{c} - \mathbf{x}\|_2$, for the basis function which has a centre $\mathbf{c}$, then several different choices for $f(.)$ are given by:

$$f(r) = r \qquad\qquad \text{the radial linear function}$$
$$f(r) = r^3 \qquad\qquad \text{the radial cubic function}$$
$$f(r) = \exp(-r^2/(2\sigma^2)) \qquad \text{the Gaussian function}$$
$$f(r) = r^2 \log(r) \qquad\qquad \text{the thin plate spline function}$$
$$f(r) = (r^2 + \sigma^2)^{0.5} \qquad \text{the multi-quadratic function}$$
$$f(r) = (r^2 + \sigma^2)^{-0.5} \qquad \text{the inverse multi-quadratic function}$$
$$f(r) = \log(r^2 + \sigma^2) \qquad \text{the shifted logarithm function}$$

for some constant σ. The chosen function influences both the modelling and the learning abilities of the network, as well as the selection of the learning rule. Of the seven functions given above, only the Gaussian and the inverse multi-quadratic functions tend to zero as r tends to infinity. Therefore only those functions which have a localised representation should be employed when the instantaneous adaptation rules given in Section 3.2.2 are used to train the weight vector, otherwise global learning interference can occur. However, it has been reported that the global functions appear to have slightly better interpolation properties than the local Gaussian ones [Buhmann and Powell, 1990]. The two-dimensional Gaussian and thin plate spline basis functions are shown in Figure 3.16. The thin plate spline function has been used extensively in the papers by Chen *et al.* [1990, 1992, 1994], and the Gaussian functions are similar to those used in a fuzzy AMN [Wang and Mendel, 1992].

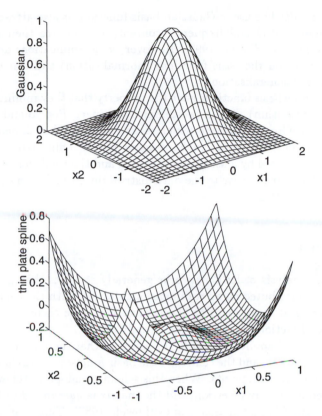

Figure 3.16 A two-dimensional representation of the Gaussian (top) and the thin plate spline (bottom) radial basis function, where their centres lie at the origin.

Theoretical Properties

Of all the different types of radial basis functions discussed above, the Gaussian and the thin plate spline in particular have some very interesting properties. The Gaussian function is *unique*, as it is the only radial basis function which can be written as a *product* of univariate functions:

$$f_i(\mathbf{x}) = \exp\left(-\frac{\|\mathbf{c}_i - \mathbf{x}\|_2^2}{2\sigma_i^2}\right) = \prod_{j=1}^{n} \exp\left(-\frac{(c_{ij} - x_j)^2}{2\sigma_i^2}\right) \qquad (3.11)$$

Thus it is easier to implement and use Gaussian RBF networks in high-dimensional input spaces where only a small number of inputs are relevant for each input [Girosi and Poggio, 1990, Poggio and Girosi, 1990]. This factorisation also makes it possible to incrementally construct Gaussian RBF networks where each node may depend on different inputs [Tresp *et al.*, 1993]. It can even be interpreted from a fuzzy viewpoint where the product operator implements a fuzzy AND, and each of the univariate Gaussian functions represent fuzzy linguistic statements such as *the error is small*. The use of Gaussian basis functions is also attractive because of their localised spatial and frequency content, as they are their own Fourier transforms [Sanner and Slotine, 1992]. However, some applications do not impose a partition of unity on the basis functions (normalisation) and this can result in poor (unexpected) generalisation.

Thin plate spline basis functions have the property that they minimise the bending energy of infinite (thin) plate in two-dimensional space [Poggio and Girosi, 1990]. The more general Green's function which is optimal in higher dimensional space and for different constraint functionals, is again *radial*. Thin plate splines do not need to be parameterised by a user-defined scalar, and the condition of the learning problem is relatively insensitive to the data distribution, which is in direct contrast to the Gaussian function.

Centre Placement

The theoretical analysis of RBF networks generally assumes that the basis functions are evenly distributed on an n-dimensional lattice, with the centre of a basis function occurring at every point on the lattice. For these theoretical results, the number of basis functions depends exponentially on n.

The first practical applications of RBFs assumed that there were as many basis functions as data points, and the centres of the basis functions were chosen to be the input vectors of these data points. When this scheme is adopted, the weight vector can be obtained by matrix inversion, and the matrix is *guaranteed* to be invertible for a wide variety of basis function shapes [Powell, 1987]. This is an *interpolation* scheme, but many of the applications require *approximation* algorithms which are capable of some degree of noise filtering. To achieve this the number of basis functions must be strictly less than the number of training examples, and so an algorithm must be devised for choosing the set centres. There are three approaches commonly found in the literature.

The first approach uses an unsupervised competitive clustering algorithm (Section 2.4.1) for positioning the centres so that they approximate the probability density of the training inputs [Chen *et al.*, 1992, Moody and Darken, 1989]. The second technique simply treats the centres of the basis functions as nonlinear parameters and attempts to train them, using a nonlinear gradient descent algorithm [Hartman and Keeler, 1991]. Finally, a third algorithm, [Chen *et al.*, 1990, Pottmann and Seborg, 1992] still assumes that the input data are used to define the basis functions' centres, but only a small subset of all the possible basis functions is actually used. An iterative orthogonal least-squares algorithm, which attempts to maximise the error reduction at each time-step, is used to select the basis functions. In comparison, the clustering algorithm learns very quickly because it is based on a linear learning law. It uses no performance feedback information, being an unsupervised learning law, although it performs well in practice. The orthogonal least squares algorithm is an efficient off-line design procedure which can be applied to polynomial term selection as well. The gradient descent learning tends to learn more slowly than the clustering algorithm, and although it uses information based on the performance of the network, practical applications indicate that clustering algorithms perform nearly as well.

Possibly the main reason for using RBF networks is to overcome the curse of dimensionality associated with the lattice AMNs. RBFs allow good models to be created for high-dimensional input vectors, with a small number of basis functions. However, the global nature of the basis functions means that they all contribute to the output, and they can no longer be interpreted as a set of linguistic rules (unless Gaussian basis functions are used).

3.3.5 Kanerva's Sparse Distributed Memory Model

Kanerva's Sparse Distributed Memory Model (KSDMM) was first proposed in 1984 [Kanerva, 1988, 1992a]. It was originally formulated as a three-layer network, and is used for high-dimensional, binary pattern classification. This section presents a very brief description of the original network, and the difficulties in using this design for on-line modelling and control tasks are described. A modified design [Prager and Fallside, 1989] is then described and parallels are drawn between these networks and the other AMNs.

The Original KSDMM

The original KSDMM is a three-layer network, defined on a high-dimensional binary space, where the number of basis functions must be specified *a priori* and each basis function has associated with it a binary vector, defined as its *centre*. A binary input vector is compared with each of the basis functions' centres, and if the Hamming distance between these two vectors is less than ρ (for a positive integer ρ), the basis function is on (has an output of 1), otherwise it is off (output is 0). The output is calculated by summing the weights corresponding to the basis functions which are turned on, or equivalently taking the dot product of the weight vector with the transformed input vector. In addition, the original Kanerva learning rule can be slightly modified as follows: *if there is an error in the network output, each weight which contributed to the output is adjusted by an amount proportional to the output error, divided by the number of active basis functions.* This is then equivalent to either of the two learning rules given in Section 3.2.2.

There are three main problems in using the KSDMM for on-line adaptive modelling and control; there is no obvious encoding of a real-valued signal to a binary representation which results in the network being able to *generalise* well, the basis function centres are chosen randomly which means that no *a priori* knowledge can be encoded within the mapping and the output of the network is piecewise constant when a binary encoding of a real-valued signal is performed. Before considering ways of resolving these difficulties, it is useful to draw a comparison between the KSDMM and the CMAC.

CMAC and Kanerva's Memory Model

The basis functions of the CMAC are defined on an n-dimensional lattice and the original basis functions are *binary*; they are either on or off depending on whether or not the input lies within its support. By representing each cell within the lattice as a binary bit, the supports of each basis function can be represented as a binary vector, where a bit is on when it lies in the support, and all the other bits are turned off. An input is then represented by a binary vector which has only one bit turned on, and this corresponds to the lattice cell containing the input. Then the CMAC's basis function is on when the Hamming distance between its binary vector and the input binary vector is $< \rho^n$, and Figure 3.17 illustrates this relationship. This would be a highly *inefficient* method for implementing the CMAC, since storing a binary vector for each basis function would be impractical. The above derivation simply shows that the original CMAC network (with binary basis functions) can be implemented within the KSDMM framework [Kanerva, 1992b].

The CMAC network has a fixed algorithm for generating the binary vectors associated with each basis function, and is *consistent* with the binary encoding of the real input signals. This illustrates two of the problems with using the KSDMM in its present form: How should the input signals be encoded, and how should the centres of the basis functions be assigned?

Modified Kanerva's Memory Model

There are many different ways in which a specific structure can be imposed on the KSDMM, so that the network can deal with real-valued signals. In this section one such approach is considered, and is related to the other AMNs described in this chapter.

In Prager and Fallside [1989], the first modification proposed for the KSDMM was to consider real-valued signals rather than a binary encoding. The basis function centres are evenly distributed in the *centre* of the real-valued n-dimensional input space. The reason for positioning the basis functions near the centre of the input space is to avoid the possibility of the support of the basis function lying mostly outside the *relevant* input space, which could occur when the basis functions are evenly distributed over the whole input space. Three distance measures were compared and the best overall measure (based on the performance of the network and the implementation cost) was the hypercubic metric. The activation radius of the basis functions was chosen so that, approximately, a constant number of basis functions were switched on for each training pattern. The basis functions were still assumed to be binary, so the overall network was very similar to the CMAC, and a

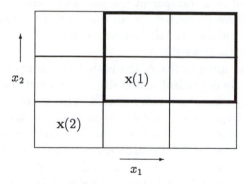

The binary encodings are given by:
$$\mathbf{c} = (0,0,0,0,1,1,0,1,1)^T$$
$$\mathbf{b}(1) = (0,0,0,0,1,0,0,0,0)^T$$
$$\mathbf{b}(2) = (1,0,0,0,0,0,0,0,0)^T$$
$$\|\mathbf{c} - \mathbf{b}(1)\|_1 = 3 < \rho^n$$
$$\|\mathbf{c} - \mathbf{b}(2)\|_1 = 5 > \rho^n$$

Figure 3.17   The encoding of the CMAC's basis functions within Kanerva's memory model. The two-dimensional CMAC has a generalisation parameter $\rho = 2$, and the support of a basis function is shown by the thick line. c denotes the binary encoding associated with the basis function, $\mathbf{b}(i)$ is the binary encoding of the ith input. x(1) lies in the support of this basis function, x(2) lies outside.

substantial improvement in the classification performance was achieved using these modifications.

When the binary behaviour of the basis functions is relaxed [Tsai *et al.*, 1990], the functional decomposition of the network is similar to most of the networks previously discussed. In particular, the definition of a distance measure on a real-valued space which compares inputs with the basis function's centre, is similar to the RBF network. Under special circumstances, the CMAC, B-spline and fuzzy networks can also be considered within this framework, defining a distance measure on the n-dimensional lattice and passing the result through a nonlinearity (see Sections 6.4 and 8.2.4), although this interpretation is not always obvious.

3.4 SUMMARY

The AMNs introduced in this chapter have been proposed for on-line adaptive modelling and control, and for static rule-based applications. AMNs are useful because their initial rate of convergence is fast, learning is local (learning about one part of the input space does not affect knowledge stored in a different area), the knowledge is stored in a transparent fashion, and convergence results can be derived. These first three properties are all due to the localised representations of the basis functions, while the last is because the network output is linearly dependent on the adjustable weight vector.

The disadvantage in using lattice AMNs is that the number of weights in these networks is exponentially dependent on the size of the input space, and an equivalent number of training examples is also required. This is not as critical for on-line adaptation, because a large amount of data are presented to the network; it is more important for the network to adapt quickly and in a stable manner, although for off-line network design, the network's structure may be too flexible for modelling certain data sets. However, the ability to initialise fuzzy rules after training has finished enables the designer to include novel network behaviour which was not contained in the training data. An MLP may be the most appropriate network to use when classifying visual data (high input space dimension and limited training data) [Wright, 1991], whereas an AMN network is generally more appropriate for guidance and control tasks [An *et al.*, 1993b].

3.4.1   Advantages in Using Local Basis Functions

It is the localised basis functions which endow the AMNs with their desirable properties. Each input activates a small number of basis functions, and it is only the knowledge stored in this region of the network which contributes to its response. The network's output is linearly dependent on these basis functions and so their contribution can easily be understood, because each basis function can be interpreted as a set of fuzzy linguistic rules. This is useful when a designer wishes to incorporate certain behaviours into the network which were not contained in the training data.

When the networks are trained using the instantaneous learning rules proposed in Section 3.2.2, only those weights which contribute to the output are updated. Thus adaptation occurs locally in the region which contributes to the network's output. Mathematically, the transformed input space is *sparse* and so the network's response and adaptation calculations occur in a sparse space. Learning is stable because the support of the basis functions is bounded, and adaptation in a different part of the input space does not cause the previously learnt behaviour to be overwritten.

The networks also have a fast initial rate of convergence, and long-term parameter convergence can be proved when certain conditions are satisfied. These results use the fact that the network's output is linearly dependent on the weight vector and the basis functions have a localised response. The rate of convergence of these networks depends on the *condition* of these basis functions, and in Section 5.5 a measure of the amount of *learning interference* is given for different shapes and sizes of basis functions. The reader should not be under the impression that parameter convergence occurs instantaneously for these networks; rather they have several properties which make them suitable for on-line, adaptive nonlinear modelling and control applications.

3.4.2 Disadvantages of Local Basis Functions

Using a sparse local internal representation of the input space in the network simplifies the learning calculation and provides network transparency. However, it was shown in Section 3.2.3 that the number of basis functions in a lattice AMN is exponentially dependent on the input space dimension, which is also true for other networks (such as RBFs), when the basis functions are evenly distributed throughout the input space. The curse of dimensionality effectively limits these networks to applications with small dimensional input spaces. Techniques such as memory hashing [Albus, 1975b] have been proposed to map a large sparse virtual memory onto a smaller physical memory, although this means that the behaviour of the network is no longer predictable. Such techniques should only be used when the number of accessed memory locations is less than the size of the physical memory, as the probability of a collision is then low.

Since the number of adjustable parameters is exponentially dependent on the dimension of the input space, a similar amount of data must be used to train the network. This is necessary when the network is required to learn the desired function across the whole of the input space. For many modelling and control applications, there may be large areas of the input space which are never visited due to the constraints which exist in the input space, and so this requirement may be partially relaxed. When this occurs, the designer should be aware that the network has not been initialised in parts of the input space and should ensure that the input vector does not enter this region during normal plant operation (for a static network).

One potential problem in using AMNs is that of over-parameterisation. This can be illustrated simply by considering the number of terms in a linear PID controller, compared with a fuzzy representation which has 7 piecewise linear rules on each axis and 343 parameters in total. Hence the following heuristic can be applied to any learning system:

The simplest acceptable adaptive system produces the best results.

When a linear model or controller gives acceptable results, there is no point in trying to use a more complex network. An adaptive network should be *parsimonious*: just flexible enough to store the required information. The generalisation abilities of an over-parameterised network will be poor, and the local representations used within AMNs means that local models of the correct form can be generated quickly, although the stored knowledge can only be extrapolated locally.

Finally, these lattice AMNs assume that it is possible, and reasonable, to represent the input using a set of basis functions, with bounded supports, evenly distributed on an n-dimensional lattice. In typical modelling and control applications the signals are bounded (due to physical effects such as actuator saturation, etc.), although these limits may be interdependent and the true input space may be much smaller than the lattice. This is especially true for higher dimensional input spaces. As an extreme example, consider the two-dimensional input space shown in Figure 3.18 where the training data lie in only a small area and because many of the basis functions are uninitialised, the generalisation ability of the network will be poor.

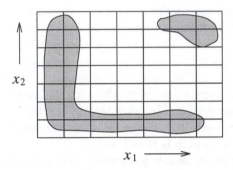

Figure 3.18   Bad input space for employing a lattice AMN. The shaded area contains training inputs.

3.4.3   Possible Solutions

For small dimensional modelling and control problems, the AMNs have many desirable properties. They allow a *qualitative* and a *quantitative* approach to be embodied within the same network. The numerical modelling and learning theory coupled with the linguistic interpretation of the network provide a very powerful toolset. These properties are a result of the network's local, internal representation of the input.

The criticisms in the previous section are because of the number of basis functions which are used by the network. However, the local representations generated by the basis functions also give the network its desirable properties. Thus, as the input space dimension grows, a more complex approach has to be adopted if these features are to be embodied within networks which can deal with larger input spaces. One such AMN is the RBF network, which distributes multivariate basis functions across the input space. A similar method has been proposed in a series of papers by Ritter *et al.* [1989], where the network's basis functions are clustered in areas of the input space where the data density is high. To ensure a local representation and a continuous network output, the support of the basis function must be finite and its output must be zero at the boundary of its support. The basis functions should also cover the input space evenly. A possible solution is to use a nearest neighbour-type approach, where a fixed number of basis functions contribute to the network's output (like the AMNs), and the centres of the basis functions are adapted using a clustering algorithm. The size and shape of the supports change as the basis function centres are adapted, but local basis functions are defined on the support, as is illustrated in Figure 3.19. The basic algorithm is

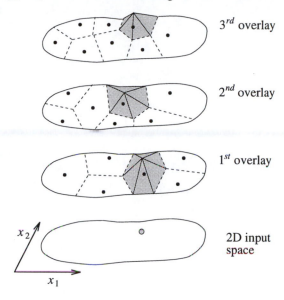

Figure 3.19 Fitting local basis functions in high-dimensional space.

similar to the approach adopted by Ritter, although the basis functions would be truly local, which in turn means that advanced tree structuring, modification and balancing algorithms would have to be employed. A related algorithm is based on *storing* all the training data and then constructing local models based on these data, [Atkeson, 1991, Specht, 1991]. These are *non-parametric* algorithms which can provide very good results from limited training data, although when the model structure changes, there are no convenient data forgetting methods currently available.

If there exists redundancy in the training data, it may be possible to additively model it using several smaller subnetworks, or the networks could be structured in a hierarchical fashion such that the output of one network is the input for another. The first approach has been used by Kavli [1994], when he design the ASMOD algorithm, which is described in Section 8.5. An ASMOD network is composed of a linear combination of several B-spline subnetworks, and the overall model is iteratively refined by introducing new input variables, combining subnetworks and increasing the flexibility of the subnetworks by introducing new basis functions. Hierarchical network structuring is a well-known technique for managing complexity both for resolving complexity at each level [Jacobs and Jordan, 1993, Moody, 1989], and for introducing new input variables at the appropriate stage [Friedman, 1991b].

There are many ways for designing networks defined on high-dimensional input spaces. The most successful approaches exploit the redundancy in the training data (whether this is based on *a priori* knowledge or learnt on-line) and produce parsimonious networks in order to both simplify the learning procedure and improve the system's transparency.

4

Adaptive Linear Modelling

4.1 INTRODUCTION

During the recent resurgence of interest in adaptive nonlinear modelling using neural networks, there have been few general theories developed which can establish convergence conditions, or which can be used to measure the rate of convergence of these adaptive networks. In contrast, over the past thirty years, there has been considerable progress (in the fields of linear adaptive control, signal processing and filter theory) in analysing the convergence and performance of adaptive linear models [Haykin, 1991, Johnson, 1988, Widrow and Stearns, 1985]. However, this work has largely been ignored by the neural network community, despite the facts that Back Propagation (BP) is simply a nonlinear generalisation of instantaneous gradient descent, and that a simple adaptive linear combiner (multiplying a weight vector by an input vector) forms part of many of the most common neural architectures, Multi-Layer Perceptrons (MLPs), the CMAC[1], etc. (see Section 2.2). It can be argued that before the BP algorithm (as a whole) can be properly understood, each of its subproblems must be understood; for instance, how does passing an error signal through a sigmoid degrade the information, or how well conditioned are the linear optimisation subproblems associated with each node? This chapter touches on both of these areas, although it is mainly concerned with basic concepts which are relevant to adaptive linear modelling. In Chapter 3, a number of neuronally inspired models were introduced in which the network's output was *linearly* dependent on the set of adjustable weights: CMAC, B-splines, etc., so the basic problem of understanding how the weights adapt and how the network behaves is simply a *linear optimisation* problem.

This chapter begins with a review of the basic concepts of adaptive linear modelling which are germane to the work presented in later parts of this book. The basic theory is not new but the issues that are pertinent to training Associative Memory Networks (AMNs) such as *singularity* and *minimum norm solutions* are described in considerable detail and the theory is illustrated with several examples.

Initially a set of common linear models found both in neural architectures and

---

[1]Cerebellar Model Articulation Controller.

conventional modelling theory is introduced. Methods for assessing the current performance of the plant are examined, and the Mean Squared Error (MSE) performance function is shown to provide computationally efficient learning algorithms with good performance. The existence of a unique solution is discussed, and this is related to the linear independence of the input vectors and the existence of a sufficiently exciting input signal. Gradient descent learning algorithms are derived and are used to find an optimal solution. It is shown that the rate of convergence depends on the *condition* of the autocorrelation matrix, which in turn is a function of the *power* and *orthogonality* of the network's input signals. Examples are given for two simple neural models, and it is shown that the AMNs internal representation produces partially orthogonal transformed input signals of similar power.

These techniques are then used to analyse why convergence is slow when MLPs are trained using gradient descent algorithms (BP). It is shown that the output of the sigmoids and the bias term are usually highly correlated, which results in slow parameter convergence. The effect of the sigmoid function's shape on the rate of convergence is also investigated and it is shown that the sigmoid transfer functions *squash* the network output error signal, and so the gradient estimates for the hidden layer weight vectors are generally too small. Heuristics are then presented for estimating the learning rates, and these are verified with an example.

Finally, the very important topic of *network stability* is examined. A network is defined to be stable if increasing the training domain does not significantly affect the previous optimal values of the weights. This property is *essential* if a network is to be used for on-line modelling and control, since the training data will be distributed in small areas of the input space for prolonged periods of time (especially if a plant is being regulated to a steady-state), and so the *local* optimal values of the weights should be close to the *global* optimal values.

4.2 LINEAR MODELS

The output of a linear model at time t is given by:

$$y(t) = \sum_{i=1}^{p} a_i(t)\, w_i(t-1) = \mathbf{a}^T(t)\,\mathbf{w}(t-1) \tag{4.1}$$

where $y(t)$ is the scalar output of the model, $\mathbf{a}(t)$ is the p-dimensional input vector and $\mathbf{w}(t-1)$ is the p-dimensional parameter vector at time t. This is illustrated in Figure 4.1.

If the weight vector is adapted after each output calculation, the above model is not strictly linear with respect to the input vector, since presenting the same input to the network at different times results in different model outputs. To resolve this apparent inconsistency, a *linear adaptive system* [Widrow and Stearns, 1985], or an *adaptive linear model* is defined as an adaptive model which becomes a linear

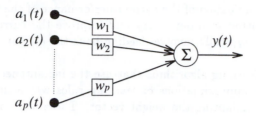

Figure 4.1 A basic linear model.

model when the adjustable parameters are held constant. In the following section, several different forms of adaptive linear models are reviewed.

There are two important interpretations of the linear model given in Equation 4.1: a *transversal* model (filter) and a *multiple* input model (filter). The multiple input model receives its inputs from a variety of different sources at the same time. Suppose that the model receives p signals at time t that are denoted by $\{x_i(t)\}_{i=1}^p$, then the inputs are given by:

$$a_i(t) = x_i(t) \qquad \text{for } i = 1, 2, \ldots, p$$

The transversal model receives a single input, and each input to the model is a time-delayed version of this original signal. For instance, suppose that the measured signal is denoted by $x(t)$, then the inputs are:

$$a_i(t) = x(t - i + 1) \qquad \text{for } i = 1, 2, \ldots, p$$

These distinctions are largely application-dependent and the learning theory can be derived without reference to which particular model is being used. However, the important points to be considered are, for instance, the independence of the input signals and the correlation of the input vectors .

If it is required to model constant functions, or to model functions which have constant and linear components, a *bias* term must be included in the above equations, otherwise the output is always zero when the input is zero. The bias term is included by setting one of the inputs equal to a constant (typically 1), and generally an extra input $a_0(t)$ is used to denote the bias term.

4.3 PERFORMANCE OF THE MODEL

In order to adapt the weight vector, an appropriate learning rule must be specified and in Chapter 2 both unsupervised and supervised learning rules were discussed, and it was shown that for supervised training, a desired network output $\hat{y}(t)$ must be given. Hence, the instantaneous network output error is:

$$\epsilon_y(t) = \hat{y}(t) - y(t) \tag{4.2}$$

which provides an indication of the current performance of the model. The weight vector is then adjusted in order to try and improve the current performance of the model and this typically involves feeding back the output error to the weight vector.

Instantaneous learning algorithms that use the instantaneous network output error are generally approximations of learning rules which use all the available training data when adapting the weight vector. These learning rules attempt to minimise a prespecified performance function, examples of which are:

$$J = \begin{cases} E\left(|\epsilon_y(t)|\right) \\ E\left(\epsilon_y^2(t)\right) \\ \max_t |\epsilon_y(t)| \end{cases}$$

where the expectation operator is taken over t for the training set $\{\mathbf{x}(t), \widehat{y}(t)\}_{t=1}^L$. Unless there exists a weight vector which enables the network to model the desired function exactly, the optimal weight vectors for each of these three performance functions are *different*, as the emphasis or importance placed on each of the (non-zero) instantaneous errors varies. The performance function also determines the complexity of the learning rule; for instance, very complex learning rules must be formulated in order to minimise the unitary performance function given by $\max_t |\epsilon_y(t)|$.

4.3.1   Why Iterative Learning?

The mean squared error performance function has a *direct*, closed-form solution, yet *iterative* gradient methods will be proposed to train the adaptive weight vector. This is justified when the size of the optimisation problem is considered, along with the real-time constraint. The memory requirements for most of the AMNs are large, and direct optimisation methods are generally based on performing a matrix inversion that has a computational cost of $O(p^3)$. The system of equations is generally both *sparse* and *singular*, so numerically stable matrix inversion routines which can exploit the problem's sparsity should be employed.

Alternatively, iterative techniques are based on refining estimates of the optimal weight vector and usually no matrix inversions are performed. This bypasses the need to invert a singular matrix, and these algorithms offer other advantages:

- **simple**, sparse iterative algorithms can be developed which operate in real-time; and
- **useful** low-accuracy solutions can be easily obtained.

The reader should not be misled by the concentration on iterative training algorithms in this book. If an *off-line* optimal weight vector needs to be calculated from a *static* batch of training data, matrix inversion routines or even the (iterative) conjugate algorithm may be appropriate. Iterative gradient descent algorithms should only be used when real-time operation is a critical factor, as occurs for *on-line* adaptation.

4.3.2 Mean Squared Error Performance

An adaptation rule typically minimises a performance function that must be specified before learning commences. The choice of performance function determines the type of learning rule, its computational complexity and the final model, although the MSE performance function is usually adopted because it gives *satisfactory* performance in most cases and the resulting learning rules are *simple* to implement. The final models produced are not always the best, rather the MSE is generally a good engineering choice.

The MSE performance function can be interpreted as a *performance surface* in $(p+1)$-dimensional space, where the scalar performance is expressed as a function of the p weights. To generate the performance surface, the MSE for a static weight vector $\mathbf{w}$ is evaluated, and this is given by:

$$J = E\left(\epsilon_y^2(t)\right) \tag{4.3}$$

where the instantaneous network output error is:

$$\epsilon_y(t) = \widehat{y}(t) - y(t) = \widehat{y}(t) - \mathbf{a}^T(t)\mathbf{w}$$

because the weight vector is now static and does not depend on t. Squaring the above expression gives:

$$\epsilon_y^2(t) = \widehat{y}^2(t) + \mathbf{w}^T\mathbf{a}(t)\mathbf{a}^T(t)\mathbf{w} - 2\widehat{y}(t)\mathbf{a}^T(t)\mathbf{w} \tag{4.4}$$

and taking the expected value of Equation 4.4 over t (assuming that $\epsilon_y(t)$, $\widehat{y}(t)$ and $\mathbf{a}(t)$ are statistically stationary):

$$J = E\left(\widehat{y}^2(t)\right) + \mathbf{w}^T E\left(\mathbf{a}(t)\mathbf{a}^T(t)\right)\mathbf{w} - 2E\left(\widehat{y}(t)\mathbf{a}^T(t)\right)\mathbf{w} \tag{4.5}$$

In order to simplify this expression, let the *autocorrelation matrix* $\mathbf{R}$ be defined as:

$$\mathbf{R} = E\left(\mathbf{a}(t)\mathbf{a}^T(t)\right) = \begin{bmatrix} E\left(a_1^2(t)\right) & E\left(a_1(t)a_2(t)\right) & \dots & E\left(a_1(t)a_p(t)\right) \\ E\left(a_2(t)a_1(t)\right) & E\left(a_2^2(t)\right) & \dots & E\left(a_2(t)a_p(t)\right) \\ \vdots & \vdots & & \vdots \\ E\left(a_p(t)a_1(t)\right) & E\left(a_p(t)a_2(t)\right) & \dots & E\left(a_p^2(t)\right) \end{bmatrix} \tag{4.6}$$

and the *cross-correlation* vector, $\mathbf{p}$, is:

$$\mathbf{p} = E\left(\widehat{y}(t)\mathbf{a}(t)\right) = \begin{bmatrix} E\left(\widehat{y}(t)a_1(t)\right) \\ E\left(\widehat{y}(t)a_2(t)\right) \\ \vdots \\ E\left(\widehat{y}(t)a_p(t)\right) \end{bmatrix} \tag{4.7}$$

Thus substituting Equations 4.6 and 4.7 into 4.5 gives:

$$J = E\left(\widehat{y}^2(t)\right) + \mathbf{w}^T\mathbf{R}\mathbf{w} - 2\mathbf{p}^T\mathbf{w} \qquad (4.8)$$

and two facts can immediately be established about J. First, the performance surface is a non-negative function because it takes the expected value of non-negative quantities. Second, it is quadratic in terms of the weight vector $\mathbf{w}$, when the inputs and the desired responses are stationary stochastic variables, and a typical MSE performance surface is shown in Figure 4.2.

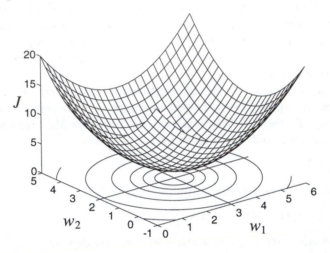

Figure 4.2   A typical MSE performance surface in two-dimensional weight space. The optimal weight vector occurs at $\widehat{\mathbf{w}} = (3,2)^T$, and a contour plot is projected onto the base of the graph.

If the (transformed) inputs, a_i, are *linearly independent* and the training data are sufficiently exciting, the MSE performance function has a single, global minimum in weight space; no local minima exist. However, when the transformed inputs are *linearly dependent* (either structurally or due to limited training data), there exists an infinite number of global minima in weight space; there are still no local minima, but the global minimum is now no longer unique. When this happens, the corresponding autocorrelation matrix is *singular* and an infinite number of solutions occur. This property is relevant to AMNs, because it is shown in Chapter 7 that the CMAC's binary basis functions are linearly dependent. Therefore there exist an infinite number of candidate optimal CMAC weight vectors, as shown in Figure 4.3.

Closed Form Pseudo-Inverse Solutions

Given a *discrete* training set $\{\mathbf{x}(t), \widehat{y}(t)\}_{t=1}^L$, the weight vector which minimises the MSE can be directly calculated using the Moore-Penrose pseudo-inverse:

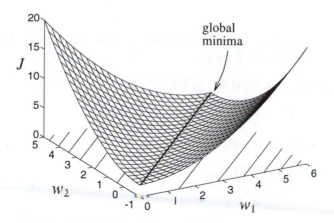

Figure 4.3 A typical *singular* MSE performance surface in two-dimensional weight space. The candidate optimal weight vectors occur along the line $w_2 = w_1 - 1$, and the contour plot is projected onto the base of the graph.

$$\widehat{\mathbf{w}} = \left(\mathbf{A}^T \mathbf{A}\right)^{-1} \mathbf{A}^T \widehat{\mathbf{y}} \tag{4.9}$$

where $\mathbf{A}$ is the solution matrix of size $(L \times p)$ whose t^{th} row is composed of the transformed input vector for the tth input, $\widehat{\mathbf{y}}$ is the vector of desired outputs (of length L) and $\widehat{\mathbf{w}}$ is the optimal weight vector. This is related to the previous work in the sense that the autocorrelation matrix and the cross-correlation vector can be calculated using:

$$\mathbf{R} = \frac{1}{L} \mathbf{A}^T \mathbf{A} \tag{4.10}$$

$$\mathbf{p} = \frac{1}{L} \mathbf{A}^T \widehat{\mathbf{y}} \tag{4.11}$$

for a discrete training set.

When there exists more weights than equations, the system is *under-determined* and an optimal weight vector can be calculated by:

$$\widehat{\mathbf{w}} = \mathbf{A}^T \left(\mathbf{A} \mathbf{A}^T\right)^{-1} \widehat{\mathbf{y}} \tag{4.12}$$

In either case, when the pseudo-inverse has zero eigenvalues, Singular-Valued Decomposition (SVD) routines can be used to select the optimal weight vector which has the smallest Euclidean norm [Haykin, 1991].

Minimum Mean Squared Error

If the autocorrelation matrix $\mathbf{R}$ is non-singular the optimal MSE weight vector, $\widehat{\mathbf{w}}$, can be found by setting the derivative of Equation 4.8 equal to zero and solving [Åström and Wittenmark, 1989, Widrow and Stearns, 1985], which gives:

$$\hat{\mathbf{w}} = \mathbf{R}^{-1}\mathbf{p} \tag{4.13}$$

The minimum MSE, $J_{\min}$, which occurs when $\mathbf{w} = \hat{\mathbf{w}}$, can now be found by substituting $\hat{\mathbf{w}}$ into Equation 4.8, to give:

$$
\begin{aligned}
J_{\min} &= E\left(\hat{y}^2(t)\right) + \hat{\mathbf{w}}^T\mathbf{R}\hat{\mathbf{w}} - 2\mathbf{p}^T\hat{\mathbf{w}} \\
&= E\left(\hat{y}^2(t)\right) + \left(\mathbf{R}^{-1}\mathbf{p}\right)^T\mathbf{R}\mathbf{R}^{-1}\mathbf{p} - 2\mathbf{p}^T\mathbf{R}^{-1}\mathbf{p}
\end{aligned}
$$

and as the autocorrelation matrix is symmetrical:

$$
\begin{aligned}
J_{\min} &= E\left(\hat{y}^2(t)\right) - \mathbf{p}^T\mathbf{R}^{-1}\mathbf{p} \\
&= E\left(\hat{y}^2(t)\right) - \mathbf{p}^T\hat{\mathbf{w}} \tag{4.14}
\end{aligned}
$$

If $J_{\min} = 0$ the model can reproduce the training data exactly, whereas if $J_{\min} > 0$ there always exists mismatch between the training data and the underlying model. Finally, by considering the case when $\mathbf{w} \equiv \mathbf{0}$, the following relationship can be derived:

$$0 \le J_{\min} \le E\left(\hat{y}^2(t)\right) \tag{4.15}$$

4.3.3   Normal Form of the Performance Surface

The MSE performance surface is simply a non-negative quadratic function of the components of the weight vector, with a minimum value occurring at $J_{\min}$ (when the weight vector takes an optimal value $\hat{\mathbf{w}}$). It can be expressed in the much simpler form:

$$J = J_{\min} + \epsilon_{\mathbf{w}}^T\mathbf{R}\epsilon_{\mathbf{w}} \tag{4.16}$$

where $\epsilon_{\mathbf{w}} = \hat{\mathbf{w}} - \mathbf{w}$ is the *current* weight vector error. The two expressions are equivalent as can be shown by substituting Equations 4.13 and 4.14 into 4.8 to give:

$$J = J_{\min} + \mathbf{w}^T\mathbf{R}\mathbf{w} + \hat{\mathbf{w}}^T\mathbf{R}\hat{\mathbf{w}} - 2\hat{\mathbf{w}}^T\mathbf{R}\mathbf{w}$$

but $\hat{\mathbf{w}}^T\mathbf{R}\mathbf{w}$ is a scalar and is equal to its own transpose. Therefore:

$$J = J_{\min} + \mathbf{w}^T\mathbf{R}\mathbf{w} + \hat{\mathbf{w}}^T\mathbf{R}\hat{\mathbf{w}} - \hat{\mathbf{w}}^T\mathbf{R}\mathbf{w} - \mathbf{w}^T\mathbf{R}\hat{\mathbf{w}}$$

and collecting terms produces Equation 4.16.

If $\epsilon_{\mathbf{w}}$ lies in the null space of $\mathbf{R}$ or is zero, $J = J_{\min}$, otherwise any error in the weight vector causes an increase in J. Clearly, the autocorrelation matrix $\mathbf{R}$ determines the shape of the performance surface, and this is further discussed in the following sections.

Example: Null Space of the Autocorrelation Matrix

A network is now described for which the basis functions are linearly dependent. The null space of the corresponding autocorrelation matrix is not empty, and so there exist an infinite number of weight vectors which can be used to form an optimal network. This shows how certain AMNs can possess an *infinite* number of global minima.

The chosen network is a univariate CMAC with a generalisation parameter $\rho = 2$. It is defined on the interval $[0, 2]$ with the single interior knot occurring at $\lambda_{1,1} = 1$, and the desired function is given by $\hat{y}(x) = 1$, for $x \in [0, 2]$. The three basis functions for this network are shown in Figure 4.4.

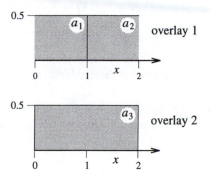

Figure 4.4 A univariate CMAC with $\rho = 2$ and one interior knot and the desired function defined on the interval $[0, 2]$.

Assuming that the input has a uniform probability density function on its domain $[0, 2]$, the autocorrelation matrix is given by:

$$\mathbf{R} = \begin{bmatrix} 0.125 & 0 & 0.125 \\ 0 & 0.125 & 0.125 \\ 0.125 & 0.125 & 0.25 \end{bmatrix}$$

The eigenvalues of $\mathbf{R}$ are: 0.375, 0.125 and 0, and its null space is spanned by the eigenvector which corresponds to its zero eigenvalue. There exists only one zero eigenvalue, so the null space of $\mathbf{R}$ is simply a line in three-dimensional space which is parallel to the orthonormal eigenvector given by $\mathbf{q} = (0.577, 0.577, -0.577)^T$. It can easily be verified that the weight vector $\hat{\mathbf{w}} = (1, 1, 1)^T$ reproduces the desired function exactly, and so *any* vector of the form:

$$\mathbf{w} = \hat{\mathbf{w}} + c\mathbf{q}$$

for some variable c, can also reproduce the desired function exactly as the error in the weight vector, $\epsilon_\mathbf{w} = \hat{\mathbf{w}} - \mathbf{w}$, is parallel to $\mathbf{q}$ and so lies in the null space of $\mathbf{R}$.

4.3.4   Normal Form of the Autocorrelation Matrix

It follows from Equation 4.16 that the properties of the performance surface, its steepness and orientation are related to the structure of the autocorrelation matrix **R**, and considerable insight into the performance surface can be obtained by expressing **R** in terms of its eigenvalues and eigenvectors. The autocorrelation matrix is by definition a real, symmetric, positive semi-definite matrix and so it can be expressed in its normal form:

$$\mathbf{R} = \mathbf{Q}\mathbf{\Lambda}\mathbf{Q}^{-1} = \mathbf{Q}\mathbf{\Lambda}\mathbf{Q}^{T} \tag{4.17}$$

where $\mathbf{\Lambda}$ is a $(p \times p)$ diagonal matrix composed of the (non-negative) eigenvalues of **R** arranged in a non-increasing order and **Q** is a unitary matrix whose columns are the corresponding orthonormal eigenvectors.

Two Examples of the Normal Autocorrelation Matrix

To illustrate these results, the autocorrelation matrices and their normal forms are derived for two networks which have identical modelling capabilities but different underlying structures. Consider a linear model of the form:

$$y(t) = a_1(x(t))w_1 + a_2(x(t))w_2 \tag{4.18}$$

where $x(t)$ has a uniform probability density function on the unit interval. The two models are given by:

$$a_1(x(t)) = 1.0, \qquad\qquad a_2(x(t)) = x(t) \tag{4.19}$$
$$a_1(x(t)) = (1 - x(t)), \qquad a_2(x(t)) = x(t) \tag{4.20}$$

The first model is simply a linear polynomial with a unity bias term, and the second a univariate B-spline network with two, order 2 basis functions. Both models can reproduce exactly any function of the form $\hat{y}(t) = a + b\,x(t)$, defined on the unit interval.

The two univariate basis functions, a_1 and a_2, which generate each model are shown in Figure 4.5, and these two models are used extensively in this chapter and Chapter 5 to illustrate many of the aspects of gradient descent and instantaneous gradient descent learning.

The conventional linear model has an autocorrelation matrix given by:

$$
\begin{aligned}
\mathbf{R} &= \begin{bmatrix} \int_0^1 1\,dx & \int_0^1 x\,dx \\ \int_0^1 x\,dx & \int_0^1 x^2\,dx \end{bmatrix} \\
&= \begin{bmatrix} 1 & 0.5 \\ 0.5 & 0.333 \end{bmatrix}
\end{aligned} \tag{4.21}
$$

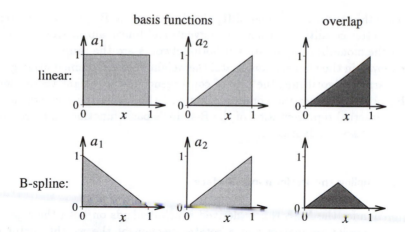

Figure 4.5   Top: the basis functions corresponding to a univariate linear model. Bottom: the basis functions for a B-spline network of order 2. Both networks have the same modelling capability on the unit interval, only the internal representation is different.

and decomposing this matrix into its eigenvalues and eigenvectors gives:

$$\mathbf{\Lambda} = \begin{bmatrix} 1.268 & 0 \\ 0 & 0.066 \end{bmatrix}$$

$$\mathbf{Q} = \begin{bmatrix} 0.882 & 0.472 \\ 0.472 & -0.882 \end{bmatrix} \tag{4.22}$$

Before the autocorrelation matrix for the B-spline network is derived, the wide range of eigenvalues produced by this simple model should be emphasised. The condition number[2] of $\mathbf{R}$, $C(\mathbf{R})$, is 19.3, which is larger than might be expected for such a simple system.

The autocorrelation matrix for the B-spline model is given by:

$$\mathbf{R} = \begin{bmatrix} \int_0^1 (1-x)^2 \, dx & \int_0^1 x(1-x) \, dx \\ \int_0^1 x(1-x) \, dx & \int_0^1 x^2 \, dx \end{bmatrix}$$

$$= \begin{bmatrix} 0.333 & 0.167 \\ 0.167 & 0.333 \end{bmatrix} \tag{4.23}$$

and the eigenvalues and eigenvectors are contained in the following matrices:

$$\mathbf{\Lambda} = \begin{bmatrix} 0.5 & 0 \\ 0 & 0.167 \end{bmatrix}$$

$$\mathbf{Q} = \frac{1}{\sqrt{2}} \begin{bmatrix} 1 & 1 \\ 1 & -1 \end{bmatrix} \tag{4.24}$$

---

[2]The condition number of a matrix is the modulus of the maximum (major) eigenvalue divided by the minimum (minor) eigenvalue.

In this case the condition number of $\mathbf{R}$, $C(\mathbf{R})$, is 3, so the B-spline autocorrelation matrix is better conditioned than the conventional linear autocorrelation matrix, although the modelling capabilities of both networks are the same.

It is shown, in the next section, that if the weight vector is trained using gradient descent learning algorithms, the rate of convergence of the weight vector depends on $C(\mathbf{R})$, and the lower the condition number, the faster the convergence rate. Hence the internal representation of the B-spline's basis functions is more suitable for gradient descent adaptation.

4.3.5   Decoupling the Performance Surface

An expression for the MSE is now derived which depends on both the eigenvalues of the autocorrelation matrix and a rotated version of the weight vector error. This is performed in order to allow the convergence of the individual weights to be studied *separately*, as the convergence of the weight vector can be deduced from the convergence of its individual translated and rotated components.

Combining Equations 4.16 and 4.17 gives:

$$J = J_{\min} + \epsilon_{\mathbf{w}}^T \mathbf{Q} \mathbf{\Lambda} \mathbf{Q}^T \epsilon_{\mathbf{w}}$$

which can be rewritten as:

$$J = J_{\min} + \mathbf{v}^T \mathbf{\Lambda} \mathbf{v} = J_{\min} + \sum_{i=1}^{p} \lambda_i v_i^2 \qquad (4.25)$$

where $\mathbf{v} = \mathbf{Q}^T \epsilon_{\mathbf{w}}$. Hence the contour projection of J is a hyperellipsoid in v-space, with a minimum occurring at the origin and with p mutually orthogonal lines which are perpendicular to all the contours of J. These lines are known as the *principal axes* of the ellipse, and they are simply the eigenvectors of the autocorrelation matrix [Widrow and Stearns, 1985]. Also $\mathbf{Q}$ is a unitary matrix, and so $\mathbf{v}$ is simply a rotated version of $\epsilon_{\mathbf{w}}$ in w-space (weight space), with an origin at $\hat{\mathbf{w}}$.

The eigenvalues of $\mathbf{R}$, which are contained in the diagonal matrix $\mathbf{\Lambda}$, represent the second derivative of J along any of the *principal axes* as:

$$\frac{\partial^2 J}{\partial v_i^2} = 2\lambda_i, \qquad \text{for } i = 1, 2, \ldots, p \qquad (4.26)$$

Thus the second partial derivatives of J are proportional to the corresponding eigenvalues of the autocorrelation matrix. They also contain *curvature* information about the performance function, and so if the eigenvalues of the autocorrelation matrix are widely spread, this is reflected in the relative steepnesses of the performance surface along the principal axes. In the following chapters, the terms *major* and *minor* are used to represent the principal axes with the largest and smallest (non-zero) eigenvalues, respectively. The performance surface is steepest along the major principal axis and is flattest along the minor principal axis, and the ratio of these two quantities (which defines the *condition number* of the autocorrelation matrix) determines the rate of convergence of the gradient descent rules.

Two Examples of the Normalised Performance Surface

In Section 4.3.4, the autocorrelation matrix and its eigenvalues and eigenvectors were calculated for a conventional linear model and a B-spline model. The conventional linear model generates an autocorrelation matrix with a condition number of 19.3, and the corresponding eigenvalues and eigenvectors are given in Expression 4.22. The MSE performance surface and its contours are plotted in Figure 4.6, and the large eigenvalue disparity (condition number) is evident in both the shape of the surface (steep side and flat bottom) and the elongated contours.

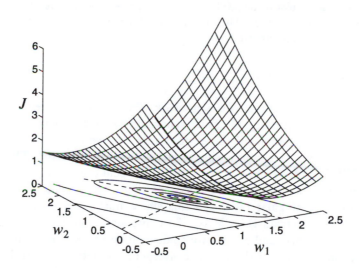

Figure 4.6 A contour plot of the MSE performance surface for a conventional linear model. The contour values are $J = 0.005, 0.015, 0.05, 0.15, 0.5, 1.5$, $\hat{w} = (1, 1)^T$ and the two dashed lines are the principal axes (eigenvectors).

The condition number of the autocorrelation matrix for the order 2 B-spline network is 3, and its eigenvalue/eigenvector decomposition is shown in Expression 4.24. The MSE performance surface and the contour plot is shown in Figure 4.7. The performance surface is more bowl shaped (smaller curvature variation) and the contours are more circular than those of the linear model's performance surface, which is due to the eigenvalues being closer.

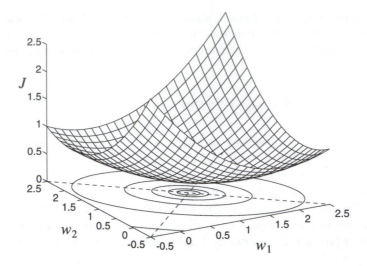

Figure 4.7   A contour plot of the MSE performance surface for an order 2 B-spline model. The contour values are $J = 0.005, 0.015, 0.05, 0.15, 0.5, 1.5$, $\hat{\mathbf{w}} = (1,1)^T$, and the two dashed lines are the principal axes (eigenvectors).

4.4 GRADIENT DESCENT

The MSE performance surface for a network which is linearly dependent on the weight vector has been analysed. This theory is now used to derive learning rules which measure the gradient of the performance surface at a particular point in weight space and use this information to adapt the weight vector. These learning rules are called *gradient descent* (or steepest descent), and they store information iteratively without any matrix inversions. The autocorrelation matrix is not required to be non-singular (although this effect must be investigated), so these rules can be used to train the CMAC. An expression is derived for their rate of parameter convergence, and it is shown to depend on the *condition* number of the autocorrelation matrix and the weight convergence in networks with lower condition numbers is faster. Finally, the domain on which a conventional linear network is defined is investigated, and it is shown that if the interval is symmetrical, the bias term and the linear term are *orthogonal functions* and the corresponding autocorrelation matrix is well conditioned. All of these results are illustrated with examples.

4.4.1 Gradient of the Performance Surface

The gradient descent learning rules, which are derived in the next section, all measure the gradient of the performance of the current network in weight space, and the weights are then updated parallel to the negative gradient of J such that the MSE is reduced. The gradient of J in w-space, is denoted by ∇, and can be obtained from differentiating Equation 4.8 with respect to $\mathbf{w}$, giving:

$$\nabla = \frac{\partial J}{\partial \mathbf{w}} = 2\mathbf{R}\mathbf{w} - 2\mathbf{p} \tag{4.27}$$

This expression can be used to find the optimal weight vector $\hat{\mathbf{w}}$, which minimises the output MSE, if the autocorrelation matrix is non-singular. The minimum MSE occurs when the gradient is zero, so:

$$\nabla = 0 = 2\mathbf{R}\hat{\mathbf{w}} - 2\mathbf{p}$$

therefore:

$$\hat{\mathbf{w}} = \mathbf{R}^{-1}\mathbf{p}$$

which is precisely the result stated in Equation 4.13.

Substituting for $\mathbf{p}$ in Equation 4.27 gives:

$$\begin{aligned}
\nabla &= 2\mathbf{R}\mathbf{w} - 2\mathbf{R}\hat{\mathbf{w}} \\
&= -2\mathbf{R}\epsilon_{\mathbf{w}}
\end{aligned} \tag{4.28}$$

where $\epsilon_{\mathbf{w}} = \hat{\mathbf{w}} - \mathbf{w}$ is the error in the weight vector. In this expression, ∇ only depends on the autocorrelation matrix and the current weight vector error.

Alternative Expression for the Gradient

An alternative expression for the gradient of the performance surface can be found from differentiating Equation 4.3 with respect to $\mathbf{w}$, which gives:

$$\begin{aligned}
\nabla &= \frac{\partial E\left(\epsilon_y^2(t)\right)}{\partial \mathbf{w}} \\
&= E\left(\frac{\partial \epsilon_y^2(t)}{\partial \mathbf{w}}\right) \\
&= -2E\left(\epsilon_y(t)\mathbf{a}(t)\right)
\end{aligned} \tag{4.29}$$

The gradient of the performance now depends on the expected output error of the network multiplied by the transformed input vector. These are *equivalent*, since:

$$\begin{aligned}
-2\mathbf{R}\epsilon_{\mathbf{w}} &= -2E\left(\mathbf{a}(t)\mathbf{a}^T(t)\right)\epsilon_{\mathbf{w}} \\
&= -2E\left(\mathbf{a}(t)\left(\mathbf{a}^T(t)\epsilon_{\mathbf{w}}\right)\right) \\
&= -2E\left(\epsilon_y(t)\mathbf{a}(t)\right)
\end{aligned}$$

and both of these expressions are used in this book.

Gradient in the Transformed Weight Space

In the following sections, the dynamical behaviour of $w(t)$ is investigated when it is updated using gradient information. This may seem a difficult task since, according to Equation 4.28, an error in *one* of the weights causes *all* of the weights to be changed. However, it was shown in Section 4.3.5 that the performance surface was "decoupled" along the principal axes in v-space, so the gradient of the performance surface with respect to these principal axes can be evaluated.

As ϵ_w and w differ by only a constant, then:

$$\nabla = \frac{\partial J}{\partial w} = -\frac{\partial J}{\partial \epsilon_w} = -2R\epsilon_w = 2(Rw - p) \tag{4.30}$$

The rotated version of the error in the weight vector is defined as $v = Q^T \epsilon_w$, where Q is a $(p \times p)$ matrix whose columns are the eigenvectors of R. From Equation 4.25, the gradient of J with respect to v is given by:

$$\frac{\partial J}{\partial v} = 2\Lambda v \tag{4.31}$$

or

$$\frac{\partial J}{\partial v_i} = 2\lambda_i v_i$$

Thus the gradient of J with respect to v_i only depends on the ith eigenvalue, λ_i and v_i, and does *not* depend on λ_j and v_j for $i \neq j$. The gradient components, in v-space, are now decoupled.

4.4.2   Gradient Descent Learning Rules

Gradient descent learning rules update the weight vector in proportion to the negative gradient of J, producing a learning rule of the form:

$$\Delta w(t-1) = -\frac{\delta}{2}\nabla(t) \tag{4.32}$$

where $\Delta w(t-1) = w(t) - w(t-1)$, and the factor of 2 has been introduced in order to "normalise" the update rule. Substituting for $\nabla(t)$ gives:

$$\begin{aligned}\Delta w(t-1) &= -\delta(Rw(t-1) - p) &&(4.33)\\ &= \delta R\epsilon_w(t-1) &&(4.34)\end{aligned}$$

Thus the weight change depends on the structure of the autocorrelation matrix and the current weight error vector.

Parameter Convergence for Different Autocorrelation Matrices

The dynamical behaviour of the gradient descent rule has been shown to depend on
the structure of the autocorrelation matrix and this relationship is now illustrated
for three different autocorrelation matrices.

In Figure 4.8, the weight path is shown superimposed on the contours of the

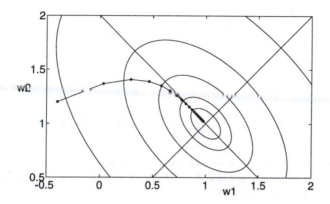

Figure 4.8   Gradient descent on the performance surface corresponding to a B-spline model,
$(C(\mathbf{R}) = 3)$, after twenty weight updates with a normalised learning rate of 0.5. The performance surface's contours correspond to $J = 0.005, 0.015, 0.05, 0.15, 0.5$.

performance surface. The condition of the autocorrelation matrix is 3 (eigenvalue
spread) and a (normalised) learning rate of $\delta = 0.5$ is used. It can clearly be seen
that the weight updates are always perpendicular to the contours, which means
that the rate of convergence depends on the shape of the performance surface. In
Figures 4.9 and 4.10, this effect is clearly illustrated. When $C(\mathbf{R}) = 1$, all the
eigenvalues are equal and convergence has occurred after five iterations, although
convergence can occur after 1 iteration if the learning rate is set equal to $1/\lambda_{\max}$. In
Figure 4.10, $C(\mathbf{R}) = 20$, and convergence has still not occurred after 100 iterations,
although *one* of the weights converged to the correct value after a small number
of iterations. The slow convergence which occurs along the minor principal axes
(corresponding to the smallest eigenvalue) is due to the learning rate being selected
to stabilise the fastest mode, which slows down convergence for the remaining
modes, and this is further discussed in Section 4.4.3.

Feedback Learning

An interesting interpretation of the gradient descent algorithm is to view it as a
feedback model, and to emphasise this point Widrow termed unsupervised learning
"open-loop adaptation" and supervised learning "closed-loop adaptation" [Widrow
and Stearns, 1985]. The feedback block diagram of Equation 4.33 is shown in

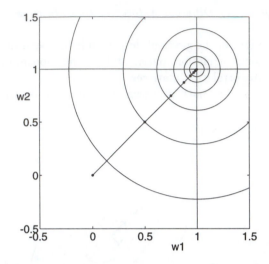

Figure 4.9   Gradient descent on a performance surface with $C(\mathbf{R}) = 1$ after ten weight updates with a normalised learning rate of 0.5. The performance surface's contours correspond to $J = 0.005, 0.015, 0.05, 0.15, 0.5, 1.5$.

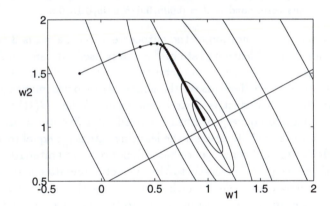

Figure 4.10   Gradient descent on the performance surface corresponding to a linear model ($C(\mathbf{R}) = 19.3$), after a hundred weight updates with a normalised learning rate of 0.5. The performance surface's contours correspond to $J = 0.005, 0.015, 0.05, 0.15, 0.5, 1.5$.

Figure 4.11 and here the weights are treated as *states* of the model with the current performance of the model being fed back to modify the weight vector. From this interpretation, it is obvious that the size of δ determines the stability of the closed-loop system, and too high a value causes unstable (un)learning.

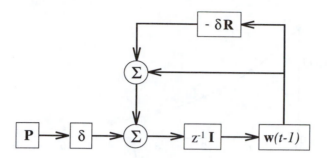

Figure 4.11 Feedback representation of gradient descent adaptation.

4.4.3 Convergence and Stability Analysis

In order to understand the dynamical behaviour of the gradient descent rule, first add and subtract $\hat{\mathbf{w}}$ from the left-hand side of Equation 4.34, to give:

$$\Delta \epsilon_{\mathbf{w}}(t-1) = -\delta \mathbf{R} \epsilon_{\mathbf{w}}(t-1) \tag{4.35}$$

or

$$\epsilon_{\mathbf{w}}(t) = (\mathbf{I} - \delta \mathbf{R}) \epsilon_{\mathbf{w}}(t-1)$$

This is a first-order recurrence relationship for the error in the weight vector. In general, this relationship is difficult to interpret, since $\mathbf{R}$ is not diagonal, although premultiplying Equation 4.35 by $\mathbf{Q}^T$ gives:

$$\Delta \mathbf{v}(t-1) = -\delta \mathbf{\Lambda} \mathbf{v}(t-1)$$

which can be rewritten in its *decoupled* components as:

$$v_i(t) = (1 - \delta \lambda_i)\, v_i(t-1) \qquad \text{for } i = 1, 2, \ldots, p \tag{4.36}$$

with closed form solutions:

$$v_i(t) = (1 - \delta \lambda_i)^t\, v_i(0) \qquad \text{for } i = 1, 2, \ldots, p \tag{4.37}$$

where $v_i(0)$ is the (rotated) i^{th} initial error in the weight vector. This is illustrated in Figure 4.12 with $(1 - \delta \lambda_i) < 1$.

Weight Vector Stability

For stable learning, it is required that:

$$|v_i(t)| \le |v_i(0)| \qquad \forall\, i = 1, 2, \ldots, p$$

or from Equation 4.37, for all non-zero eigenvalues:

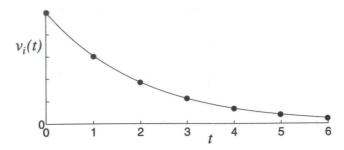

Figure 4.12   Convergence of the ith natural mode, $v_i(t)$, which is adapted using a gradient descent rule with $(1 - \delta\lambda_i) < 1$ (an exponential curve has been fitted to the discrete points).

$$|1 - \delta\lambda_i| < 1$$

Therefore learning is stable if and only if the learning rate, δ, satisfies:

$$0 < \delta < \frac{2}{\lambda_{\max}} \qquad\qquad (4.38)$$

If the learning rate satisfies Expression 4.38, the weight vector tends to an optimal value as $t \to \infty$, because (for a non-zero i^{th} eigenvalue of $\mathbf{R}$) the right-hand sides of Equation 4.37 all tend to zero as time increases. The error in the weight vector therefore tends to zero as t increases.

When the ith eigenvalue of $\mathbf{R}$ is zero, the error along the ith principal axis does not decay. This does not mean that the weight vector does not converge, but rather when $\mathbf{R}$ is singular, there exist an *infinite* number of global minima in weight space (see Figure 4.3), and the errors which exist along the principal axes (corresponding to the zero eigenvalues) *constrain* the learning algorithm so that there exists a unique minimum, which depends on the initial values of the weights. For a particular candidate optimal weight vector, $\hat{w}$, if there exists a non-zero, initial, rotated, weight error vector, $v_i(0)$, and the ith eigenvalue of $\mathbf{R}$ is zero, the weight vector does not converge to this optimal weight vector. When the weight vector is trained using gradient descent, it converges to the optimal weight vector for which $v_i(0) = 0$ for *all* the zero eigenvalues.

4.4.4   Minimum Norm Solutions

The previous example considered an AMN which has a singular autocorrelation matrix and it was shown that the weight vector still converged to a weight vector which minimised the MSE, although there exist an infinite number of different optimal weight vectors. It was also demonstrated that the final weight vector depends on the initial weight vector, and so it is pertinent to investigate what initial weight vector should be used when the autocorrelation matrix is singular. For a network to locally generalise sensibly the final weight values, and the corresponding final weight vector $\|\hat{\mathbf{w}}\|$, should be relatively small. This constraint on the weight vector's magnitude now makes the solution to the optimisation problem unique, although it depends on the specific vector norm. In most applications the Euclidean norm is used, as the resulting optimal weight vector has a closed-form solution which can be simply calculated using an SVD algorithm [Haykin, 1991]. This section establishes that a *zero* initial weight vector always converges to the minimum (Euclidean) norm solution when it is updated using a gradient descent algorithm.

If the autocorrelation $\mathbf{R}$ is singular and $r = \text{rank}(\mathbf{R})$, there exists $(p - r)$ zero eigenvalues and the minimum (Euclidean) norm solution is given by:

$$\hat{\mathbf{w}}^* = \sum_{i=1}^{r} \frac{\mathbf{q}_i^T \mathbf{P}}{\lambda_i} \mathbf{q}_i \tag{4.39}$$

where λ_i is the ith eigenvalue of $\mathbf{R}$ (the eigenvalues are arranged in a non-increasing order), and $\mathbf{q}_i$ is the corresponding orthonormal eigenvector [Haykin, 1991]. The optimal minimum norm solution is a linear combination of the eigenvectors which have non-zero eigenvalues. Denoting the error between this minimum norm solution and a zero initial weight vector by:

$$\epsilon_{\mathbf{w}}(0) = \hat{\mathbf{w}}^*$$

then the rotated initial error in v-space is:

$$
\begin{aligned}
\mathbf{v}(0) &= \mathbf{Q}^T \epsilon_{\mathbf{w}}(0) \\
&= \mathbf{Q}^T \hat{\mathbf{w}}^* \\
&= \sum_{i=1}^{r} \frac{\mathbf{q}_i^T \mathbf{P}}{\lambda_i} \mathbf{Q}^T \mathbf{q}_i
\end{aligned}
$$

The quantity $\mathbf{q}_j^T \mathbf{q}_i$ is zero when $j > r$, so $v_j(0)$ is zero for all values which correspond to the zero eigenvalues ($j > r$), and the remaining values ($v_j, j \leq r$) all converge to zero when they are trained. Therefore a zero initial weight vector always converges to the minimum norm solution when it is trained using the gradient descent learning algorithm.

If the weight vector is initialised with a non-zero value when the autocorrelation matrix is singular and the weight vector does not lie in the space spanned by the basis vectors $\{\mathbf{q}_i\}_{i=1}^{r}$, it converges to a solution which has a Euclidean norm greater than $\|\hat{\mathbf{w}}^*\|_2^2$. The difference between these two norms gives an indication of the ability of the network to generalise. For example, consider the binary CMAC described in Section 4.3.3, and used in the previous example. One optimal weight vector is:

$$\hat{\mathbf{w}} = (-99, -99, 100)^T$$

whereas the minimum norm solution is given by:

$$\widehat{\mathbf{w}}^* = (0.67, 0.67, 1.33)^T$$

Both weight vectors reproduce the desired function exactly, although the Euclidean norms are given by $\|\widehat{\mathbf{w}}\|_2 = 172$ and $\|\widehat{\mathbf{w}}^*\|_2 = 1.633$. If the information is extrapolated to either side of the network, the normalised network output is -99 for the first network and 0.67 for the second. It is anticipated that the extrapolated value should be close to 1, and so the second network's output is preferable.

4.4.5   Rate of Convergence

For stable adaptation to occur, the learning rate must satisfy Expression 4.38 which can be achieved by considering a new learning rate defined by $\delta_\lambda = \delta/\lambda_{\max}$, then stable learning is assured if and only if $0 < \delta < 2$. Performing this normalisation of the learning rate allows the rate of convergence for different networks to be compared (for the same value of δ), as was shown in Section 4.4.2.

Decay Constants

Referring back to the update Equation 4.37, an exponential decay curve can be fitted to the discrete values of $v_i(t)$, as shown in Figure 4.12. This curve has an initial value $v_i(0)$ and a decay constant τ_i, defined by:

$$v_i(0) \exp\left(-\frac{t}{\tau_i}\right) = v_i(0) \left(1 - \delta_\lambda \lambda_i\right)^t$$

solving for τ_i gives:

$$
\begin{aligned}
\tau_i &= \frac{-1}{\ln\left(1 - \delta_\lambda \lambda_i\right)} \\
&= \frac{-1}{\ln\left(1 - \delta \lambda_i / \lambda_{\max}\right)}
\end{aligned}
\tag{4.40}
$$

The decay constant, τ_i, represents the time taken for the amplitude of the signal, $v_i(t)$, to decay to e^{-1} of its initial value, and so it can be used to compare the rate of convergence of different learning modes and different networks. The largest time constant $\tau_{\max}$ corresponds to the minor (smallest) eigenvalue, and it is given by:

$$
\begin{aligned}
\tau_{\max} &= \frac{-1}{\ln\left(1 - \delta \lambda_{\min}/\lambda_{\max}\right)} \\
&= \frac{-1}{\ln\left(1 - \delta/C(\mathbf{R})\right)}
\end{aligned}
\tag{4.41}
$$

where the condition of the the autocorrelation matrix is defined by $C(\mathbf{R}) = \lambda_{\max}/\lambda_{\min}$. Networks with large condition numbers (large major/minor eigenvalue ratios) have components of the weight vector which possess very large time constants, and they learn the information which lies along the minor principal axis very slowly. It should be re-emphasised that the minor eigenvalue has been defined to be the smallest *positive* eigenvalue. The condition of the matrix is not allowed to be infinite, as the zero eigenvalues do not influence the rate of convergence; they only influence the actual values of the weights, as shown by the above example.

Weight Convergence

To compare the rate of convergence of the individual weights in a network, it is necessary to transform the above equations from v-space to w-space, by noting that $\mathbf{Q}^T c_{\mathbf{w}}(t) = \mathbf{v}(t)$, hence:

$$
\begin{aligned}
\mathbf{w}(t) &= \hat{\mathbf{w}} - \mathbf{Q}\mathbf{v}(t) \\
&= \hat{\mathbf{w}} - \sum_{j=1}^{p} \mathbf{q}_j v_j(t)
\end{aligned}
$$

where $\mathbf{q}_j$ is the jth column eigenvector of $\mathbf{Q}$. Using Equation 4.37 and considering the dynamical behaviour of each weight separately gives [Haykin, 1991]:

$$
w_i(t) = \hat{w}_i - \sum_{j=1}^{p} q_{j,i} v_j(0) \left(1 - \delta \frac{\lambda_j}{\lambda_{\max}}\right)^t \tag{4.42}
$$

for $i = 1, 2, \ldots p$, where $q_{j,i}$ is the ith element of the jth column eigenvector.

This formulation allows the rate of convergence of the weight vector to be investigated. The rate of convergence of each weight depends on a sum of exponentials of the form $(1 - \delta \lambda_i / \lambda_{\max})^t$. As time increases, this sum becomes dominated by the term which is decaying the slowest. Assuming a non-zero rotated initial error $v_i(0)$, and that the minimum eigenvalue $\lambda_{\min}$ is unique, the rate of convergence of the slowest mode is given by:

$$
\left(1 - \delta \frac{\lambda_{\min}}{\lambda_{\max}}\right)^t = \left(1 - \frac{\delta}{C(\mathbf{R})}\right)^t \tag{4.43}
$$

When the condition number of the autocorrelation matrix is large, or equivalently the eigenvalue spread is large, Expression 4.43 is close to 1 and convergence occurs slowly. Alternatively, if $C(\mathbf{R}) = 1$ and $\delta = 1$, which happens when all the eigenvalues are equal, parameter convergence occurs after one iteration.

To compare the suitability of two different networks for gradient descent learning, all that needs to be investigated is the condition number of the autocorrelation matrix. The parameter convergence of networks with lower condition numbers is in general faster, and they are better able to generalise.

Performance Function

From Equations 4.25 and 4.37, the MSE performance function at time t can be expressed as:

$$J(t) = J_{\min} + \sum_{i=1}^{p} \lambda_i \left(1 - \delta' \lambda_i\right)^{2t} v_i^2(0) \qquad (4.44)$$

For stability, the learning rate must satisfy $\delta' < 1/\lambda_{\max}$, so redefining δ' as $\delta/\lambda_{\max}$, where $0 < \delta < 2$, the above becomes:

$$J(t) - J_{\min} = \sum_{i=1}^{p} \lambda_i v_i^2(0) \left(1 - \delta \frac{\lambda_i}{\lambda_{\max}}\right)^{2t} \qquad (4.45)$$

the performance of the current model depends on a sum of decaying exponentials, and the steady-state rate of convergence is dominated by the term with the largest decay constant, $\tau_{\max}$, as shown in Figure 4.13. The minor (smallest) eigenvalue

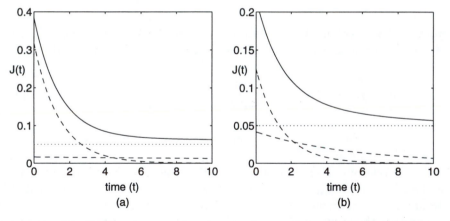

Figure 4.13   The MSE performance functions for the two-dimensional linear (a) and B-spline (b) networks. Each performance function (solid line) is formed from the sum of two decaying exponentials (dashed line) and the minimum value of the performance function (dotted line). After a certain period of time, the rate of convergence of the performance function is dominated by the slowest mode.

determines the slowest rate of decay, and the corresponding decay constant is given by:

$$\tau_{\max} = \frac{-1}{2 \ln \left(1 - \delta / C(\mathbf{R})\right)}$$

This is simply *half* of the corresponding weight decay time constant, as the MSE performance function is the sum of *squared* errors in the individual weights. The slow final convergence does not generally affect the initial rate of convergence, as the magnitude of the initial error with the largest time constant, $\lambda_{\min} v_{\min}^2(0)$, is small. It is only significant when *all* the other initial errors have decayed sufficiently. This term decays slowly as J is not sensitive with respect to these errors, and the performance function is nearly flat along the corresponding principal axis.

Example: Convergence in Singular Autocorrelation Matrices

The dynamical behaviour of the weights trained using gradient descent is illustrated using the univariate CMAC network described in Section 4.3.3. The autocorrelation matrix, its eigenvalues and eigenvectors are given by:

$$\mathbf{R} = \begin{bmatrix} 0.125 & 0 & 0.125 \\ 0 & 0.125 & 0.125 \\ 0.125 & 0.125 & 0.25 \end{bmatrix}$$

$$\mathbf{\Lambda} = \begin{bmatrix} 0.375 & 0 & 0 \\ 0 & 0.125 & 0 \\ 0 & 0 & 0 \end{bmatrix}$$

$$\mathbf{Q} = \begin{bmatrix} 0.408 & 0.707 & 0.577 \\ 0.408 & -0.707 & 0.577 \\ 0.816 & 0 & -0.577 \end{bmatrix}$$

This example considers the evolution of the weight vector from an initial condition $\mathbf{w}(0) = (0, 1, 0)^T$, and the weight vector to which the network converges is given by $\hat{\mathbf{w}} = (1, 1, 1)^T$. An identically zero initial weight vector is not used as this produces an error along only one of the principal axes and so its dynamical behaviour is not very instructive.

The initial weight vector error is $\epsilon_{\mathbf{w}}(0) = (1, 0, 1)^T$, and the rotated version in v-space is $\mathbf{v}(0) = (1.225, 0.707, 0.0)^T$. The rotated error v_3, which corresponds to the zero eigenvalue is zero, since the weight vector converges to $\hat{\mathbf{w}}$ but v_3 does not decay (due to its zero eigenvalue). If another *optimal* weight vector which reproduces the desired function without modelling error (for example $\hat{\mathbf{w}}' = (0, 0, 2)^T$) is considered, the initial error weight vector is $\epsilon_{\mathbf{w}}'(0) = (0, -1, 2)^T$, and the rotated weight error is given by $\mathbf{v}' = (1.225, -0.707, 1.732)^T$. The first two components of $\mathbf{v}'$ converge to zero as $t \rightarrow \infty$, because the corresponding eigenvalues are non-zero. However, v_3' is constant, so there exists a non-zero error in the weight vector for all time, indicating that the network converges to a *different, optimal* solution.

The dynamical evolution of the weight vector $\mathbf{w}$ and the rotated weight vector error $\mathbf{v}$ are shown in Table 4.1. A learning rate of $\delta = 1/(2\lambda_{\max})$ was used, which is reflected in the rate of decay of $v_1(t)$ (corresponding to the maximum eigenvalue) whose value halves at each iteration. The rate of decay of v_2 is also as predicted by Equation 4.36; the value reduces by 5/6 $(= (1 - \lambda_2/(2\lambda_{\max})))$ after each update. Parameter convergence in v-space is therefore very regular, as illustrated in Figure 4.14. In weight space however, the error in w_2 increases during

t	$w_1(t)$	$w_2(t)$	$w_3(t)$	$v_1(t)$	$v_2(t)$	$v_3(t)$	J
0	0.0	1.0	0.0	1.225	0.707	0.0	0.623
1	0.333	1.167	0.5	0.612	0.589	0.0	0.184
2	0.528	1.222	0.75	0.306	0.491	0.0	0.065
3	0.648	1.227	0.875	0.153	0.409	0.0	0.030
4	0.728	1.210	0.938	0.077	0.341	0.0	0.017
5	0.783	1.185	0.969	0.038	0.284	0.0	0.011
6	0.825	1.160	0.984	0.019	0.237	0.0	0.007

Table 4.1   The dynamical evolution of the weight vector **w**, the rotated weight vector error **v**, and the MSE performance function J.

the first few iterations. The output error is decreasing, but the error in one of the parameters increases! The theory derived in this chapter does *not* predict that all of the individual weights converge monotonically to their optimal values, it only shows that $\|\epsilon_\mathbf{w}(t)\|_2^2$ and $E\left(\epsilon_y^2(t)\right)$ decrease when the weight vector is updated.

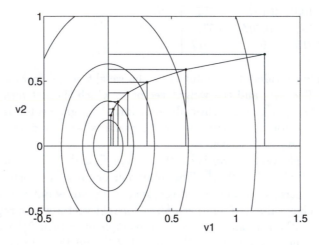

Figure 4.14   Parameter convergence in two-dimensional v-space from an initial value $\mathbf{v} = (1.061, 0.707)^T$. The projection of the learning path onto each of the principal axes is regular.

4.4.6   Condition of Linear Models

It has been shown that the rate of convergence of networks which are linear in their weight vector directly depends on the condition number of the autocorrelation matrix. The condition of the conventional linear network is now examined for different input domains, and it is shown that when the bias term and the linear term are *orthogonal* and the magnitude of the bias term is *reduced*, then $C(\mathbf{R}) = 1$.

Initially consider a linear model, which consists of a linear and a bias term, such that the network output is given by:

$$
\begin{aligned}
y(t) &= a_1(t)w_1(t-1) + a_2(t)w_2(t-1) \\
a_1(t) &= 1, \qquad\qquad a_2(t) = x(t)
\end{aligned}
$$

and it is defined on the interval $[a, b]$. The corresponding autocorrelation matrix is:

$$
\mathbf{R} = \begin{bmatrix} \int_a^b p(x)\,dx & \int_a^b x\,p(x)\,dx \\ \int_a^b x\,p(x)\,dx & \int_a^b x^2\,p(x)\,dx \end{bmatrix}
$$

for a given probability density function $p(x)$ on $[a, b]$. The eigenvalues of $C(\mathbf{R})$ vary according to the values of a and b and the associated probability density function, and so to investigate the conditioning of the autocorrelation matrix, various realistic combinations are evaluated. The simplest probability density function is a uniform distribution on the interval $[a, b]$ given by:

$$
p(x) = \begin{cases} 1/(b-a) & \text{if } x \in [a, b] \\ 0 & \text{otherwise} \end{cases}
$$

When the data are discrete, it is assumed that the training examples are evenly spread and there are sufficient to *approximate* a uniform distribution on $[a, b]$.

Linear Models defined on Positive Intervals

The condition of a linear model (linear term + bias term) is now derived for training data which have a uniform probability density function on the interval $[0, b]$. This produces an autocorrelation matrix of the form:

$$
\mathbf{R} = \begin{bmatrix} b^{-1}\int_0^b 1\,dx & b^{-1}\int_0^b x\,dx \\ b^{-1}\int_0^b x\,dx & b^{-1}\int_0^b x^2\,dx \end{bmatrix}
$$

On the interval $(0, b]$ all of the integrands are positive and when b is a small number (≈ 1.5) the matrix coefficients are approximately the same magnitude, so it would be expected that the eigenvalues of $\mathbf{R}$ are quite widely spread. This turns out to be true as shown in Figure 4.15, where $C(\mathbf{R})$ is plotted for different values of b; the graph has a single minimum at $b = \sqrt{3}$.

The shape of the graph can be easily explained. When b is close to zero, the power of the bias term dominates the other terms in $\mathbf{R}$, and the network is slow to learn the information contained in the linear term. As b increases, the power of the linear term increases and the eigenvalues move closer together until $b = \sqrt{3}$. At this point the power of the bias term is equal to the power of the linear term, so $\mathbf{R}$ is symmetrical with equal diagonal elements (although the off-diagonal elements are significant). Then as b increases still further, the power of the linear term starts

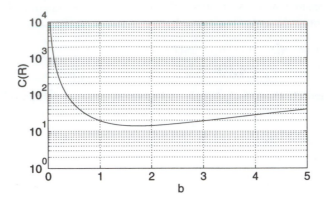

Figure 4.15 Condition of the autocorrelation matrix for the linear model, when the input is defined on the interval $[0, b]$.

to dominate the other elements of $\mathbf{R}$, and the network is slow to learn the bias term.

A surprising result is the condition number of $\mathbf{R}$ when $b = 1$, which was calculated in Section 4.3.4 to be 19.3, larger than might be expected. This is because the power of the bias input dominates, and the off-diagonal terms are greater than the power of the linear input. By reducing the output of the bias term to $1/\sqrt{3}$, the matrix becomes better conditioned as $C(\mathbf{R}) = 13.9$, although this is still large and is due to the size of the off-diagonal terms, which reflects the significant cross-coupling between the linear and the bias terms.

Linear Models Defined on Symmetric Intervals

Let the conventional linear model be defined on a domain $[-b, b]$ for some positive number b. The probability density function is assumed to be uniform on the interval $[-b, b]$, and so the autocorrelation matrix is given by:

$$\mathbf{R} = \begin{bmatrix} (2b)^{-1} \int_{-b}^{b} 1 \, dx & (2b)^{-1} \int_{-b}^{b} x \, dx \\ (2b)^{-1} \int_{-b}^{b} x \, dx & (2b)^{-1} \int_{-b}^{b} x^2 \, dx \end{bmatrix}$$

The off-diagonal elements are now all zero, because the integrands are anti-symmetric functions, therefore the integral over a symmetric interval is zero, and the above expression for $\mathbf{R}$ simplifies to:

$$\mathbf{R} = \begin{bmatrix} (2b)^{-1} \int_{-b}^{b} 1 \, dx & 0 \\ 0 & (2b)^{-1} \int_{-b}^{b} x^2 \, dx \end{bmatrix}$$

In Figure 4.16, the condition number of this autocorrelation matrix is plotted for different values of b. There is a unique minimum at $b = \sqrt{3}$ where $C(\mathbf{R}) = 1$; the two eigenvalues are equal, since $\mathbf{R}$ is diagonal with *equal* diagonal terms. Therefore to make a network as well conditioned as possible, the autocorrelation matrix should be nearly diagonal with diagonal terms of equal magnitude.

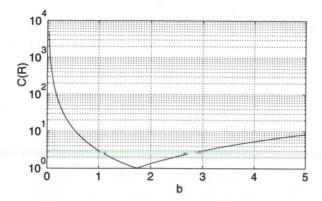

Figure 4.16 Condition of the autocorrelation matrix for a linear model, when the input is defined on the interval $[-b, b]$.

If it is required that $C(\mathbf{R}) = 1$, then instead of integrating over $[-\sqrt{3}, \sqrt{3}]$, the bias output could be set to be $1/\sqrt{3}$ and the integral taken over the interval $[-1, 1]$, since this produces the same result. It should also be noted that for *any* input probability density function the power of a unity bias term is *always* greater than or equal to the power of the linear term, which is defined on the interval $[-1, 1]$. Hence, the condition of these basic models can be drastically improved by choosing the size of the bias term appropriately and by altering the domain of the linear term.

4.4.7   Orthogonal Basis Functions

In the previous section, it was shown that a network is well conditioned if the diagonal elements of the autocorrelation matrix are approximately equal and the off-diagonal terms are almost zero, as this results in a condition number of almost unity (by Gerschgorin's circle theorem). The requirement that the off-diagonal terms be zero means that $\int_a^b a_i(x)\, a_j(x)\, p(x)\, dx$ must also be zero for all $i \neq j$. Similarly, the requirement that all the diagonal terms be equal means that $\int_a^b a_i^2(x)\, p(x)\, dx$ is equal to some constant, for all i.

Before generalising these concepts, some notation is introduced. Let the *inner product*, (a_i, a_j), of two functions $a_i(x), a_j(x)$, be defined as:

$$(a_i, a_j) = \int_a^b a_i(x)\, a_j(x)\, p(x)\, dx$$

Requiring the off-diagonal terms to be zero means that:

$$(a_i, a_j) = 0 \qquad \text{for } i \neq j \qquad\qquad (4.46)$$

and the requirement that the diagonal terms have equal power means that:

$$(a_i, a_i) = c \qquad \text{for } i = 1, 2, \ldots, p \qquad\qquad (4.47)$$

where c is a constant.

Two functions whose inner product is identically zero are said to be *orthogonal* on the interval $[a, b]$. If in addition, $(a_i, a_i) = 1$, $\forall i$, the functions are said to be *orthonormal*. Therefore, when the set of basis functions $\{a_i\}_{i=1}^p$ forms an orthonormal set (to within a multiplicative constant), the network is well conditioned for gradient descent learning and one-shot learning can be achieved by setting $\delta = 1/(a_i, a_i)$. Various sets of orthogonal polynomial basis functions have been derived: Legendre Polynomials, Chebyshev Polynomials [Burden and Faires, 1993] and these basis functions can reduce the computational burden when gradient descent learning rules are used. It has been shown that the bias function $a_1(t)$ and the linear function $a_2(t)$ are orthogonal on the interval $[-1, 1]$, which is not surprising since they are simply the first two terms in the set of Legendre polynomials.

4.5 MULTI-LAYER PERCEPTRONS AND BACK PROPAGATION

Multi-Layer Perceptrons trained using back propagation are currently the most widely used artificial neural networks. This is for a variety of reasons: the biological implications were emphasised when the training algorithm was reinvented [Rumelhart and McClelland, 1986]; the network is a universal nonlinear approximator, [Hornik *et al.*, 1989]; and the network seems capable of learning poorly understood, high-dimensional functions from only a small amount of training data [Wright, 1991]. However, MLPs trained using gradient descent and instantaneous gradient descent algorithms have serious shortcomings; convergence cannot be proved except for trivial cases, the learning process is poorly understood and few heuristics exist for choosing the learning rates.

This section considers two factors that influence the rate of convergence of MLPs: the basic condition of the linear optimisation subproblems associated with each of the nodes, and the effect of the sigmoid on the weight updates. The results derived are valid for an MLP trained using gradient descent, and their applicability to instantaneous gradient descent algorithms is currently under evaluation. It is shown that the linear optimisation subproblems are inherently ill conditioned, although this can be improved by using sigmoids which have outputs lying in the range $(-1, 1)$ and by reducing the output of the bias term. The effect the nonlinear transfer function has on the back-propagated error is then investigated, and it is shown that when an error is passed back through the sigmoid, the error term must be multiplied by approximately the inverse squared derivative in order to preserve its magnitude. For example, the scaling factor for a sigmoid given by $f(u) = 1/(1 + \exp(-u))$ is approximately 36, and the output error must be multiplied by this term for each *layer* of sigmoids. Equivalently, the learning rate for each layer in an MLP should be of this magnitude for responsive learning.

4.5.1 Architecture and Learning Rules

Interest in ANNs wained after the limitations of the types of logical functions which a single perceptron could reproduce became apparent [Minsky and Papert, 1969]. It was well known during the sixties that multi-layer networks were capable of reproducing these mappings, but no learning algorithm capable of training these more complex structures existed. Even when the BP training rule was reinvented in the mid-eighties, it is doubtful whether it would have been useful during the sixties, due to the limited computing power available at that time. The BP algorithm is a gradient descent algorithm which has become very popular because of its efficient way for calculating the network's sensitivity derivatives. MLP networks are composed of perceptron "type" units or *nodes*, which are arranged into layers where the outputs of the nodes in one layer constitute the inputs to the nodes in the next layer. The signals received by the first layer are the training inputs and the network's response is the outputs of the last layer, as shown in Figure 2.10. Each of the nodes has associated with it a weight vector and a transfer (or activation) function, where the dot product of the weight vector and the incoming input vector is taken, and the resultant scalar is transformed by the activation function. For a suitable arrangement of nodes and layers, and for appropriate weight vectors and activation functions, it can be shown that this class of networks can reproduce *any* logical function exactly and can approximate any continuous nonlinear function to within an arbitrary accuracy [Ellacott, 1994].

The description given above is very general and in order to analyse the behaviour of the weight adaptation rules, the network's architecture (the number of layers and nodes and the activation functions) must be specified. The results given in this section apply to multi-input, multi-output networks, although the theory is derived for a multi-input, single-output network with one hidden layer.

Consider therefore an n-input, single-output network with p nodes in the hidden layer, as shown in Figure 4.17. The $(p + 1)$-dimensional weight vector associated with the output node is denoted by $\mathbf{w}_0$, and the $(n + 1)$-dimensional weight vector associated with the ith node in the hidden layer is given by $\mathbf{w}_i$ for $i = 1, 2, \ldots, p$. Each node has connections with all the nodes in the previous layer plus a bias term. The $(n + 1)$-dimensional network's input vector at time t is composed of the bias term plus the n-dimensional input vector and is denoted by $\mathbf{x}(t)$, and the network's output is $y(t)$. The output of the nodes in the hidden layer is denoted by the $(p+1)$-dimensional vector $\mathbf{a}(t)$, and the output of the ith node in the hidden layer is given by:

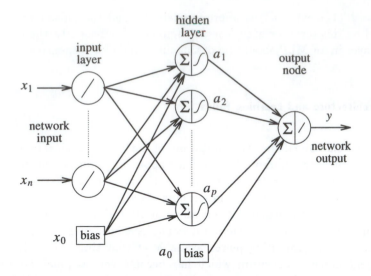

Figure 4.17 Three-layer network with linear input nodes and a linear output node.

$$a_i(t) = \begin{cases} 1 & \text{if } i = 0 \\ f(u_i(t)) & \text{otherwise} \end{cases} \tag{4.48}$$

where $u_i(t) = \sum_{j=0}^{n} x_j(t)w_{i,j} = \mathbf{x}^T(t)\mathbf{w}_i$ for $i = 1, 2, \ldots, p$. In this section, the dependence of the weights on time is neglected to simplify the notation, and when the network has an output node with a linear transfer function, its output is:

$$y(t) = \sum_{j=0}^{p} a_j(t)w_{0,j} = \mathbf{a}^T(t)\mathbf{w}_0 \tag{4.49}$$

Output Layer Learning

Gradient descent training rules adapt the output layer and the hidden layer weight vectors using the current gradient of the MSE performance function. Since the network is linearly dependent on the output layer weight vector, the learning rule is of the form given in Equation 4.33:

$$\begin{aligned} \Delta\mathbf{w}_0 &= -\delta_0 E\left(\epsilon_y(t)\mathbf{a}(t)\right) \\ &= -\delta_0\left(\mathbf{R}\mathbf{w}_0 - \mathbf{p}\right) \end{aligned}$$

where $\mathbf{R}$ is the hidden layer autocorrelation matrix, $\mathbf{p}$ is the cross-correlation vector, δ_0 is the learning rate for the output layer and $\Delta\mathbf{w}_0$ denotes the change in the weight vector.

Hidden Layer Learning

The network output depends *nonlinearly* on the hidden layer weight vectors, so using the local gradient information gives an update rule of the form:

$$\Delta \mathbf{w}_i = -\delta_i w_{0,i} E \left(\frac{df\left(u_i(t)\right)}{du} \epsilon_y(t)\mathbf{x}(t) \right) \qquad \text{for } i = 1, 2, \ldots, p \qquad (4.50)$$

where δ_i is the learning rate for the ith node in the hidden layer, and $w_{0,i}$ is the ith element of the output layer weight vector. In this expression, the network output error has been back propagated (hence the name of the learning algorithm) through the output layer, multiplying it by the derivative of the ith sigmoid in the hidden layer and the ith weight in the output layer. The error signal passed to the weight vectors in the hidden layer therefore depends on both these quantities, and their effects on the learning rule are considered in Section 4.5.3.

The designer has little control over the internal states (weight values, size of derivatives) of the adaptive network, as they are determined by the training data. The only parameters which can be set are the *learning rates*, the *type of activation function* and the size of the *bias* term. Heuristics for choosing all of these quantities are presented in the following sections, and reasons why improved convergence rates should be obtained (improved conditioning of each node, full error back propagation) are presented.

4.5.2 Condition of the Linear Optimisation Problems

It has been argued that gradient descent learning in MLP networks is composed of many *linear* optimisation subproblems. The network output is linearly dependent on the output layer weights, and a gradient descent learning rule is used to train the hidden layer weights. Thus if the linear optimisation subproblems are badly conditioned, the overall nonlinear adaptation strategy is also badly conditioned, although even if they are well conditioned, other factors may reduce the rate of convergence.

The rate of convergence of the weight vector of the output node depends on the condition of the hidden layer autocorrelation matrix, $C(\mathbf{R})$, and the rate of adaptation of the hidden layer weights. If the weights in the hidden layer are fixed, there exists a stationary global minimum in the output layer's weight space; whereas if the hidden layer weights adapt, the global minimum of the output layer is non-stationary and its position in weight space changes as the structure of the hidden layer is updated. When the global minimum is stationary, the rate of convergence is directly related to $C(\mathbf{R})$, which depends on the interaction between p sigmoidal transfer functions and a single bias node.

If the hidden layer is composed of a single node and a bias term, the condition of the output layer optimisation problem is identical to the condition of the linear models defined in Section 4.4.6, where x now denotes the output of the hidden layer node. Therefore, when the output of the hidden layer node is defined on a symmetrical interval ($[-b, b]$ for some positive number b) and the corresponding probability density function is symmetrical, the resulting optimisation problem is far better conditioned than if the output of the hidden layer node were to lie in the interval $[0, 1]$. In addition, if the output of the hidden layer node lies in the interval $[-1, 1]$ and has a uniform probability density function, and the output of the bias term is $1/\sqrt{3}$, a gradient descent learning algorithm can converge after one iteration when the learning rate is chosen appropriately. Finally, when the output of the hidden layer node lies in $[-1, 1]$, the power of the bias term is always greater than or equal to the power of the hidden layer node, irrespective of the characteristics of the probability density function. Even when the probability density function is non-symmetric, improved rates of convergence can be obtained by simply reducing the output of the bias node and increasing the learning rate to compensate.

When there is more than one hidden layer sigmoidal node, the analysis becomes more difficult, as the autocorrelation matrix is no longer diagonal, although some of the results can be generalised. The power of the bias term is always greater than or equal to the power of any term in the hidden layer when the outputs of the sigmoids lie in the interval $[-1, 1]$ or $[0, 1]$. Thus, by reducing the output of the bias node, the condition of the optimisation problem can be improved. Also when the output of the sigmoid lies in the interval $[-1, 1]$ (for a symmetrical probability density function), each sigmoidal node is orthogonal to the bias node. Thus the optimisation problem can again be improved. However, convergence is still generally slow, due to the considerable cross-coupling between different sigmoidal nodes, which produces significant off-diagonal elements in the autocorrelation matrix.

4.5.3 Backwards Error Propagation Through a Sigmoid

Considering the amount of research which has been aimed at understanding how MLPs learn using gradient descent algorithms, it is surprising that so little effort has been directed at learning rate selection and the effect of the sigmoid transfer functions on the overall optimisation strategy. When the output error is uncorrelated with the derivative of the sigmoid, the hidden layer adaptation can be decomposed into a set of linear optimisation subproblems, and the rate of convergence is a function of the condition of these linear subnetworks, but the *strength* of the output error signal back propagated through the network has not been considered. This is directly related to the form of (nonlinear) transfer function used in each node, so initially the two most popular sigmoidal transfer functions are investigated and compared. A simple example then motivates the remainder of the work developed in this section and demonstrates that the learning rates for the basic gradient descent learning algorithm are substantially *underestimated*. It is therefore not surprising that conventional MLPs converge very slowly when trained using gradient descent algorithms; the linear optimisation subproblems are generally ill conditioned and the learning rates are usually far too small.

Sigmoidal Transfer Functions

Consider the form of the two most common nonlinear transfer functions:

$$f_1(u) = \frac{1}{1 + \exp(-u)} \qquad \in (0,1) \qquad (4.51)$$

$$f_2(u) = \tanh(u) = \frac{1 - \exp(-2u)}{1 + \exp(-2u)} \qquad \in (-1,1) \qquad (4.52)$$

which are shown in Figure 4.18. These functions are monotonic, bounded and infinitely differentiable everywhere, so they are suitable candidates for replacing the original discontinuous binary threshold function. The output of f_1 lies in the interval $(0,1)$ whereas the output of f_2 is contained in $(-1,1)$. Therefore from a conditioning viewpoint, the tanh(.) function is preferable, as it is potentially orthogonal to the bias term. These transfer functions are related in the sense that:

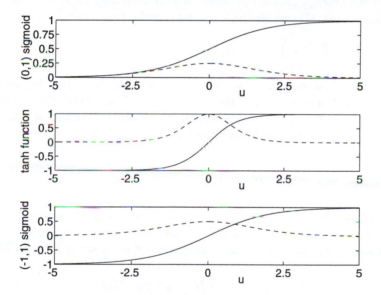

Figure 4.18   Three nonlinear sigmoidal transfer functions (solid lines) and their derivatives (dashed lines).

$$f_2(u) = 2f_1(2u) - 1 \qquad (4.53)$$

showing that the tanh(.) function is *not* a linearly transformed sigmoid. The modelling abilities of networks which include these *different* transfer functions are equivalent when the input is multiplied by two (or the weights in the previous layer are twice their size), and it is also worthwhile comparing their derivatives in order to understand how this influences the error back propagated to the hidden layer.

The derivative of f_1 is given by:

$$\frac{df_1(u)}{du} = \frac{\exp(-u)}{(1+\exp(-u))^2} = f_1(u)(1-f_1(u)) \qquad \in (0,1/4] \qquad (4.54)$$

and is shown in Figure 4.18. For a uniform probability density function on the output of the node, its expected value is 1/6, and square of the derivative of this function lies in the interval $(0,1/16]$ with an expected value of 1/30. For the tanh(.) function, its derivative lies in the interval $[0,1]$, with an expected value of 2/3 for a uniform probability density function on the output of the node, and the corresponding expected value of the square of the derivative is 2/15.

From Equation 4.50, the size of the weight update is influenced by the size of the derivative of the transfer function. Intuitively it would be expected that it should take a value close to unity, otherwise learning could be slow or unstable. From this viewpoint, the tanh(.) is again the most desirable transfer function, as its derivatives are closer to unity.

The modelling abilities of each transfer function are similar, although in order to compare the storage abilities of different networks they must have the *same* basic structure. Hence, consider the following transfer function:

$$f_3(u) = \left(\frac{2}{1+\exp(-u)}\right) - 1 = 2f_1(u) - 1 \qquad \in (-1,1) \qquad (4.55)$$

which is a scaled and translated version of the sigmoid $f_1(u)$ (see Figure 4.18). This ensures that the output lies in the interval $(-1,1)$, and the derivative is twice that of f_1. For the remainder of this section, this transfer function is referred to as a sigmoid whose output lies in the interval $(-1,1)$.

Example: Single Node Linear Network

To illustrate how the output errors are *squashed* when they are passed back through a transfer function whose derivative is less than unity, consider updating a single-layered network which has a single linear output node, as shown in Figure 4.19. The

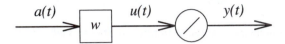

Figure 4.19 Single-layered network with a linear output node.

gain of the linear output node is 1/2, hence the gradient descent, back propagation algorithm is given by:

$$\Delta w = -\delta/2E\left(\frac{d\epsilon_y^2(t)}{dw}\right) = \delta E\left(\frac{dy(t)}{du(t)}\frac{du(t)}{dw}\epsilon_y(t')\right) \qquad (4.56)$$

For this example, the derivative of the activation function is constant $(= 1/2)$, and the weight update is:

$$\Delta w = \delta/2E\left(\epsilon_y(t')a(t')\right)$$

In order to assess the effect of this weight update, the expected *a posteriori* network output, $\underline{y}(t)$, must be evaluated, and is simply equal to:

$$
\begin{aligned}
E\left(\underline{y}(t')\right) &= E\left(\frac{a(t')w}{2} + \frac{a(t')\Delta w}{2}\right) \\
&= E\left(y(t')\right) + E\left(\frac{\delta\epsilon_y(t')}{4}\right)
\end{aligned}
$$

when the input signal is identically unity ($a(t) \equiv 1$). Thus it is possible to express the *a posteriori* output error in terms of the *a priori* output error as:

$$
\begin{aligned}
E\left(\epsilon_{\underline{y}}(t')\right) &= E\left(\hat{\underline{y}}(t')\right) - E_{t}\left(\underline{y}(t')\right) \\
&= E\left(\hat{y}(t')\right) - E\left(y(t')\right) - \frac{\delta\left(E\left(\hat{y}(t')\right) - E\left(y(t')\right)\right)}{4} \\
&= \left(1 - \frac{\delta}{4}\right)E\left(\epsilon_y(t')\right)
\end{aligned}
$$

The network output error is reduced by $\delta/4$, since the plant output error is multiplied by the derivative of the transfer function when the weight is updated, and the weight update is *again* multiplied by the derivative of the transfer function when the *a posteriori* network output is calculated. Thus, the *a priori* plant output error is being scaled by a factor of $E\left(dy/du\right)^2$ and, unless the learning rate δ is chosen so as to compensate for this quantity, adaptation can be very slow (for a derivative < 1) or unstable (derivative > 1). This approach was partially developed by Kalman and Kwasny [1992] and Rigler *et al.* [1991], where the learning rate, δ, (or equivalently the gain of the transfer function) was chosen so that it compensated for the value of $E\left(dy/du\right)$. This quantity was proposed because only the effect of the derivative of the transfer function on the weight update was considered, whereas its effect on the network evaluation also needs to be taken into account. Hence, the learning rate should be set to be (approximately) the inverse of the *square* of the expected value of the derivative of the transfer function.

Output Layer Gradient Descent Adaptation

In deriving the above results, a number of assumptions were made; the network only had one input node and the input signal was identically unity. It is now shown that these basic results can be generalised to the more realistic problem where the network has more than one hidden layer, the input signal is a vector whose magnitude is not required to be unity, and the output layer can have a nonlinear transfer function.

When the input signal is multi-dimensional and/or the magnitude of the input signal is not equal to 1, the selection of the learning rate depends on the size of this signal (see Section 4.4.5) and stable learning is assured if:

$$0 < \delta < \frac{2}{\lambda_{\max}}$$

where $\lambda_{\max}$ is the maximum eigenvalue of the autocorrelation matrix. This condition *assumes* that the complete network output error signal is used to update the weight vector, and so it should be used in conjunction with the factor necessary to compensate for the derivative of the transfer function. The previous example can therefore be extended to cope with non-constant, multivariate inputs, and stable learning results if:

$$0 < \delta < \frac{8}{\lambda_{\max}}$$

Thus the condition of the underlying *linear* optimisation problem does not affect the theory developed for choosing the factor with which to scale the learning rate in order to compensate for the derivative of the transfer function.

When the output layer has a nonlinear transfer function, its gradient descent learning rule is given by:

$$\Delta \mathbf{w}_0 = \delta_0 E \left(f'(u(t)) \epsilon_y(t) \mathbf{a}(t) \right)$$

where the size of the weight update depends on the derivative of $f()$. So the learning rate δ_0 must be selected so that it incorporates information about the size of both the derivative of f and the output of the hidden layer, $\mathbf{a}(t)$. When the transfer function is nonlinear, no exact scaling factor for the learning rate can be determined to counteract the size of $f'()$. However, if it is assumed that the output error is not correlated with the derivative of the sigmoid, to a first approximation the learning rate should be multiplied by the *square* of the inverse of the expected value of the derivative of the transfer function. When a $(0, 1)$ sigmoid is employed as the transfer function, the learning rate should be multiplied by 36, whereas it should be multiplied by 9 when the output of the sigmoid lies in $(-1, 1)$. If it cannot be assumed that the output error is not correlated with the derivative of the sigmoid, the learning rate should (at least) be multiplied by the minimum value of $(f'(u(t)))^{-2}$. For the $(0, 1)$ sigmoid this value is 16, and for the $(-1, 1)$ sigmoid the scaling factor is 4.

For any of these cases, it can be seen that the sigmoid function squashes the output error as it is propagated backwards through the network. Requiring that the learning rate be selected so as to counteract this scaling effect means that, for gradient descent learning, δ should be substantially greater than 1. The learning rate should also take into account the size of the input vectors, and should also be scaled by the maximum eigenvalue of the autocorrelation matrix.

An alternative strategy was proposed by Biegler-König and Bärmann [1993], where the desired network output is passed through the *inverse* sigmoid function, and the difference between this quantity and $(u(t))$ is used to train the weight vector, using a direct matrix inversion technique or by using a standard gradient descent algorithm. This overcomes the problem of slow learning due to the slope of the sigmoid, but it should be noted that now the network output MSE is *not* being minimised, but rather the algorithm minimises the quantity:

$$E\left(\epsilon_u^2(t)\right)$$

where $\epsilon_u(t) = f^{-1}(\widehat{y}(t)) - u(t)$. This is a *standard* technique for finding an approximate solution [Burden and Faires, 1993], although it should be emphasised that when $E\left(\epsilon_u^2(t)\right)$ is small, it does not follow that $E\left(\epsilon_y^2(t)\right)$ is small, especially when the modelling error is large (see Section 2.2.2). Other schemes have been proposed, such as minimising alternative performance functions [Ooyen and Nienhuis, 1992], which produce learning rules that do not depend on the derivative of the sigmoid.

Hidden Layer Gradient Descent

The heuristics for choosing the learning rates in the hidden layer are similar to those which have just been derived for choosing the learning rates in the output layer, as the learning rates have to be chosen to account for the condition of input layer autocorrelation matrix, the values of the derivatives of both the output and hidden layer transfer functions and also the value of the appropriate output layer weight. Denoting the output layer and hidden layer transfer functions by $f'(.)$ and $a'(.)$, respectively, the ith learning rate should be multiplied by the factor:

$$\left[w_{0,i}E\left(f'(t)a_i'(t)\right)\right]^{-2}$$

in order to account for the squashing effects of the nonlinear transfer functions. It has also been implicitly assumed that the learning rate is scaled by the inverse of the maximum eigenvalue of the input layer correlation matrix, as in the previous section. Finally, it is undesirable for different nodes in the hidden layer to have different learning rates, as the network would be biased, so generally the minimum learning rate is calculated, and this is used for each node in the hidden layer.

When a $(0,1)$ sigmoid is used in the output and in the hidden layer, the learning rate should have a value of at least $(16 * 16/w_{0,i})^2$. Considering the size of this quantity, it is not surprising that slow convergence is reported in many MLPs which have unity learning rates and a small weight initialisation strategy.

The learning rates for the output layer and the hidden layers should not be chosen independently, as this can result in unstable learning for networks with three or more layers. These learning rates should again be scaled so that the output layer learns faster than the hidden layer, and these scaling factors should be chosen such that they are normalised (sum to one), as this implies a full output error correction across the whole network. Thus the learning rates for the output layer and hidden layer should be scaled by a factor of approximately 0.7 and 0.3, respectively.

4.5.4   Summary and an Example

There have been many heuristics proposed for choosing the type of nonlinear transfer functions and the learning rates for use in MLPs which are trained using gradient descent algorithms, and these are now summarised.

The use of sigmoidal transfer functions, whose output lies in the interval $(-1, 1)$ rather than $(0, 1)$, generates a *better* conditioned network for training using gradient descent rules. This is because the sigmoidal nodes and the bias nodes become orthogonal functions for symmetrical probability density distributions. Also the power of the bias node is *always* greater than the power of the sigmoidal node, and so improved convergence can be obtained by reducing the output of the bias node. Again, the optimal value of the output of the bias node depends on the range of the sigmoidal outputs.

Several heuristic rules have also been proposed for choosing the value of the learning rate, and under certain conditions, it depends on the maximum eigenvalue of the appropriate autocorrelation matrix, the square of the expected value of the subsequent forward path of the signal (sigmoid derivatives, output layer weights, etc.) and the number of layers in the network. The use of the tanh(.) sigmoidal function is therefore preferred, because it improves the conditioning of the linear optimisation subproblems associated with each node and the derivative of this function lies in the interval $[0, 1]$. Both properties should improve the rate of convergence of MLPs trained using gradient descent algorithms.

For the output layer, a learning rate of the form:

$$\delta_0 = \frac{0.7 * \delta}{\lambda_{\max} \left[E \left(f'(u(t)) \right) \right]^2}$$

was proposed, where $\lambda_{\max}$ is the maximum eigenvalue of the hidden layer autocorrelation matrix. A similar derivation for the hidden layer learning rate produced:

$$\delta_1 = \min_{i=1}^{p} \frac{0.3 * \delta}{\lambda_{\max} \left[w_{0,i} E \left(a'_i(t) f'(u(t)) \right) \right]^2}$$

where $\lambda_{\max}$ is the maximum eigenvalue of the input layer autocorrelation matrix.

When these heuristics are used to scale the learning rates (for $0 < \delta < 2$), they produce learning rates for the hidden layer substantially larger than commonly used. This is because the "squashing" effect of the sigmoid is generally not taken into account, and hence learning is very slow. These heuristics are now illustrated with a simple example.

Example: MLP Trained Using Gradient Descent

Consider a desired function that can be modelled exactly by a three-layered network, as shown in Figure 4.20. The single input, single output network has linear input and output node transfer functions, there exists one sigmoidal node in the hidden layer whose output lies in the interval $(0,1)$, plus a unity output bias node in each layer and the desired value of each weight is one. Therefore a network of the same structure has the potential to learn this function *exactly*.

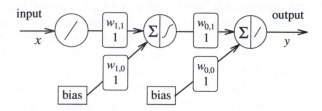

Figure 4.20   The three-layered network realisation of the desired function, where $(0,1)$ sigmoids are employed with unity bias terms.

Two networks are compared. The first one has the same structure as the desired function, with sigmoids whose output lies in $(0,1)$ and unity bias terms. Therefore all the desired weights are equal to unity. The simulations are started with the hidden layer weight vectors being correct and the output layer weights being set to zero. The reason for this is because the rate of convergence of the weight vector is being studied, and setting too large an initial error in the hidden layer weight vector may cause the weights to converge to a local minimum. The training samples are chosen so that a uniform probability density function on the output of the hidden layer sigmoid is produced.

The second network has the same basic structure as the desired network, except that the output of the sigmoid lies in the interval $(-1,1)$, and the outputs of the bias nodes are reduced, so that the power of the bias and the linear/sigmoidal nodes are the same. The training data are uniformly distributed on the output of the hidden layer node and so the output of the hidden layer bias node is $1/\sqrt{3}$. Assuming a uniform distribution on the output of the sigmoidal node, the probability density distribution of the training data in the input layer is:

$$p(x) = \frac{\exp\left(-x+1\right)}{\left(1+\exp\left(-x+1\right)\right)^2}$$

which is simply the (shifted) derivative of the sigmoid transfer function. Thus the power of the linear input node is given by:

$$\int_{-\infty}^{\infty} x^2 \, \frac{\exp\left(-x+1\right)}{\left(1+\exp\left(-x+1\right)\right)^2} \, dx$$

Integrating by parts this expression evaluates to $\pi^2/3$, and so the output of the input layer bias node is set equal to $\sqrt{(3+\pi^2)/3}$. Thus the nodes in each layer have the same power, and in the output layer they are orthogonal. The output layer desired weight vector is given by $\hat{\mathbf{w}}_0 = \left(1.5\sqrt{3}, 0.5\right)^T$, and the desired hidden layer weight vector is $\hat{\mathbf{w}}_1 = \left(\sqrt{3/(3 + \pi^2)}, 1.0\right)^T$. The initial weight vectors again have the correct values for the hidden layer weights and a zero output layer weight vector.

For each network the learning rate for the output layer is set equal to $0.7/\lambda_{\max}$, because the output layer node is linear with a unity gain, where $\lambda_{\max}$ is the largest eigenvalue of the hidden layer autocorrelation matrix. The learning rate for the hidden layer is set equal to $0.3 * 36/\lambda_{\max}$, because for each network, the desired linear weight in the output layer, multiplied by the derivative of the hidden layer node, is a constant and the inverse of the square of the expected value is 36. Here $\lambda_{\max}$ is the largest eigenvalue of the input layer autocorrelation matrix.

The learning histories for these two networks are shown in Figures 4.21 and 4.22. The first figure shows the initial stages of learning, when the learning is slightly

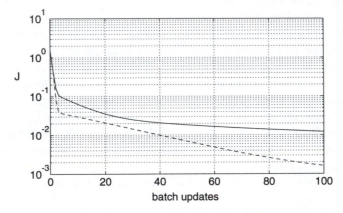

Figure 4.21   MLP initial learning history. The solid line is generated by a standard $(0, 1)$ sigmoidal network, whereas the dashed line corresponds to a symmetrical $(-1, 1)$ sigmoidal network.

erratic, although it can clearly be seen that the modified network learns faster initially. The second figure gives a clearer impression of the comparative rates of output convergence; the "steady-state" rate of convergence for the scaled and translated MLP is much quicker.

Similar results are observed when the rates of convergence of the individual weights in the two networks are compared. In Table 4.2, the normalised errors in each of the weights are displayed for $t = 0, 10, 100$ and 1000. It is clear that the weights in each of the two networks are converging at approximately the same rate and that convergence occurs at a faster rate in the MLP with $(-1, 1)$ sigmoids. In this example, the learning rates for both networks were selected, using substantial *a priori* knowledge, so that both networks stored the information quickly. The relative rates of convergence reflect the conditions of the MLPs, although in general, the convergence rate in MLPs using BP is slower, because the learning rates are substantially *underestimated*.

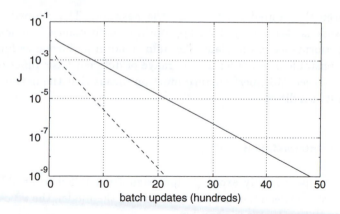

Figure 4.22 MLP final learning history. The faster rate of convergence (dashed line) corresponds to a symmetrical $(-1, 1)$ sigmoidal network, whereas the slower rate of convergence (solid line) is generated by a $(0, 1)$ sigmoidal network.

	Normalised Absolute Weight Errors							
	$(0, 1)$ sigmoidal MLP				$(-1, 1)$ sigmoidal MLP			
t	$w_{0,0}$	$w_{0,1}$	$w_{1,0}$	$w_{1,1}$	$w_{0,0}$	$w_{0,1}$	$w_{1,0}$	$w_{1,1}$
0	1.0	1.0	0.0	0.0	1.0	1.0	0.0	0.0
10	$1.30e^{-1}$	$2.51e^{-1}$	$9.44e^{-2}$	$1.09e^{-1}$	$4.86e^{-2}$	$4.08e^{-2}$	$3.37e^{-1}$	$5.27e^{-2}$
10^2	$4.60e^{-2}$	$8.25e^{-2}$	$1.36e^{-1}$	$1.65e^{-1}$	$1.94e^{-3}$	$6.81e^{-3}$	$3.07e^{-2}$	$1.38e^{-2}$
10^3	$1.41e^{-3}$	$3.44e^{-3}$	$8.19e^{-3}$	$6.28e^{-3}$	$8.44e^{-7}$	$1.74e^{-5}$	$4.18e^{-5}$	$3.42e^{-5}$

Table 4.2 The absolute value of the normalised errors in the individual weights for a $(0, 1)$ and a $(-1, 1)$ sigmoidal MLP. The superior rate of weight convergence for the latter network is obvious.

4.6 NETWORK STABILITY

A network is defined to be *stable* if the learnt weight values are *minimally* affected when the training domain is increased. This is an important concept for on-line adaptive modelling and control as if training data are biased towards certain areas of the input space for prolonged periods only relevant information should be stored. When the learnt weight values are *not* close to the globally optimal values, significant retraining takes place after the plant has stopped operating in this region: learning is neither local nor stable. Therefore parameter convergence needs to be investigated when the input signal is both *locally* and *globally* persistently exciting.

If a locally persistently exciting input signal is applied to a model which can reproduce the desired function exactly, the appropriate weights converge to their correct values. Therefore network stability needs to be assessed when a mismatch exists between the desired function and the network. This immediately implies that a network is stable only if an appropriate performance function is used to assess the performance of the plant. An infinity norm performance function could result in an unstable network for a wide range of different basis functions, whereas the commonly used MSE performance function means that the network's stability depends on its condition.

4.6.1   Matrix Interpretation

When a locally persistently exciting input signal is applied to an AMN where a mismatch exists between the desired function and the model, the ith component of the cross-correlation vector is given by:

$$p_i = E\left(\widehat{y}(t)a_i(t)\right) \tag{4.57}$$

Extending the training data means that the desired function is defined over a larger input domain, but p_i generally remains *unchanged* (at least to within a multiplicative constant) if the desired function was previously completely specified in the support of the ith basis function. Changing the desired function does not affect the optimal value of w_i directly, but it is different, due to the effects of neighbouring weight changes. The amount that w_i is altered depends on the type of desired function and the interaction or *condition* of the set of basis functions (see Section 8.3.2). Basis functions with little or no overlap (low condition numbers) are more stable than the basis functions of a network with a large amount of generalisation.

It is also worth noting that if the inner product of the desired function with every basis function (the integral over the respective domain of the product of the desired function with the basis function) is zero, the AMN is completely unable to model the desired function; the optimal weight vector is also zero. This well-known result was stated in the context of the CMACs modelling abilities by Cotter and Guillerm [1992] and is used in Chapter 7 to construct a set of desired functions which are orthogonal to the set of the CMAC's binary basis functions.

4.6.2   B-spline Example

Consider the desired quadratic function $\widehat{y}(x) = x^2$, which is defined on three input domains:

$$
\begin{aligned}
I_1 &= [-1, 1] \\
I_2 &= [-1, 2] \\
I_3 &= [-1, 3]
\end{aligned}
$$

In each case, the input is defined to have a uniform probability density function on the appropriate interval, and an order 2 (piecewise linear) B-spline network, Figure 4.23, is used as the basic model.

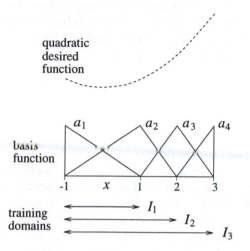

Figure 4.23 Desired function defined on the extended input domain and the B-spline model.

When the input lies within the interval I_1, only two weights are set and their optimal values are given by:

$$\widehat{w}_1 = 0.333, \qquad \widehat{w}_2 = 0.333 \tag{4.58}$$

Extending the input domain to $[-1, 2]$, so that there now exist three B-splines of order 2 defined on this input space and one extra weight is introduced. The new optimal weights values are:

$$\widehat{w}_1 = 0.25, \qquad \widehat{w}_2 = 0.5, \qquad \widehat{w}_3 = 4 \tag{4.59}$$

The new optimal value for w_1 is 25% less than the value given in Equation 4.58, which is a reasonable estimate, because the input is being extended to an interval immediately adjacent to the support of the basis function. The new optimal value for w_2 is 50% greater than that given in Equation 4.58, which is not too great considering that the new input domain lies within the support of the basis function and therefore p_2 is altered. The weight was previously estimated when the desired function was only defined on half of the basis function's support, and a similar situation occurs at the edges of the training domain.

Finally consider the input interval I_3, where four B-splines of order 2 are defined on the input space. The calculated weights are:

$$\widehat{w}_1 = 0.242, \qquad \widehat{w}_2 = 0.515,$$
$$\widehat{w}_3 = 3.924, \qquad \widehat{w}_4 = 8.788 \tag{4.60}$$

and compared to Equation 4.59, the value of w_1 decreases by 3% and the value of w_2 increases by about 3%. Hence these parameters only change slightly (p_1 and p_2 are unchanged) and the network can be said to be truly stable.

To understand exactly how extending the input domains changes the parameters already stored, the structure of the above matrices must be investigated. It is worth noting that they are real, symmetric and banded of bandwidth $(2k - 1)$ and this area remains a topic for further research.

4.7   CONCLUSION

Gradient descent learning algorithms and their instantaneous counterparts, which are described in Chapter 5 are probably the most popular methods for training supervised artificial neural networks. There has also been a vast amount of theory derived over the past thirty years which can be used to explain the behaviour of the learning networks, and this chapter has summarised the most relevant results and described how they may be applied to both linear and nonlinear adaptive systems. This separation of the AMN's parameters into a set of linear weights that may be adapted on-line and a group of nonlinear terms (knots, input variables, etc.) which must be determined before learning commences (either from an expert or using a network construction learning algorithm as described in Section 8.5) means that the behaviour of the algorithms can be predicted and convergence and stability results derived.

Various aspects of parameter convergence have been examined and, in particular, this chapter has concentrated on issues such as rate of convergence, singularity, minimum norm solutions and the relationship between parameter and output errors. Parameter convergence is essential if a network is to be expected to generalise sensibly and the rate of convergence must be rapid in order for networks to be useful, especially if on-line adaptation is important. Higher order learning algorithms can increase the rate of parameter convergence and if the computational cost of training the weights is not important, these training rules should be used. However, one of the reasons for studying (batch) gradient descent in some detail is to understand the behaviour of the instantaneous learning algorithms and to gain insight into how the network's structure may determine whether a particular training rule is successful.

The rate of convergence of the parameter vector has been shown to depend on the *condition* of the autocorrelation matrix. This in turn is determined by the shape and size of the network's basis functions and by the distribution and amount of training data. The size and shape of the basis functions (fuzzy membership functions) can be determined by the user (if they have sufficient *a priori* knowledge), hence biasing the network can improve its convergence characteristics, as long as it is sufficiently flexible to adequately model the data. The AMNs considered in this book are all generally defined on some lattice covering the input space, so singularity is a prime concern as the training data are unlikely to cover the whole of the input space adequately, being rich in both frequency and spatial content. Network construction algorithms which automatically determine the partitioning of the input space can improve this situation, although singularity and the resulting minimum norm solutions still need to be considered.

These results were all derived for the linear set of adjustable parameters, although this was generalised in Section 4.5, where it was shown how they could be applied to MLPs. Heuristics were developed for choosing the learning rates associated with each layer and they were shown to depend on the condition of the local autocorrelation matrix and the strength of the derivative of the transfer function. The adaptive linear theory cannot be used to prove convergence, but it has a significant role to play in understanding the behaviour of these more complex models.

5

Instantaneous Learning Algorithms

5.1 INTRODUCTION

The instantaneous learning rules derived in this chapter have been likened to the following scenario [Johnson, 1988]:

> *Imagine standing on the side of a valley on a moonless night and wanting to reach the village at the bottom of the valley. Choose a search direction at random and take a small number of steps along the direction which leads downhill. From this new starting position, choose another direction of motion and keep repeating this procedure until the ground is locally flat.*

If the valley is bowl-shaped and the set of search directions has sufficient variety, and if it is possible to determine when the ground is locally flat, this *instantaneous* gradient descent algorithm approximates the true gradient descent algorithm, and the shorter each step is, the more accurate the approximation. When the valley is not bowl-shaped, it is possible to stop on a plateau (or in a local valley) which is not the lowest point of the valley. If the direction of motion does not have sufficient variety, there is not enough freedom in the direction of the descent in order to reach the bottom of the valley. Finally, when the "flatness" of the ground cannot be determined accurately, the most likely final behaviour is to wander around the bottom of the valley close to the village.

Many neural net training rules are simply instantaneous learning algorithms (together with nonlinear variations) and in recent years there has been a growing interest in trying to understand *what* and *how* neural networks learn. This is crucial in areas such as adaptive modelling and control, where the algorithms are designed to operate with the minimum of human intervention. In the field of linear modelling and control, a vast body of literature has been generated over the past forty years [Åström and Wittenmark, 1989, Sastry and Bodson, 1989] and one advantage offered by Artificial Neural Networks (ANNs) is that they provide a method for extending this work to a more general nonlinear framework. Many ANNs can approximate arbitrarily well any continuous nonlinear function, with the network's architecture and learning algorithm determining how easily learning proceeds. For on-line modelling and control, it is unrealistic to assume that the input signal will excite the whole of the state space, and so the effects of a reduced input signal on the overall functional approximation must be investigated. Hence learning must be *local*, in that the parameters adapted should only affect the output of the network locally.

Instantaneous learning rules are formulated by minimising instantaneous estimates of a performance function, which is generally the Mean Square output Error (MSE), and the parameters are updated using gradient descent rules. The Associative Memory Networks (AMNs) considered in this book all depend *linearly* on a set of weights, and it is these parameters which are updated using the basic learning rules. Prior to this linear layer, the algorithms perform a fixed, nonlinear transformation of the input, mapping it to a high dimensional sparse internal representation. Instantaneous learning rules can take advantage of this structure as they only update those weights which contribute to the output, and learning is local because dissimilar inputs are mapped to different weight sets. Therefore, it appears that instantaneous learning laws are appropriate for training these weight vectors, and this chapter considers their performance in a variety of situations.

To begin, the basic Least Mean Square (LMS) algorithm is derived and its ability to store the current training sample assessed. This motivates the development of the Normalised Least Mean Square (NLMS) rule, which always stores the training pair exactly. The directions of the weight updates (search directions) are the same for each learning rule, it is only the distance travelled (step size) which distinguishes them. These rules have a specific geometrical interpretation which can be used to assess their convergence properties. They both possess the *minimal disturbance principle*; the weight update is the smallest that causes the current information to be stored. It is proved that when there does not exist any modelling error or measurement noise, the network learns to interpolate the data exactly. Even if the optimisation problem is underspecified, the weight vector converges to the *minimum norm solution*. When modelling error exists, output dead-zones and stochastic learning rates are proposed to counteract the inaccurate gradient estimates, which cause the weights to converge to a *minimal capture zone* rather than to a single value. An expression for the rate of parameter convergence is then derived and fast parameter convergence is shown to occur when the transformed input vectors are orthogonal, and slow convergence if the they are highly correlated.

An important issue in this work is the relationship between instantaneous and true gradient descent. The instantaneous estimates made of the gradient introduces *noise* into the updating algorithm, causing erratic weight updates, and when modelling error exists, the parameters converge to a minimal capture zone. The size of the gradient noise is related to the *condition* of the autocorrelation matrix, providing a direct link with the rate of convergence of the true gradient descent algorithm. Examples are presented both for the "transient" gradient noise and the "steady-state" behaviour within the minimal capture zones. Then a measure of the amount of *learning interference* present in various associative memory networks is derived. Learning interference can be defined as the information lost due to neighbouring weight updates, and provides another insight into the effect of gradient noise.

Finally, a set of higher order learning algorithms is derived which address some of the deficiencies of the basic LMS training rules. They use the information contained in *several* training samples to increase the rate of parameter convergence and to reduce the amount of gradient noise and are particularly useful for *on-line* learning where the training data are usually highly correlated and contains a significant amount of measurement noise.

5.2   INSTANTANEOUS LEARNING RULES

In this section, several instantaneous linear learning laws are derived from a variety of different viewpoints: gradient descent, error correction, stochastic approximation, etc. They are shown to have the same basic structure; the change in the weight vector is equal to the (scaled) output error multiplied by the transformed input vector, and their performance is thoroughly analysed.

In order to adapt the set of linear parameters (weight vector), the current performance of the network must be evaluated and generally the network's MSE is used. Other measures can be employed instead of the Euclidean norm, such as the one norm, which places a low emphasis on the data outliers, or the infinity norm which minimises the largest output error. The (weighted) MSE *usually* leads to acceptable models and the criterion has been thoroughly studied. It generates simple instantaneous learning laws and when the data have been corrupted by additive Gaussian noise, the trained model often corresponds to the maximum likelihood estimate.

The instantaneous learning rules considered in this section can be formulated as *line search* optimisation methods [Fletcher, 1987], since these algorithms generate a *search path*, along which the weights are updated, then calculate a *step size* which ensures stable learning. Both the search path and the step size are determined using *limited* training information (often only a single input/output pair), and the effect of using these estimations is thoroughly investigated.

5.2.1   Instantaneous Gradient Descent

The MSE of a set of training examples for a model which is linear in its parameter vector forms a quadratic hyperbowl in p-dimensional weight space (see Figure 4.2), which has a unique minimum at $\hat{w}$ if the set of training signals is sufficiently rich and the basis functions are linearly independent. When the parameters are trained using gradient descent rules, they converge (in the limit) to this unique weight vector. The choice about using gradient descent rules, and if so, whether to train by example or batch depends on the particular application. It is argued in this section that instantaneous gradient descent rules are appropriate for AMNs, rather than the more complicated recursive least-squares algorithm, and as such they are suitable for on-line learning.

The MSE of the network is given by:

$$J = E\left(\epsilon_y^2(t)\right) \tag{5.1}$$

where $\epsilon_y(t)$ is the *a priori* output error $(\hat{y}(t) - y(t))$, and $E(.)$ is the expectation operator. The instantaneous estimate of the MSE at time t is:

$$J_i(t) = \epsilon_y^2(t) \tag{5.2}$$

and the instantaneous estimate of the gradient of the performance function (in weight space) at time t is given by:

$$\frac{\partial J_i(t)}{\partial \mathbf{w}} = -2\epsilon_y(t)\mathbf{a}(t)$$

It can easily be established that this is an unbiased estimate of the gradient of the true performance function given in Equation 5.1 [Widrow and Stearns, 1985]. Instantaneous gradient descent training rules update the weight vector in proportion to the negative instantaneous gradient, producing a learning rule of the form:

$$\Delta\mathbf{w}(t - 1) = \delta\epsilon_y(t)\mathbf{a}(t) \tag{5.3}$$

where $\Delta\mathbf{w}(t - 1) = \mathbf{w}(t) - \mathbf{w}(t - 1)$ and δ is the learning rate. The *search path* is parallel to the transformed input vector and the *step size* is equal to the learning rate (a constant chosen before learning commences) multiplied by the instantaneous output error. This is the well-known LMS instantaneous gradient descent rule, which has the basic form *weight update = scalar * transformed input vector*. The computational simplicity of this learning rule is also obvious, especially when the transformed input vector $\mathbf{a}(t)$ is sparse; only those weights which contribute to the output are updated and the size of their modification is in proportion to the output of the basis function.

Now if any of the weights in an AMN have not been initialised, Equation 5.3 becomes:

$$w_i(t) = \begin{cases} \hat{y}(t) : \text{if the } i^{th} \text{ weight was not initialised and } a_i(t) > 0 \\ w_i(t - 1) + \delta\epsilon_y(t)a_i(t) : \text{otherwise} \end{cases} \tag{5.4}$$

If a weight has not been initialised, it is set equal to the desired output. This heuristic is justified, as the basis functions have a compact support (local representation) and because the outputs of the basis functions are normalised.

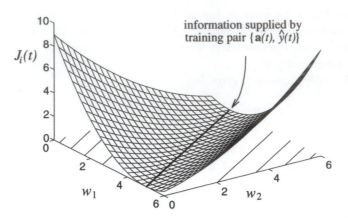

Figure 5.1   An instantaneous performance surface in two-dimensional weight space. At time t, the information supplied to the network is contained in the data pair: $\mathbf{a}(t) = (0.6, 0.4)$, $\widehat{y}(t) = 3$.

Performance Surface

An instantaneous performance surface generated by a single training is shown in Figure 5.1. Each training pair presented to the network produces *one* equation which the p weights must satisfy:

$$\mathbf{a}^T(t)\,\mathbf{w} = \widehat{y}(t)$$

in order to store the data exactly. Hence the performance surface is *singular*, as many weight vectors exist which satisfy this equation and the instantaneous optimisation problem is underconstrained. When the training information is rich enough, or the input is *sufficiently exciting*, the global minima of successive performance surfaces (these are termed solution hyperplanes in Section 5.2.3) intersect at a unique point which is the global minimum of the true MSE performance function, as illustrated in Figure 5.2.

From Equation 4.5, the instantaneous autocorrelation matrix $\mathbf{R}(t)$ associated with $J_i(t)$ is of the form:

$$\mathbf{R}(t) = \mathbf{a}(t)\mathbf{a}^T(t) \tag{5.5}$$

This is a $(p \times p)$ matrix which has only one non-zero eigenvalue $(\|\mathbf{a}(t)\|_2^2)$ and the corresponding eigenvector is $\mathbf{a}(t)$. Therefore using Equation 4.26, the curvature (second derivative) of the instantaneous performance surface along the major axis, $\mathbf{a}(t)$, is $2\|\mathbf{a}(t)\|_2^2$ and it is flat along all the remaining axes. Hence the batch steepest descent algorithm and the instantaneous version are equivalent for the instantaneous performance surface, and learning stability is assured (see Section 4.4.3) when:

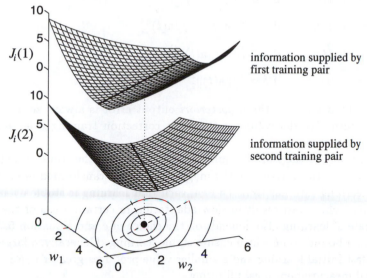

Figure 5.2 The performance surfaces generated by the two training pairs: $\mathbf{a}(1) = (0.6, 0.4)$, $\widehat{y}(1) = 3$ and $\mathbf{a}(2) = (0.2, 0.8)$, $\widehat{y}(2) = 3$. The global minima of each of these performance surfaces (bold lines) are projected onto the base of the figure (dashed lines) where they intersect at a unique point. The contour plot of the average or expected MSE is also shown.

$$0 < \delta < \frac{2}{\|\mathbf{a}(t)\|_2^2}$$

This relationship can also be derived by considering the reduction in the output error after the weights have been updated, as will now be described.

Output Error Reduction

After adaptation using the LMS learning rule, the *a posteriori* output of the network, $\underline{y}(t)$, is given by:

$$
\begin{aligned}
\underline{y}(t) &= \mathbf{a}^T(t)\mathbf{w}(t) \\
&= \mathbf{a}^T(t)\mathbf{w}(t-1) + \mathbf{a}^T(t)\delta\epsilon_y(t)\mathbf{a}(t) \\
&= \delta\|\mathbf{a}(t)\|_2^2\widehat{y}(t) + \left(1 - \delta\|\mathbf{a}(t)\|_2^2\right)y(t)
\end{aligned}
$$

where $\|\mathbf{a}(t)\|_2^2 = \mathbf{a}^T(t)\mathbf{a}(t)$, and the *a posteriori* output error $\epsilon_{\underline{y}}(t)$ is given by:

$$
\begin{aligned}
\epsilon_{\underline{y}}(t) &= \widehat{y}(t) - \underline{y}(t) \\
&= \left(1 - \delta\|\mathbf{a}(t)\|_2^2\right)\epsilon_y(t)
\end{aligned}
\tag{5.6}
$$

For a non-zero $\epsilon_y(t)$, the following relationships between the *a priori* and the *a posteriori* output errors can be then established for different values of δ.

$$
\begin{array}{rcll}
|\epsilon_{\underline{y}}(t)| & > & |\epsilon_y(t)| & \text{if } \delta \notin \left[0, 2/\|\mathbf{a}(t)\|_2^2\right] \\
|\epsilon_{\underline{y}}(t)| & = & |\epsilon_y(t)| & \text{if } \delta = 0 \text{ or } \delta = 2/\|\mathbf{a}(t)\|_2^2 \\
|\epsilon_{\underline{y}}(t)| & < & |\epsilon_y(t)| & \text{if } \delta \in \left(0, 2/\|\mathbf{a}(t)\|_2^2\right) \\
\epsilon_{\underline{y}}(t) & = & 0 & \text{if } \delta = 1/\|\mathbf{a}(t)\|_2^2
\end{array}
\tag{5.7}
$$

When $\delta = 1/\|\mathbf{a}(t)\|_2^2 \ \forall \ t$, the *a posteriori* output error is always zero and this provides an alternative derivation of the error correction training rule, which is described in the next section.

The dependency of the LMS learning rule on the size of the transformed input vector is obvious. If the variance in the magnitude of the transformed input vector is large, the learning rate has to be chosen, such that learning is stable even when $\|\mathbf{a}(t)\|_2^2$ is large, which can result in slow adaptation in certain areas of the input space. The rate of learning also depends on the choice of δ and Equation 5.7 gives upper and lower bounds on δ which ensure stable learning. Generally, a large value of δ ensures fast initial learning and a smaller value produces greater noise (model mismatch and measurement noise) filtering.

5.2.2 Error Correction Rule

Instantaneous error correction rules, as the name suggests, update the network weight vector to reduce the error in the output after each training pair is presented to the network. The weight update $\Delta \mathbf{w}(t-1)$ is calculated, so that the search direction is always parallel to the transformed input vector (similar to the instantaneous gradient algorithm), but the step size is calculated at each time-step, such that the input/output information at time t is stored *exactly* by the network. This is achieved if the *a posteriori* weight vector satisfies:

$$
\widehat{y}(t) = \mathbf{a}^T(t)\mathbf{w}(t)
\tag{5.8}
$$

Updating the weight vector parallel to the transformed input vector also means that the learning rule incorporates the principle of minimal disturbance (described in Section 5.2.3), meaning that the weight change recommended is the *smallest* possible which still stores the desired output. The learning rule at time t can therefore be written as:

$$
\Delta \mathbf{w}(t-1) = c(t)\mathbf{a}(t)
\tag{5.9}
$$

for the step size $c(t)$. It is now possible to solve for $c(t)$ by substituting Equation 5.9 into 5.8 which gives:

$$
\widehat{y}(t) = \mathbf{a}^T(t)(\mathbf{w}(t-1) + c(t)\mathbf{a}(t))
$$

Noticing that the first term on the right-hand side is simply $y(t)$ and rearranging in order to find $c(t)$ gives:

$$c(t) = \frac{\epsilon_y(t)}{\mathbf{a}^T(t)\mathbf{a}(t)}$$

resulting in a learning rule of the form:

$$\Delta\mathbf{w}(t-1) = \frac{\delta\epsilon_y(t)\mathbf{a}(t)}{\|\mathbf{a}(t)\|_2^2} \tag{5.10}$$

where δ is the learning rate. This NLMS weight update rule is again of the form *scalar * transformed input vector* except that, in this case, the magnitude of the weight change is *normalised* by the magnitude of the transformed input vector. It should also be emphasised that the NLMS learning rule is equivalent to the LMS algorithm when $\delta = \delta'/\|\mathbf{a}(t)\|_2^2 \ \forall t$. The weight update is in the *same* direction (parallel to the transformed input vector), it is only the step size that changes.

For the AMNs considered in this book, if any of the basis functions have not been initialised, Equation 5.10 becomes:

$$w_i(t) = \begin{cases} \hat{y}(t) : \text{if the } i^{th} \text{ weight was not initialised and } a_i(t) > 0 \\ w_i(t-1) + \frac{\delta\epsilon_y(t)}{\|\mathbf{a}(t)\|_2^2}a_i(t) : \text{otherwise} \end{cases} \tag{5.11}$$

As with the LMS adaptation rule, if a weight has not been initialised, it is set equal to the value of the desired function and in Appendix A it is shown that the modified NLMS rule (above) stores the desired value exactly when $\delta = 1$.

These error correction and gradient descent laws have a very interesting geometrical interpretation originally noticed by Kaczmarz, [1937], who derived the error correction algorithm; the change in the weight vector is perpendicular to the hyperplane which consists of the set of the solutions of the equation $\hat{y}(t) = \mathbf{a}^T(t)\mathbf{w}$. This enables the rate of convergence of this algorithm to be explored when the input signals have different properties (i.e. orthogonal, ill conditioned, inconsistent) and this is expanded on in Section 5.3.

Arbitrary Search Directions

In deriving the NLMS rule, it was assumed that the search direction is parallel to the transformed input vector and a theoretical justification of this is presented in Section 5.2.3. As an alternative, consider updating the weights along an arbitrary search direction s(t), where the only restriction is that $\mathbf{a}^T(t)\mathbf{s}(t) \neq 0$. Then the learning rule given by:

$$\Delta\mathbf{w}(t-1) = \frac{\delta\epsilon_y(t)\mathbf{s}(t)}{\mathbf{a}^T(t)\mathbf{s}(t)}$$

stores the current training pair exactly when $\delta = 1$, as the *a posteriori* output error is zero. Setting the search direction equal to the transformed input vector works well in many cases, although sometimes an alternative, *orthogonal* search direction needs to be calculated to increase the rate of parameter convergence (see Section 5.6.2). The NLMS algorithm can be easily modified to incorporate this extension.

Output Error Reduction

The *a posteriori* output of a network trained using the NLMS rule is:

$$
\begin{aligned}
\underline{y}(t) &= \mathbf{a}^T(t)\mathbf{w}(t) \\
&= \mathbf{a}^T(t)\mathbf{w}(t-1) + \mathbf{a}^T(t)\frac{\delta\epsilon_y(t)\mathbf{a}(t)}{\|\mathbf{a}(t)\|_2^2} \\
&= \delta\hat{y}(t) + (1-\delta)y(t)
\end{aligned}
$$

and the *a posteriori* output error is given by:

$$
\begin{aligned}
\epsilon_{\underline{y}}(t) &= \hat{y}(t) - \underline{y}(t) \\
&= (1-\delta)\epsilon_y(t)
\end{aligned}
\tag{5.12}
$$

For a non-zero $\epsilon_y(t)$, a set of relationships between the *a priori* and the *a posteriori* output errors can again be derived for different values of δ:

$$
\begin{array}{llll}
|\epsilon_{\underline{y}}(t)| & > & |\epsilon_y(t)| & \text{if } \delta \notin [0,2] \\
|\epsilon_{\underline{y}}(t)| & = & |\epsilon_y(t)| & \text{if } \delta = 0 \text{ or } \delta = 2 \\
|\epsilon_{\underline{y}}(t)| & < & |\epsilon_y(t)| & \text{if } \delta \in (0,2) \\
\epsilon_{\underline{y}}(t) & = & 0 & \text{if } \delta = 1
\end{array}
$$

Here the NLMS adaptation rule is self-normalising, unlike the LMS rule, since it does *not* depend on the size of the transformed input vector. Stable learning (for a single training example) occurs if the learning rate lies in the interval $(0,2)$, and if $\delta = 1$, the training example is stored after just one iteration (the learning algorithm was formulated to incorporate this constraint). However, the weight update depends on the size of the transformed vector, and so it does not minimise the output MSE, as shown in the next section.

Projection Algorithm

The NLMS learning rule can be interpreted as a *projection algorithm* when $\delta = 1$, as the weight vector only changes once when the *same* training pair is sequentially presented to the network. Specifically, a projection algorithm [Ellacott, 1994] is a linear mapping $\mathbf{P} : R^p \to R^p$ that satisfies:

$$
\mathbf{P}(\mathbf{P}\mathbf{w}) = \mathbf{P}\mathbf{w} \qquad \forall\, \mathbf{w} \in R^p
\tag{5.13}
$$

Clearly the NLMS algorithm satisfies this condition, as the *a posteriori* output error is zero if $\delta = 1$, so repeated application of the NLMS rule does not alter the weight vector. To derive the matrix that represents the NLMS projection mapping, note that Equation 5.10 may be expressed as:

$$
\mathbf{w}(t) = \left(\mathbf{I} - \frac{\mathbf{a}(t)\mathbf{a}^T(t)}{\|\mathbf{a}(t)\|_2^2}\right)\mathbf{w}(t-1) + \frac{\mathbf{a}(t)\hat{y}(t)}{\|\mathbf{a}(t)\|_2^2}
$$

Assuming there exists no modelling error (i.e. $y(t) = \mathbf{a}^T(t)\hat{\mathbf{w}}$), then the error in the weight vector, $\epsilon_{\mathbf{w}}(t) = \hat{\mathbf{w}} - \mathbf{w}(t)$ evolves according to the following first-order relationship:

$$\epsilon_{\mathbf{w}}(t) = \left(\mathbf{I} - \frac{\mathbf{a}(t)\mathbf{a}^T(t)}{\|\mathbf{a}(t)\|_2^2} \right) \epsilon_{\mathbf{w}}(t-1) \tag{5.14}$$

where the projection matrix has the form:

$$\mathbf{P}(t) = \left(\mathbf{I} - \frac{\mathbf{a}(t)\mathbf{a}^T(t)}{\|\mathbf{a}(t)\|_2^2} \right)$$

This matrix has one zero eigenvalue with a corresponding eigenvector $\mathbf{a}(t)$ and the other $(p-1)$ eigenvalues are all equal to one.

Normalised MSE Performance Function

The output error reduction does not depend on the magnitude of the transformed input vector when the weight vector is trained using the NLMS learning rule. However, the dependence of the learning rate on the size of the transformed input vector means that the NLMS rule no longer minimises the MSE. This can be seen by noticing that the right-hand side of Equation 5.10 is a scaled (by -0.5) instantaneous estimate of the gradient of the *normalised* MSE performance function given by:

$$J = E\left(\frac{\epsilon_y^2(t)}{\|\mathbf{a}(t)\|_2^2} \right) \tag{5.15}$$

If there exists a unique weight vector for which $\epsilon_y(t) = 0 \; \forall t$ (no modelling error) or if $\|\mathbf{a}(t)\|_2 = constant \; \forall t$ (as occurs in the original binary CMAC), the optimal weight vectors which minimise the performance functions given in Equations 5.1 and 5.15 are equivalent. Otherwise the minima of the two performance functions occur at different locations.

One way of visualising how the NLMS learning algorithm affects the MSE cost function is to consider the instantaneous performance functions described in Section 5.2.1. The instantaneous autocorrelation matrix for the NLMS rule has only one non-zero eigenvalue, and its value is *one*. The corresponding eigenvector is still $\mathbf{a}(t)$, hence every instantaneous performance function is replaced by a *normalised* version that has the same set of global minima, but its curvature along the major axis is constant (2); it does not depend on the magnitude of the input vector. Clearly when model mismatch exists, the contributions made by the instantaneous performance function and the normalised version to the MSE are different, unless all the input vectors have the same magnitude.

The normalised performance function is generated from the normalised training set given by $\{\mathbf{a}(t)/\|\mathbf{a}(t)\|_2, \hat{y}(t)/\|\mathbf{a}(t)\|_2\}_{t=1}^L$ and the distance between the two global minima is greatest when both the modelling error and the variance of $\|\mathbf{a}(t)\|_2$ are large. This point was essentially noted by Widrow and Lehr [1990], where they showed that the NLMS adaptation rule can be written as:

$$\mathbf{w}(t) = \mathbf{w}(t-1) + \delta \left(\frac{\widehat{y}(t)}{\|\mathbf{a}(t)\|_2} - \frac{\mathbf{a}^T(t)}{\|\mathbf{a}(t)\|_2} \mathbf{w}(t-1) \right) \frac{\mathbf{a}(t)}{\|\mathbf{a}(t)\|_2} \qquad (5.16)$$

which is an LMS adaptation rule with a *normalised* training set. It is shown in Section 5.4.1 that the noise in the gradient estimates in the LMS rule is directly proportional to the condition of the autocorrelation matrix. Hence it may be conjectured that the gradient noise in NLMS rule is proportional to the condition of the *normalised* autocorrelation matrix, generated from the normalised training set. In fact, the rates of convergence of these two algorithms have been compared [Slock, 1993] and it has been shown that the NLMS rule generally converges at a faster rate.

The interpretation of the NLMS rule as a gradient descent rule for a normalised performance function illustrates a potential problem in using this training rule. When the size of the input is small, $\|\mathbf{a}(t)\|_2 < \epsilon$, the size of the weight change is inversely proportional to ϵ. If the measurement noise does *not* depend on the size of the transformed input vector, the magnitude of the weight change can be unbounded. For AMNs, the size of the transformed input vector is bounded above by 1, and is always strictly greater than 0. Hence, the size of the weight change is bounded above, although more emphasis is placed on the output errors when the corresponding transformed input vectors are small.

5.2.3   Geometric Interpretation of the LMS Rules

Historically, this geometrical interpretation of the learning rules was first noticed by Kaczmarz [1937], when he proposed the NLMS rule for solving a consistent set of simultaneous linear equations. It was proved that the linear parameters converged to their true values, given that a unique solution existed, and convergence occurred for $0 < \delta < 2$. The training data were interpreted as a set of *solution hyperplanes* which intersected at a unique point in weight space. The weight vector was *projected orthogonally* onto each solution hyperplane and it spiralled in towards the unique value. This interpretation was re-derived by Nagumo and Noda [1967], and was also employed by Parks and Militzer [1989] where it was used to investigate the convergence of the CMAC's weight vector.

Solution Hyperplanes

The LMS and NLMS learning rules described by Equations 5.3 and 5.10 have a simple basic form:

$$\Delta \mathbf{w}(t-1) = c(t)\mathbf{a}(t)$$

and it is only the scalar quantity $c(t)$ that makes the LMS and the NLMS rules different. For the LMS rule, the scalar is simply the output error multiplied by a learning rate, whereas the NLMS rule multiplies the output error by the learning rate and divides this quantity by the "size" of the transformed input vector. In both cases, the weight change is *parallel* to the current transformed input vector.

The desired output of the network at time t is $\hat{y}(t)$ and thus the aim is to update the weight vector $\mathbf{w}(t-1)$ so that:

$$\hat{y}(t) = \mathbf{a}^T(t)\mathbf{w}(t)$$

This specifies one equation in p variables (the set of linear weights), and so the solutions to this equation lie on the $(p-1)$-dimensional *solution hyperplane* $\mathbf{h}(t)$ (in weight space), which is composed of the set of solutions to the following equation:

$$0 = \mathbf{a}^T(t)\mathbf{w} - \hat{y}(t) \tag{5.17}$$

Each solution hyperplane corresponds to the global minima of the corresponding instantaneous performance surface, as shown in Figure 5.1.

The normal to the solution hyperplane is the vector $\mathbf{a}(t)$, and the unit normal is given by $\mathbf{a}(t)/\|\mathbf{a}(t)\|_2$. In a two-dimensional weight space, the solutions of such an equation lie on a line (one equation with two variables), as illustrated in Figure 5.3.

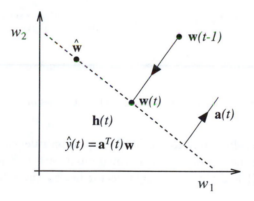

Figure 5.3   A solution hyperplane (dashed line) and perpendicular learning in two-dimensional weight space.

The search direction for the LMS rules is parallel to the transformed input vector and the transformed input vector is perpendicular to the solution hyperplane, which means that the weight change is also perpendicular to the solution hyperplane, as shown in Figure 5.3.

The effect of different values of δ being used in the NLMS adaptation rule can be illustrated geometrically. When $\delta = 1$, the weight update is such that the weight vector is projected onto the solution hyperplane. If $0 < \delta < 1$, the weight update means that the weight vector does not reach the solution hyperplane (overdamped), whereas if $1 < \delta < 2$, the weight update vector crosses the solution hyperplane and lies on the opposite side of the hyperplane compared to $\mathbf{w}(t-1)$ (underdamped). If $\delta < 0$, the weight update moves perpendicularly away from the solution hyperplane and if $\delta > 2$, the weight update crosses the solution hyperplane and the new weight vector lies further away on the opposite side (unstable learning). This is illustrated in Figure 5.4.

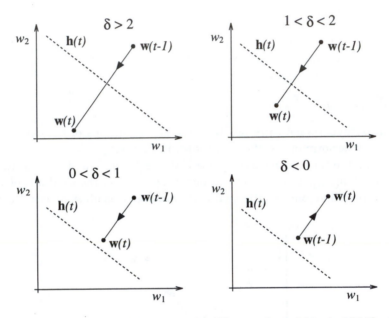

Figure 5.4   The effect on the weight updates for different values of δ in the NLMS adaptation rule.

These comments also apply to the LMS adaptation rule, except that the learning rate depends on the size of the transformed input vector. Stable learning occurs if $\delta \in \left(0, 2/\|\mathbf{a}(t)\|_2^2\right)$ and the new weight vector lies on the solution hyperplane if $\delta = 1/\|\mathbf{a}(t)\|_2^2$.

Principle of Minimal Disturbance

The LMS family of learning rules embody the principle of *minimal disturbance*, that is to say, the change made to the weight vector is the smallest which causes the new desired output to be stored. Since the weight change is the smallest of all the possible weight changes which cause the new weight vector to store the new desired output, it interferes *minimally* with the learnt information.

If the network is required to store the new output after adaptation, the new weight vector must lie on the solution hyperplane generated by $0 = \mathbf{a}^T(t)\mathbf{w} - \hat{y}(t)$. The minimal distance, $d(t)$, of this hyperplane from the point $\mathbf{w}(t-1)$ is:

$$
\begin{aligned}
d(t) &= \frac{\left|\mathbf{a}^T(t)\mathbf{w}(t-1) - \hat{y}(t)\right|}{\|\mathbf{a}(t)\|_2} \\
&= \frac{|\epsilon_y(t)|}{\|\mathbf{a}(t)\|_2}
\end{aligned}
$$

This distance is measured by dropping a perpendicular from $\mathbf{w}(t-1)$ to a point on the solution hyperplane in weight space. However, the LMS family of learning rules do exactly this; they update weight vectors perpendicular to the solution hyperplane. Hence the LMS family of learning rules embody the principle of minimal disturbance and the size of the weight change which stores the desired output is given by:

$$
\|\Delta\mathbf{w}(t-1)\|_2 = d(t) = \frac{|\epsilon_y(t)|}{\|\mathbf{a}(t)\|_2}
$$

and this is illustrated in Figure 5.5.

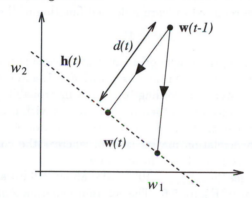

Figure 5.5 The principle of minimal disturbance for LMS learning rules.

The NLMS learning rule can be derived by viewing it as a constrained optimisation problem, where the constraint ensures that the new weight vector lies on the solution hyperplane, and the problem is to generate the new weight vector such that $\|\Delta\mathbf{w}(t-1)\|_2$ is minimised. This means that the minimal disturbance principle is automatically incorporated into the learning law. The optimisation problem can then be solved using Lagrange multipliers [Haykin, 1991], and (not surprisingly) the recommended change in the weight vector is simply parallel to the transformed input vector. It is also possible to use Singular-Valued Decomposition (SVD) to solve the single training example optimisation problem, and the recommended weight change is simply that calculated using the NLMS training rule. The SVD algorithm automatically calculates the minimum norm solution to singular optimisation problems.

5.2.4   Example: Weight Update Using NLMS

In Chapter 4 it was shown that an order 2, B-spline network and a conventional linear network had the same modelling capabilities, although the ability of the two networks to store information, when trained using steepest descent, was very different. The B-spline network was better conditioned and learned the desired function much quicker.

The LMS and NLMS learning rules, which have just been derived, attempt to approximate gradient descent adaptation using instantaneous estimates of the performance function. As a result, it is necessary to compare how well the information is stored when instantaneous training rules are used. The modelling capabilities of the two networks are the same and, if the NLMS rule is used to train the weight vectors, the *a posteriori* output errors are also the same. Therefore, to compare the networks, the errors in the weight vector must be examined. A thorough theoretical comparison is very difficult, because the rate of convergence is highly dependent on the input sequence presented to the network, although some interesting insights can be obtained by considering specific input distributions (see Section 5.4.1). Thus this comparison is strictly simulation based.

The networks are required to learn a desired function of the form:

$$\widehat{y}(t) = -0.5 + x(t) \qquad\qquad x \in [0, 1]$$

and in both cases the optimal networks can reproduce this function exactly. For both networks an NLMS learning rule is used with $\delta \equiv 1$, and the input is randomly drawn from the unit interval according to a uniform probability density function, so the weight vectors converge to their optimal values with probability 1.

For the conventional linear model described in Section 4.3.4, the condition of the normalised autocorrelation matrix is 18.6, whereas the condition of the true autocorrelation matrix is 19.3. From an initial value of $\mathbf{w}(0) = (-1.0, 0)^T$, the weight vector converges ($\|\epsilon_{\mathbf{w}}(t)\|_\infty < 10^{-6}$) after about 270 iterations, and a typical weight path is shown in Figure 5.6. The solution hyperplane is vertical at $x = 0$, and rotates anticlockwise as x is increased until $x = 1$, when it forms an angle of $-45°$ with the horizontal axis, as shown in Figure 5.6. This zone of convergence, which is spanned by the solution hyperplanes, limits the rate of convergence of the weight vector, as the weight updates change by only (at most) $45°$ at each iteration.

The order 2 B-spline model is again described in Section 4.3.4, and the condition of the normalised autocorrelation matrix is 3.66, compared with a condition number of 3 for the unnormalised version. A typical trajectory in weight space is shown in Figure 5.7. This time the initial weight vector is $\mathbf{w}(0) = (0, -0.5)^T$, as the previous value now lies on a principal axis, and is closer to the optimal solution than in the previous case. The B-spline weight vector converges ($\|\epsilon_{\mathbf{w}}(t)\|_\infty < 10^{-6}$) after fifty iterations, which is approximately five times faster than the conventional linear model. It is interesting to note that the ratio of the condition numbers of the normalised autocorrelation matrices is also close to 5. This increase in the rate

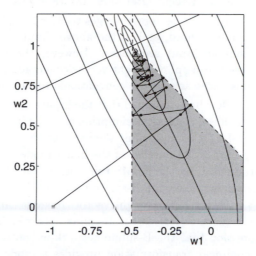

Figure 5.6   Weight convergence for a linear model trained using the NLMS rule. The contours are given by $\xi = 0.001, 0.003, 0.01, 0.03, 0.1, 0.3$. The zone of convergence (spanned by the solution hyperplanes) is the shaded area.

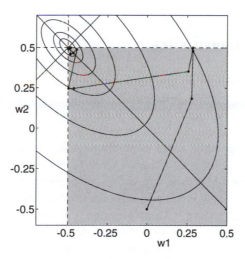

Figure 5.7   Weight convergence for a B-spline model trained using the NLMS rule. The contours are given by $\xi = 0.001, 0.003, 0.01, 0.03, 0.1$, and the shaded area shows the zone of convergence.

of convergence can be explained by considering the zone of convergence generated by the solution hyperplanes. At $x = 0$, the solution hyperplane is vertical and rotates anticlockwise as x is increased until $x = 1$, when the solution hyperplane is horizontal. The zone of convergence is larger, which allows perpendicular weight updates and convergence could occur after only two iterations.

The rate at which the weight vector converges to its optimal value depends on the particular input sequence used, and so the above simulation was run several times with different random input sequences. However the results varied only slightly, and in each case the B-spline weight vector converged after about 20% of the number of iterations necessary for the conventional linear model.

In this comparison, it has been assumed that the inputs were evenly distributed in the interval $[0, 1]$, which is a common assumption in many implementations. But it was shown in Section 4.4.6 that by scaling the input so that it lies in the interval $[-1, 1]$, this reduces the condition number of the linear model from 19.3 to 3. The new linear model is trained using the NLMS rule and the weight vector converges at approximately the same rate as the B-spline network. When the bias term is also reduced to $1/\sqrt{3}$, $C(\mathbf{R}) = 1$, the weight vector converges after about twenty-five iterations, which is twice as fast as the B-spline network. The reader should not get the false impression that the B-spline network is "optimal" in any sense, but rather that the nonlinear transformation provides a model structure that is well conditioned for (instantaneous) gradient descent learning, is computationally efficient and provides a convenient set of basis functions with which to produce more complex piecewise polynomial models.

5.3   PARAMETER CONVERGENCE

In this section, it is shown that the LMS rules converge to a weight vector which minimises the MSE performance function. Initially, the case when there exists a unique solution to the training data is considered, and it is proved that the weights converge to their optimal values. This assumption is generally unrealistic and it is extended by considering when there exists more than one solution to the supplied data (singular systems), and also the problem of modelling error or noise. For singular networks, the LMS rules are shown to converge to the minimum norm solution if the initial weight vector is zero, and output dead-zones are proposed as a simple method for coping with measurement errors. The instantaneous estimates made of the gradient in the presence of modelling error cause the weights to converge to a domain rather than a fixed point, and the output dead-zones cause adaptation to cease once the weights enter this zone.

5.3.1   Convergence

Suppose that there is a set of training examples $\{\mathbf{x}(t), \widehat{y}(t)\}_{t=1}^{L}$, which is cyclically presented to an AMN trained using the NLMS rule. Also assume that there exists a unique weight vector $\widehat{\mathbf{w}}$ such that:

$$\hat{y}(t) = \mathbf{a}^T(t)\,\hat{\mathbf{w}} \qquad\qquad \forall\, t \qquad\qquad (5.18)$$

It is immediately acknowledged that this assumption is in general unreasonable, but the derivation is performed so as to provide an insight into the important topics of output and parametric convergence. Similar descriptions have also been presented in Åström and Wittenmark [1989], Johnson [1988] and Kaczmarz [1937].

Consider using a learning rule of the form:

$$\Delta\mathbf{w}(t-1) = \frac{\delta\epsilon_y(t)}{\|\mathbf{a}(t)\|_2^2}\mathbf{a}(t) \qquad\qquad (5.19)$$

where the transformed input vectors possess the property that:

$$0 < a_\epsilon \leq \|\mathbf{a}(t)\|_2^2 \leq 1 \qquad\qquad \forall\, t$$

for some positive constant a_ϵ. Then the weights converge to their optimal values as $t \to \infty$.

Proof. Let the error in the weight vector be defined as:

$$\epsilon_{\mathbf{w}}(t) = \hat{\mathbf{w}} - \mathbf{w}(t)$$

and an energy function $V(t)$ be defined as the square of the modulus of this error vector at time t. Then:

$$V(t) = \|\epsilon_{\mathbf{w}}(t)\|_2^2 = \epsilon_{\mathbf{w}}^T(t)\epsilon_{\mathbf{w}}(t)$$

and the output error is given by:

$$\begin{aligned}
\epsilon_y(t) &= \mathbf{a}^T(t)\hat{\mathbf{w}} - \mathbf{a}^T(t)\mathbf{w}(t-1) \\
&= \mathbf{a}^T(t)\epsilon_{\mathbf{w}}(t-1) \qquad\qquad (5.20)
\end{aligned}$$

Now subtracting $\hat{\mathbf{w}}$ from both sides of Equation 5.19 gives:

$$-\epsilon_{\mathbf{w}}(t) = -\epsilon_{\mathbf{w}}(t-1) + \frac{\delta\epsilon_y(t)}{\|\mathbf{a}(t)\|_2^2}\mathbf{a}(t)$$

Taking the Euclidean norm of both sides of this expression and rearranging terms gives:

$$\begin{aligned}
V(t) - V(t-1) &= \frac{\delta^2\epsilon_y^2(t)\mathbf{a}^T(t)\mathbf{a}(t)}{\|\mathbf{a}(t)\|_2^4} - \frac{2\delta\epsilon_y(t)\mathbf{a}^T(t)\epsilon_{\mathbf{w}}(t-1)}{\|\mathbf{a}(t)\|_2^2} \qquad (5.21) \\
&= \frac{\delta^2\epsilon_y^2(t) - 2\delta\epsilon_y^2(t)}{\|\mathbf{a}(t)\|_2^2} \\
&= \frac{\delta\epsilon_y^2(t)}{\|\mathbf{a}(t)\|_2^2}(\delta - 2) \qquad\qquad (5.22)
\end{aligned}$$

In Equation 5.22 if $\delta < 0$ or $\delta > 2$ and $\epsilon_y(t) \neq 0$ then:

$$V(t) > V(t-1)$$

so the error in the weight vector increases and the weight vector diverges. If $\delta = 0$ or $\delta = 2$, then no learning occurs, as the weight vector stays the same distance away from the solution hyperplane (see Section 5.2.3), and if $0 < \delta < 2$:

$$0 \leq \|\hat{\mathbf{w}} - \mathbf{w}(t)\|_2 \leq \|\hat{\mathbf{w}} - \mathbf{w}(t-1)\|_2 \leq \|\hat{\mathbf{w}} - \mathbf{w}(0)\|_2 \tag{5.23}$$

After each iteration, the weight vector gets closer to its optimal value (for a non-zero output error). Convergence of the weight vector to its optimal value has *not* been proved; it has merely been shown that energy function $V(t)$ is a non-increasing function of time. Now:

$$V(L) = V(0) + \sum_{t=1}^{L} (\delta - 2)\delta \frac{\epsilon_y^2(t)}{\|\mathbf{a}(t)\|_2^2} \tag{5.24}$$

and setting $\delta' = \delta(2 - \delta)$, which is positive if $0 < \delta < 2$, means that the above expression can be rewritten as:

$$\sum_{t=1}^{L} \frac{\epsilon_y^2(t)}{\|\mathbf{a}(t)\|_2^2} = \frac{1}{\delta'} (V(0) - V(L))$$

$$\leq \frac{V(0)}{\delta'}$$

Letting $L \to \infty$, then $0 \leq V(L) \leq V(0)$ and because of the fact that $\|\mathbf{a}(t)\|_2^2$ is positive and bounded above and below, the positive sequence $\epsilon_y^2(t)/\|\mathbf{a}(t)\|_2^2$ must tend to zero as $t \to \infty$, implying that:

$$\epsilon_y^2(t) \to 0 \qquad\qquad \text{as } t \to \infty$$

So if a set of training examples is presented to the network in a cyclic fashion, the above result shows that the output error for each training sample tends to 0 as time increases. It has been assumed that there exists a unique weight vector which satisfies Equation 5.18 and so, if the training set is sufficiently rich, output convergence also implies parameter convergence.   □

This is illustrated in Figure 5.8, where the weight vector is projected orthogonally onto each solution hyperplane and because the solution hyperplanes intersect at a unique point (consistent data), convergence of the learning algorithm is obvious.

This derivation assumed that the inputs are bounded which is necessary to ensure that the network is attempting to learn a one-to-one or a many-to-one function. If the inputs were unbounded, the desired mapping would no longer be a function, because the network is defined over a compact space. Proving the boundedness of the input signals is a major problem in deriving stability theories for these AMNs within control loops, although some initial attempts have assumed that there is a fixed, stabilising controller surrounding these AMNs, which ensures that the input never stays outside the AMN's support for more than a finite time [Sanner and Slotine, 1992].

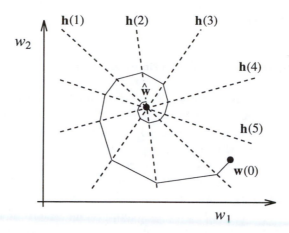

Figure 5.8   Parameter convergence for a consistent set of training data. Each training pair generates a solution hyperplane (dotted line) and because the set is consistent, they intersect at a unique point. The weight vector is projected orthogonally onto each solution hyperplane at each training iteration.

Weight and Output Convergence

It is important to understand the relationship between the result which has just been proved (weight convergence) and the widely used mean squared output error measure of system performance. For each weight vector update, Equation 5.23 shows that the size of the weight vector error does not increase, and given sufficient training examples, the weight vector converges to its optimal solution. The NLMS learning rules are derived, based on the assumption that the instantaneous output error is fully corrected, so for each update, the size of the weight vector error and the instantaneous output error decrease. However, this does *not* imply that the overall MSE also decreases. This is illustrated in Figure 5.7 where the second weight update causes the output error to increase, although the weight vector error is reduced and the *a posteriori* instantaneous output error is zero.

The MSE increases, due to the effect of the update on the information already stored in the weight vector. Storing one training example causes a significant reduction in the information retained about neighbouring data pairs, and this effect is known as *learning interference*. After cyclically presenting the data to the network, the MSE is reduced, although it is quite easy to construct simple examples where the overall performance of the system is reduced, simply by storing one piece of data. This occurs when the weight vector is projected from a region where the performance function has a low curvature into a part of the weight space where it has a large curvature.

True (batch) gradient descent learning causes the MSE *and* the parameter error to decrease at each update iteration. The instantaneous estimates which are being formed of the gradient introduce *noise* into the weight updates, and this can cause the overall MSE to increase.

5.3.2   Singular Systems and Minimum Norm Solutions

When a unique solution exists to the set of linear equations generated by the AMN training data, the parameter vector converges to this value if an NLMS training rule is used and the learning rate satisfies $0 < \delta < 2$. It is now proven that when the training data are *consistent* (there exists no modelling error or measurement noise) but the rank of the solution matrix $\mathbf{A}$ is lower than the number of parameters and the weights are trained using an NLMS learning rule, the weight vector converges to a value which exactly reproduces the training data and whose length has a *minimum Euclidean norm*, if the initial weight vector is zero or lies in the space spanned by the transposed rows of the solution matrix. So consider trying to solve the consistent set of *linear* equations given by:

$$\mathbf{Aw} = \widehat{\mathbf{y}} \tag{5.25}$$

where $\mathbf{A}$ is the solution matrix of size $(L \times p)$ whose i^{th} row is the transposed basis function vector $\mathbf{a}^T(i)$, $\mathbf{w}$ is the unknown p-dimensional coefficient vector and $\widehat{\mathbf{y}}$ is the L-dimensional vector composed of the desired network responses. By assumption, the equations are consistent and letting $r = \text{rank}(\mathbf{A})$, then $r < p$ and $r = \dim(\text{span}\{\mathbf{a}(1), \dots, \mathbf{a}(L)\})$.

Proof. By a minor modification to the proof described above, it can be shown that when more than one weight vector exists which can reproduce the training data exactly, the weight vector converges to a solution which lies in the space spanned by:

$$\{\mathbf{a}(1), \mathbf{a}(2), \dots, \mathbf{a}(L), \mathbf{w}(0)\}$$

when it is trained using an NLMS learning algorithm with $0 < \delta < 2$. This follows directly from the definition of the training rule, coupled with the fact that the output error tends to zero for each of the training samples as time increases. If $\mathbf{w}(0)$ is a linear combination of the above basis function vectors (or is identically zero), the final weight vector $\widehat{\mathbf{w}}^1$ lies in:

$$\widehat{\mathbf{w}}^1 \in \text{span}\{\mathbf{a}(1), \mathbf{a}(2), \dots, \mathbf{a}(L)\} \tag{5.26}$$

Note: $\mathbf{w}(0)$ can be uniquely decomposed into a vector which lies in the null space of $\mathbf{A}$, $\mathbf{w}^N(0)$, and a vector which lies in the space spanned by the basis function vectors. If $\mathbf{w}^N(0)$ is non-zero, it does not affect the rate of convergence, because $\mathbf{a}^T(t)\mathbf{w}^N(0) = 0 \ \forall\, t$, and the network output errors and the rate of convergence do not depend on $\mathbf{w}^N(0)$, but it influences the final value of the weight vector. This is analogous to the behaviour of the gradient descent training algorithm.

It can easily be shown that $\widehat{\mathbf{w}}^1$ is unique by assuming that there is another distinct optimal weight vector $\widehat{\mathbf{w}}^2$, which lies in the same space. If this occurs:

$$\mathbf{A}\left(\widehat{\mathbf{w}}^1 - \widehat{\mathbf{w}}^2\right) = 0$$

which implies that the vector $(\widehat{\mathbf{w}}^1 - \widehat{\mathbf{w}}^2)$ lies in the null space of $\mathbf{A}$, but by assumption both vectors lie in the row space which is orthogonal to the null space, and this provides a contradiction. Hence $\widehat{\mathbf{w}}^1$ is unique.

Finally, it can be established that the minimum norm solutions found using SVD [Haykin, 1991] all lie in the row space of the solution matrix. Hence they are unique and the weight vector, trained using the NLMS rule, converges to the minimum norm solution. ☐

The convergence proof for the NLMS algorithm can be simply modified and parameter convergence can be shown to occur when the weight vector is trained using the LMS rule, if the learning rate satisfies $0 < \delta < \max_t 2/\|\mathbf{a}(t)\|_2^2$. From Figure 5.9, it can clearly be seen that the initial value of the weight vector only affects the final solution, but does not influence the rate of convergence.

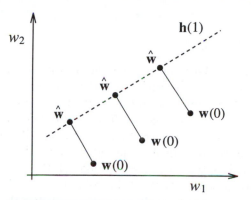

Figure 5.9　A singular set of training data (one equation in two unknowns). The weight vector is projected orthogonally onto the solution hyperplane, so the orthogonal component of the initial value is never overwritten.

5.3.3　Model Mismatch and Output Dead-Zones

When there is modelling error or measurement noise, the instantaneous output error can be linearly decomposed as:

$$\epsilon_y(t) = \mathbf{a}^T(t)\left(\widehat{\mathbf{w}} - \mathbf{w}(t-1)\right) + \left(\widehat{y}(t) - \mathbf{a}^T(t)\widehat{\mathbf{w}}\right) \tag{5.27}$$

$$= \epsilon_y^o(t) + \epsilon_y^n(t) \tag{5.28}$$

where $\epsilon_y^o(t)$ is due to the difference between the current and the optimal network and $\epsilon_y^n(t)$ is the optimal modelling error or measurement noise. If the original output error is used in the LMS training rules, the network is unable to determine when the weight vector is optimal, as the term $\epsilon_y^n(t)$ causes the gradient estimate to be non-zero.

Output dead-zones can be used to counteract this behaviour partially and to ensure that the energy function $V(t)$ is a non-increasing function of time, even when there is modelling error or noise. The technique does not update the weight vector when the output error is less than some predefined value ζ, and if it is greater, the difference is used in place of the output error. This can be written as:

$$\epsilon_y^d(t) := \begin{cases} 0 & \text{if } |\epsilon_y(t)| \leq \zeta \\ \epsilon_y(t) + \zeta & \text{if } \epsilon_y(t) < -\zeta \\ \epsilon_y(t) - \zeta & \text{if } \epsilon_y(t) > \zeta \end{cases} \quad (5.29)$$

where ϵ_y^d is the modified output error used in place of the normal output error in the LMS learning rules. It is hoped that $\epsilon_y^d(t)$ is a computationally cheap estimate of $\epsilon_y^o(t)$.

Generally ζ is selected to be either:

$$\zeta = \max_t \left| \epsilon_y^n(t) \right| \quad (5.30)$$

or

$$\zeta = E\left(\left| \epsilon_y^n(t) \right| \right) \quad (5.31)$$

For the former case, it is shown below that the weights converge to a fixed value, after which the output errors lie within a ball of radius ζ [Chen and Lin, 1993]. However, this is a very conservative estimate for the overall modelling error and it often causes slow rates of convergence. The second definition of ζ is often used in practice as the rate of convergence is faster, although it cannot be shown that the parameters converge to a fixed value and in general they move about in a p-dimensional minimal capture zone (see Section 5.4.3).

Consider when Equations 5.29 and 5.30 define the modified output error used in the NLMS learning algorithm. If the output error is less than or equal to ζ, the weights are not adapted and the energy function $V(t)$ does not increase. Now let $\epsilon_y^d(t)$ be positive (an analogous derivation can be performed if it is negative), then from Equation 5.21:

$$\begin{aligned} V(t) - V(t-1) &= \frac{\delta^2 \epsilon_y^d(t) \|a(t)\|_2^2}{\|a(t)\|_2^4} - \frac{2\delta \epsilon_y^d(t)\epsilon_y^o(t)}{\|a(t)\|_2^2} \\ &\leq \frac{\delta(\delta-2)\epsilon_y^d(t)}{\|a(t)\|_2^2} \end{aligned}$$

as $\epsilon_y^d(t) \leq \epsilon_y^o$. But $\delta(\delta-2)$ is negative for $\delta \in (0,2)$ and $\epsilon_y^d(t)$ is positive, sp $V(t)$ is a non-increasing function of time. Using the arguments expounded in Section 5.3.1, it can be shown that $\epsilon_y^d(t) \to 0$ as $t \to \infty$, proving that the output errors lie within a ball of radius ζ about zero, and that the weights converge to a fixed value.

Output dead-zones provide a quick fix to the LMS rules when modelling mismatch exists. However, a successful application requires substantial *a priori* knowledge about ϵ_y^o, in order to choose ζ. If the value is too large, the system may terminate learning prematurely, with a resulting loss in performance, whereas if it is too small its overall effect is negligible.

Parameter Convergence

Given that the output error lies within a ball of radius ζ centred on the origin, it is possible to generate a worst case estimate of the final weight vector error based on the condition of the autocorrelation matrix. Suppose that the weight vector in the above algorithm converges to a value $\mathbf{w}$, then the magnitude of the weight vector error related to the output error is given by:

$$\frac{\|\epsilon_{\mathbf{w}}\|}{\|\mathbf{w}\|} \leq C(\mathbf{R})\frac{\|\epsilon_{\mathbf{y}}\|}{\|\widehat{\mathbf{y}}\|} \tag{5.32}$$

where $\epsilon_{\mathbf{y}}$ is the vector of output errors, $\widehat{\mathbf{y}}$ is the desired output vector, $\|.\|$ is any natural norm and the condition number $C(\mathbf{R})$ is defined as $\|\mathbf{R}\|\,\|\mathbf{R}^{-1}\|$ for the appropriate matrix induced norm. This definition of the condition number of a matrix is *closely* related to the more common expression: as the ratio of the largest and the smallest eigenvalues.

   If the condition of the model is poor (highly correlated solution hyperplanes), output error tests are not sufficient to determine when to stop training, as the parameter errors could be very large. The network's ability to generalise is closely related to its parameter convergence, and using output dead-zones which are too large can result in the adaptation process terminating prematurely when the parameter errors are still large. Again, the condition of the basic model is an important factor in its overall ability to generalise sensibly.

Stochastic Approximation

As an alternative to counteracting model mismatch using output dead-zones, it is possible to *filter out* the gradient noise using a *stochastic approximation* LMS learning rule. Therefore consider making two simple changes to the learning rate:

- **assign** an individual learning rate to each basis function; and
- **reduce** δ_i through time as the confidence in a particular weight increases.

These modifications retain the fast initial convergence rate while in the long term filtering out measurement and modelling noise.

   Therefore, let the ith basis function have a learning rate $\delta_i(t)$ associated with it and assume that it is also a function of time. The necessary conditions on $\delta_i(t)$ for the weight vector to converge [Luo, 1991, Robbins and Monro, 1951] are given by:

$$\begin{aligned}
\delta_i(t) &> 0 \\
\sum_{t=1}^{\infty} \delta_i(t) &= \infty \\
\sum_{t=1}^{\infty} \delta_i^2(t) &< \infty
\end{aligned} \tag{5.33}$$

where the third condition implies that $\delta_i(t) \to 0$. One such function which satisfies these constraints is:

$$\delta_i(t) = \frac{\delta_1}{1 + t_i/\delta_2} \tag{5.34}$$

where δ_1 and δ_2 are positive constants which denote the initial learning rate and the rate of decay, respectively, and t_i is the number of times that the ith basis function has been updated.

Equations 5.3 and 5.10 now become:

$$w_i(t) = w_i(t-1) + \delta_i(t)\epsilon_y(t)a_i(t) \tag{5.35}$$

$$w_i(t) = w_i(t-1) + \delta_i(t)\epsilon_y(t)\frac{a_i(t)}{\|\mathbf{a}(t)\|_2^2} \tag{5.36}$$

where $\delta_i(t)$ is of the form given by Expression 5.33. Equivalent stochastic approximation learning rules exist when any of the basis functions have not been initialised.

If the plant is subject to sudden parameter changes, it may be necessary to reset $\delta_i(t)$ to a nominal value. This is needed because, while the constraints given in Expression 5.33 guarantee parameter convergence (subject to the usual conditions), convergence may occur very slowly. Hence, when a significant change is detected, a reasonable counteraction is to make the model more plastic.

5.3.4 Rate of Convergence

The rate of convergence of the NLMS learning algorithm is now shown shown to depend directly on the *orthogonality* of the training patterns. From Equations 5.20 and 5.24, the size of the parameter error after one learning iteration is given by:

$$V(1) = V(0) - \delta' \frac{\left(\mathbf{a}^T(1)\epsilon_\mathbf{w}(0)\right)^2}{\|\mathbf{a}(1)\|_2^2} \tag{5.37}$$

$$= V(0)\left(1 - \delta' \cos^2(\theta(1))\right) \tag{5.38}$$

where δ' is defined above and is positive, and $\theta(t)$ is the angle between $\mathbf{a}(1)$ and $\epsilon_\mathbf{w}(0)$ [Nagumo and Noda, 1967]. Setting $\delta' = 1$, the above expression can be written as:

$$V(1) = V(0)\sin^2(\theta(1)) \tag{5.39}$$

and after L updates the size of the weight vector error is:

$$V(L) = V(0)\prod_{t=1}^{L} \sin^2(\theta(t)) \tag{5.40}$$

where $\theta(t)$ is the angle between $\mathbf{a}(t)$ and $\epsilon_\mathbf{w}(t-1)$. Therefore the rate of parameter convergence depends on the relative orientations of the successive weight vector errors and the transformed input vectors, and if the weight update is nearly perpendicular to the weight vector error, parameter convergence occurs very slowly.

The LMS family of learning rules embody the principle of minimal disturbance, although this does not mean that the current knowledge stored in the network is not overwritten, it only ensures that the new weight vector is *close* to the previous weight vector. This section considers the conditions under which the weight updates do not interfere with each other and the implications that this has for the type of nonlinear mapping used to transform $\mathbf{x}(t)$ to $\mathbf{a}(t)$.

Fast Convergence for Orthogonal Inputs

Consider the case when the AMN has two weights and the two transformed input vectors $\mathbf{a}(1)$ and $\mathbf{a}(2)$ are mutually orthogonal. The normals to the solution hyperplanes are also mutually orthogonal, as are the weight updates. When an NLMS adaptation rule is used with $\delta \equiv 1$, the size of the weight vector error after two iterations can be calculated using Equation 5.40:

$$V(2) = V(0) \prod_{t=1}^{2} \sin^2\left(\theta(t)\right)$$

where $\theta(t)$ is the angle between $\mathbf{a}(t)$ and $\epsilon_\mathbf{w}(t-1)$. After one iteration, the error in the weight vector lies along $\mathbf{h}(1)$, is perpendicular to $\mathbf{a}(1)$ and is therefore parallel to $\mathbf{a}(2)$. Therefore $\theta(2) = 0$ and:

$$V(2) = 0$$

So after two iterations, the error in the parameter vector is zero if the transformed input vectors are orthogonal, and parameter convergence occurs in a finite number of iterations, as illustrated in Figure 5.10. When the transformed input vectors are orthogonal, the new information stored does *not* interfere with the (orthogonal) information already stored. This can be seen by noting that:

$$\mathbf{w}(2) = \mathbf{w}(0) + \epsilon_y(1)\frac{\mathbf{a}(1)}{\|\mathbf{a}(1)\|_2^2} + \epsilon_y(2)\frac{\mathbf{a}(2)}{\|\mathbf{a}(2)\|_2^2}$$

and the *a posteriori* output error is zero for the second input. Taking the dot product with the first transformed input vector gives:

$$\mathbf{a}^T(1)\mathbf{w}(2) = \mathbf{a}^T(1)\mathbf{w}(1) = \hat{y}(1)$$

and so the information stored after the first iteration has not been corrupted. If all the transformed input vectors are orthogonal, but the number of vectors is less than the number of weights, the network stores the data exactly in a finite time,

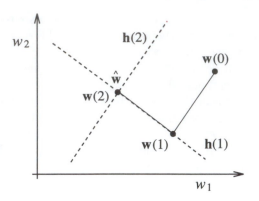

Figure 5.10 Orthogonal solution hyperplanes in two-dimensional weight space.

but the overall system is singular and there are an *infinite* number of possible optimal parameter vectors.

The fast initial learning that occurs in AMNs is because the transformed input vectors are *sparse*. Hence many of the transformed input vectors are mutually orthogonal and learning about one part of the input space does not interfere with information stored in another region.

If two transformed input vectors are mutually orthogonal, there is no learning interference between the two patterns. However, if the two inputs have common non-zero transformed input elements, using the LMS family of rules ensures that the updates interfere with each other. In Parks and Militzer [1992], Shao *et al.* [1993] and Section 5.6.2, learning rules are derived which produce weight changes perpendicular to the previous L weight changes (where L is a user-defined parameter) and so learning is orthogonal for the last L training examples (Gram-Schmidt procedure).

Slow Convergence With Correlated Data

In contrast, now consider the case when the transformed input vectors are nearly parallel (i.e. when the set of the transformed input vectors are ill conditioned) as illustrated in Figure 5.11.

Consider when the training set consists of two inputs which map to ill-conditioned transformed input vectors $\mathbf{a}(1)$ and $\mathbf{a}(2)$, an NLMS updating rule is used with $\delta = 1$, and at time 0 the weight vector lies on the second solution hyperplane. This last condition means that $\mathbf{a}^T(2)\mathbf{w}(0) = \mathbf{a}^T(2)\hat{\mathbf{w}} = 0$. Let θ be the angle between the two solution hyperplanes, then the angle between the current weight vector error and current solution hyperplane is always $\theta' = (90° - \theta)$. Hence from Equation 5.40, the relative reduction in the magnitude of the parameter error is given by:

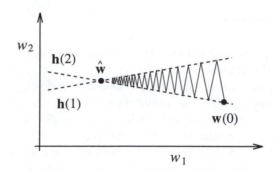

Figure 5.11 Ill-conditioned solution hyperplanes in a two-dimensional weight space.

$$\frac{V(2)}{V(0)} = \sin^4(\theta') = \cos^4(\theta) \qquad (5.41)$$

θ is close to zero, therefore $\cos^4(\theta)$ is almost unity, and after two learning iterations the size of the weight vector error is nearly equal to its initial value.

As in Section 5.2.3, the LMS family of instantaneous rules drop the weight vector perpendicularly onto the next solution hyperplane. This is illustrated in Figure 5.11 for a two-dimensional weight space with two ill-conditioned trans- formed input vectors. The ill-conditioned solution hyperplanes result in the weight updates being nearly perpendicular to the *optimal* weight updates which always point directly to the optimal weight vector.

The Basis Functions and the Solution Hyperplanes

These last two sections may suggest to the reader the misleading conclusion that it is always desirable to have the transformed input vectors orthogonal to each other. This can only be achieved if the weight set mapped to by each input is unique, so there is at most one input lying in the support of each basis functions. However, in order to generalise sensibly, each basis function needs an input lying in its support, and if the training samples are corrupted by noise, it is both necessary and desirable to have several samples lying in the support of each basis function, even though this produces non-orthogonal transformed input vectors. Also, if the basis functions have a very small support, the memory requirements would be enormous. The size and shape of the basis functions are design parameters which influence the modelling capabilities, the computational cost and the rate of convergence of these algorithms.

5.3.5   On-Line Adaptation

This geometrical interpretation illustrates a potential problem with *over-correlated* training data. If the training pairs produce transformed input vectors that are all nearly parallel, parameter convergence occurs very slowly, whereas if some of the training data are rejected, the condition of the optimisation problem can be improved. As an extreme example consider a two-weight optimisation problem, where one training set contains two orthogonal transformed input vectors and the second training set has 18 training vectors, which have equal angles of 10° between them. This can be generated by a first-order single input, single output recurrent system subject to a sinusoidal-type control signal, and the first set forms a subset of the second. The weights converge to their correct values after one pass through the first training set (see Section 5.3.4), whereas after one cyclic pass through the second training set, $V(18) \approx 0.58V(0)$ when the initial weight vector lies on the first solution hyperplane, as shown in Figure 5.12. Nearly parallel transformed input vectors result in very slow parameter convergence, as described in Sections 5.3.4.

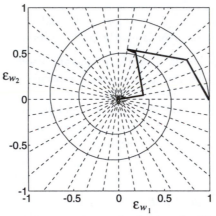

Figure 5.12   Parameter convergence when trained using the NLMS learning algorithm and the input data are correlated. The solution hyperplanes are shown as the dashed lines, and the sequential and random weight updates are the solid and bold lines, respectively.

This problem occurs with LMS rules, since the current search direction is always set equal to the current transformed input vector. This generally works well and is cheap to implement, although when the network receives correlated input data, modified training algorithms must be formulated. A simple way to counteract this phenomenon is, when the network receives two pieces of correlated training data $\{\mathbf{a}(t), \widehat{y}(t)\}_{t=1}^{2}$, first train the network on both pieces of data and then train it on the *difference*, $\{\mathbf{a}(2) - \mathbf{a}(1), \widehat{y}(2) - \widehat{y}(1)\}$. The new transformed input vector $\mathbf{a}(3) = \mathbf{a}(2) - \mathbf{a}(1)$ is nearly orthogonal to both $\mathbf{a}(2)$ and $\mathbf{a}(1)$, and the subsequent rate of parameter convergence is increased. The idea of maintaining a set of past training examples and developing higher order LMS-type learning rules is further discussed in Section 5.6.

This insight can explain why it is necessary to randomise a training set before presenting it to a network trained using a batch LMS learning algorithm. Randomising the data tends to reduce the amount of correlated data which are successively presented to the network, leading to increased rates of convergence. This has been noted in [Parks and Militzer, 1992, Weiss and Kulikowski, 1991], and is illustrated in Figure 5.12.

Instantaneous Momentum

Momentum is a much abused technique in the neural network literature. Often LMS-type learning rules are augmented by a term that keeps the weights moving in the same direction, and the LMS rule would be reformulated as:

$$\Delta \mathbf{w}(t-1) = \delta_1 \epsilon_y(t) \mathbf{a}(t) + \delta_2 \Delta \mathbf{w}(t-2) \tag{5.42}$$

where the learning rates δ_1 and δ_2 control the relative contributions of the current and past estimates of the the change in weights, respectively. Momentum may be used to increase the rate of convergence in plateau areas of the performance function, it can reduce oscillatory behaviour in batch training algorithms, and can also be used to compensate for badly chosen learning rates δ_1. However, to reduce the oscillatory behaviour (and to increase the rate of convergence), both δ_1 and δ_2 must be chosen carefully, otherwise they may *cause* oscillations near the optimal solutions. Plateau areas do not occur in the quadratic performance functions considered in this book, and often increased rates of convergence can be obtained by simply choosing the learning rate δ_1 carefully. Consider the two weight optimisation problem just described where the momentum term would cause even slower parameter convergence as the weight updates become even more *correlated*. A final important point which is pertinent to the AMN networks is that gradually the optimisation process becomes non-sparse when momentum is used. *All* the weights are updated at each time-step, with little or no increase in the rate of convergence. Thus momentum should only be used for non-sparse, nonlinear optimisation problems and the learning rates should be chosen carefully.

5.4 THE EFFECTS OF INSTANTANEOUS ESTIMATES

A potential problem that affects both the transient and the steady-state convergence of the weight vector is the *noise* associated with taking an instantaneous estimate of the true gradient. The *gradient noise* results in the instantaneous weight updates approximately following the path of the true gradient, although there exists a seemingly random component which is due to the instantaneous gradient estimates, as illustrated in Figures 5.6 and 5.7. An expression for the covariance of the gradient noise is derived in Section 5.4.1, and this can be *directly* related to the condition of the autocorrelation matrix, showing that, if the model is badly conditioned, the noise in the gradient estimates is large. These results extend the work of Widrow and Stearns [1985], as the covariance expression is valid for any weight errors; it is not restricted to small perturbations about its optimal value. When there is modelling mismatch or measurement error, the weights never settle to a fixed value when they are trained using LMS-type rules (unless the learning rule is modified as described in Section 5.3.3) but converge to a bounded area known as a *minimal capture* domain whose size and shape depend on the relative orientations of the solution hyperplanes, which in turn determine the condition of the autocorrelation matrix. Hence both (related) aspects of instantaneous learning are influenced by the condition of the autocorrelation matrix, which is influenced by the structure of the network and the type and the distribution of the training data.

5.4.1 Noisy Instantaneous Gradient Estimates

The minimal capture zones, which have been described, are simply due to the noisy estimates of the true gradient when instantaneous learning rules are used. The noise in the gradient estimate makes it impossible to determine when the current set of weights are optimal (in the MSE sense) when modelling error exists between the desired function and the network, and so adaptation does not stop, even if the current weight vector is optimal. To examine the gradient noise in greater detail, let $\epsilon_\nabla(t)$ denote the error or *noise* in the current estimate of the true steepest descent path, i.e.:

$$\epsilon_\nabla(t) = \nabla(t) - (-\epsilon_y(t)\mathbf{a}(t)) \tag{5.43}$$

The performance of the LMS learning rule and hence the noise in its instantaneous estimates depend on the particular input sequence presented to the network. A measure of how well instantaneous gradient descent approximates the batch steepest descent algorithm is the instantaneous covariance of the gradient noise, given by:

$$\mathbf{N} = \mathrm{cov}\left(\epsilon_\nabla\right) = E\left(\epsilon_\nabla \epsilon_\nabla^T\right) \tag{5.44}$$

where the expected value is taken over all possible training pairs, for a weight vector *fixed* at its current value.

When the network is updated using the LMS rule, the weight change is related to the true gradient and the gradient noise as follows:

$$\Delta\mathbf{w} = -\Delta\epsilon_\mathbf{w} = -\delta\left(\nabla - \epsilon_\nabla\right)$$

Hence the covariance of the weight change (change in weight vector error) can be expressed as:

$$\text{cov}(\Delta \mathbf{w}) = \text{cov}(\Delta \epsilon_{\mathbf{w}})$$
$$= E\left((\Delta \epsilon_{\mathbf{w}} - E(\Delta \epsilon_{\mathbf{w}}))(\Delta \epsilon_{\mathbf{w}} - E(\Delta \epsilon_{\mathbf{w}}))^T\right)$$

but the instantaneous gradient is an unbiased estimate of the true gradient, hence:

$$\text{cov}(\Delta \mathbf{w}) = E\left((\delta(\nabla - \epsilon_\nabla) - \delta\nabla)(\delta(\nabla - \epsilon_\nabla) - \delta\nabla)^T\right)$$
$$= \delta^2 E\left(\epsilon_\nabla \epsilon_\nabla^T\right) = \delta^2 \mathbf{N} \tag{5.45}$$

As expected, the gradient noise directly influences the weight updates (and hence the weight vector error), although this effect can be made acceptably small by reducing the learning rate, which produces greater averaging.

The covariance of the gradient noise is given as:

$$\mathbf{N} = E\left((\epsilon_y \mathbf{a} - E(\epsilon_y \mathbf{a}))(\epsilon_y \mathbf{a} - E(\epsilon_y \mathbf{a}))^T\right)$$
$$= E\left(\mathbf{a}\epsilon_y \epsilon_y \mathbf{a}^T\right) - E(\epsilon_y \mathbf{a}) E\left(\epsilon_y \mathbf{a}^T\right) \tag{5.46}$$

It is only possible to gain any insight into this expression if certain assumptions are made about the network inputs and/or the correlation between the random variables.

Local Estimate

To simplify Equation 5.46, Widrow assumed that the weight vector was close to its optimal value, meaning that the true gradient is almost zero and the output error is approximately uncorrelated with the transformed input vector. This produces the following *local* expression for the gradient noise covariance:

$$\mathbf{N} \approx J_{\min} \mathbf{R} \tag{5.47}$$

where $J_{\min}$ is the minimum mean square output error. The corresponding expression for the covariance of the LMS learning algorithm shows that if there exists modelling error ($J_{\min} > 0$), the weight updates are not zero when they assume their optimal values. Hence convergence to a fixed value does not occur, although this effect can be made negligible by choosing a sufficiently small learning rate, and this is discussed further in Section 5.4.3.

Global Expression

If the transformed input vector $\mathbf{a}(t)$ and the output errors $\epsilon_y(t)$ are assumed to be zero-mean Gaussian random variables (this assumption is relaxed further on), the Gaussian moment factoring theorem [An *et al.*, 1994b, Haykin, 1991] can be applied to the fourth moment term in Equation 5.46 giving:

$$E\left(\mathbf{a}\epsilon_y\epsilon_y\mathbf{a}^T\right) = E\left(\epsilon_y\mathbf{a}\right)E\left(\epsilon_y\mathbf{a}^T\right) + E\left(\epsilon_y\mathbf{a}\right)E\left(\epsilon_y\mathbf{a}^T\right) + E\left(\epsilon_y^2\right)E\left(\mathbf{a}\mathbf{a}^T\right)$$

Thus the gradient noise covariance is:

$$\mathbf{N} = \mathbf{R}\epsilon_{\mathbf{w}}\epsilon_{\mathbf{w}}^T\mathbf{R} + \mathbf{R}\left(J_{\min} + \epsilon_{\mathbf{w}}^T\mathbf{R}\epsilon_{\mathbf{w}}\right) \tag{5.48}$$

This shows that the orientation and magnitude of the gradient noise covariance is a function of the current weight vector error, the structure of the autocorrelation and the minimum MSE. Also, when the weight vector error is small, Equation 5.48 reduces to 5.47, as expected.

When $\mathbf{a}$ and ϵ_y are non-zero mean Gaussian variables, they can be linearly decomposed into constant (denoted by an overscore) and zero-mean Gaussian terms (shown by an underscore), ensuring that the Gaussian moment factoring theorem is still valid. After some algebraic manipulation, the expression for the gradient noise covariance is given by:

$$\mathbf{N} = \mathbf{R}\epsilon_{\mathbf{w}}\underline{\epsilon}_{\mathbf{w}}^T\underline{\mathbf{R}} + \mathbf{R}\left(J_{\min} + \underline{\epsilon}_{\mathbf{w}}^T\mathbf{R}\underline{\epsilon}_{\mathbf{w}}\right) + \underline{\mathbf{R}}\overline{\epsilon_y^2} + \underline{\mathbf{R}}\epsilon_{\mathbf{w}}\overline{\epsilon_y\mathbf{a}}^T \tag{5.49}$$

and when the bias terms are zero, this expression reduces to Equation 5.48. It is now possible to calculate the gradient noise covariance for the linear network introduced in Section 4.3.4 and, assuming that the Gaussian variables have a sufficiently small variance, it can also be applied to the order 2, B-spline model.

Extensive simulations have been performed and it has been found that the gradient noise condition number is approximately related to the autocorrelation matrix number, as $C(\mathbf{N}) \approx 2C(\mathbf{R})$. Similarly, it has also been been found that the principal axes of $\mathbf{N}$ are related to the principal axes of $\mathbf{R}$, and that the noise component along each axis is approximately proportional to the corresponding eigenvalues. This is illustrated in Figure 5.6, although in Figure 5.7, the gradient noise is less correlated as the eigenvalues are closer and the condition of the optimisation problem is better. This occurs in the local result given in Equation 5.47, although it is more difficult to prove anything about the eigenvalue/eigenvector decomposition of global gradient noise covariance expression.

5.4.2   Gradient Noise Examples

To illustrate the gradient noise covariance, consider a univariate, order 2, B-spline network which consists of two basis functions, that is defined on the interval $[-3, 3]$. The outputs of the basis functions are:

$$a_1 = \frac{3-x}{6}, \qquad a_2 = \frac{3+x}{6}, \qquad\qquad x \in [-3, 3]$$

The input x is a zero-mean Gaussian random variable with a standard deviation of 1, so the output of each basis function is (approximately) a biased Gaussian random variable, neglecting the fact that each basis function is only defined on a compact interval. The autocorrelation matrix has a condition number of 9, and is given by:

$$\mathbf{R} = \begin{bmatrix} 0.2778 & 0.2222 \\ 0.2222 & 0.2778 \end{bmatrix}$$

The corresponding eigenvectors for this network are oriented at $\pm45°$ relative to the horizontal, which is the same as the autocorrelation matrix generated from a uniform input probability density function.

The desired function is:

$$\hat{y}(x) = 1 + x, \qquad\qquad x \in [-3, 3]$$

which the network can reproduce exactly with an optimal weight vector $\hat{\mathbf{w}} = (-2, 4)^T$, therefore $J_{\min} = 0$. The weights are trained using an LMS rule with a learning rate of $\delta = 0.4$. This ensures stable learning as $0.5 \le \|\mathbf{a}(t)\|_2^2 \le 1.0$.

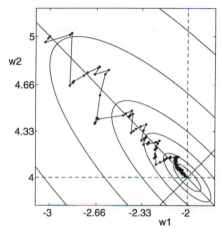

Figure 5.13 Gradient noise in the B-spline's weight vector when the initial weight error lies along the minor principal axis. The contours correspond to $J = 0.001, 0.003, 0.01, 0.03, 0.1, 0.3$.

Figures 5.13, 5.14 and 5.15 show the learning histories for three different initial weight vectors. In the first figure, the initial error in the weight vector lies along the minor principal axis (lowest curvature), and the gradient noise is highly correlated with the major (steepest) principal axis as the weight vector "jitters" along the performance surface valley. The size of the noise is also proportional to the (square) of the weight vector, error as predicted in Equation 5.49 and, as the weight vector tends to its optimal value, the gradient noise tends to zero. When the initial weight vector error lies along the major principal axis (largest curvature), the gradient noise only weakly depends on the information contained along the minor principal axis, and the weight vector goes straight to its optimal value (see Figure 5.14). This effect can also be explained in terms of solution hyperplanes, as the majority lie nearly parallel to the minor principal axis (Gaussian distribution about this line), so the weight updates are mostly nearly parallel to the major principal axis. Finally,

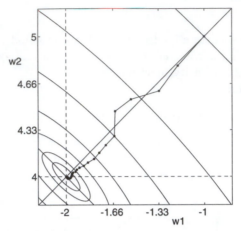

Figure 5.14 Gradient noise in the B-spline's weight vector when the initial weight error lies along the major principal axis. The contours correspond to $J = 0.001, 0.003, 0.01, 0.03, 0.1, 0.3, 1$.

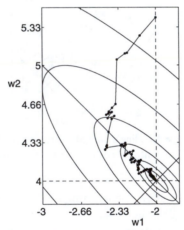

Figure 5.15 Gradient noise in the B-spline's weight vector when the initial weight error lies between the major and minor principal axes. The contours correspond to $J = 0.001, 0.003, 0.01, 0.03, 0.1, 0.3$.

when the learning process is started with a weight vector which lies somewhere between the major and minor principal axes (as generally occurs in practice), the gradient noise is correlated with the principal axis and its magnitude depends on the weight vector error.

When the modelling error is non-zero, the steady-state dynamics of the weight vector is determined by the interplay between the terms which repel the weight vector from its optimal value (gradient noise) and those which drive it towards its optimal value, based on instantaneous gradient estimates. The parameters converge to some domain in weight space, and Equation 5.32 gives an expression for its size when output dead-zones are used to cease training once the output error is sufficiently small. In general, it is difficult to imagine the size and shape of these domains, although when the network is trained using an NLMS rule, it is possible to interpret the steady-state weight behaviour geometrically, as will now be described.

5.4.3   Minimal Capture Zones

When no optimal weight vector exists which can reproduce the training data exactly, the weights converge to a domain rather than to a unique value. The size and shape of this domain depends on the size of the modelling error and the internal representation used by the network. This section describes the structure of these domains for models trained using the NLMS learning rule and an example is given which employs both a linear and a B-spline network.

When the weight vector is close to its optimal value it is simpler to consider the dynamical behaviour of w relative to the current solution hyperplanes rather than considering the overall gradient noise covariance. Each time the weight vector is updated using the NLMS rule with $\delta \leq 1$, it moves towards the solution hyperplane, but *never* crosses.

If the training samples are drawn randomly from a possibly infinite training set and the NLMS learning rate lies in the interval $(0, 1]$, the original definition of a *minimal capture zone* [Parks and Militzer, 1989] can be modified to:

> *The smallest p-dimensional domain which contains the intersection of all the solution hyperplanes such that the orthogonal projection of every point (in the domain) onto every solution hyperplane is completely contained in this domain.*

A two-dimensional minimal capture zone is shown in Figure 5.16.

The size and shape of these minimal capture zones depends on the training samples, their order of presentation, the adaptation rule and the learning rate, δ. The smaller the learning rate, the greater will be the noise filtering (or equivalently the modelling error filtering), although the capture zone is not always smaller. However, if the learning rate satisfies the stochastic approximation conditions given in Section 5.3.3 and an LMS updating rule is used, convergence of the weight vector to the optimal MSE value is assured. Similarly, if the stochastic approximation NLMS learning rule is used, the weight vector converges towards the optimal weight vector which minimises the normalised performance function given in Equation 5.15.

This can be simply illustrated by considering the instantaneous and average performance functions generated by a network with a single, linear adaptable parameter w. The training set consists of two data pairs $\{1, 1\}$ and $\{2, 4\}$, which

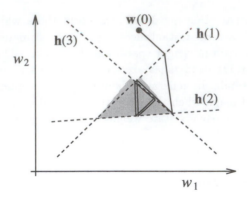

Figure 5.16   A two-dimensional minimal capture zone (shaded area) generated by three inconsistent solution hyperplanes. The weights no longer converge to an optimal value.

the network is unable to store exactly, as the optimal weight value for each case is different (1 and 2). Inspecting Figure 5.17, it can be seen that the optimal weight value (with respect to the MSE performance function) is $w = 1.8$, however the instantaneous gradient estimates in the interval $[1, 2]$ provide conflicting information and the weights never settle down to a constant value. Hence the minimal capture zone is the univariate interval $[1, 2]$.

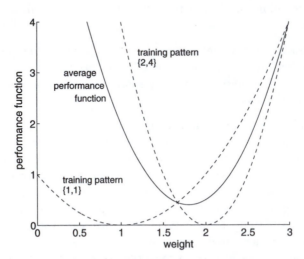

Figure 5.17   The instantaneous (dashed lines) and average performance functions (solid line) generated by a linear network with a single weight and two inconsistent training samples.

5.4.4   Example: Linear and B-spline Minimal Capture Zones

An example is now given which generates and compares the minimal capture zones of a conventional linear and a B-spline network. Both networks have identical modelling capabilities and the effect of modelling mismatch is illustrated when both networks attempt to reproduce the squared function $\widehat{y}(x) = x^2$ on the unit interval $[0, 1]$, as shown in Figure 5.18.

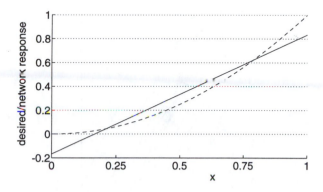

Figure 5.18   Approximating the quadratic function $\widehat{y}(x) = x^2$ (dashed line) with an optimal linear network (solid line).

Linear Network

Initially consider using a conventional linear network of the form:

$$y = w_1 + w_2 x \qquad x \in [0, 1]$$

to try and store this relationship. The optimal weight vector (for a MSE performance function and a uniform probability density function) is:

$$\widehat{w}_1 = -0.167, \qquad \widehat{w}_2 = 1.0$$

whereas when the desired function is sampled at the three points $\{0.25, 0.5, 0.75\}$, the optimal weights become:

$$\widehat{w}_1 = -0.208, \qquad \widehat{w}_2 = 1.0$$

The minimal capture zones for the conventional linear network trained using the NLMS rule can be generated by examining the solution hyperplanes that are given by:

$$x^2 = w_1 + w_2 x \qquad x \in [0, 1]$$

The network and the desired function are initially sampled at the three points $\{0.25, 0.5, 0.75\}$, and the training samples are randomly presented to a linear network that is trained using an NLMS rule, with $\delta \leq 1$. The intersection of the solution hyperplanes and the minimal capture zone are shown in Figure 5.19. This minimal capture zone is fairly small, and is *stretched* in the direction of w_2 (the

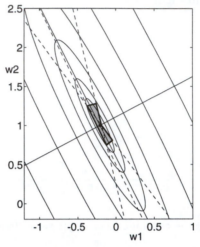

Figure 5.19   Minimal capture zone (shaded area), when the input is sampled at three points (three solution hyperplanes) and a conventional linear model is employed. The contour values are given by: $J = J_{\min} + \{0.005, 0.015, 0.05, 0.15, 0.5, 1.5\}$.

weight which multiplies the linear term). The power of the linear input is less than the power of the bias term, and so the size of the minimal capture zone is greater in the direction of w_2.

Now imagine the case when the input is randomly sampled in the domain $[0, 1]$, which is a more realistic situation for a network learning on-line. The overall minimal capture zone is the union of all the possible minimal capture zones corresponding to a finite training set, and a computer-generated approximation to the minimal capture zone is shown in Figure 5.20. The optimal weight vector is contained within the minimal capture zone, but the domain is much larger than that shown in Figure 5.19, and for certain points in the minimal capture zone $\|\widehat{\mathbf{w}} - \mathbf{w}(t)\|_\infty = 1.3$.

When the weights are trained using the stochastic version of the NLMS rule, they do not converge to the optimal MSE values given above, but to the slightly biased weights:

$$w_1 = -0.147, \qquad w_2 = 0.952$$

This can be obtained from the *normalised* autocorrelation matrix and the *normalised* cross-correlation vector, where each element is divided by $\|\mathbf{a}(t)\|_2^2$ before taking the expected value.

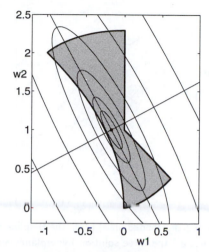

Figure 5.20 Minimal capture zone (shaded area), when the input is randomly sampled in $[0,1]$ and a conventional linear model is employed. The contour values are given by: $J = J_{min} + \{0.005, 0.015, 0.05, 0.15, 0.5, 1.5\}$.

B-spline Network

The two corresponding minimal capture zones for the B-spline network are now generated. The output of an order 2 B-spline network with two basis functions is expressed as:

$$y(x) = w_1(1-x) + w_2 x \qquad x \in [0,1]$$

and the network is again trained using an NLMS rule with $\delta \leq 1$. The optimal MSE weight vector (for a uniform probability density function on the unit interval) is given by:

$$\hat{w}_1 = -0.167, \qquad \hat{w}_2 = 0.833$$

and the optimal weight vector, when the training set consists of just three training examples, is:

$$\hat{w}_1 = -0.208, \qquad \hat{w}_2 = 0.792$$

The solution hyperplanes are now given by:

$$x^2 = w_1(1-x) + w_2 x \qquad x \in [0,1]$$

and the first minimal capture zone is again generated when the input is sampled at the three points $\{0.25, 0.5, 0.75\}$. The intersection of the solution lines and the minimal parameter zone are shown in Figure 5.21. This region is fairly small and

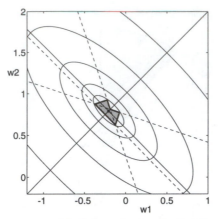

Figure 5.21   Minimal capture zone (shaded area) when the B-spline model is sampled at three input points. The dashed lines are the three solution hyperplanes and the contour values are given by: $J = J_{\min} + \{0.005, 0.015, 0.05, 0.15, 0.5\}$.

more regular than the corresponding linear minimal capture zone (Figure 5.19) as the two transformed inputs now have the same power.

Similarly, if the training input data are now generated by randomly sampling the domain $[0, 1]$, the new minimal capture zone is shown in Figure 5.22. For this minimal capture zone $\|\hat{\mathbf{w}} - \mathbf{w}(t)\|_{\infty} \leq 0.833$, which is smaller than the corresponding value for the linear network. This is to be expected because the output errors

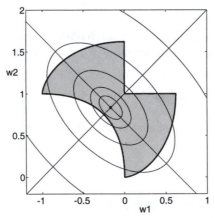

Figure 5.22   Minimal capture zone (shaded area) for a random input to a B-spline model. The contour values are given by: $J = J_{\min} + \{0.005, 0.015, 0.05, 0.15, 0.5\}$.

are the same for the two networks, but the condition of the linear network is higher, so from Equation 5.32, the expected error in the weight vector is larger.

Also if the weights for the B-spline network are trained using the stochastic approximation version of the NLMS rules, they converge to the slightly biased values:

$$w_1 = -0.182, \qquad w_2 = 0.818$$

Comparing these values with the optimal linear network model shows that the optimal networks are different. This occurs because the NLMS rule produces an optimal model, which minimises the MSE performance function normalised by $\|a(t)\|_2^2$. For each network the internal representation and hence the strength of the signal is different, as are the relative weightings on the output errors.

When Figure 5.19 is compared with Figure 5.21, it is immediately obvious that the set of three solution lines is more ill conditioned for the conventional linear network, and the variation with respect to the linear term, w_2, is greater. These points are illustrated better when the minimal capture zones for random inputs are compared. For the linear network, the minimal capture zone is stretched in the direction of the linear term, so although the two representations have the same modelling capabilities, the estimated value of w_2 is worse for the linear case. The B-spline representation gives equal emphasis to both w_1 and w_2, so the minimal capture zone is more regular. It should also be emphasised that the models formed by either the linear or the B-spline network are equivalent for weight vectors which lie within the respective minimal capture zones. However, if these networks form part of a larger model structure (using three or more order 2 B-splines, for instance), the rate of parameter convergence is faster for networks with more regular minimal capture zones, due to a smaller parameter error being propagated along the weight vector.

Finally, the inclusion of dead-zones within the adaptation algorithm makes it possible to reduce the size of the minimal capture zone significantly when the input is randomly sampled. Ill conditioned input sequences, coupled with modelling mismatch, drag the weights far away from their optimal values, so if adaptation ceases when the output error is small, the effect of this phenomenon is reduced.

5.5   LEARNING INTERFERENCE IN ASSOCIATIVE MEMORY NETWORKS

Within the context of AMN adaptation, *learning interference* is defined as the weight updates which degrade knowledge stored at a particular point. As has been demonstrated in Section 5.3.4, if the solution hyperplanes are all mutually orthogonal no learning interference occurs, as updating at one point has no effect on the knowledge stored at the other points. However, in order to filter out measurement noise, or to produce a best fit (in some sense), or simply to model the underlying structure of the desired function, there generally exists more than one data point in the support of each basis function, and when this occurs, the recommended weight updates may partially cancel each other out reducing the rate of convergence.

   This section derives a measure of the amount of learning interference for different univariate basis function shapes. The theory can be directly applied to both univariate CMAC and B-spline (fuzzy logic) networks, as they both arrange the univariate basis functions on the lattice in an identical manner. For each network, moving to the next interval means that one basis function is dropped from, and introduced to, the output calculation. Hence the basic network structure is the same; only the shape of the basis functions are different. Each input maps to ρ basis functions and ρ is the width of the support of each basis function.

   Section 5.5.1 gives a precise definition of learning interference, fixing the type of input signal and learning rules, etc., so that only the type of basis function varies. It is then shown how this definition can be interpreted using matrices, and how the problem of measuring the learning interference reduces to finding the spectral radii of the generated matrices. It is also shown that these matrices have a characteristic equation which has a special form, and so the learning interference measures are equivalent to finding the largest root of a characteristic equation.

5.5.1 Defining Learning Interference in Associative Memory Networks

The learning interference that occurs between weights is very difficult to predict without a full knowledge of the training set. If the training set is completely specified, it is possible to construct the matrices which iterate the weight vector from one batch training cycle to the next. In this case, a measure of the learning interference (or rate of convergence) would be the spectral radius of the iteration matrix, with a value close to 1 meaning that there was substantial learning interference occurring and a value close to zero indicating very rapid convergence of the weight vector [Parks and Militzer, 1989]. However, such a measure depends on the statistics of the input signal (when and where the input occurred), the learning rule used (and the adaptation rate) as well as the size and shape of basis functions defined on the input space. The objective of this section is to measure the learning interference which is due to using different shapes of basis functions in the "hidden layer", so the remaining terms which affect the learning interference must be *fixed*. This results in the following definition of the learning interference which depends on the type of basis functions employed:

> *Assume that the model and the desired function have the same basic structure, and that at time 0 all the weights are correctly initialised apart from those mapped to by an input which lies at the origin. A measure of learning interference is then the size of the eventual relative change in the local error weight vector, where the input sequence moves away from the origin one cell at a time and an NLMS learning rule is employed with $\delta = 1$.*

Here the local error weight vector is defined as the vector of weight updates (errors) which influence the network output for a particular input. An NLMS learning rule is used, so that the output error reduction does *not* depend on the size of the transformed input vector. Finally, the input sequence is assumed to move away from the origin one cell at a time, ensuring that only one new weight is introduced into the output/update calculations at each iteration and when the basis functions generalise, the weight which was previously correct is corrupted due to the previous local error weight vector.

As $t \to \infty$, the magnitude of the ratio of the current weight change over the previous weight change gives a measure of how fast the parameter error is decaying, and hence measures how *local* the adaptation is. It is expected that the larger the amount of generalisation, the less local the learning and the greater the corresponding learning interference measure.

5.5.2 Matrix Interpretation of Learning Interference

Consider the case illustrated in Figure 5.23, a univariate B-spline network with order 2 basis functions defined on simple knots. Suppose that the desired function

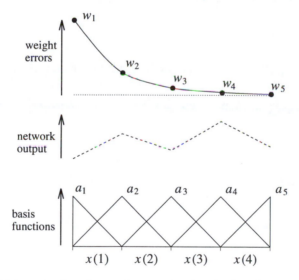

Figure 5.23 Learning interference for an order 2, univariate B-spline network.

can be reproduced exactly by the AMN with a weight vector $\hat{w}$, and at time zero the weight vector of the AMN is defined by:

$$w_i(0) = \begin{cases} \hat{w}_i + \varepsilon & \text{if } i = 1 \\ \hat{w}_i & \text{otherwise} \end{cases} \tag{5.50}$$

where ε is the error in the weight mapped to by an input at $x = 0$. For a different input lying in the tth interval, the local weight vector consists of two weights $w_t(t-1)$, $w_{t+1}(t-1)$ and by induction $w_{t+1}(t-1) = \hat{w}_{t+1}$. The input lies in the tth interval, then $x_t \in I_t = [\lambda_{t-1}, \lambda_t)$ and:

$$
\begin{aligned}
y(t) &= a_t(t)w_t(t-1) + a_{t+1}(t)w_{t+1}(t-1) \\
\hat{y}(t) &= a_t(t)w_t(t-2) + a_{t+1}(t)w_{t+1}(t-1) \\
\epsilon_y(t) &= \hat{y}(t) - y(t) = -a_t(t)\Delta w_t(t-2)
\end{aligned}
\tag{5.51}
$$
$$\tag{5.52}$$

where $x(t) \in I_t = [\lambda_{t-1}, \lambda_t)$. For simplicity, let $a_1 = a_t(t)$ and $a_2 = a_{t+1}(t)$, and the dependence on time and position is now implicit for the outputs of the basis functions. Therefore the change in $w_{t+1}(t-1)$ caused by applying the NLMS rule with $\delta = 1$ is:

$$
\begin{aligned}
\Delta w_{t+1}(t-1) &= \frac{\epsilon_y(t)a_2}{a_1^2 + a_2^2} = -\frac{a_1 a_2}{a_1^2 + a_2^2}\Delta w_t(t-2) \\
&= -c(t)\Delta w_t(t-2) = \prod_{i=1}^{t}(-c(i))\varepsilon
\end{aligned}
\tag{5.53}
$$

where the decay factor $c(t)$ is defined to be $a_1 a_2/(a_1^2 + a_2^2)$. Hence the ratio of the errors in the adapted weights is given by:

$$
\left| \frac{\Delta w_{t+1}(t-1)}{\Delta w_t(t-2)} \right| = c(t)
\tag{5.54}
$$

For each t, $c(t) \in [0, 0.5]$, and so the absolute value of the errors in the weight vector decays exponentially as the input moves away from the origin. The expected value of $c(t)$ is 0.285 (for a uniform input probability density function on each unit interval) and a plot of $c(t)$ against $x(t)$ is given in Figure 5.24.

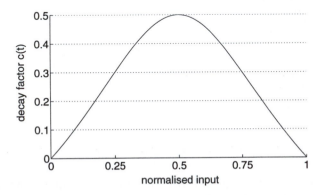

Figure 5.24 Decay factor for piecewise linear univariate B-splines.

In this case, the local error weight vectors (scalars) are related by a scalar, and a similar phenomenon happens for the CMAC's univariate binary basis functions with $\rho = 2$, where $c(t) \equiv 0.5 \ \forall \ t$. However, as the width of the basis functions increases the local error weight vectors are related by learning interference matrices, as is now shown.

Basis Functions of Width 4

Now consider a univariate lattice AMN, which has basis functions of width 4. If the weight vector has the correct values apart from an error in the local weight vector at the origin initially, the network output error, at time t, depends on six terms: $\Delta w_t(t-2)$, $\Delta w_{t+1}(t-2)$, $\Delta w_t(t-3)$, $\Delta w_{t+2}(t-2)$, $\Delta w_{t+1}(t-3)$ and $\Delta w_t(t-4)$. Defining the local error weight vector as $\epsilon'_w(t) = (\Delta w_{t+1}(t-1), \Delta w_{t+2}(t-1),$ $\Delta w_{t+1}(t-2), \Delta w_{t+3}(t-1), \Delta w_{t+2}(t-2), \Delta w_{t+1}(t-3))^T$, then it is related to the previous one by:

$$\epsilon'_w(t) = \mathbf{B}(t)\epsilon'_w(t-1)$$

where the learning interference matrix $\mathbf{B}(t)$ has the form:

$$
\begin{bmatrix}
-\delta a_1 a_2 & -\delta a_2^2 & -\delta a_1 a_2 & -\delta a_2 a_3 & -\delta a_2^2 & -\delta a_1 a_2 \\
-\delta a_1 a_3 & -\delta a_2 a_3 & -\delta a_1 a_3 & -\delta a_3^2 & -\delta a_2 a_3 & -\delta a_1 a_3 \\
0 & 1 & 0 & 0 & 0 & 0 \\
-\delta a_1 a_4 & -\delta a_2 a_4 & -\delta a_1 a_4 & -\delta a_3 a_4 & -\delta a_2 a_4 & -\delta a_1 a_4 \\
0 & 0 & 0 & 1 & 0 & 0 \\
0 & 0 & 0 & 0 & 1 & 0
\end{bmatrix}
\tag{5.55}
$$

and $a_1 = a_t(t)$, $a_2 = a_{t+1}(t)$, $a_3 = a_{t+2}(t)$, $a_4 = a_{t+3}(t)$ with $\delta = 1/\|\mathbf{a}\|_2^2$. The top-left element in this matrix is the learning interference measure for basis functions of width two. Similarly, the top-left submatrix of size (3×3) is the learning interference matrix for basis functions of width 3. It can also be seen that this is the top-left submatrix of the learning interference matrix for basis functions of width 5 (size (10×10)).

For an initial local weight vector error of magnitude:

$$\|\epsilon'_w(0)\| = \varepsilon$$

the induced local weight vector error depends on the size of the learning interference matrix, $\|\mathbf{B}(t)\|$. Taking a vector induced matrix norm does not give a sensible answer, due to the presence of the rows containing a single unity element, which results in the vector induced matrix norms being greater than 1, and this can be interpreted as meaning that the initial error in the weight vector *increases*. This occurs because these norms are derived by considering the worst possible case for one iteration, whereas this work is directed towards measuring the *average* or *eventual* change in the weight vector error.

5.5.3 A Measurement of Learning Interference

The measurement of the size of the matrices derived in the previous section should *not* reflect the initial transients, which are due to slightly different local error weight vectors at the origin. Rather it should reflect the *eventual* ratio as $t \to \infty$. The spectral radius (which provides a sharp lower bound for the space of matrix norms) is one such measure and is defined as the absolute value of the largest eigenvalue of the matrix, or equivalently the absolute value of the largest root of the corresponding characteristic equation.

It can be shown that the corresponding characteristic equations of the learning interference matrices generated by basis functions of orders 2, 3 and 4, are given by:

$$\lambda + \delta a_1 a_2 = 0 \tag{5.56}$$

$$\lambda^2 + \lambda\delta(a_1 a_2 + a_2 a_3) + \delta a_1 a_3 = 0 \tag{5.57}$$

$$\lambda^3 + \lambda^2 \delta(a_1 a_2 + a_2 a_3 + a_3 a_4) + \lambda\delta(a_1 a_3 + a_2 a_4) + \delta a_1 a_4 = 0 \tag{5.58}$$

respectively. It is worth noting that for each matrix, many of the eigenvalues are zero and that the characteristic equations possess a regular structure that holds for higher order basis functions (larger ρ) as the matrix construction and reduction methods are identical. Hence the characteristic equation for a univariate AMN with basis functions of width ρ is:

$$\lambda^{\rho-1} + \delta \sum_{i=1}^{\rho-1} \lambda^{\rho-1-i} \left(\sum_{j=1}^{\rho-i} a_j a_{j+i} \right) = 0 \tag{5.59}$$

The problem of measuring the amount of learning interference reduces to finding the absolute value of the largest root of this equation.

Examples: Hat and Binary Univariate Basis Functions

To illustrate the theory developed in this section, the learning interference measures will be calculated for four different univariate basis function shapes. The aim is to illustrate how the learning interference measures change for different set shapes and widths. The shapes of the basis functions are chosen as rectangles and triangular functions which have supports of both 2 and 4 units wide, and these four basis functions are shown in Figure 5.25. When the basis functions are rectangles, which are also called binary basis functions because their output is either 0 or a positive constant, they are equivalent to the original basis functions used in the CMAC (see Chapter 6). Similarly, the hat functions of width 2 are equivalent to B-splines of order 2, and the hat functions of width 4 are just dilated B-splines of order 2 and width 4 (see Section 8.2).

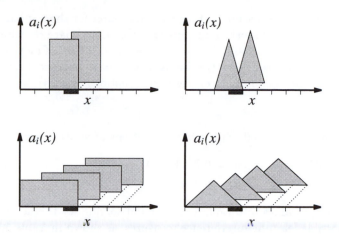

Figure 5.25 Binary and hat univariate basis functions of width 2 and 4.

Without loss of generality, these basis functions are assumed to form a partition of unity ($\sum_{i=1}^{\rho} a_i \equiv 1$), so their outputs are:

$$a_1 = a_2 = 0.5 \qquad\qquad\qquad \text{width 2}$$
$$a_1 = a_2 = a_3 = a_4 = 0.25 \qquad\qquad \text{width 4}$$

which generate the following characteristic equations:

$$\lambda + \frac{1}{2} = 0$$
$$\lambda^3 + \frac{3}{4}\lambda^2 + \frac{2}{4}\lambda + \frac{1}{4} = 0$$

Solving for the largest root gives the learning interference measures shown in Table 5.1.

Width	Learning Interference
2	0.5
4	0.642

Table 5.1 Univariate binary basis functions learning interference measures.

These measures support the work developed in Section 5.3.4, where it was shown that parameter convergence was slow if the transformed input vectors were ill conditioned. The angle between two univariate binary transformed input vectors (which differ by one cell parallel to one of the axes) is given by $\cos^{-1}(1 - 1/\rho)$. Hence when the generalisation parameter is increased by one, the angle between the two new transformed input vectors is $\cos^{-1}(1 - 1/(\rho + 1))$, which is closer to zero. Thus, the transformed input vectors become more ill conditioned as ρ increases, and parameter convergence is correspondingly slower (see Section 5.3.4), and increasingly noisy (Section 5.4.1).

In order to derive the learning interference measures for the hat basis functions, a number of assumptions must be made about the input space on which the basis functions are defined. Without loss of generality, the basis functions are again assumed to be normalised, and the input space to be divided into equi-spaced unity intervals. Then the outputs of the basis functions of width 2 are given by:

$$a_1 = (1 - x), \qquad a_2 = x$$

for $x \in [0, 1]$, and the outputs of the basis functions of width 4 are:

$$a_1 = 0.25(1 - x), \qquad a_2 = 0.25(2 - x)$$
$$a_3 = 0.25(x + 1), \qquad a_4 = 0.25x$$

Substituting these values into Equation 5.59 and solving for the largest root gives a range of possible learning interference measures, as shown in Table 5.2. Intervals

Width	Learning Interference	Expected Value
2	$[0.0, 0.5]$	0.285
4	$(0.370, 0.409)$	0.389

Table 5.2 Univariate hat basis functions learning interference measures.

are shown because different learning interference measures are generated for different values of $x \in [0, 1]$, so to produce a single measure (which can then be compared with the binary basis functions) it is assumed that x is a random variable with a uniform probability density function on $[0,1]$, and these results are also shown in Table 5.2.

Comparing these values, it can be clearly seen that for the same-shaped basis functions, as their widths increase the corresponding learning interference measures also increase. When the basis functions of the same width are compared, it is possible to see the difference in the learning interference measures for different shapes. Binary basis functions produce a greater amount of learning interference compared to triangular representations, because the output of the binary basis functions only weakly depends (two states) on the distance from the centre of the basis function, whereas the triangular functions tend towards zero as the distance increases. Therefore the tapered basis functions, which are defined over a small support, produce the smallest amount of learning interference.

This work shows that choosing a larger binary basis functions increases the learning interference and it also indicates that using tapered basis functions (i.e. B-splines) reduces the amount of learning interference. However, it would be *incorrect* to draw the conclusion that using basis functions for which learning interference is larger always means that learning is slower. The rate of convergence also depends on the choice of the model for a particular problem. For instance, if the function is constant and the input space is divided into a large number of intervals, the best strategy is to choose ρ as large as possible, since this results in very fast convergence, and in this case learning interference is desirable. If the desired function varies rapidly, a large amount of learning interference would cause slow parameter convergence as the gradient noise would be large.

In practice, the number of intervals on each axis is small and these networks are generally employed to learn highly nonlinear functions. Thus it would seem that local learning is *essential* if these networks are to be used in on-line applications.

5.6   HIGHER ORDER LEARNING RULES

For some applications, the gradient noise associated with taking an instantaneous estimate of the true gradient may be undesirable, or the rate of convergence may be too slow, especially when correlated inputs are sequentially presented to the network. These criticisms of the LMS-type learning algorithms can be partially overcome by using *higher order* training rules which use more than one data pair for each parameter update. No improved training algorithm should increase the computational complexity too much, and in particular it is desirable to formulate rules which exploit the sparsity of the transformed input vectors. These higher order learning rules use more than one data example in order to update the weight vector, so a common feature is that a *store* of L training pairs must be maintained. It is argued that maintaining this information store properly is as critical as the choice about which learning algorithm to use, so this is described first. Several higher order training rules are then described, and a comparison is made both of the method for calculating the weight update and also the implementation cost.

An alternative higher order learning rule is derived in Section 5.6.4, where the past training data are used to filter out non-Gaussian noise from the training signal [Mukhopadhyay and Narendra, 1993]. It is assumed that the noise signal is generated by an *immeasurable*, finite order, linear, stable, recurrence relationship and LMS-type instantaneous gradient descent rules have been used to model the nonlinear time series described in Section 6.6 when the data have been corrupted by sinusoidal noise [Wang *et al.*, 1994a].

5.6.1   Training Data Store

It is necessary to maintain a store of L training examples for the higher order learning rules, to provide information about the input/output relationship of the plant. These data must be current to model and control slowly time-varying plants on-line and to filter out any modelling error or measurement noise, but it is also generally required that the training inputs should be partially orthogonal, so that higher order LMS-type rules can be employed. The simplest way of maintaining the store is to adopt the *first in, last out* principle, which replaces the oldest training example with the most recent. This always ensures that the store's information is current, although when a plant is operating at a steady-state value for prolonged periods of time, the store will only contain very correlated input data. Another method is to search the store for the training sample which most closely matches the current input. The matched sample is then replaced by the new training example. The potential difficulty with this approach is that very old data which are no longer representative of the plant input/output mapping could be maintained in the store and the network's noise filtering abilities would be greatly reduced as the data would generally be consistent. The third possible method for maintaining the store would be to combine the two approaches by *weighting* each training pair according to both the length of time it has been a member of the store and the degree of correlation with the current training sample. Then either the training example with the highest score is replaced by the current sample, or else a data pair is chosen and replaced at random, with its weight influencing how likely it is to be deleted.

The selection of L depends on various factors such as the size of the parameter vector, the desired rate of convergence, the noise level, etc., although it should be stressed that when $L = 1$, all of these algorithms reduce to the standard LMS rules.

5.6.2   Improved Training Schemes

The aim of these improved learning algorithms is to increase the rate of *parameter* convergence, and this is achieved by choosing or constructing search directions which are nearly orthogonal and choosing the step size, such that the training rule filters out measurement noise or modelling error. The instantaneous output error is not always fully corrected (as occurs in the standard NLMS rule with $\delta \equiv 1$), so the instantaneous performance may be slightly degraded, but overall the parameters converge at a much faster rate.

Random Training Algorithm

This algorithm selects the current training example from the store at random and trains the weight vector, using the standard LMS rules. By selecting the training pair at random, it is hoped that it is uncorrelated with the previous learning example, and it is an extremely cheap algorithm to implement; the only overheads are those associated with updating the store. In practice, the algorithm generally performs better than the instantaneous LMS rules [Parks and Militzer, 1992], although increased rates of parameter convergence can be obtained through a more "intelligent" use of the data in the store and by choosing a step size which dampens the weight changes made during the final stages of learning.

Partially Optimised Step Length Algorithm

In this learning algorithm, the weight vector is updated in the direction of the current input, $\mathbf{a}(t)$, and so retains the minimal disturbance principle, although the size of the update $c(t)$ in Equation 5.60 is not chosen solely to correct the current instantaneous output error; rather it minimises the sum of the squared perpendiculars to the hyperplanes for the L training samples in the store:

$$\sum_{i=t-L+1}^{t} \left(\hat{y}(i) - \mathbf{a}^T(i)\mathbf{w}(t-1)\right)^2$$

where (for notational convenience) the store consists of the last L training samples. The weight vector is updated according to the following rule:

$$\Delta\mathbf{w}(t-1) = \delta c(t)\mathbf{a}(t)$$

$$c(t) = \frac{\sum_{i=t-L+1}^{t} \left(\hat{y}(i) - \mathbf{a}^T(i)\mathbf{w}(t-1)\right)\mathbf{a}^T(i)\mathbf{a}(t)}{\sum_{i=t-L+1}^{t} \left(\mathbf{a}^T(i)\mathbf{a}(t)\right)^2} \qquad (5.60)$$

where $\delta \in (0,2)$. This adaptation strategy provides a computationally efficient generalisation of the instantaneous NLMS error correction rule. Optimising the weight vector with respect to the last L pieces of data ensures that the gradient noise (jittering and minimal capture zones) is reduced, although always choosing the search direction to be parallel to the current transformed input vector means that parameter convergence is still slow. A possible improvement to this algorithm could be to choose the "current" training sample at random from the store and then minimise the local MSE or to choose the one with the maximum absolute output error. The computational cost would still depend linearly on L, although L network outputs must be calculated. This is the only algorithm that attempts to choose an "intelligent" step size, using all of the noisy data contained in the store.

Batch Normalised Least-Mean Squares

This algorithm provides a computationally efficient generalisation of the NLMS learning rule by setting the search direction to:

$$\mathbf{s}(t) = \frac{1}{L} \sum_{i=t-L+1}^{t} \frac{\epsilon_y(i)\mathbf{a}(i)}{\|\mathbf{a}(i)\|_2^2}.$$

A fixed step size δ is taken along the search direction given by the moving average of the normalised steepest descent directions associated with each member of the training store. This allows for a better approximation to the true gradient descent path and hence reduces the amount of gradient noise, although no attempt is made to optimise the choice of step size.

A similar learning algorithm which stores the last L training examples and simply performs L NLMS updates at each learning iteration was also proposed by Nagumo and Noda [1967].

Maximum Error Training Algorithm

The maximum error training algorithm selects the training sample from the store with the largest output error (with respect to the current weight vector) and uses this as the "current" learning pair. Denoting the output error of the ith training example at time t by $\epsilon_y(t,i)$:

$$\epsilon_y(t,i) = \hat{y}(i) - \mathbf{a}^T(i)\mathbf{w}(t-1)$$

an index j must be selected such that:

$$|\epsilon_y(t,j)| > |\epsilon_y(t,i)|$$

for $i,j = t - L + 1, \ldots, t$, $i \neq j$. If j is not unique, i.e. there exist several different maximum errors, j is selected according to some predetermined criteria such as choosing the first or last maximum error. The weight vector $\mathbf{w}(t-1)$ is then updated using the LMS or NLMS learning rule and the training pair $\epsilon_y(t,j)$.

This updating algorithm has been shown to perform better than the previous two for a wide variety of test problems [Parks and Militzer, 1992], although because only one training sample is used to update the weight vector, there is a significant amount of gradient noise, especially within the minimal capture zone. The required computational effort is slightly less than the previous algorithm and possibly the best option would be to combine the two approaches, as described previously.

Partial Gram-Schmidt Training Algorithm

The success of all of the previous higher order learning rules (except the difference training method) depends on the store maintaining a sufficiently varied set of training examples (inputs uncorrelated). It may be that there *is* sufficient information contained in the store, only that the inputs are highly correlated, and this results in a slow convergence rate. The difference training algorithm attempts to overcome this deficiency by constructing nearly orthogonal search directions (still in the same transformed input space), and this can be placed in a firm theoretical framework using Gram-Schmidt orthogonalisation methods [Monzingo and Miller, 1980, Parks and Militzer, 1992, Shao *et al.*, 1993].

The Gram-Schmidt training algorithm is based on calculating successive orthogonal search directions, $\mathbf{d}(t)$, for the last L training samples. Each weight update is made in a direction parallel to a vector $\mathbf{d}(t)$, which lies in the subspace spanned by the set $\{\mathbf{a}(i)\}_{i=t-L+1}^{t}$ and is orthogonal to all the transformed input vectors except $\mathbf{a}(t)$. The corresponding step size is chosen so that the new weight vector $\mathbf{w}(t)$ lies on the tth solution hyperplane, $\mathbf{h}(t)$ and the weight update direction $\mathbf{d}(t)$ is generated using the standard Gram-Schmidt orthogonalisation procedure; if $\mathbf{a}(t)$ is a linear combination of the set $\{\mathbf{a}(i)\}_{i=t-L+1}^{t-1}$, the weight update direction can either be set equal to $\mathbf{a}(t)$, or else no update is performed. This produces the following update rule:

$$\Delta \mathbf{w}(t-1) = \delta \frac{\epsilon_y(t)\mathbf{d}(t)}{\mathbf{a}^T(t)\mathbf{d}(t)} \tag{5.61}$$

where $\delta \in (0,2)$. The computational cost of implementing the algorithm is *linearly* dependent on L, *if* the new search direction is not a linear combination of input data in the store [Shao *et al.*, 1993]. Otherwise, if the store is constructed using the first in, last out principle and a complete set of orthogonal search directions must be calculated, the computational cost is $O(L^2)$ [Parks and Militzer, 1992].

When L is large and the data do not contain modelling errors, this algorithm performs very well and if L is equal to the number of training examples, the learning algorithm converges in a finite number of steps. If the data contain measurement noise or significant modelling error, the gradient noise is *amplified* especially close to the optimal weight vector, although stochastic learning rates and output dead-zones can help to minimise this effect.

As a computationally cheaper alternative, possibly the simplest way for introducing orthogonality into the current training data ($L = 2$) is first to adapt the weight vector using the LMS rules on the two most recent examples and then to train it on the *difference*. If the two transformed input vectors are almost orthogonal, little effort is wasted when the network is trained on the difference, although, when they are highly correlated, the training pair given by $\{\mathbf{a}(2)-\mathbf{a}(1), \widehat{y}(2)-\widehat{y}(1)\}$ is almost orthogonal to both $\mathbf{a}(1)$ and $\mathbf{a}(2)$, and faster parameter convergence occurs [Shao *et al.*, 1993]. However, this can have a detrimental effect on the convergence rate as the two training samples are consistent (assuming that the two input vectors are different) and any measurement noise or modelling error in the process will be stored, resulting in rapid parameter drift during the final stages of learning.

5.6.3 Example

In order to visualise the dynamical evolution of the weight vector, it is necessary to consider an AMN with only two adaptable parameters. Therefore a univariate linear B-spline network with only two basis functions and no interior knots is used to model the following linear single input, single output difference equation:

$$x(t+1) = 0.9x(t) + 0.1$$

initialised to be $x(0) = 0.01$. This generated a training set containing 50 data pairs that becomes increasingly correlated as x approaches 1.0. The value of x always lies in the interval $[0,1]$ hence the B-spline network realisation of the above difference equation is:

$$y(t) = (1 - x(t))w_1(t-1) + x(t)w_2(t-1)$$

and the optimal weight vector is $\widehat{\mathbf{w}} = (0.1, 1.0)^T$. A training store with 5 locations was used, and this was maintained by replacing the oldest piece of data that was

sufficiently correlated with the current sample, and if there were no similar input
vectors, the closest match was discarded. For this example, similarity was defined
by a correlation threshold set to be 7°. This is a very simple method for updating
the store and in large-scale optimisation problems *time* must be considered more
carefully; an area for future research.

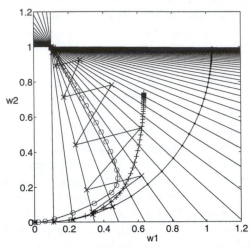

Figure 5.26   Solution hyperplanes and weight trajectories for the learning rules: $\cdot$ = NLMS,
$\circ$ = batch NLMS, + = partially optimised step length and $\times$ = maximum error.

The dynamic behaviour of the weight vector is shown in Figure 5.26, and these
results clearly show how using a training store can partially overcome the problem
of slow parameter convergence due to correlated training data. The NLMS rule
projects the weight vector orthogonally onto each of the solution hyperplanes,
which produces search directions nearly perpendicular to the desired one (pointing
directly at the optimal solution). Improved parameter convergence occurs when the
partially optimised step length algorithm is used, but the search direction is still
correlated with past values, so convergence is still quite slow. Training the weight
vector using the batch NLMS rule improves the rate of convergence and the weight
updates approximate the steepest descent path, as would be expected. Finally,
the maximum error training algorithm produces an erratic weight update path,
but parameter convergence occurs very quickly, confirming the results reported
by Parks and Militzer [1992]. For this learning rule, the maximum output error
corresponds to the most uncorrelated piece of training data in the store, hence the
search directions become increasingly perpendicular.

    This simulation is unrealistic in the sense that the network could model the
desired function exactly and no measurement noise exists, therefore the rate of
convergence depends only on how orthogonal the search directions are. In order to
consider other situations, the above problem was repeated and an additive uniform
noise signal in the range $[0, 0.1]$ was added to the process at each time-step. A linear

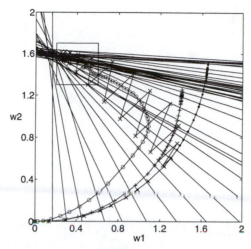

Figure 5.27 Solution hyperplanes and weight trajectories for the learning rules: · = NLMS, ○ = batch NLMS, + = partially optimised step length and × = maximum error. The process is subject to a random disturbance signal.

B-spline network with two weights was again used as the model, except that it was defined on the interval $[0, 1.6]$ to account for the noise perturbations. The solution hyperplanes and initial convergence characteristics of the higher order learning rules shown in Figure 5.27 are similar to the noise-free case, although close to the optimal solution the maximum training rule produces erratic weight updates in the minimal capture zone formed by the training samples in the store, whereas the batch NLMS learning algorithm filters out the measurement noise and parameter convergence is significantly smoother (see Figure 5.28).

These two simple examples are illustrative, in the sense that they show the importance of using orthogonal search directions and of using several pieces of data for calculating the step length. However, to assess the higher order learning algorithm's true performance, they must be applied to much higher dimensional real-world problems, the method for updating the store should be carefully considered and heuristics for estimating the store size must be developed. These are future research topics.

5.6.4   Immeasurable Disturbances

The learning rules which have been proposed in the previous sections are generally based on minimising MSE performance functions. This generally produces good results if the training examples contain unbiased Gaussian measurement noise or if the modelling errors are small and evenly distributed. When the measurement noise has a specific structure, however (additive step noise, for instance), these

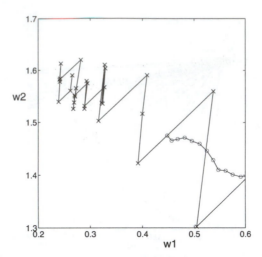

Figure 5.28   Weight trajectories for the batch NLMS (○) and the maximum error (×) rules when the weights are close to their optimal values. This forms an expanded view of Figure 5.27.

training rules produce biased models and do not learn the underlying structure of the desired function.

This section considers a modified learning rule for rejecting a specific type of *immeasurable* additive measurement noise or disturbance [Wang *et al.*, 1994a]. Denoting the disturbance signal at time t by $\eta(t)$, the dynamic behaviour of a nonlinear time series can be expressed as:

$$\tilde{y}(t) = f(\tilde{y}(t-1), \dots, \tilde{y}(t-n)) + \eta(t) \tag{5.62}$$

and it is assumed that the disturbance signal satisfies the following unforced, finite order, stable recurrence relationship:

$$g(\eta(t), \dots, \eta(t-L)) = \eta(t) + d_1\eta(t-1) + \cdots + d_L\eta(t-L) = 0 \tag{5.63}$$

where $\{d_i\}_{i=1}^{L}$ are the disturbance coefficients, L is a parameter which depends on the type of disturbance and the roots of the polynomial $\lambda^L + d_1\lambda^{L-1} + \cdots + d_L$ are called the modes. It is worth noting that the *size* of the disturbance is not required, it can be arbitrarily large, the only *a priori* knowledge assumed is its structure, in the form of the set of coefficients and the time-delay parameter L.

For example, when the disturbance is a step signal (of unknown magnitude), Equation 5.63 becomes:

$$g(\eta(t), \eta(t-1)) = \eta(t) - \eta(t-1) = 0 \tag{5.64}$$

Similarly if the immeasurable disturbance is a sinusoidal signal of unknown magnitude but known frequency, ω_0, it satisfies the following equation:

$$g(\eta(t), \eta(t-1), \eta(t-2)) = \eta(t) - 2\eta(t-1)\cos(\omega_0) + \eta(t-2) = 0 \tag{5.65}$$

The learning rules for such additive disturbance signals are now generated and in Wang *et al.* [1994] they are used to extract the underlying dynamics from a nonlinear time series which is subject to an additive *sinusoidal* signal of unknown magnitude but known frequency.

Performance Function

Using the notation just developed, consider training an AMN to reproduce the underlying dynamics of $f(.)$ if only $\tilde{y}(t)$ is measurable. Define the network output error as:

$$\epsilon_y(t) = (\hat{y}(t) - y(t))$$

where $\hat{y}(t) = \tilde{y}(t)$ and $y(t)$ is the network output at time t. If the AMN is trained using any of the rules proposed in this chapter, it learns to approximate the unknown function $f(.)$ plus the additive disturbance $\eta(.)$. Minimising the MSE performance function $E(\epsilon_y(t))$ does not generally result in the network being able to approximate the underlying dynamics of $f(.)$. Instead an alternative training strategy must be developed.

When the AMN can reproduce the desired function exactly, $y(t) = \tilde{y}(t) - \eta(t)$, $\forall\, t$, and Equation 5.63 becomes:

$$g(\epsilon_y(t), \ldots, \epsilon_y(t-L)) = \epsilon_y(t) + d_1\epsilon_y(t-1) + \cdots + d_L\epsilon_y(t-L) = 0 \qquad (5.66)$$

If the network output errors satisfy the above polynomial expression, it is able to model the underlying dynamics of the desired function successfully. Therefore consider the following performance function:

$$J_1 = E\left(g^2\left(\epsilon_y(t), \ldots, \epsilon_y(t-L)\right)\right) \qquad (5.67)$$

When $J_1 = 0$, the network can model the underlying dynamics exactly, otherwise it attempts to produce a MSE solution. The form of Equation 5.67 immediately suggests using gradient descent rules to train the AMN's weight vector, and these are now derived.

Training Rules

Instantaneous gradient descent rules attempt to minimise J_1, using an instantaneous estimate of the true performance function. Therefore:

$$\Delta\mathbf{w}(t-1) = -\frac{\delta}{2}\frac{\partial J_1}{\partial\mathbf{w}(t-1)}$$

where the weight vector has been held *constant* for the previous L training samples (or else the network output error has been recomputed for the previous L samples at time t). Now:

$$\frac{\partial J_1}{\partial \mathbf{w}} = -2g\left(\epsilon_y(t), \ldots, \epsilon_y(t-L)\right)\left(\mathbf{a}(t) + d_1\mathbf{a}(t-1) + \cdots + d_L\mathbf{a}(t-L)\right)$$

and combining both expressions gives a learning rule of the form:

$$\Delta\mathbf{w}(t-1) = \delta g\left(\epsilon_y(t), \ldots, \epsilon_y(t-L)\right)\left(\mathbf{a}(t) + d_1\mathbf{a}(t-1) + \cdots + d_L\mathbf{a}(t-L)\right) \quad (5.68)$$

This learning rule is equivalent to an Lth-order LMS training strategy and a similar NLMS training rule can be derived by dividing the right-hand side of Equation 5.68 by:

$$\|\mathbf{a}(t) + d_1\mathbf{a}(t-1) + \cdots + d_L\mathbf{a}(t-L)\|_2^2$$

However, the NLMS rule may have to be slightly modified in order to take into account the possibility of this *new* input vector being equal to zero and this is generally achieved by the addition of a positive term to the divisor $\|.\|_2^2$.

5.7   DISCUSSION

This chapter has derived, described and measured the suitability of using instantaneous learning laws for training AMNs. The class of AMNs includes the CMAC, B-splines and in Chapters 10 and 11 it is shown that these learning rules can also be used to train a certain class of fuzzy systems. It is worth emphasising this output equivalence between fuzzy and neural (B-spline) networks, as it allows not only the output of the fuzzy models to be analysed, but also convergence and the rate of convergence to be proved and measured, respectively. In the past, many self-organising fuzzy models and controllers have been simulated, but convergence has never been proved. The relationships established in these chapters and the theory described in this chapter allows the rate of convergence to be interpreted in terms of the type of basis (membership) functions and their condition.

The instantaneous gradient descent LMS rule and the error correction rule have both been shown to minimise an MSE performance function, and the instantaneous storage ability of the learning rules has been assessed. An instructive geometric interpretation of these learning algorithms, which relates their rate of convergence to the distribution of the input signals and the type of basis functions, has also been derived. Parameter convergence is the ultimate aim of these training rules and this insight shows how they adopt the minimal disturbance principle, and explains why parameter convergence is slow (fast) when the training signals are highly correlated (orthogonal). On-line learning generally produces correlated training data, and higher order learning algorithms (as described in Section 5.6) may have to be used to introduce some variety into the sequentially presented training samples.

Much of the theory described in this chapter has been aimed at showing how instantaneous gradient descent differs from the true algorithm described in Chapter 4. The concept of gradient noise has been described and its affect on the rate of parameter convergence during the transient and steady-state phases of learning has been illustrated. When there is modelling error and the instantaneous learning rules are used to train the networks, the parameters converge to a minimal capture zone (rather than to a unique value) whose size and shape depends on the condition of the data and the network. These ideas are also related to measuring the amount of learning interference which occurs in AMNs. All this work measures the deviation from some ideal training behaviour, and the amount of learning interference relates the incorrect parameter updates directly to the size and shape of the basis functions (fuzzy membership functions). Each piece of work provides insights into the compromises made when the basis functions are chosen to be sufficiently wide for good generalisation (interpolation/extrapolation) and low computational requirements, but also sufficiently narrow so that much of the training data are orthogonal.

Future research should be aimed at developing new higher order learning algorithms which take into account the network's structure that is being considered here: large-dimensional, sparse, non-negative input spaces, and possibly develop their ability to model time-varying plants [Gardner, 1987]. Another line of investigation which appears very fruitful is to view the LMS algorithm as a feedback control scheme, where the control gain is simply the learning rate, δ. The classical stability results can be derived using this approach, although it is also possible to use frequency-based design algorithms to improve the tracking ability at low frequencies [Dabis and Moir, 1991].

6

The CMAC Algorithm

6.1 INTRODUCTION

Intelligence within a network could be defined as its ability to learn from its past experiences, generalising the stored information so that it influences the network's output (response) when similar inputs (stimuli) are received. These two important factors:

- **potential** to learn from past interaction with the environment; and
- **ability** to store the new information locally

inspired Albus' development of the Cerebellar Model Articulation Controller (CMAC) in the mid-seventies, [Albus, 1975a, b]. The algorithm was based on Albus' understanding of the functioning of the cerebellum, but it was *not* proposed as a biologically plausible model. Albus [1975a] wrote

> *there is good reason to believe, however, that it may be possible to dupli-cate the* functional *properties of the brain's manipulator control system without necessarily modelling the* structural *characteristics of the neuronal substrate.*

In this respect, the CMAC is different from other neural networks, which attempt to *model* the characteristics of the individual neurons in order to endow the composite networks with the brain's functional properties, such as local generalisation and learning.

The CMAC was originally developed for adaptively controlling robotic manip-ulators *on-line*. This required an algorithm which could deal with large input spaces, operate in real-time and learn in a stable manner. The formulation of the CMAC reflected the necessity of an efficient implementation as well as a desire to model the functioning of the cerebellum (but not the individual neurons). This and Chapter 7 do not describe the biological theory behind the development of the CMAC as this can be found in Albus [1979a, b, c, 1989] and Tolle and Ersü [1992], rather they concentrate solely on the computational aspects of the CMAC that make it appealing for adaptive nonlinear modelling and control. It should also be emphasised that the description of the CMAC given in these two chapters is very different from that first described by Albus. This description is more geometric and gives the reader a greater understanding of the network's properties, although it can be verified that they are equivalent (see a later paper written by Albus [1979b], where a more geometric description is given).

In the past fifteen years since the CMAC was first proposed, a number of researchers have continued to use modified versions of the basic algorithm for adaptive modelling and control [Miller, 1993, Tolle and Ersü, 1992], signal processing and classification tasks [Miller *et al.*, 1990a, Miller *et al.*, 1993]. Indeed when DARPA [1988] performed a survey of the neural network field, the CMAC was one of the few algorithms for which successful real-world applications could be reported. These applications preceded the theory which explained how the CMAC generalises and adapts, which has only been thoroughly explained in the past few years, [Brown *et al.*, 1993, Ellison, 1988, Parks and Militzer, 1989, 1991a, 1992] (the first reference forms the basis for Chapter 7) and Chapter 5. The theory has confirmed many of the experimental results which indicated that the CMAC could initially learn very quickly and that the weights would tend towards their optimal values (given sufficient training examples). It can also be used to determine the class of nonlinear functions that the CMAC can model exactly and those functions which it is unable to model, as described in Chapter 7. Many theoretical and experimental results are now known about this network, although this knowledge is not widely distributed. To compensate, this chapter describes the basic CMAC algorithm, as well as several extensions, some applications are also reviewed and a particular software implementation scheme is described in Appendix C.

6.2 THE BASIC ALGORITHM

The CMAC algorithm is a descendant of Rosenblatt's Perceptron (Section 3.2.1), and is classed as a lattice AMN. Like the Perceptron, the CMAC can be decomposed into a fixed, nonlinear and an adaptive linear mapping, although there are some important differences:

- The network output is linearly dependent on the weight set, unlike the Perceptron, where the output is given by passing the weighted transformed input through a hard limiter, giving a binary output.
- The initial CMAC nonlinear map is determined by a set of parameters which are provided by the designer, unlike the Perceptron, where sparse, random connections are formed.

Another influence in the design of the CMAC was the adaptive look-up table. As described in Chapter 2, an adaptive look-up table is unable to distribute (generalise) the stored knowledge and the memory requirements are not feasible for large input spaces. However, the associated learning algorithm is extremely simple, it is temporally stable (due to the lack of generalisation) and convergence to a global minimum is guaranteed, even in the presence of random noise.

Like a look-up table, output information is stored in a small region of the network, although in a CMAC these regions are not disjoint, with similar inputs mapping to similar regions, while dissimilar inputs map to completely different regions of the network. The output is formed from a linear combination of nonlinearly transformed inputs, and it is this initial nonlinear mapping which determines the topology preserving features and the modelling abilities of the network.

6.2.1   Notation

The CMAC algorithm can be decomposed into two separate mappings. The first is a nonlinear, topology conserving transformation that maps the network's input into a higher dimensional space, in which only a small number of the variables have a non-zero output. Thus the CMAC produces a *sparse* internal representation of the input vector. The designer must specify a *generalisation parameter*, ρ, which determines the number of non-zero variables in the hidden layer, and also specifies the size of the network's internal region that influences its response.

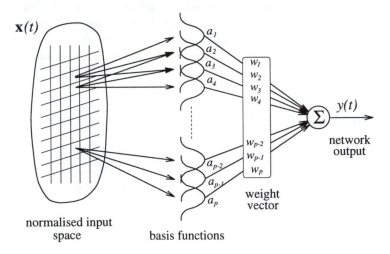

Figure 6.1 Schematic illustration of the basic CMAC algorithm.

The network's n-dimensional input vector is denoted by $\mathbf{x}$, and the network's sparse internal representation is denoted by the p-dimensional vector $\mathbf{a}$; this vector is called the *transformed input vector* or the *basis function output vector*. The transformed input vector, $\mathbf{a}$, has as elements the outputs of the basis functions in the hidden layer (see Figure 6.1) and the output, y, of the CMAC network is formed from a linear combination of these basis functions. The network output is therefore given by:

$$y(t) \quad = \quad \sum_{i=1}^{p} a_i(t) \, w_i(t-1) \tag{6.1}$$

$$= \quad \sum_{i=1}^{\rho} a_{ad(i)}(t) \, w_{ad(i)}(t-1) \tag{6.2}$$

where $ad(i)$ is a function which returns the address of the ith non-zero basis function. When the generalisation parameter is much less than the number of basis functions ($\rho \ll p$), Equation 6.2 illustrates the large reduction in the CMAC implementation cost, if there exists an efficient algorithm for generating the addresses of the non-zero basis functions, as the input space parameters satisfy the following relationship:

$$n < \rho \ll p \ll p'$$

where p' is the number of lattice cells.

The key to understanding how the CMAC operates is in the construction of the initial nonlinear map. The CMAC is a lattice AMN, hence the basis functions are defined on a lattice which the designer must specify (these basis functions are termed association cells by Albus [1975a, b]). Each basis function has an associated *support* (or receptive field), which is defined as the set of network inputs for which the output of the basis function is non-zero (see Figure 3.7). The CMAC algorithm operates by *sparsely* and *evenly* distributing these supports across the lattice, such that any network input lies in the supports of exactly ρ basis functions. This lattice and the set of basis functions determine the topology conserving features of the network (i.e. how the network generalises in n-dimensional space) and, in order to discuss its properties, a precise definition of the algorithm is now presented.

6.2.2   Partitioning of the Input Space

As well as supplying the generalisation parameter, ρ, the designer must also specify an n-dimensional lattice which *normalises* the input space. As explained in Section 3.2.3, this involves giving a minimum and a maximum value for each input, which generates the domain over which the CMAC is defined: a bounded hyperrectangle in n-dimensional space. A set of r_i knots lying between x_i^{min} and x_i^{max} must also be specified whose positions determine the sensitivity of the CMAC to particular regions of the input space. These knots generate the n-dimensional lattice in the input space. The lattice can also be generated by assuming that it is defined on the hypercube $[0, 1]^n$ (or equivalently $[-1, 1]^n$), all the interior knots are evenly spaced and a linear function is used to map the real values to the network's domain [Tolle and Ersü, 1992].

In either case, the lattice consists of $p' = \prod_{i=1}^{n}(r_i+1)$ n-dimensional cells or unit hypercubes. This discretisation provides a normalisation of the space on which the basis functions are defined, and generates a mapping from the original network input vectors, x, to a normalised input vector x'. In this new normalised input space, $\mathbf{X}' \subset \Re^n$, the position of a knot is simply its integer index and the reason for explaining this map is because it is often necessary to measure the distance between two input vectors in order to compare their similarity. Measuring the distance between two inputs in the original input space would not allow a true comparison to be made, although it is possible to generate a useful measure in the normalised input space.

As with look-up tables, the position of the knots is important in determining the ability of a CMAC to reproduce a function to within a desired tolerance. If the value of the desired function is changing rapidly over a certain domain, then more knots should be assigned to this region. Similarly, if the desired function changes very slowly over a certain domain, then the knots can be distributed so that only a small number are assigned to this region. The definition of the lattice influences the network's memory requirements, its modelling capabilities and the rate of convergence of the learning rules.

6.2.3   The Topology Conserving Map

The local generalisation which occurs in the CMAC is uniquely specified by the initial nonlinear mapping, as each *basis function* has associated with it a *support*, which is a hypercube of volume ρ^n defined on the lattice. Hence the generalisation parameter ρ not only specifies the number of basis functions that contribute to the network output, but also determines the *size* of their supports. As ρ increases, the size of the supports also increase and learning becomes *less local*. This is an important point, because the CMAC generates a sparse internal representation of the input lattice, where the number of basis functions is significantly less than the number of cells defined on the lattice ($p \ll p'$). When ρ increases, learning becomes less local and the modelling capabilities of the network generally decrease, as shown in Chapter 7.

By definition, the output of a basis function is non-zero if and only if the input lies in its hypercubic support. These hypercubes are distributed across the lattice, so that the supports of ρ basis functions cover each lattice cell. Thus the network's output is calculated by summing over only ρ weights. In order to ensure that exactly ρ supports cover each cell, a set of ρ *overlays* is formed, where each overlay consists of the union of *adjacent non-overlapping* supports. Each input lies in the support of one and only one basis function on each overlay, and overall the input lies in the supports of ρ basis functions (one on each overlay). For a one-dimensional CMAC with $\rho = 3$, this arrangement is shown in Figure 6.2, and the overlays are displaced by one cell relative to each other. A two-dimensional arrangement is also shown in Figure 6.3 where $\rho = 3$, and the supports are squares of size 3^2. For a multivariate CMAC, the overlays are displaced relative to each other in *every* dimension, and this is how the network sparsely encodes the input

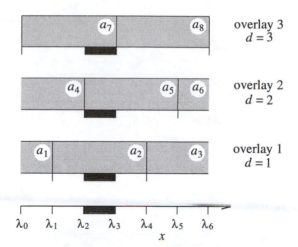

Figure 6.2 Univariate CMAC overlays with $\rho = 3$. Each overlay is composed of adjacent non-overlapping supports, and the input (which lies in the third interval) lies in the support of a_2, a_5 and a_7.

lattice. An n-dimensional integer *overlay displacement vector* $\mathbf{d}$ is generated and each overlay is displaced by an amount $\mathbf{d}$ relative to the previous overlay. The elements of the displacement vector satisfy $1 \leq d_i < \rho$, with d_i co-prime to ρ.

The major feature of this algorithm is that the number of basis functions, ρ, mapped to for a particular input does *not* depend on the dimension of the input space n. Thus the computational cost is *linearly* dependent on n, and although the memory requirements are exponentially dependent on n (see Section 6.5), memory hashing schemes can be used to provide a trade-off between approximation accuracy and available memory.

In order to achieve smooth, local generalisation, the overlays are displaced relative to each other, so that they always have a *uniform projection* onto each of the input axes. As an input moves parallel to one axis by one cell, the weight set mapped to has $(\rho - 1)$ weights in common with the previous weight set. Inputs which are more than $(\rho - 1)$ cells apart (with respect to the lattice and using an $\|.\|_\infty$ metric) have no weights in common and learning about either of these points will not affect the output of the network for the other; hence the CMAC generalises locally. This occurs because the algorithm ensures that each of the $(n-1)$-dimensional lattice hyperplanes, which passes through a knot on one of the axes and is parallel to the remaining $(n-1)$ input axes, forms part of one and only one of the overlays. The CMAC is said to obey the *uniform projection principle*: as the input moves one cell parallel to one input axis, the number of new weights mapped to is a constant, and for the CMAC this constant is 1.

Finally, a CMAC is said to be *well defined* if the generalisation parameter satisfies:

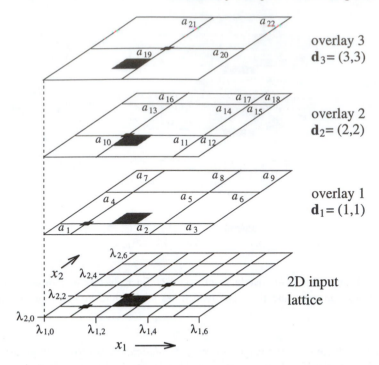

Figure 6.3   2-dimensional CMAC overlays with $\rho = 3$. The overlays are displaced relative to each other in *both* dimensions, and the input lies in the support of a_5, a_{11} and a_{19}.

$$1 \le \rho \le \max_{i=1}^{n}(r_i + 1)$$

Each basis function in a well defined CMAC makes a *relevant* contribution to the output. If this expression is not satisfied, there are two or more supports which completely cover the input lattice, and the corresponding binary basis functions are indistinguishable.

Generalisation in a Two-Dimensional CMAC

Consider the two-dimensional CMAC shown in Figure 6.3; the generalisation parameter is 3, 5 interior knots exist on each axis, the number of lattice cells is 36 and the network uses 22 basis functions to cover the input space. The 3 overlays, which are formed from the union of the non-overlapping supports, are successively displaced relative to each other one interval along each axis, and on each overlay the input lies in the support of one and only one basis function.

Each input to the binary CMAC activates a *unique* set of ρ ($= 3$) basis functions, and in order to understand how the CMAC generalises, it is instructive to consider the relationship between neighbouring sets of active basis functions. The two-dimensional input space is shown in Figure 6.4 and a reference input is denoted

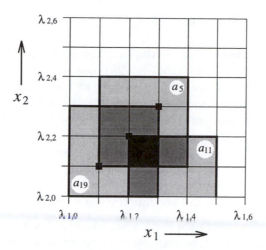

Figure 6.4 CMAC generalisation in a two-dimensional input space, where the generalisation parameter is 3. The black cell shows the input and is used to denote the cells which map to three common active basis functions. The dark grey, light grey and white areas show those cells which have two, one and zero active basis functions in common with the original input. The thick lines denote the support of the input's active basis functions.

by a black cell. Inputs which are close activate *similar* sets of basis functions, and the degree of generalisation is shown by plotting the number of common ones. The reference input is the only one which maps to the same three basis functions, so this cell is coloured black. Cells which are coloured dark and light grey share two and one active basis functions, respectively, with the original input, and cells which are blank map to totally different sets of active basis functions. If the weights are updated using the LMS learning rules, only those which contributed to the output are changed, and the network's response is modified only in the region corresponding to the union of the active supports (shaded area). Hence the CMAC stores and learns information locally. It might be supposed that the response regions should be almost circular, and while this is discussed further below, it is informative to investigate the placement of the supports in greater detail. As the input moves away from its original location parallel to any of the input axes, one common active basis function is dropped from the output calculation and inputs which are more than $(\rho - 1)$ cells (parallel to any axis) do not access any common basis functions. When inputs move diagonally away from the original cell, the pattern is less symmetrical, and this is due to the placement of the overlays. Overlay displacements which lie solely along the main diagonal (as occurs in the original Albus binary CMAC) may not be desirable.

Overlay Displacement Strategies

The problem of distributing the overlays on the lattice has been considered by a number of researchers. Lane *et al.* [1992] proposed to increase the number of basis functions mapped to for each input and to make this number depend on the input space dimension, which is an undesirable property. However, An *et al.* [1991] and Parks and Militzer [1991a] approached the problem by searching for more efficient ways of distributing the existing ρ overlays on the lattice. Once one corner of the overlay has been fixed, the overlay placement is uniquely defined. So the overlay displacement problem reduces to finding ρ displacement vectors (one for each overlay) in the first hypercube of dimension ρ^n, such that they obey the uniform projection principle and are "distributed evenly". This was defined as:

> *For points $\mathbf{x}'_k$ lying within a certain neighbourhood of a chosen point $\mathbf{x}'_j$ the Hamming distance between the corresponding binary transformed input vectors $\mathbf{a}_k$ and $\mathbf{a}_j$ should be linearly proportional to the Euclidean distance between the points $\mathbf{x}'_k$ and $\mathbf{x}'_j$.*

The original Albus scheme placed the overlay displacement vectors on the main diagonals of the lattice, so the displacements of the overlays are given by $(1, 1, \ldots, 1)$, $(2, 2, \ldots, 2)$, $\ldots$, $(\rho, \rho, \ldots, \rho)$, where each vector has n elements. This distribution obeys the uniform projection principle, but it lies solely on the main diagonal, so the network output may not be as smooth as can be achieved using other strategies. In Parks and Militzer [1991a], a vector of elements $(d_1, d_2, \ldots, d_n)$, where $1 \leq d_i < \rho$ and d_i and ρ are coprime, was computed, and the overlay displacement vectors are defined to be:

$$
\begin{aligned}
\mathbf{d}_1 &= (d_1, d_2, \ldots, d_n) \\
\mathbf{d}_2 &= (2d_1, 2d_2, \ldots, 2d_n) \\
&\vdots \\
\mathbf{d}_{\rho-1} &= ((\rho-1)d_1, (\rho-1)d_2, \ldots, (\rho-1)d_n) \\
\mathbf{d}_\rho &= (\rho, \rho, \ldots, \rho)
\end{aligned}
$$

where all the calculations are performed using modulo ρ arithmetic (with 0 being mapped to ρ on the final overlay). As d_i and ρ are coprime, the set $\{1, 2, \ldots, \rho\}$ is generated for each input using the above algorithm (d_i is a generator of this set) and the principle of uniform projection is preserved, as each $(n-1)$-dimensional hyperplane which generates the lattice lies in one and only one overlay. The original displacement vector is calculated using an exhaustive search which maximises the minimum distance between the supports, in order to ensure that the basis functions are evenly distributed throughout the lattice rather than only lying on the diagonals. Tables are given by Parks and Militzer [1991a, b] for $1 \leq n \leq 15$ and $1 \leq \rho \leq 100$, and these tables are reproduced in Appendix B. An *et al.* [1991] adopted a similar scheme, except that three heuristics are used in place of the exhaustive search to generate the original displacement vector.

Example of the Improved Overlay Displacement

In order to illustrate the improvement that results from using this generalised overlay displacement scheme, a CMAC network was trained to reproduce the desired
function $\hat{y}(x_1, x_2) = \sin(x_1) * \sin(x_2)$ over the input domain $[0°, 360°] \times [0°, 180°]$.
The generalisation parameter chosen was 5 and the knots were (uniformly) placed
20° apart on each axis. Multivariate basis functions were formed from univariate
"hat" basis functions, combined using the *product* operator, and their outputs were
normalised to form a partition of unity (all these terms are explained in the following sections). Figure 6.5 shows the network output when the original displacement
vector, $(1, 1)$, is used and Figure 6.6 shows the network output when the improved
displacement vector, $(1, 2)$, is used. This dramatic improvement is also reflected

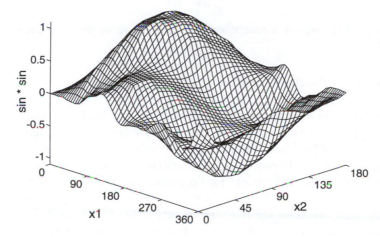

Figure 6.5   CMAC network output using the original overlay displacement scheme with d =
$(1, 1)$.

in the RMS error values after training: 0.101 and 0.015 and the peak absolute
errors: 0.276 and 0.042 for the original and improved overlay displacement algorithms, respectively. Using the improved displacement vectors results in nearly a
seven fold increase in the quality of the final model. The reasons for this dramatic
improvement are explained in Section 7.3.1 where it is shown that the unity displacement vector produces a network with large *diagonal* areas (of width ρ) where
the output is locally additive. The CMAC with the improved displacement vector
has *smaller* locally additive areas, which are more evenly spaced. Hence, the ridges
in Figure 6.5 do not appear in the improved network's surface.

   It would be incorrect to say that using this improved overlay displacement
scheme always produces a *better* model. Changing the displacement vector alters the basic CMAC model (as shown in Chapter 7), although this modification
generally produces a much *smoother* network output.

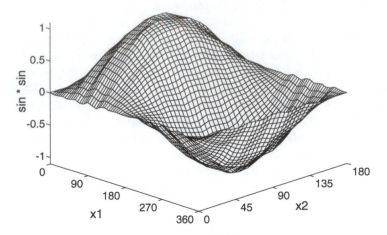

Figure 6.6   CMAC network output using the improved overlay displacement scheme with $\mathbf{d} = (1, 2)$.

6.2.4   Output of the Basis Function

Having defined the distribution of the supports on the lattice, it only remains to specify the nonlinear transformations that generate the basis functions' outputs.

The original CMAC used *binary* basis functions; if an input lies in the support of the basis function, its output is 1, otherwise it is 0. So the sum over all of the (unnormalised) basis functions for each input is a constant and equal to ρ, as the same number of basis functions are activated for each input. In this book, the binary basis functions are modified to output $1/\rho$ if the input lies in the corresponding support and 0 otherwise (see Figure 6.7). This simple modification is made so that the basis functions form a *partition of unity*, $\sum_{i=1}^{\rho} a_{ad(i)}(t) \equiv 1$, which is consistent with the nomenclature employed elsewhere in this book. The result of this modification is simply to scale each weight by a factor of ρ; it affects neither the output of the network nor the weight adaptation algorithms.

Using binary basis functions means that the network output is *piecewise con-*

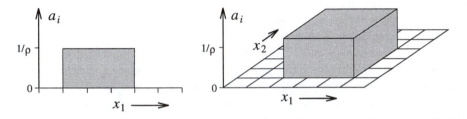

Figure 6.7 One and two-dimensional CMAC binary basis functions, $\rho = 3$.

stant. It has discontinuities along each $(n-1)$-dimensional hyperplane which passes through a knot lying on one of the axes, and is constant otherwise. While this is very similar to a look-up table, it is incorrect to think that the CMAC has the same modelling capabilities as a look-up table. When $\rho = 1$, the CMAC is a look-up table, with one overlay consisting of unit hypercubic supports; no generalisation occurs and the output of the network is simply the stored weight which corresponds to the single basis function with a non-zero output. When ρ increases, the number of basis functions required by the CMAC decreases (for $n > 1$) and the flexibility of the network also decreases. The theory that measures the flexibility of the CMAC network and describes which functions the CMAC can and cannot model is found in Chapter 7. Binary basis functions can only exactly reproduce piecewise constant functions and, while this may seem a very severe limitation, it should be noted that since the output of each basis function is equal, and the same number of basis functions contribute to the output, then $\|\mathbf{a}\|_1 \equiv 1$ and $\|\mathbf{a}\|_2^2 \equiv 1/\rho$. Therefore the LMS and the NLMS learning rule are *equivalent* when they are used to train the network, as the factor $\|\mathbf{a}\|_2^2$ is a constant and is simply absorbed into the learning rate δ. However, when trying to model continuous functions, the model mismatch between the CMAC and the desired function causes the weights to converge to the minimal capture zone, as described in Section 5.4.3, rather than to a fixed value. An obvious extension to the basic CMAC is to consider basis functions (defined on the same support) that generate a (piecewise) continuous output, and various algorithms are proposed and assessed in Section 6.4.

6.2.5 Memory Hashing

As with look-up tables, the memory requirements of the CMAC are exponentially dependent on n. In particular, for a multi-dimensional input ($n \geq 2$), the number of basis functions p can be approximated by (see Section 6.5.1):

$$p \approx \frac{\prod_{i=1}^{n} r_i}{\rho^{n-1}}$$

The number of basis functions is still considerably fewer than the number of cells in a look-up table ($p \ll p'$), although the exponential dependence on n means that the algorithm may only be suitable for small- to medium-sized input spaces, unless memory hashing is used. Memory hashing is a computer science technique which is used to store large, *sparse* virtual memories in a small physical memory. A pseudo-random mapping is constructed, whose input is the address of the original virtual location and whose output is the address in the physical memory, as shown in Figure 6.8. In this illustration it is possible to see the major problem with hashing techniques; information is overwritten when *collisions* occur. A collision happens when two distinct memory addresses in the virtual memory map to the *same* location in the physical memory, and because the virtual memory is much larger than the physical memory, collisions occur. In his original paper, Albus separated the collisions into two distinct categories:

- those occurring when the same physical memory address is accessed by two or more active virtual memory addresses for the **same** network input; and
- collisions occurring when the same physical memory address is accessed by two or more virtual memory addresses corresponding to **different** network inputs

and both of these cases are illustrated in Figure 6.8. In theory, by suitable choice of the size of the physical memory, the probability of a collision can be kept low. It is also possible to employ collision detection routines within the hashing algorithm, which detect if the physical memory address has already been accessed from a different virtual memory address. When this occurs, the original virtual memory address is modified and the process is repeated for a fixed number of iterations or until the correct (empty) physical memory address is found [Tolle and Ersü, 1992].

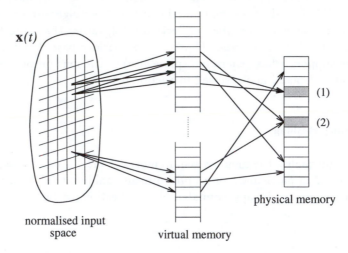

Figure 6.8   Memory hashing showing two different kinds of collisions (shaded memory locations); (1) illustrates a collision between two memory addresses mapped to by the *same* input, whereas (2) shows a collision between memory addresses mapped to by different inputs.

These hashing algorithms can only be used if the hashing address mapping is truly pseudo-random and also if the size of the physical memory is chosen appropriately [Miller *et al.*, 1987]. If the size of the physical memory is too small, then too many collisions (of both types) occur, the approximation ability of the network is poor and the computational cost increases, as the network iteratively tries to find the correct location in the physical memory. The requirement for the hashing address mapping to be truly pseudo-random is very important [Ellison, 1991], and efficient ones which have been found to perform well transform the virtual memory address into a binary word, and bit manipulation routines are used to scramble the words [Tolle and Ersü, 1992].

Finally, it is worth emphasising when hashing is appropriate. It works well when the number of accessed locations in the virtual memory *approximately equals* the size of the physical memory. This property may hold when the desired function is being approximated over only a small part of the input domain. Hashing should *not* be employed when attempting to perform global functional approximation (accessing every location in the virtual memory) if the size of the virtual memory is much greater than that of the physical memory. It is also *not* required when the size of the virtual memory is equal to (or less than) the physical memory size. Memory hashing is generally used when the network is learning on-line, although it introduces a form of temporal instability into the network, with the output being dependent on inputs which lie far away in the input space. If memory hashing is to be used in static networks, the global approximation ability of the network is degraded and the network must be carefully analysed to see how much the output is affected over the *whole* of the input space.

In the remainder of this chapter, the CMAC networks considered do not employ memory hashing, as this makes theoretical analysis and useful network comparisons very difficult. Generally, the performance of the studied networks provides an upper bound for the CMACs which use hash coding techniques.

6.2.6 Algorithm Summary

It is worthwhile summarising the basic CMAC algorithm and its various properties before discussing possible extensions and implementation issues.

The basic CMAC algorithm is decomposed into a nonlinear *static* mapping and an *adaptive* linear mapping. Therefore the output of the CMAC network is formed from a linear combination of basis functions. These basis functions, which are generated in the initial nonlinear topology conserving map, have the property that only a small number ρ are active at any one time, so only a small number of adaptive parameters contribute to the network output. Information is stored *locally* in this network. These basis functions are defined on the lattice supplied by the designer and the corresponding supports are hypercubes of size ρ^n, which are arranged on the lattice so that the *uniform projection* property holds. The *sparse* internal representation of the network means that instantaneous learning rules can be used to train the network very efficiently. Section 6.2.3 described an improved overlay displacement scheme. Whilst this was not part of the original algorithm, it distributes the hypercubic supports more evenly across the lattice and generally results in a smoother output. It is the authors' opinion that it should be integrated into the standard CMAC formulation, and so the optimal displacement tables have been reproduced in Appendix B. Finally, it should not be forgotten that the output of the basis functions and the network is piecewise constant, similar to a look-up table.

The main advantages which the CMAC has over a conventional look-up table are the fast initial learning and the huge memory reductions which are due to using a sparse set of basis functions that locally generalise. However, as is shown in Chapter 7, the modelling capabilities of the CMAC are generally not as flexible as those of a look-up table.

6.3 ADAPTATION STRATEGIES

The original CMAC algorithm has been shown to provide an efficient, tabular memory scheme that stores information and generalises locally. However, a network can only be used for on-line adaptive modelling if the training strategies are *appropriate* for the network. It is not sufficient to have a general-purpose network if it cannot efficiently adapt and store new information.

This section describes the relationship between the original Albus learning rule and those proposed in Chapter 5, and it is shown that the algorithm is simply an NLMS rule with an output dead-zone. The structure of the autocorrelation matrix is investigated and an upper bound for the number of non-zero elements is derived, which shows that it is *sparse*. Finally, a set of learning interference measures are derived for univariate, binary basis functions, which shows that as the generalisation parameter increases, learning becomes more difficult.

6.3.1 Albus Training Algorithm

The original binary CMAC training scheme proposed by Albus [1975b] can be expressed as:

$$\Delta \mathbf{w}(t-1) = \begin{cases} \delta \, \epsilon_y(t) \, \mathbf{a}(t) & \text{if } \epsilon_y(t) > \zeta \\ \mathbf{0} & \text{otherwise} \end{cases} \tag{6.3}$$

and only those weights which contributed to the output are updated, so knowledge is distributed *locally*. It can easily be seen that Equation 6.3 is simply the LMS learning rule derived in Chapter 5 with $\delta \equiv \rho$ (the inverse of $\|\mathbf{a}\|_2^2$ which is *constant*) and an output dead-zone. The learning algorithm is also equivalent to the NLMS rule (with output dead-zones) when $\delta \equiv 1$. This interpretation of the original learning algorithm allows many simple generalisations to be made, such as using the NLMS learning rule and choosing $\delta \in (0, 2)$, or having non-binary basis functions, or even using higher order learning algorithms for improved rates of convergence.

6.3.2 Autocorrelation Matrix Structure

The maximum number of non-zero elements per row in the autocorrelation matrix can easily be calculated. An element in the autocorrelation matrix is non-zero only if the supports of the corresponding basis functions have a non-null intersection. Consider the support of an arbitrary basis function. It has a non-null intersection with only one support on its own overlay (itself) and on each of the other $(\rho - 1)$ overlays, 2^n basis functions have a non-null intersection with it. Hence at most $(2^n(\rho - 1) + 1)$ non-zero elements exist in each row of the autocorrelation matrix. Both the solution matrix (which has ρ non-zero elements in each row) and the autocorrelation matrix are *sparse*.

6.3.3 Binary CMAC Learning Interference

The local learning which occurs in the CMAC is often cited as one of its most desirable features, as the network's initial rate of convergence is generally very fast. In this respect, it is desirable to have a fairly large generalisation parameter, because more weights are set during the initial stages of learning and the corresponding output MSE is in general smaller. As time increases this generalisation can cause *unlearning* in certain parts of the input space, when adaptation at one point causes a weight to move away from its optimal value. This phenomenon is called *learning interference* and as described in Section 5.5, it can be measured. For univariate basis functions of width ρ, a measure of the amount of learning interference is given by finding the largest root of:

$$\lambda^{\rho-1} + \delta \sum_{i=1}^{\rho-1} \lambda^{\rho-1-i} \left(\sum_{j=1}^{\rho-i} a_j a_{j+i} \right) = 0$$

For binary basis functions, the output of each active basis function a_j is $1/\rho$ and the learning rate is $\delta = \rho$. Therefore when $\rho = 2, \ldots, 5$, the equations become:

$$\lambda + \frac{1}{2} = 0$$

$$\lambda^2 + \frac{2}{3}\lambda + \frac{1}{3} = 0$$

$$\lambda^3 + \frac{3}{4}\lambda^2 + \frac{2}{4}\lambda + \frac{1}{4} = 0$$

$$\lambda^4 + \frac{4}{5}\lambda^3 + \frac{3}{5}\lambda^2 + \frac{2}{5}\lambda + \frac{1}{5} = 0$$

These equations can be solved exactly, giving the learning interference measures shown in Table 6.1. It is also possible to find an interval in which the maximal root lies and it has been shown by Pryce and Parks [1993] that the largest root $\lambda_{\max}$ satisfies:

$$\left(\frac{1}{2\rho + 1} \right)^{1/\rho} \leq |\lambda_{\max}| \leq 1$$

ρ	CMAC	ρ	CMAC
1	0.0	7	0.760
2	0.5	10	0.821
3	0.577	15	0.874
4	0.642	20	0.903
5	0.692	40	0.950

Table 6.1   Univariate CMAC learning interference measures. The values for $\rho > 5$ are generated using numerical methods.

Hence, as ρ increases, the lower bound tends to 1, as does the learning interference measure. When this occurs, the initial parameter error is *totally* propagated along the weight vector; learning is no longer local and correcting the output error for one input may cause a similar output error for a different network input.

Example: Rate of Convergence

These measures indicate that as ρ increases learning becomes less local and that for well structured problems, the rate of convergence will be slower. This is illustrated in Figure 6.9, where a univariate CMAC was trained to reproduce the discrete function $\sin(x)$ on the domain $[0°, 360°]$ with $r_1 = 35$. Each subinterval was 10°

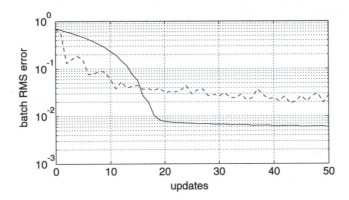

Figure 6.9   Learning histories for two univariate CMACs with the generalisation parameter, ρ, equal to 2 (solid line) and 15 (dashed line).

wide, the desired value of the function was obtained by sampling the function at the centre of each subinterval and a maximum training scheme (see Section 5.6.2) was used to train each network. Fast initial learning and slow long-term convergence can be seen when ρ is large, and when ρ is small the initial convergence is slower as fewer weights are being initialised, although long-term convergence is generally quicker due to the reduced amount of learning interference.

Higher order (tapered) basis functions (discussed in the next section) can reduce the amount of learning interference for basis functions of the same width, and these measures are calculated in Section 8.3.1.

6.4 HIGHER ORDER BASIS FUNCTIONS

In the original CMAC, the basis functions are binary; they are either *on* or *off*. The output of a basis function is $1/\rho$ if the input lies in its support and 0 otherwise. Therefore the output of the network is piecewise constant, as within each lattice cell the output is constant, and at a knot the output is discontinuous (analogous to a look-up table). While a piecewise constant approximation may be sufficient for some applications, a continuous network output (and higher orders if a continuous derivative needs to be calculated) is required in many systems, and so several researchers have proposed using non-binary basis functions in order to generate a continuous network output [An *et al.*, 1991, Brown and Harris, 1993, Lane *et al.*, 1992, Mischo, 1992, Moody, 1989]. These modified basis functions are defined on the same supports, ensuring that the network still generates a sparse encoding of the input space. For the network output to be continuous, the basis function output *must* be zero on the boundary of its support, and the method for calculating the output of these basis functions naturally partitions the proposed approaches into two categories; those which define a distance measure on the normalised input space, X', and pass this distance through a univariate, nonlinear function, and those which transform each input variable and combine these quantities. Both methods can be shown to reduce to the binary CMAC under certain conditions and they are both related to other networks:

- Defining a distance measure on the input space and nonlinearly transforming these quantities is similar to the approach used in radial basis function neural networks (see Section 3.3.4) and the CMAC sparse coding algorithm has been proposed as a method for positioning the centres [Mason and Parks, 1992].
- Passing each input variable through a univariate function and then combining the results is exactly what happens when multivariate B-splines and Gaussian radial basis functions are calculated (Chapters 8 and 10).

Both approaches are described in more detail in the following sections, although it should be stressed that most of the algorithms described give a satisfactory network output, and it is also required that *all* of the basis functions form a partition of unity. Another important property of the basis functions which are proposed is that their localised shape results in the learning being more local and the measured learning interference being lower.

When using higher order basis functions, it may be necessary to define an extra set of exterior knots on each input axis. Typically these extra knots are all equispaced or all defined to be equal to the input's maximum and minimum values, as discussed in Section 3.2.3.

6.4.1   Metric Basis Functions

The output of a metric basis function is calculated by measuring the distance between the input to the network and the centre of the corresponding support and passing this quantity through a nonlinear function. The output of the ith basis function is given by:

$$a_i(t) = f(\|\mathbf{c}'_i - \mathbf{x}'(t)\|)$$

where $\mathbf{c}'_i$ and $\mathbf{x}'(t)$ are the centre of the support of the ith basis function and the network input at time t, respectively, and are defined on the normalised input space X', $\|.\|$ represents the distance measure defined on the normalised input space and $f(.)$ is the univariate nonlinear function. Therefore, different choices of $\|.\|$ and $f(.)$ generate different basis function shapes.

Infinity Norm Basis Functions

When the infinity norm, $d = \|.\|_\infty$ is chosen and the univariate nonlinear function

$$f(d) = \begin{cases} 1/\rho & \text{if } d < \rho/2 \\ 0 & \text{otherwise} \end{cases}$$

is used, these metric basis functions are equivalent to the original binary basis functions. Another function which uses the $\|.\|_\infty$ distance measure is given by:

$$f(d) = \begin{cases} 1/\rho - 2d/\rho^2 & \text{if } d < \rho/2 \\ 0 & \text{otherwise} \end{cases}$$

This function is zero for inputs lying on the boundary of the support, has a value $1/\rho$ if the input is lying at the centre, and in one-dimensional space provides linear interpolation in between. For a two-dimensional input space, the basis function is a "pyramidal" shape, and for higher order basis functions, only changes in the input for which $\left|c'_{ij} - x'_{ij}\right|$ is the greatest affects the output of the basis function. This distance measure is appropriate to use, because the contour shapes (set of inputs for which the basis functions have the same output) *exactly* match the shape of the support, as shown in Figure 6.10.

Euclidean and One Norm Basis Functions

When the distance measure is the Euclidean norm $d = \|.\|_2$ and the nonlinear function is represented by:

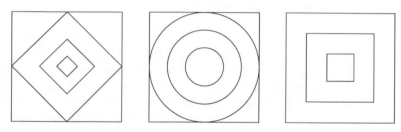

Figure 6.10 The supports and two-dimensional contour plots for metric basis functions. From left to right: $\|.\|_1$, $\|.\|_2$ and $\|.\|_\infty$.

$$f(d) = \exp\left(d^2/c\right)$$

for some positive constant c, the CMAC is similar to the Gaussian radial basis function network, when the centres of the basis functions are chosen to cover the input space sparsely in a regular fashion. The two algorithms are not strictly identical, because the output of the Gaussian function is not identically zero on the boundary of the hypercube. However, this effect can be made negligible by choosing c appropriately or by thresholding the basis functions so that they contribute to the output only if $f(d)$ is greater than some value α; this is similar to the α-cut used in fuzzy logic [Linkens and Shieh, 1992].

Another distance measure that can be used is the $\|.\|_1$, which sums $\left|c'_{ij} - x'_{ij}\right|$ for $j = 1, \ldots, n$ [Mischo, 1992]. This technique shares the same problem as the Euclidean norm; the contour shapes do *not* match the hypercubic shape of the supports, although the effect can be made arbitrarily small by choosing an appropriate nonlinear function.

Superspheres

The CMAC's basis function contour shapes must be hypercubic at the boundary of the supports, but it is desirable to have basis functions which depend on all the inputs at its centre. This can be achieved using a Euclidean distance measure at the centre, an infinity norm distance measure at the boundary and in between the measure is continuously deformed from $\|.\|_2 \to \|.\|_\infty$. This is termed a *supersphere* by An *et al.* [1991] and although it is an elegant solution to the problem, the computational cost makes the approach not feasible for real-time operation. In the next section, much simpler basis functions are developed which approximately satisfy this design constraint.

6.4.2   Combined Univariate Basis Functions

Another method for generating the membership of a multivariate basis function
is to combine the members of a set of univariate basis functions. The memberships of each variable in the univariate basis functions are determined, and the n
values combined using an appropriate operator to generate the membership of the
multivariate basis function. Therefore if the ith multivariate basis function a_i is
generated from the set of univariate basis functions $\{a_j^i\}_{j=1}^n$, it may be expressed
as:

$$a_i(\mathbf{x}(t)) = \bigwedge_{j=1}^n a_j^i(x_j(t)) \tag{6.4}$$

In this equation, the univariate and multivariate dependency has been made *explicit*
and the symbol chosen to represent the possible operators, $\wedge$, has been "borrowed"
from the fuzzy logic literature, where it is used to denote the fuzzy AND (see Figure 3.9). This notation was chosen because of the close links between this operator
and the fuzzy conjunction. For example, if the set of univariate basis functions,
$\{a_j^i(.)\}_{j=1}^n$, is used to define the univariate fuzzy membership functions and the *min*
operator is used to combine the univariate basis functions, Equation 6.4 represents
the most common method for implementing the fuzzy conjunction of several univariate fuzzy sets. Another operator which has been proposed to implement the
fuzzy AND is the *product* operator which has a close resemblance to the multivariate B-splines commonly found in surface fitting algorithms (see Section 8.2.4).
When the supports of the univariate basis functions are compact and their outputs
are zero at either end, the support of the resulting multivariate basis function is a
hypercube in n-dimensional space, and its output is zero at the boundary. Hence
these basis functions possess the desired properties.

Binary Multiplicative Functions

Before discussing the univariate basis function shapes in more detail, it is important
to show how the original binary basis functions are a special case of this more
general formulation. If the *min* operator is used and the univariate basis functions
are defined by:

$$a_j^i(x_j(t)) = \begin{cases} 1/\rho & \text{if } \left|c_{ij}' - x_j'(t)\right| < \rho/2 \\ 0 & \text{otherwise} \end{cases} \qquad \text{for each } j = 1, \ldots, n$$

this is equivalent to the original binary basis functions. Alternatively, if the *product*
operator is used, and the univariate basis functions are given by:

$$a_j^i(x_j(t)) = \begin{cases} \rho^{-1/n} & \text{if } \left|c_{ij}' - x_j'(t)\right| < \rho/2 \\ 0 & \text{otherwise} \end{cases} \qquad \text{for each } j = 1, \ldots, n$$

then this is also equivalent to the original binary basis functions.

B-spline Univariate Basis Functions

One important set of candidates for generating the univariate basis functions are B-splines [Brown and Harris, 1993, Moody, 1989], and the more general dilated B-splines [Lane *et al.*, 1992], both of which are fully described in Chapter 8. The basic univariate B-spline basis functions are piecewise polynomials of order k, whose supports are k units wide. At the edge of each basis function's support, its outputs and its first $(k-2)$ derivatives are zero, so across each knot the network's output is C^{k-2}. Hence smooth network surfaces can be produced. There exist efficient and numerically stable recurrence relationships for generating, differentiating and integrating the basis functions (given in Chapter 8), therefore they are suitable for real-time implementation.

The problem with using these basis functions in the CMAC is that their order is *directly* related to the width of the support. If the width of the support is large, the order of the basis function is also large, and the resulting set membership would be computationally expensive to evaluate (due to the high order). High-order networks also have a tendency to overfit the training data, and an algorithm for generating low-order, wide basis functions is required. This problem can be overcome using dilated B-splines [Lane *et al.*, 1992], which are generalised basis functions, where the width of the support is only loosely coupled to the order of the basis function (see Figure 6.11). It is assumed that the size of the support is

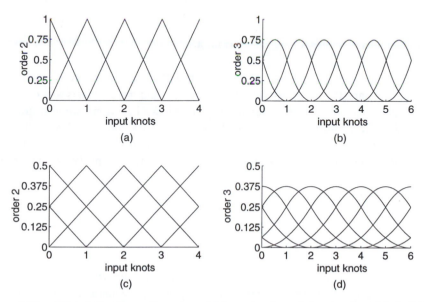

Figure 6.11   Dilated B-splines of varying widths and orders: (a) order 2, width 2; (b) order 3, width 3; (c) order 2, width 4; (d) order 3, width 6.

an *integer* multiple of the order of the basis function, i.e. $\rho = mk$ for some positive integer m. When $m = 1$ the original B-spline basis functions are recovered, and when $k = 1$ and $m = \rho$, the original CMAC binary basis functions are generated by the algorithm. Again, there are stable recurrence relationships for evaluating, differentiating and integrating the basis functions, as described in Chapter 8.

In graphical applications, the multivariate basis functions are formed by applying the *product* operator to each univariate basis function. This is because the *product* operator retains the smoothness of the univariate basis functions and also ensures that the normalisation (sum over all the basis functions is unity) still holds for regular multivariate B-splines. These properties are discussed in more detail in Chapter 8. The *min* operator has also been proposed, but this introduces discontinuities along the main diagonals of the multivariate basis functions. It should also be noted that the multivariate basis function formed from the minimum of identical, univariate dilated B-splines (defined on equispaced knots) is *equivalent* to a multivariate metric basis function, where the infinity norm is used and the univariate nonlinear function is the dilated B-spline. Hence, under certain conditions, there is an equivalence between these two kinds of basis functions.

Infinitely Differentiable Functions

Another basis function recently proposed is similar to the Gaussian function, but it is defined on a compact support [Werntges, 1993]. This is an important contribution as the basis function is *infinitely* differentiable, and is generated by:

$$a_j^i(x_j) = \begin{cases} \exp\left(-\frac{(\lambda_2-\lambda_1)^2/4}{(x_j-\lambda_1)(\lambda_2-x_j)}\right) & \text{if } x_j \in (\lambda_1, \lambda_2) \\ 0 & \text{otherwise} \end{cases}$$

when the support of a_{ij} is (λ_1, λ_2). This function has a maximum at $x_j = (\lambda_2 + \lambda_1)/2$, where its value is $\exp(-1)$, and if the univariate basis are combined using a tensor product, the resulting multivariate basis functions are also infinitely differentiable. A two-dimensional basis function and its contour plot is shown in Figure 6.12, and it is interesting to note that the contour shapes are almost circular (Euclidean norm) close to its centre and almost square (infinity norm) close to the edge of its support. Therefore these functions are similar to the superspheres previously described.

Example: Minimum and Product Operator

These points are illustrated in the following example, where a two-dimensional CMAC was trained to reproduce the function $\hat{y}(x_1, x_2) = \sin(x_1) * \sin(x_2)$ on the input domain $[0°, 360°] \times [0°, 180°]$. The interval between knots was 20° on each axis and the generalisation parameter was chosen to be 5. Hat functions (modified

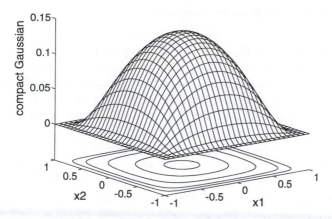

Figure 6.12   A two-dimensional Gaussian-type basis function defined on a compact support and its contour plot.

dilated B-splines of order 2) were defined on each axis, the output was normalised, which imposed a partition of unity on the network, and the weights were calculated using a numerically stable matrix inversion algorithm. The overlay displacement vector was $(1, 2)$, as recommended by Parks and Militzer [1991a] and obtained from Appendix B. In Figure 6.13 a *min* rule is used to combine the univariate basis

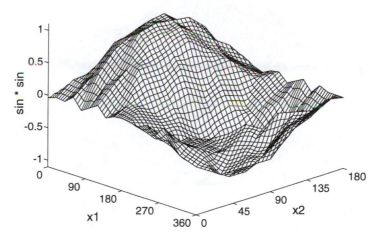

Figure 6.13   Using the *min* operator to combine univariate hat functions for a CMAC attempting to learn a two-dimensional sin $*$ sin function.

functions and in Figure 6.6 the *product* operator is used instead. The *min* operator produces derivative discontinuities in the multivariate basis functions and in the output of the network, whereas the *product* operator produces a much smoother surface. Both networks used a total of 57 weights.

6.4.3   Higher Order Basis Function Summary

Many different ways of extending the CMAC's original binary basis functions have
been proposed in recent years, although few guidelines have appeared on what is
actually useful. It is the authors' experience that low-order dilated B-splines (generally order 2 is sufficient), or hat functions, combined using the *product* operator
result in a smooth output surface with only a small increase in the computational
cost. The supports of these basis functions are the *same* as those of the original
binary basis functions and when the *product* operator is used, there are no derivative discontinuities within each of the cells, as occurs when the *min* operator or
an infinity norm-based metric basis function are used. Another important point
to note is that these (unnormalised) higher order basis functions no longer form
a partition of unity, although this can be overcome as described in Section 3.2.3
and this normalisation technique is *essential* when a smooth network output is
required. This is illustrated in Figure 6.14, for the output of an unnormalised two-
dimensional CMAC network that uses triangular basis functions combined using
the product operator. Finally, if higher order basis functions are used, the gener-

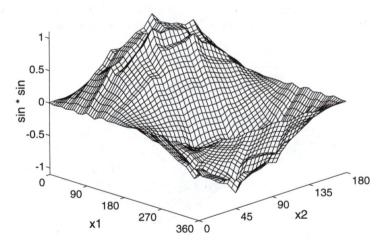

Figure 6.14   Using the *product* operator to combine univariate hat functions, without a
normalised output for a CMAC attempting to learn a two-dimensional sin * sin function.

alisation parameter must be chosen to be greater than the input space dimension
$(\rho > n)$. When this relationship is not satisfied, there exist points in the input
space for which the output of every basis function is zero. This occurs because
the normalised input lies on the boundary of *every* multivariate basis function, so
their outputs are all zero.

6.5 COMPUTATIONAL REQUIREMENTS

Expressions for the memory requirements and the arithmetic cost for a CMAC implementation are now derived. Hash coding is not considered in these networks; they are assumed to be multi-input, single-output and use the generalised overlay displacement algorithm described in Section 6.2.3.

6.5.1 Memory Size

For a multi-input, single-output CMAC network, the number of weights is simply equal to the number of basis functions, whereas for an m-dimensional network output, the number of weights that need to be stored is simply $m * p$. It is shown that for a multivariate CMAC, a substantial reduction in memory requirements is achieved compared to a look-up table, i.e. $p \ll p'$, and the relationship between p and ρ is described.

An expression for the number of basis functions can be *easily* calculated using the geometric construction which has been described in previous sections. The number of basis functions on each overlay is calculated, then p is simply the sum of the total number of basis functions on each overlay. The number of basis functions on each overlay is given by multiplying together the number of univariate intervals of length ρ on each of the lattice axes. On the ith input lattice axis, there are $(r_i + 1)$ unit intervals and one basis function exists before d_{ij}, where d_{ij} is the ith component of the displacement vector for the jth overlay. There remain $(r_i + 1 - d_{ij})$ unit intervals on the axis, so the number of intervals ρ units long that can be generated is:

$$\left\lceil \frac{(r_i + 1 - d_{ij})}{\rho} \right\rceil$$

where $\lceil . \rceil$ is the ceil function which rounds up to the nearest integer. So the total number of basis functions is given by:

$$p = \sum_{j=1}^{\rho} \prod_{i=1}^{n} \left(\left\lceil \frac{(r_i + 1 - d_{ij})}{\rho} \right\rceil + 1 \right) \tag{6.5}$$

If $n = 1$, Equation 6.5 reduces to:

$$p = r_1 + 1 + (\rho - 1)$$

but in most implementations, $r_i + 1$ is several times the size of ρ and $n > 1$, so Equation 6.5 approximates to:

$$p \approx \frac{\prod_{i=1}^{n} r_i}{\rho^{n-1}}$$

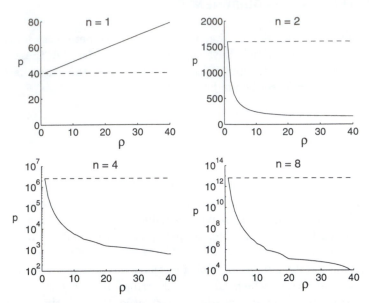

Figure 6.15   CMAC memory requirements for different-sized input spaces with 40 intervals on each axis. The dashed line shows the number of cells in the lattice; note the log scaling for $n = 4, 8$.

Hence doubling the size of the generalisation parameter ρ means that the memory requirements will be reduced by a factor of approximately 2^{n-1}. A set of graphs plotting p against ρ is given in Figure 6.15.

When $\rho = 1$, the CMAC is simply a look-up table with no generalisation occurring, but as ρ increases, the memory requirements *and* the modelling flexibility (number of weights) of the multivariate network are substantially reduced.

6.5.2   Arithmetic Operations

The computational cost of the original CMAC algorithm is dominated by the calculation of the addresses of the set of ρ non-zero basis functions. A floating point multiplication or division operation is denoted by float m & d, and a floating point addition or subtraction is represented by float a & s. A similar notation is used for integer operations.

The address calculation algorithm is similar to the expression for the number of basis functions, which has just been derived. The address calculation on each overlay is performed *separately*, and the address calculations on each overlay are *equivalent*, so consider the problem of calculating the address of the active basis function on the jth overlay. This addressing scheme is illustrated in Figures 6.2 and 6.3 and naturally extends to higher dimensional spaces.

Once the index of the subinterval in which the input lies, I_i, has been determined, the address of the active basis function on the jth overlay is denoted by $ad(j)$, and can be calculated from:

$$ad(j) = \sum_{i=2}^{n} \left(\left\lceil \frac{I_i - d_{ij}}{\rho} \right\rceil * \prod_{k=1}^{i-1} \left(\left\lceil \frac{r_k + 1 - d_{kj}}{\rho} \right\rceil + 1 \right) \right) + \left\lceil \frac{I_1 - d_{1j}}{\rho} \right\rceil$$

The arithmetic cost for calculating the ρ addresses is therefore:

$\rho(2n-1)$ int m & d
ρn int a & s

Once the ρ addresses have been calculated, the output is formed from the sum of ρ weights which (obviously) takes $\rho - 1$ float a & s, and dividing the output by ρ (which is equivalent to having each basis function output $1/\rho$) takes 1 float m & d. Therefore the total arithmetic cost of calculating the output of the binary CMAC (without memory hashing) is given by:

1   float m & d
$\rho - 1$ float a & s
$\rho(2n-1)$ int m & d   (6.6)
ρn int a & s

Hence, for a fixed generalisation parameter, the computational cost depends *linearly* on the input space dimension. Similarly, for a fixed input space dimension, the computational cost is *linearly* dependent on ρ.

The arithmetic cost of the various extensions to the basic algorithm are now considered. If the multivariate basis functions are formed from the product over each univariate basis functions, and the cost of finding the membership of a univariate input is k_1 float m & d and k_2 float a & s, the cost of forming the output of the ρ multivariate basis functions is given by:

$\rho(nk_1 + n - 1)$ float m & d
$\rho n k_2$ float a & s

The normalised output is formed using:

ρ float m & d
$2(\rho - 1)$ float a & s

which results in a total arithmetic cost of:

$\rho n(k_1 + 1)$ float m & d
$2(\rho - 1)$ float a & s
$\rho(2n - 1)$ int m & d   (6.7)
ρn int a & s

The arithmetic cost is still linear with respect to ρ and n, although the floating point multiplications and divisions are dominant for this implementation.

The original binary CMAC is computationally very cheap to implement; the address calculations use only integer operations and the output involves only a single floating point division. Introducing higher order basis functions results in a smoother output surface, but at a cost of making the number of floating point operations linearly dependent on ρ and n.

It has been assumed that the subintervals in which the input lies are already available, possibly from specific hardware. If this is not possible, a search must be performed for each input variable, to determine which intervals the input lies within. Constructing binary trees, the cost of performing the searches are $O\left(\sum_{i=1}^{n} \log_2(r_i + 1)\right)$, although this can be reduced using certain techniques (i.e. using the last input as the starting value, rather than the root).

6.6   NONLINEAR TIME SERIES MODELLING

In this section, the CMAC is required to learn the nonlinear dynamics of a simulated time series from a set of noisy observations. This modelling example is quite informative, as the true dynamics of the process are known (a globally stable limit cycle with an unstable equilibrium) and because other similar networks have been applied to the same benchmark problem, B-splines in Chapter 8 and RBF and MLP neural networks in [An *et al.*, 1993c, Chen *et al.*, 1992, Chen and Billings, 1994]. After training, the CMAC model is examined using a variety of techniques. Some of these are only applicable because the true solution is known, although general data modelling validation tests are also used. The chosen time series has only two inputs (smallest multivariable test possible), but this was specifically chosen because the network output surface can be plotted, which enables its smoothness to be assessed, and is informative, in that it reflects the type of multivariate basis function used.

6.6.1   The Nonlinear Time Series

Consider the two-input, single-output nonlinear difference equation given by:

$$
\begin{aligned}
y(t) = & \left(0.8 - 0.5\exp\left(-y^2(t-1)\right)\right) y(t-1) - \\
& \left(0.3 + 0.9\exp\left(-y^2(t-1)\right)\right) y(t-2) + 0.1\sin(\pi y(t-1)) + \eta(t) \quad (6.8)
\end{aligned}
$$

where $\eta(t)$ denotes the additive noise at time t. When $\eta(t) \equiv 0$, this difference equation has an unstable equilibrium at the origin and a globally attracting limit cycle as shown in Figure 6.16. The "surface" of the time series which the above difference equation represents is also shown in Figure 6.17. From this figure and

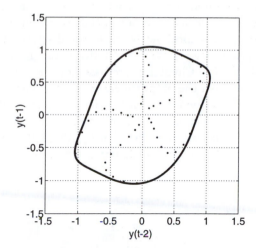

Figure 6.16 Iterated time series mapping from the initial condition $(0.1, 0.1)^T$. The time series "spirals" out from the unstable equilibrium at the origin towards the globally attracting limit cycle.

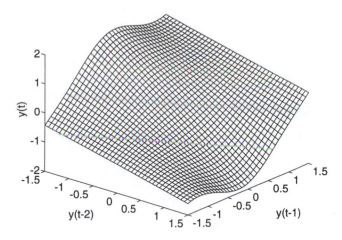

Figure 6.17 Time series output surface $y(t)$ v. $y(t-1) \times y(t-2)$, where $y(t)$ lies in the interval $[-1.872, 1.872]$.

Equation 6.8, it can clearly be seen that when $y(t-1)$ is held constant, $y(t)$ is *linear* in $y(t-2)$ and so the basic structure of the desired function with respect to the two inputs is different. The B-spline representation makes it possible to incorporate this kind of *a priori* knowledge into the network design, as shown in Section 8.6.

Defining the two-dimensional input vector $\mathbf{x}(t) = (y(t-1), y(t-2))^T$ and using $y(t)$ as the (scalar) desired output, a set of noisy training samples can be collected by iterating Equation 6.8 from an initial condition $\mathbf{x}(1)$. The data set then represents the noisy dynamical behaviour of the discrete time series.

The stochastic approximation LMS and NLMS learning algorithms for which convergence was established in Chapter 5, are based on the assumption that zero-mean Gaussian noise corrupts the output measurements. Therefore, let $\eta(t)$ be a zero-mean Gaussian white sequence which has a variance of 0.01. A set of noisy training samples from Equation 6.8 can be collected, and used to train a CMAC network. From an initial condition $\mathbf{x}(1) = (0,0)$, the 300 training inputs that represent the noisy iterated dynamics of Equation 6.8 are shown as a scatter plot in Figure 6.18.

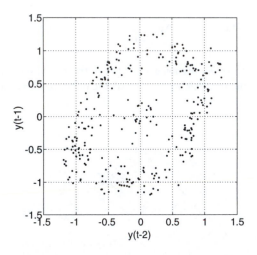

Figure 6.18   The noisy iterated dynamics of the time series from an initial position at the origin.

6.6.2   Network Design

The network's input domain is chosen to be the two-dimensional space $[-1.5, 1.5] \times [-1.5, 1.5]$, with 17 equispaced intervals defined on each axis ($r_1 = r_2 = 16$). The generalisation parameter is also selected to be $\rho = 17$, and the reason for choosing such a large value is that the desired output surface is smooth and the larger generalisation parameter generates a smaller network with a greater degree of noise filtering. Piecewise linear univariate basis functions (hat functions) are defined on each axis and these are combined using the *product* operator and the multivariate basis functions are normalised so that the network forms a partition of unity. The overlay displacement vector is set to $\mathbf{d} = (1,7)$, which is not the same as the "optimal" displacement vector ($\mathbf{d} = (1,4)$) proposed in Appendix B.

This is because the selected displacement vector gives a slightly smoother output surface, although the performance of a network which uses either is much better than that of a network using Albus' original displacement vector, $\mathbf{d} = (1,1)$.

Thus the network has a total of 65 weights, 60 of which are used in the time series prediction process. The weight vector is trained using an SANLMS rule with a learning rate of the form:

$$\delta_i(t) = \frac{1}{1 + t_i/30}$$

where t_i is the number of times that the ith basis function has been updated. The network was trained for 20 cycles and the learning convergence during the first training cycle is shown in Figure 6.19.

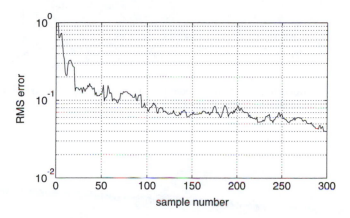

Figure 6.19 Learning history of the CMAC network during the first training cycle.

6.6.3 Network Evaluation

The trained CMAC model is evaluated using a variety of tests. The first scheme simply tests how well the network reproduces the training data. Denoting the network output error for the tth training sample as $\epsilon_y(t)$, the normalised output error autocorrelation function $\hat{\phi}(\tau)$ for the training set is:

$$\hat{\phi}(\tau) = \frac{\sum_{t=1+\tau}^{L} \left(\epsilon_y(t)\epsilon_y(t-\tau)\right)}{\sum_{t=1}^{L} \epsilon_y^2(t)} \tag{6.9}$$

where τ is the time lag and $L \ (= 300)$ is the training set size. When the network can reproduce the underlying function exactly, $\hat{\phi}(\tau)$ is equal to 1 if $\tau = 0$ and approximately 0 otherwise, and any modelling error results in the one step ahead prediction error being correlated with past values [Billings and Voon, 1986]. Generally, when the training set is large, the standard deviation of the correlation estimate is $1/\sqrt{L}$, and the 95% confidence limits are $\pm 1.96/\sqrt{L}$. The autocorrelation plot for the CMAC network after the 20th training cycle is shown in Figure 6.20, and the values lie within the 95% confidence bands.

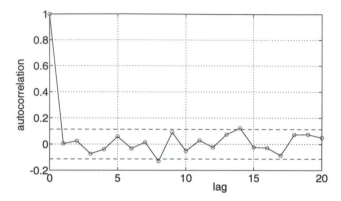

Figure 6.20   The autocorrelation of the prediction errors after the 20th training cycle. The 95% confidence band is shown by the dashed lines.

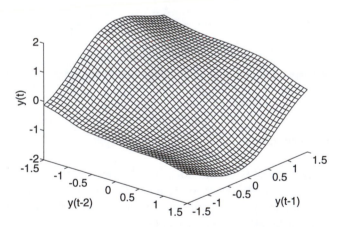

Figure 6.21   The CMAC network approximation to the true time series surface. The output lies in the interval $[-1.64, 1.74]$.

The output of the CMAC is plotted in Figure 6.21 and the piecewise linear nature of the basis functions is reflected in the network's surface. Many of the training examples lie around the attracting limit cycle, and the network's surface is very smooth about this area. However, little *a priori* knowledge is encoded in the design of the basis functions, so the network's ability to locally extrapolate is limited, especially about the points $(1.5, -1.5)$ and $(-1.5, 1.5)$.

Another method for assessing the network's performance is to iterate the CMAC network from an initial condition $\mathbf{x}(1) = (0.1, 0.1)$, then plot the resulting dynamical behaviour. At time t, the output of the network is calculated, and the input vector at the next time instant is set equal to $\mathbf{x}(t + 1) = (y(t), x_2(t))$. A scatter plot illustrating the network's dynamical behaviour is shown in Figure 6.22 and a good approximation to Figure 6.16 is clearly achieved.

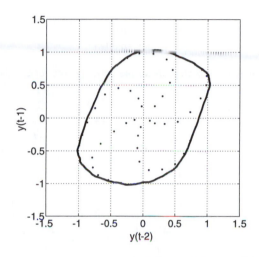

Figure 6.22 The iterated dynamical behaviour of the CMAC network from the initial condition $\mathbf{x}(1) = (0.1, 0.1)^T$. The basic shape of the limit cycle is a good approximation to the true one and the five armed interior spiral can be seen clearly.

The last two methods for assessing the network's performance are based on subjective visual comparisons with the true result. One test for which the exact information is not necessary (although it is used for this example) and which produces a quantitative answer is to measure the variance of a k-step-ahead prediction error for a given test set. For this time series prediction problem, the test set consisted of 900 noiseless iterated input/output points from an initial condition which lay on the limit cycle, and the prediction accuracy for up to 20 steps in the future was assessed, as shown in Figure 6.23. The CMAC performs well, and when a noisy test set is used for evaluating the network's performance, a similar result is obtained, except that the noise offset is higher. This result is confirmed by comparing the iterated network and true time history from an initial condition $\mathbf{x}(1) = (0.5, 1.0)^T$ for up to 50 steps into the future. These two plots are shown in Figure 6.24, and it can be seen that there is good agreement between the two time histories.

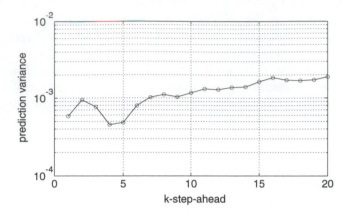

Figure 6.23 Variance of the k-step-ahead prediction errors for a noiseless test set.

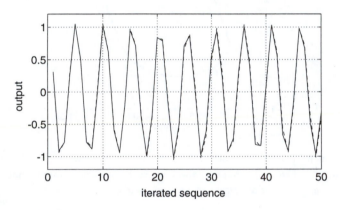

Figure 6.24   The CMAC and the true iterated mapping from an initial condition $x(1) = (0.5, 1.0)^T$. The solid line denotes the CMAC time history and the dashed line represents the true time history.

6.6.4   Time Series Conclusions

The CMAC network has been applied to the problem of nonlinear time series prediction and it has been demonstrated that it can model the training data well, filtering out the additive Gaussian noise and producing a model which closely approximates the desired dynamical behaviour. A variety of network tests have been used to validate the network's performance and *any* nonlinear learning algorithm should be subject to an extensive range of tests, once learning has ceased.

One of the CMAC's most desirable properties is that learning is local. However, in this example the generalisation parameter ρ was set equal to 17, so for each input approximately a quarter of the networks total weights contributed to the output.

The reason for this is that the CMAC was being used for *off-line* modelling and the most important criterion was the accuracy of the final model. Choosing such a large value of ρ results in a smoother (less flexible) network with fewer adjustable parameters, which is in keeping with the form of the desired function. Also the network's displacement vector was not the one recommended in Appendix B. The displacement vectors proposed in this table are generated by choosing the vector which best distributes the overlays on the lattice. This generally results in a smoother network output, although other displacement vectors may give slightly better performance measured with respect to the MSE. Despite this, the values contained in these tables generally improve the performance of the CMAC (see Section 6.2.3).

6.7 MODELLING AND CONTROL APPLICATIONS

The CMAC is a flexible learning module which can be used widely in intelligent control applications. In its basic form, it is a distributed look-up table which locally generalises learnt information, such that it affects the network's response for nearby inputs. The computational aspects of the algorithm also make it suitable for on-line training. Some of the applications which have been reported in the literature are reviewed, and the reasons why this algorithm may be preferred over others are discussed.

6.7.1 Process Control

Since the late seventies, a research group at the Technische Hochschul, Darmstadt, Germany have been actively engaged in research projects which have investigated both the theoretical and practical aspects of CMAC networks applied in on-line learning control loops [Tolle and Ersü, 1992]. These learning controllers have been applied to a wide range of different plants: process control, air-conditioning, robotic grippers and collision avoidance path planning [Tolle *et al.*, 1994]. However, only the general learning control loop architecture will be described.

In order to make a learning control loop *intelligent*, Ersü identified four key elements that must be included:

- **performance function** for evaluating the state of the process;
- **learning** to construct a predictive model;
- **exploration** of different control strategies; and
- **remember** previous control actions.

These ideas were translated into a learning control architecture shown in Figure 6.25, that was called LERNAS. The detailed operation of this structure can be found in Tolle and Ersü [1992], although its basic operation is now summarised.

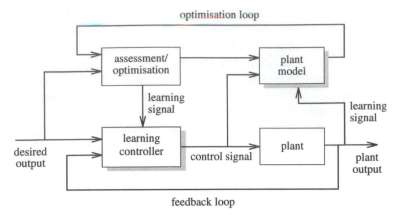

Figure 6.25   The LERNAS predictive control architecture where the learning modules have grey shadows.

The overall aim of the learning control elements is to generate the best possible control strategy which can track the desired response of the system. The basic optimisation strategies plan only one step ahead, although once the process model is sufficiently accurate, multi-step ahead optimisation procedures can be adopted if the computation is feasible. Optimisation routines search the space of possible control actions, evaluating the performance on the process model. Once a satisfactory control action has been calculated, it is applied to the plant and used to train the learning control module, and the true plant output can be used to adapt the plant model. At the beginning of the next optimisation calculation, the output of the control module can be used as an initial estimate of the control action in the optimisation routines.

The CMAC is suitable for use within this learning control loop because it initially learns quickly and in a stable fashion. The learning module must adapt locally, so that prolonged training in one part of the input space (regulating a plant to a steady-state for instance) does not corrupt the knowledge stored in a different region. The process industry is an ideal application area for this control architecture because of the severe nonlinearities and the long control time-scales.

A computationally simpler predictive control loop, called MINILERNAS, has also been proposed by Ersü [Tolle and Ersü, 1992]. Instead of forming a plant model, an AMN is trained to produce the cost of a particular control action for the current plant state, learning whether a certain situation is advantageous. This both reduces the memory requirements and the computational cost of the overall control loop, as a plant model is no longer formed.

6.7.2 Robotics Control

Since the mid-eighties, Miller and his research group at the University of New Hampshire have used the CMAC to control, amongst other things, multi-degree of freedom robotic manipulators [Miller, 1987, 1989, Miller *et al.*, 1990b]. The learning network is run in parallel with a predesigned, linear feedback controller, which stabilises the plant during the initial learning stages and provides a training signal for the CMAC. This control architecture is shown in Figure 6.26.

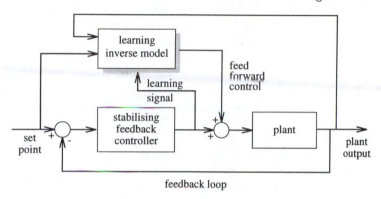

Figure 6.26 A stabilising linear feedback learning control architecture.

This control system has been used to control real five degree of freedom robotic manipulators in real-time. The use of a linear feedback controller to generate the training signal restricts the type of plants for which this approach can be used, although the initial rate of convergence is generally very quick and greatly reduces the computational cost of implementing the control loop. A CMAC network has been implemented in hardware using programmable CMOS logic cell arrays and response times of below 1 millisecond were achieved for a 32 input, 8 output system with up to 1 million weights available for memory hashing [Miller *et al.*, 1991].

This work is motivated by the desire to develop general purpose learning control schemes which can be used to control complex plants and require a minimal amount of *a priori* information. The stable, local learning of the CMAC, coupled with its low computational cost and local generalisation make it eminently suitable for this learning control task. Kraft and Campagna [1990] compared this learning control scheme with two traditional adaptive linear control algorithms: a self-tuning regulator and a Lyapunov model reference adaptive controller. Not surprisingly, the linear adaptive controllers performed best when the plant was *linear*, although when a significant nonlinear term was added, the CMAC performed best. The CMAC's learning convergence was the slowest, and this is due to the fact that little is assumed about the structure of the desired controller. The more *a priori* knowledge assumed about the plant, the faster the rate of convergence, although the control scheme is correspondingly less flexible; another manifestation of the bias/variance dilemma described in Section 2.5.1.

6.7.3   Reinforcement Learning

When reinforcement learning systems were first investigated [Barto *et al.*, 1983], the input space was assumed to be divided up into a number of lattice cells, and a parameter was associated with each bin. This is equivalent to a look-up table approach, and it is therefore surprising that it was not until 1991 that the CMAC was used for reinforcement learning control [Lin and Kim, 1991]. The local generalisation and learning which occur in this network increase the initial rate of convergence of the adaptive search element and the adaptive critic element, which in turn improves the performance of the overall control scheme. Also the learning algorithm is a *direct* generalisation of the conventional associative reward-penalty training scheme. A similar implementation has also been used to design a control algorithm which can reverse a standard trailer truck [Shelton and Peterson, 1992]. The low computational cost of the calculations make it feasible to train the CMAC modules *on-line*.

Watkins [1989] also used the CMAC to coarse code the world model used in a planning system that was represented by a grid. The coarse coding strategy allowed the information to be locally distributed and improved the performance of the system. However, the CMAC has to learn discontinuities in the surface (obstacles) and apart from some very trivial examples, the network has difficulties in performing this task (see Chapter 7). Prescott and Mayhew [1992] have successfully applied the CMAC to a similar reinforcement learning problem, although Dayan [1991] proposed a more intelligent strategy that allocated resources where the planning algorithm needed them. This motivates the hypothesis that the original coarse coding algorithm may not be appropriate for all cases.

6.7.4   Other Applications

There are many other ways in which the CMAC has been included in a learning control loop. In Ersü and Wienand [1986], the CMAC is trained to output two linear control gains: proportional and integral, and a predictive optimisation strategy similar to LERNAS is used to search the space of linear control gains. The CMAC is taught to emulate a rule-based controller in Handelman *et al.* [1990]. For this application, the network learns to swing a two-degree of freedom link manipulator to try and hit a baseball. A static CMAC network has also been used to synthesise a set of PID control parameters from a set of labels (rise time, peak overshoot, etc.) which describe the closed-loop response of the system [Lawrence and Harris, 1992]. This network can also be used higher up in the control hierarchy, for robotic manipulator trajectory planning [El-Zorkany *et al.*, 1985], and coordination tasks [Ersü and Tolle, 1988].

6.8 CONCLUSIONS

The CMAC was originally developed as an adaptive look-up table which locally generalises, and was proposed for on-line learning manipulator control. It models the high-level functionality of human reasoning, storing information locally, with similar inputs activating topologically similar regions of the network. However, its suitability for modelling and control applications depends on the performance of the algorithm; its modelling capabilities, the efficiency of the learning algorithm and the implementation cost. It has been shown to be a very flexible network with many desirable attributes for on-line, nonlinear, learning control tasks.

The CMAC network requires a number of parameters to be specified by the designer; the position of the knots, which in turn generates an n-dimensional lattice, the generalisation parameter and the shape of the basis functions. These parameters determine the modelling capabilities of the network, and a bad choice results in the network performing poorly. Fortunately the performance of the network appears to be robust with respect to most sensible choices, although it should be ensured that the network is *not* over-parameterised (high variance). The relationship of the network's output to the parameter set is very complex and is explored further in Chapter 7.

This relationship may be very complicated, however the static nonlinear map where the input is transformed to a higher dimensional input space simplifies the adaptation process. The generalisation parameter determines the number of non-zero basis functions for any input and only loosely depends on the dimension of the input space. Thus the computational cost of the algorithm (response and learning) is *not* exponentially dependent on the input space dimension. The basic instantaneous learning task occurs in a *sparse linear* space, and one of the problems with applying lattice AMNs to medium- and high-dimensional spaces is that the number of active basis functions, and hence the number of adapted weights, depends exponentially on the input space dimension. This does not occur in the CMAC network, although the total number of basis functions is exponentially dependent on the input space dimension.

There is a wide range of possible applications, from nonlinear learning modelling and control modules to static networks which have been trained to reproduce a human's response. For on-line modelling and control applications, there is no reason for using these algorithms if the plant is linear. The simplest possible control strategy should always be employed and it would be more "intelligent" to employ classical (adaptive) control techniques in this case. For nonlinear adaptive modelling, the algorithm has many desirable features: fast initial learning, long-term convergence, flexible modelling capabilities. Theoretical work on the CMAC has focused on the basic modelling capabilities of the network and its associated learning laws. The latter topic is generic to all the associative memory networks (see Chapter 5), although it is extremely important to understand what type of functions the basic network is capable of modelling exactly, and what influence its parameters (generalisation parameter, displacement vector and knot vector) has on the network's output. It is these features which distinguish the CMAC from the other AMNs and they are considered in Chapter 7.

7

The Modelling Capabilities of the Binary CMAC

7.1 MODELLING AND GENERALISATION IN THE BINARY CMAC

Chapter 6 described the CMAC algorithm and reported on some of the many successful applications. This network is often used because it is believed to be a general nonlinear approximation algorithm and because of its apparent rapid initial rate of convergence and long-term convergence properties, although none of this had been rigorously established. As with adaptive control in the fifties and sixties, practical applications have preceded the theory and in Albus' first papers [1975a, b], he admitted that "at the present time there exists no formal proof of convergence of the procedure". The modelling capabilities of many of the neural network algorithms are also poorly understood. Recent theoretical results [Cotter, 1990, Ellacott, 1994, Girosi and Poggio, 1990, Hornik *et al.*, 1989] have established that some networks are universal nonlinear approximators; given potentially infinite resources, they can approximate *any* continuous nonlinear function arbitrarily closely. However, in practice the network designer takes a fixed structured network and then has to resolve the important questions:

> *For a particular network structure, what functions can the network learn to reproduce* exactly, *and which mappings are the network* unable *to learn?*

For an MLP, the structure of the network is given by the number of layers in the network, the number of nodes in each layer and the internode connections. The structure of the binary CMAC is specified by the generalisation parameter, the position of the knots and the overlay displacement vectors; other parameters could include the size of the physical memory, type of basis function, etc. In this context the above questions could be rewritten as:

> *For a particular network structure, is there a weight vector which enables the network to reproduce a particular data set exactly, and for what type of (non-zero) functions is the optimal network output identically zero?*

It is important to answer these questions, because neural networks are *soft* modelling algorithms, and it is necessary to understand what assumptions have been made before a complete theoretical analysis of the learning properties can be performed. The network's *structure* determines how it *generalises* in n-dimensional space.

Therefore, theoretical research into the CMAC has been directed towards the following two areas:

- **investigating** parameter convergence and deriving new learning algorithms; and
- **establishing** the modelling capabilities of the network and analysing how it generalises in n-dimensional space.

The convergence of the CMAC weight vector has been analysed by a number of authors in recent years [Ellison, 1988, Parks and Militzer, 1989, 1992], and the main results are presented in Chapter 5 and Section 6.3. Other contributions, such as Ellison [1991] and Wong and Sideris [1992], have analytical flaws, due to incorrect assumptions which have been made about the modelling capabilities of the multivariate network, and this chapter addresses this topic.

Many different results can be generated about the binary CMAC's modelling capabilities. In Section 7.2, a theorem is derived which shows that its modelling ability depends on the number of basis functions, which in turn depends on the generalisation parameter. For a multivariate CMAC, the number of basis functions is generally considerably less than the number of lattice cells in a look-up table and so there must exist multivariate look-up tables which the network *cannot* exactly model. An expression is given for the dimension of the range of the CMAC, and it is shown that when the CMAC generalises ($\rho > 1$), the solution and autocorrelation matrices are always singular. Thus there exist an *infinite* number of optimal solutions to the weight optimisation problem.

In Section 7.3, a set of simple relationships is derived which explains how the CMAC generalises in n-dimensional space. These *consistency equations* must be satisfied for the CMAC to reproduce the data set exactly, and they are complete, in the sense that any look-up table which satisfies these relationships can be stored exactly by the network. They explain how the CMAC produces locally additive models, and allows the modelling abilities of different CMACs to be compared. In particular, networks with different generalisation parameters and overlay displacement vectors cannot model the same type of functions. It then follows that the *only* functions which an arbitrary multivariate CMAC can reproduce are the set of additive look-up tables.

A set of local *orthogonal functions* is then derived which the CMAC is unable to model in a least-squares sense. It is important to investigate the class of orthogonal functions, because any mapping can be *uniquely* decomposed into two parts: one which the optimal network can reproduce exactly and another which it is completely unable to model in a least-squares sense. Thus when a data set depends on these orthogonal functions, the network is not able to model the data set exactly.

The consistency equations are local relationships which must be satisfied by the training data, and the orthogonal functions are local mappings which the network cannot model. These local orthogonal functions are also complete, as any orthogonal function can be expressed as a linear combination of these mappings. This is illustrated when a *multiplicative* function is also shown to be a local orthogonal function, proving that there exist simple mappings which the CMAC is unable to reproduce.

Thus, any look-up table can be uniquely decomposed into one part which satisfies the consistency equations and one formed from a linear combination of the local orthogonal functions. This interpretation enables a simple and cheap lower bound of the CMAC's output error to be calculated *without* finding the optimal weight vector, using only local operations. It relates the output error to the local variations in the training data, and provides a firm theoretical basis for the knot placement design strategy, which sparsely populates areas that have low output variations and densely populates regions possessing large local variations. The lower bound for the modelling error is then used to analyse how well the CMAC can model a general multiplicative function, and it is shown that it is unable to model *any* non-trivial multiplicative function exactly. This result is not surprising, since the CMAC produces a locally additive model when it generalises.

Finally, the accuracy of the coarse coding mapping is investigated. A well-known result is given, which expresses the accuracy of the coarse coding map in terms of the memory requirements and the width of the basis functions, and an interpretation is presented for the CMAC. This expression shows that the accuracy of the coarse coding does *not* depend on the generalisation parameter, which is slightly surprising because, as the generalisation parameter increases, the internal representation of the input is distributed amongst a greater number of cells, and it might be thought that an internal error would have less of an effect when the original input was reconstructed. However, the CMAC's internal representation is not *truly* distributed, and any loss of information causes two or more inputs to be indistinguishable.

7.2 MEASURING THE FLEXIBILITY OF THE BINARY CMAC

The original binary CMAC, described in Section 6.2, is formed from a linear combination of binary basis functions defined on a lattice. Therefore, the range of functions that can be reproduced (or modelled exactly) forms a subset of the space of look-up tables defined on the same lattice. It is instructive to consider whether the CMAC has the same modelling flexibility as a look-up table (this is trivial for the case $\rho = 1$, as the CMAC *is* a look-up table) and if not what type of look-up tables can be realised. The modelling capabilities of the network depend on the number, position and size of its basis functions, and in Section 6.5.1, an expression for the number of basis functions, p, is given and its dependency on ρ is examined.

The positioning of basis functions' supports on the overlays, and positioning of the overlays on the lattice was discussed in Section 6.2.3. Both these aspects of the CMAC algorithm are used to generate a measure of its modelling capabilities.

It has already been established that the number of basis functions in a multi-variate CMAC is generally a lot less than the number of lattice cells, p', each of which contains a single training pair. Therefore using elementary matrix algebra, the flexibility of the CMAC is bounded above by the number of basis functions, p, and the overall network cannot reproduce an arbitrary look-up table simply because $p < p'$.

In this section, it is shown that the dimension of the set of p' binary basis function vectors (transformed input vectors) of length p is $p - (\rho - 1)$. Thus when the basis functions generalise between cells defined on the lattice ($\rho > 1$), the dimension of the space spanned by the basis functions is strictly less than the number of basis functions. The result can then be used to find a sharp upper bound for the rank of the solution and the autocorrelation matrices which shows that they are always singular for $\rho > 1$. This should not be surprising since it was shown by Powell [1987] that although radial basis function matrices are always non-singular, infinity norm basis function matrices can be singular.

7.2.1 The Dimension of the Set of Basis Function Vectors

Theorem 7.1 The dimension of the set of p' binary basis function vectors (trans-formed input vectors) of length p of a well defined CMAC is $p - (\rho - 1)$.

Proof. Initially consider the case when $\rho = 1$. The CMAC is simply a look-up table and the output is identically zero if and only if all the p weights are zero. Hence the theorem holds for $\rho = 1$.

Now consider the case of $\rho > 1$. Theorem 7.1 will be proved using the fact that the dimension of a set of basis function vectors is equal to p—*the dimension of its null space*. Hence all that needs to be found is the dimension of the null space of the basis function vectors. This is achieved by dividing the weight vector of the CMAC $\mathbf{w}$ into ρ sets, denoted by $\mathbf{w}^1$, $\mathbf{w}^2$, ..., $\mathbf{w}^\rho$, where each set consists of those weights whose supports lie on the same overlay (ρ overlays). Therefore:

$$\mathbf{w} = \bigcup_{j=1}^{\rho} \mathbf{w}^j \quad \text{and} \quad \mathbf{w}^i \cap \mathbf{w}^j = \emptyset \quad \forall i \neq j$$

Now for $j = 1, 2, \ldots, \rho - 1$ assign:

$$w_i^j = c^j \qquad \text{for each } i \tag{7.1}$$

where c^j is an arbitrary constant. Then set:

$$w_i^\rho = -\sum_{j=1}^{\rho-1} c^j \qquad \text{for each } i \tag{7.2}$$

as shown in Figure 7.1. Every input maps to one and only one basis function on each overlay, and the network output is formed from the (scaled by $1/\rho$) sum of each of the corresponding weights. By Equations 7.1 and 7.2, the output of the network is identically zero, that is, a set of $(\rho - 1)$ non-zero coefficients has been constructed such that the summed output of the ρ basis functions is identically zero. Hence the dimension of the null space is $\geq (\rho - 1)$.

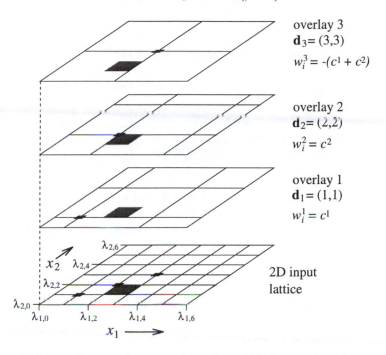

overlay 3
$\mathbf{d}_3 = (3,3)$
$w_i^3 = -(c^1 + c^2)$

overlay 2
$\mathbf{d}_2 = (2,2)$
$w_i^2 = c^2$

overlay 1
$\mathbf{d}_1 = (1,1)$
$w_i^1 = c^1$

2D input
lattice

Figure 7.1   Weight assignment for a two-dimensional CMAC network with $\rho = 3$. This (non-zero) weight set ensures that the network output is identically zero.

Now it is shown that when the CMAC's output is identically zero, every weight whose support lies on the same overlay must be equal. Therefore at most $(\rho - 1)$ parameters can be chosen arbitrarily. If the output is zero for a particular input, moving one cell parallel to one axis means that one new weight is introduced into the output calculation. Since the output is formed from a linear combination of weights, if one of the weights is changed and the output is the same, the new weight must be equal to the old weight. Since the new weight and the old weight lie on the same overlay and because the whole input space can be paved by moving in this manner, every weight whose support lies on the same overlay must be equal. Finally, in a well defined CMAC, ρ parameters cannot be chosen arbitrarily because this would give a non-zero output, so the dimension of the null space is $\leq (\rho - 1)$.

Combining the two results means that the dimension of the null space is $(\rho - 1)$ and hence the dimension of the set of basis function vectors is $p - (\rho - 1)$. $\square$

An obvious corollary of this theorem is Corollary 7.1.

Corollary 7.1 When the basis function vectors are generated by a well defined binary CMAC which generalises, the dimension of the space spanned by the basis function vectors is (strictly) less than the number of basis functions.

Proof. This is easily shown by noting that for a well defined binary CMAC to generalise, ρ must be > 1 and so the dimension of the set of basis function vectors is $< p$.   □

7.2.2   Singular Solution and Autocorrelation Matrices

The solution matrix of a CMAC is defined to be the matrix, whose t^{th} row consists of the vector output of the basis functions $(\mathbf{a}^T(t))$ for the tth training example $(\mathbf{x}(t))$. So given a set of training examples $\{\mathbf{x}(t), \widehat{y}(t)\}_{t=1}^{L}$, a weight vector must be found which solves the system of linear equations given by:

$$\mathbf{A}\mathbf{w} = \widehat{\mathbf{y}} \tag{7.3}$$

where $\mathbf{A}$ has dimension $(L \times p)$ and $\widehat{\mathbf{y}}$ is a vector of length L which consists of the desired outputs of the training set. The corresponding autocorrelation matrix $\mathbf{R}$ of size $(p \times p)$ is defined by $\mathbf{R} = \mathbf{A}^T\mathbf{A}$, and as shown in Chapter 4, the above system of equations can be interpreted as finding the weight vector which solves:

$$\mathbf{R}\mathbf{w} = \mathbf{p} \tag{7.4}$$

where $\mathbf{p}$ is the cross-correlation vector given by $\mathbf{p} = \mathbf{A}^T\widehat{\mathbf{y}}$. Both the autocorrelation and the solution matrices have the same rank [Horn and Johnson, 1985], thus the following corollaries can be established.

Corollary 7.2 The rank of the solution (autocorrelation) matrix of a well defined binary CMAC is $\leq p - (\rho - 1)$.

Proof. This can be established immediately, as the rank of $\mathbf{A}$, is equal to the dimension of the row space of $\mathbf{A}$ which is simply the dimension of the set of basis function vectors mapped to for the given training set. The set of basis function vectors generated by the training data is contained within the set of all possible basis function vectors and so the rank$(\mathbf{A}) \leq p - (\rho - 1)$.   □

When it was established that the dimension of the set of basis function vectors was $p - (\rho - 1)$, the proof consisted of two parts. The first part constructed a lower bound for the nullity of the basis function vectors, and this still holds for Corollary 7.2. However, the second part of the proof breaks down because it assumes that every possible basis function vector is generated, whereas in practice only a subset of the basis function vectors are formed from a training set. This result disproves the related theory developed by Ellison [1991]. The second corollary about the solution matrix is now given.

Corollary 7.3 If a well defined binary CMAC generalises, the corresponding solution (autocorrelation) matrix is always singular.

Proof. Corollary 7.2 established that the rank of the solution matrix is $\leq p-(\rho-1)$ and when the basis functions generalise, the generalisation parameter ρ is > 1. Hence $\text{rank}(\mathbf{A}) < p$, and the solution matrix is singular. $\square$

A singular solution matrix means that an infinite number of weight vectors which satisfy Equations 7.3 and 7.4 exist. Therefore an infinite number of global minima for the MSE cost functions investigated in Chapters 4 and 5 exist.

7.2.3 The Rank of the Solution and Autocorrelation Matrices

In the previous two sections it was established that the $\text{rank}(\mathbf{A}) \leq p - (\rho - 1)$, and that equality holds if a set of training examples is given such that at least one training input lies in each cell defined on the input lattice (i.e. the desired function is completely specified). The effect of varying ρ on this quantity is now investigated for the univariate and the multivariate CMAC.

By Equation 6.5 and the above corollaries, the rank of the solution matrix for a completely specified desired function is given by:

$$\text{rank}(\mathbf{A}) = \sum_{j=1}^{\rho} \prod_{i=1}^{n} \left(\left\lceil \frac{(r_i + 1 - d_{ij})}{\rho} \right\rceil + 1 \right) - (\rho - 1) \qquad (7.5)$$

where r_i is the number of interior knots defined on the ith univariate axis.

When $n = 1$, the rank of the univariate CMAC's solution matrix is:

$$\text{rank}(\mathbf{A}) = r_1 + 1$$

which is independent of ρ. A completely specified univariate look-up table has $(r_1 + 1)$ training pairs, one for each interval. Therefore the univariate CMAC is as flexible as a look-up table for any ρ, although the network requires $(\rho - 1)$ more parameters to store the same information. The set of equations in 7.3 is *under-determined* (see Section 4.3.2), and at least one weight vector exists that can model the data exactly [Ellison, 1988].

For a more realistic multivariate CMAC network where r_i is several times larger than ρ, Equation 7.5 approximates to:

$$\text{rank}(\mathbf{A}) \approx \frac{\prod_{i=1}^{n} r_i}{\rho^{n-1}}$$

Hence, not only does the training data produce an over-determined set of equations, but the autocorrelation matrix is also singular. When ρ is doubled, $\text{rank}(\mathbf{A})$ decreases by a factor of approximately 2^{n-1}, and referring back to Figure 6.15 shows the *significant* decrease in $\text{rank}(\mathbf{A})$ when ρ is increased. Therefore as ρ increases, the size of the space of the functions which the multivariate CMAC can model decreases exponentially. However, it may be possible that a multivariate CMAC with a generalisation parameter ρ, can exactly model a function, whereas an equivalent network with a generalisation parameter $(\rho - 1)$ cannot. This is because the space of functions which the former can model is *not* a subset of the space of functions of the latter, and this is discussed further in Section 7.3.3.

7.3 CONSISTENCY EQUATIONS

It has been shown that the rank of the solution matrix is in general considerably lower than the size of the desired function vector (look-up table values). Therefore some relationships must exist between the elements in this vector for the CMAC to be able to model the desired function exactly. A univariate CMAC can exactly reproduce the training data for any value of ρ, but a multivariate network that generalises produces an over-determined system of equations. For these equations to be consistent, the training data must satisfy the *consistency equations* which result from reducing the CMAC's augmented solution matrix $\mathbf{A}$ to an upper trapezoidal matrix. The consistency equations are then those relationships in the augmented matrix below the diagonal submatrix. When the desired function is fully specified p' input/desired output training pairs exist. The rank of the solution matrix is $p - (\rho - 1)$ and so there are $p' - (p - (\rho - 1))$ non-redundant consistency equations which specify the relationships that must exist in the training data when the CMAC can model them exactly.

Initially, a set of local consistency equations are shown to exist, which the training data must satisfy for the CMAC to have the potential to store the data exactly. This produces a set of *necessary* consistency equations. The structure and position of these relationships shows that the CMAC forms *locally additive models* in n-dimensional space, thus explaining how the network generalises. This set of local consistency equations is then shown to be *sufficient*, or complete, in the sense that any function which satisfies these relationships can be exactly modelled by the appropriate CMAC, enabling the modelling abilities of various networks to be compared. In particular, it is shown that the only mappings which every CMAC can learn are the class of additive functions.

7.3.1 Geometrical Interpretation of the Consistency Equations

Consider an n-dimensional input lattice (as described in Section 6.2 with $n > 1$) which is composed of n-dimensional unit hypercubes. Now imagine a slab being formed from four of these n-dimensional unit hypercubes, as shown in Figure 7.2, and numbering them 1 to 4. Each input represents the vectors:

$$1: \quad (x_1, \ldots, x_n)$$

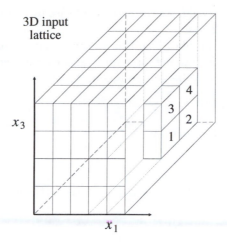

Figure 7.2 A (2×2) consistency equation slab in a three-dimensional lattice.

$$2: \quad (x_1, \ldots, x_{i-1}, x_i + 1, x_{i+1}, \ldots, x_n)$$
$$3: \quad (x_1, \ldots, x_{j-1}, x_j + 1, x_{j+1}, \ldots, x_n)$$
$$4: \quad (x_1, \ldots, x_i + 1, \ldots, x_j + 1, \ldots, x_n)$$

where the indices i and j satisfy $1 \leq i < j \leq n$. Also let $\mathbf{h}_i$ and $\mathbf{h}_j$ denote the $(n-1)$-dimensional lattice hyperplanes separating x_i from $x_i + 1$ and x_j from $x_j + 1$, respectively. The indexed unit hypercubes represent inputs to a well defined binary CMAC, and when the n-dimensional slab is placed completely within the lattice, two situations can occur:

1. $\mathbf{h}_i$ and $\mathbf{h}_j$ from part of the same overlay; or
2. $\mathbf{h}_i$ and $\mathbf{h}_j$ lie on different overlays.

For the former case, nothing can be said. However, for the latter case a relationship exists between the network's outputs, corresponding to the inputs which lie on the slab. This can easily be derived by noting that, as the two $(n - 1)$-dimensional hyperplanes form part of different overlays, the weight change in moving from $1 \rightarrow 2$ is the same as moving from $3 \rightarrow 4$, due to the fact that the same weights are being dropped from, and introduced to, the output calculation. Hence:

$$y(2) - y(1) = y(4) - y(3) = \frac{w^J(2)}{\rho} - \frac{w^J(1)}{\rho} = \frac{w^J(4)}{\rho} - \frac{w^J(3)}{\rho}$$

where $w^k(i)$ represents the weight on the kth overlay, which is non-zero for an input i. For this example, the hyperplane and the weight change occur on the Jth overlay, although considering only the relationship between the network outputs gives:

$$y(1) + y(4) = y(2) + y(3) \tag{7.6}$$

Equation 7.6 can also be derived from the equivalent case by considering the weight change moving $1 \rightarrow 3$ and $2 \rightarrow 4$.

The above relationship specifies how the neighbouring outputs of the CMAC network are related, and the training data must also contain equivalent relationships in order for the CMAC to be able to model it exactly, i.e.:

$$\hat{y}(1) + \hat{y}(4) = \hat{y}(2) + \hat{y}(3) \tag{7.7}$$

and a consistency relationship such as:

$$0 = \hat{y}(1) + \hat{y}(4) - \hat{y}(2) - \hat{y}(3)$$

would occur in the reduced augmented solution matrix, as the following relationship must hold:

$$\mathbf{a}(1) + \mathbf{a}(4) - \mathbf{a}(2) - \mathbf{a}(3) = \mathbf{0}$$

Thus wherever the n-dimensional slab is placed completely within the input lattice, such that the two $(n-1)$-dimensional hyperplanes which cut it do not form part of the same overlay, then Equation 7.7 *must* be satisfied for the CMAC to be able to model the data exactly. This generates a necessary, and generally redundant, set of consistency equations which the training data must satisfy, or else the CMAC is unable to reproduce the desired function exactly.

Locally Additive Models

These consistency equations mean that the CMAC generalises using locally *additive* models. An n-dimensional additive model, $\hat{y}(\mathbf{x})$, is formed from the linear sum of n univariate functions:

$$\hat{y}(\mathbf{x}) = \sum_{i=1}^{n} \hat{y}_i(x_i) \tag{7.8}$$

where $\hat{y}_i(x_i)$ are its univariate components. Consider the $n = 2$ case where $\hat{y}_i(x_i)$ are both univariate look-up tables, then:

$$\hat{y}(i,j) = \hat{y}_1(i) + \hat{y}_2(j)$$

for $i, j \in \{1, 2\}$. It can easily be verified that this (2×2) mapping satisfies the CMAC's local consistency equation and that two (non-unique) univariate functions which satisfy this relationship can always be found. Hence across a slab, the CMAC produces a locally additive model. When a group of the slabs forms a consistency equation cube (three-dimensional) or a hypercube (n-dimensional), the network's output in this part of the input space is identical to an n-dimensional additive model. Similarly, when these neighbouring (sharing two or more inputs) consistency equation slabs form part of larger *monotonic* regions, the CMAC's output is still locally equivalent to an additive model across this domain, as illustrated in Figure 7.3.

The consistency equations build up areas in the input space where the CMAC's output is locally additive and the original Albus strategy of making the overlay displacement vectors lie on the main diagonal produces large (diagonal) areas in the input space where the CMAC reduces to an additive model (see Figure 6.5). Using the improved displacement vectors described in Section 6.2.3 means that the locally additive areas are generally *smaller* and more evenly distributed, so the network's surface is much smoother, as shown in Figure 6.6.

Lattice-Based Consistency Equations

The set of consistency equations which the training data must satisfy has been derived with specific reference to the CMAC, although the proof does not depend on a particular network; it is applicable to all those which have basis functions sparsely defined on a lattice with a binary output. If no basis function has a corner at the centre of the consistency equation slab, the consistency equations *must* hold for the network to accurately model the data. This can easily be proved simply by modifying the above definition to take into account the possibility that one or more basis functions may be dropped from, and introduced to, the output calculation as the input crosses a lattice hyperplane.

This implies that if a lattice-based binary network is to model a look-up table *exactly*, the number of basis functions in the network must be greater than or equal to the number of lattice cells ($p \geq p'$). Any coarse coding of the input space, as occurs in the CMAC, reduces the modelling capabilities of the network when compared with a look-up table.

Example of the Consistency Equations

The two-dimensional CMAC's parameters are given by $r_1 = r_2 = 4$, $\rho = 3$ and $d_1 = d_2 = 1$, the output of the network is shown in Figure 7.3 and the weights on each overlay are given in Figure 7.4; this network is used in all remaining examples in this chapter. Hence the CMAC's lattice has $p' = 25$ unit squares and the network uses $p = 17$ weights, so $p' - (p - (\rho - 1)) = 10$, which is the number of local consistency equations whose slabs are centred on the circles in this figure. This relationship always holds for a two-dimensional input space, as the local consistency equations are *not* redundant.

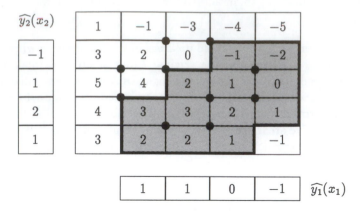

Figure 7.3 The output of a two-dimensional CMAC with $\rho = 3$. The solid circles are positioned at the centres of the consistency equation slabs and the univariate functions, that correspond to the CMAC's locally additive model in the shaded region, are shown along the axes.

7.3.2 A Complete Set of Consistency Equations

The local consistency equations must be satisfied by the training data if the CMAC is to model the function exactly. They form a *necessary* set of constraints which determine the type of mappings that a CMAC can model. This section shows that they are also *sufficient*, as any function which satisfies these constraints can be reproduced by a CMAC, and the set of local consistency equations is said to be complete. The proof of this theorem is quite long and involved, although a brief outline is given to guide the interested reader through it. It can be omitted, as it provides little extra insight into the generalisation abilities of a multivariate network. However, the local consistency equations that have just been described should be thoroughly understood, as much of the remaining work in this chapter is based on these simple relationships.

Theorem 7.2 Let $\mathrm{CMAC}(\rho, \mathbf{d})$ represent the class of functions that a well defined multivariate CMAC, with generalisation parameter ρ and overlay displacement vector $\mathbf{d}$, can reproduce. Consider an n-dimensional look-up table represented by the function $\hat{y}(\mathbf{x})$, which satisfies all the CMAC's local consistency equations. Then the CMAC can model this function exactly.

The theorem is proved in two parts; a network is constructed which can reproduce the desired function exactly at certain *training points*, then it is shown that the desired function is equivalent to the network.

As the CMAC's and the look-up table's structures are identical, all that needs to be determined is an appropriate weight vector that can store the desired values

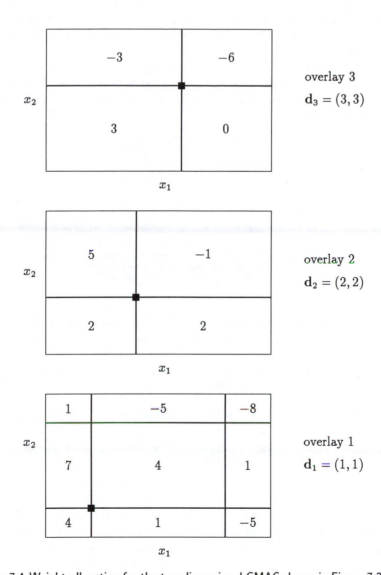

Figure 7.4 Weight allocation for the two-dimensional CMAC shown in Figure 7.3.

at the training points. The fact that nullity($\mathbf{A}$) = $\rho - 1$ means that ($\rho - 1$) weights can be (arbitrarily) set to zero without affecting the modelling capabilities of the network. These are chosen to be the ($\rho - 1$) weights mapped to by the first cell in the lattice, and the remaining weight is set equal the desired value at this point. This ensures that the network's output for the first input is correct. The weight-initialising algorithm then proceeds by moving through the input space in a pre-defined fashion, one lattice cell at a time. Therefore, at most one new weight is introduced (and correspondingly a weight dropped from) the network output calculation. If a new weight is used in the output calculation, the input lattice cell is known as a training point and the weight is set so that the network's output is correct for this input. The whole of the weight vector can be initialised in this manner, so it only remains to verify that the network generalises appropriately for inputs which do not correspond to training points. This initialisation algorithm is illustrated in Figure 7.5.

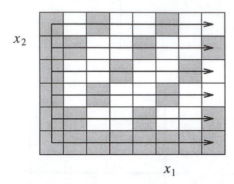

Figure 7.5   The weight initialisation algorithm for a two-dimensional CMAC with $\rho = 3$ and $d = (1, 1)$, where the shaded lattice cells represent training points and the arrow illustrates how the weight initialisation algorithm moves through the input space. The input is incremented in the direction x_1 until the edge is reached, then x_1 is reset to 1, x_2 is incremented one cell, and the input is again moved parallel to the first axis.

Induction is then used to show that the output of the CMAC and the desired function are equivalent. If it is assumed that the output is correct in an n-dimensional hyperrectangular region of the lattice, it only needs to be proved that the desired function and the CMAC agree, as an input moves one lattice cell parallel to any axis because the whole input space can be paved in this manner. The weight-initialisation algorithm results in the network output being correct in the lattice's first cell, and the final part of the proof constructs two parallel paths along which the local consistency equations hold, relating the change in the value of network's output to the desired function, showing that the CMAC and the desired function are equivalent. The theorem is now formally proved.

Proof. To begin some notation will be defined. Without loss of generality, the input $\mathbf{x}$ is assumed to be discrete and each element x_i lies in the set $\{1, 2, \ldots, r_i + 1\}$, so the input space and the lattice are analogous. $y(x_1, x_2, \ldots, x_n)$ is the CMAC's output for an input $(x_1, x_2, \ldots, x_n)$, $w^j(x_1, x_2, \ldots, x_n)$ refers to the weight on the jth overlay whose support contains the input $(x_1, x_2, \ldots, x_n)$ and x_i^j denotes the minimum value of x on the ith axis that causes x_i and x_i^j to map to the same weight on the jth overlay, i.e. $w^j(x_1, x_2, \ldots, x_n) = w^j(x_1^j, x_2^j, \ldots, x_n^j)$. It should also be noted that in this proof, the CMAC's output is generated by summing over every active weight, i.e. the output of an active basis function is 1. While this may seem to be inconsistent with a network maintaining a partition of unity, the weights are simply a multiple of $1/\rho$ greater than the networks described in Chapter 6.

It may also be assumed that on the ith input axis, $\hat{y}(.)$ has r_i internal discontinuities which lie on the integers $\{1, 2, \ldots, r_i\}$ and that it is defined on the domain $[0, r_i + 1]$. Hence, the CMAC's knots also occur at these points and a generalisation parameter is chosen, such that $1 \leq \rho \leq \max_{i=1}^{n}\{r_i + 1\}$, as well as an overlay displacement vector **d** that satisfies the conditions described in Section 6.2.3. The binary CMAC is then well defined and its output is piecewise continuous on each unit hypercube defined on the lattice.

To begin, the trivial case of $\rho = 1$ is considered; then $p = p'$ and the CMAC has only one overlay consisting of supports which are unit hypercubes. The network is a general look-up table, as setting:

$$w^1(\mathbf{x}) = \hat{y}(\mathbf{x}), \qquad \forall\, \mathbf{x} \tag{7.9}$$

means that the CMAC can store the desired function (an arbitrary look-up table). Hence the theorem holds for $\rho = 1$.

Now when $\rho > 1$, the network has $(\rho - 1)$ free parameters which can be chosen arbitrarily. For this theorem they are:

$$w^1(\mathbf{1}) = w^2(\mathbf{1}) = \cdots = w^{\rho-1}(\mathbf{1}) = 0 \tag{7.10}$$

and then setting:

$$w^{\rho}(\mathbf{1}) = \hat{y}(\mathbf{1}) \tag{7.11}$$

means that:

$$y(\mathbf{1}) = \sum_{j=1}^{\rho} w^j(\mathbf{1}) = \hat{y}(\mathbf{1}) \tag{7.12}$$

and the network has the correct output for the first input. The remaining weights are then initialised using Algorithm 7.1.

Algorithm 7.1

1. set $x_i = 1,$ $\qquad \forall\, i = 1, 2, \ldots, n$
2. find the minimum k such that $x_k \neq r_k + 1$

   increment x_k by one

   set $x_i = 1,$ $\qquad \forall\, 0 < i < k$

3. if any weight mapped to is uninitialised, set:

$$w^J(\mathbf{x}) = \hat{y}(\mathbf{x}) - \sum_{j=1, j \neq J}^{\rho} w^j(\mathbf{x})$$

where the new weight is defined to lie on the Jth overlay.

4. if $(x_i = r_i + 1, \; \forall i = 1, 2, \dots, n)$ stop
   else goto 2.

This algorithm iterates $(p' - 1)$ times and is well defined in the sense that all the weights are initialised for the first input (**1**) and moving one cell parallel to the kth axis means that at most one weight is uninitialised (due to the uniform projection property) and this weight is defined to lie on the Jth overlay, $J \in \{1, 2, \dots, \rho\}$.

Since the CMAC is well defined, all p weights have now been initialised (by Algorithm 7.1 and the initial conditions) at the $p - (\rho - 1)$ *training points*, where a training point is defined as the input which causes a new weight to be initialised. At these training points, the CMAC's output and the desired function agree, due to the weight initialisation in Algorithm 7.1, and the question is now: Does the CMAC generalise correctly, in the sense that the output is still correct for the inputs which are not training points?

By Equation 7.12, the network has the correct output for an input (**1**). Now assume that:

$$y(x_1, \dots, x_n) = \hat{y}(x_1, \dots, x_n), \qquad \text{for each } 1 \le x_i \le c_i \le r_i + 1, \quad \forall i \qquad (7.13)$$

then it must be shown that:

$$\hat{y}(x_1, \dots, c_L + 1, \dots, x_n) = y(x_1, \dots, c_L + 1, \dots, x_n) \qquad (7.14)$$

for $1 \le L \le n$, $c_L \ne r_L + 1$, i.e. given that the network and the desired function agree on a hyperrectangular region in the input space, show that the output is also correct when any of the inputs is incremented by 1. If this result can be established, Theorem 7.2 is true by induction, as the whole of the discrete input space can be paved in this manner.

In moving one step parallel to the Lth axis, the weight on the Jth overlay changes. Now suppose that $w^J(x_1, \dots, c_L + 1, \dots, x_n)$ is initialised by Algorithm 7.1 when the input is incremented by one cell parallel to the kth axis $(1 \le k \le n)$. It immediately follows that $k \le L$, because if it is assumed that $k > L$, then by Algorithm 7.1 $x_i = 1 \; \forall 0 < i < k$, hence $c_L + 1 = 1$. But $c_L \ge 1$ contradicting the original assumption.

Consider when $1 \le k \le L$ and a weight on the Jth overlay changes as the c_L is incremented by 1, then by assumption and by Algorithm 7.1 the network's output is correct at the points:

$$\begin{aligned}
\mathbf{x}(3) &= (x_1^J, \dots, x_{L-1}^J, c_L + 1, x_{L+1}^J, \dots, x_n^J) \\
\mathbf{x}(1) &= (x_1(3), \dots, x_{L-1}(3), c_L, x_{L+1}(3), \dots, x_n(3))
\end{aligned}$$

where $x_i^J = 1$ for all $i < k$ and the input's index refers to its position on a consistency equation slab rather than time. Now let l be the minimum index, such that $x_l(1) < x_l$, $l \ne L$, then by assumption the output is also correct for an input:

$$\mathbf{x}(2) = (x_1(1), \dots, x_{l-1}(1), x_l(1) + 1, x_{l+1}(1), \dots, x_n(1))$$

By assumption, the hyperplanes which pass through the lth and L^{th} axes must be distinct (otherwise the support's corner would occur at a different position), so a consistency equation is produced, which is satisfied by the network's output at:

$$\mathbf{x}(4) \;=\; (x_1(3), \ldots, x_{l-1}(3), x_l(3) + 1, x_{l+1}(3), \ldots, x_n(3))$$

but the desired function also satisfies all the consistency equations, so it has the same output as the network at this point. The above procedure can be repeated (by the redefining $\mathbf{x}(3)$ and $\mathbf{x}(1)$ as $\mathbf{x}(4)$ and $\mathbf{x}(2)$, respectively, and by assumption the output for new input $\mathbf{x}(3)$ is also correct), and the consistency equations pave a path along each axis, until it is shown that the network's output and the desired function satisfy Equation 7.14.

Therefore, the result holds by induction and the CMAC can store any function exactly which satisfies all of its consistency equations.   □

The local consistency equations are both necessary and sufficient conditions, and for 3 and higher dimensional networks, they are also redundant.

7.3.3   Relationships Between Different CMACs

These local consistency equations make it very easy to prove that various relationships do *not* exist between different CMACs. The theory developed in Section 7.2 might have suggested to the reader the misleading conclusion that it is *always* better to have a smaller generalisation parameter, because the modelling capabilities of such a network are greater, and correspondingly the number of consistency equations which must be satisfied by the training data is smaller. This is broadly true, but it does not follow that a certain CMAC with a small generalisation parameter can store exactly the information contained in another network which has a larger generalisation parameter. The former network has been shown to have greater modelling flexibility, although the modelling capabilities of the two networks are *different*, due to the distributions of the basis functions.

Similarly, these consistency equations can also be used to show that the modelling capabilities of two CMAC networks with different overlay displacement vectors are not the same. This is because the overlays are distributed differently across the lattice, so the consistency equations must also occur at different positions where the CMAC produces locally additive models. Hence the modelling capabilities are generally different.

CMACs With Different Generalisation Parameters

Theorem 7.3 For a well defined multivariate binary CMAC with $\rho > 2$ and $\min_{i=1}^{n} r_i > \rho$, the following expression holds:

$$\mathrm{CMAC}(\rho, \mathbf{d}) \not\subset \mathrm{CMAC}(\rho - 1, \mathbf{d})$$

Proof. It can be shown that it is always possible to find a basis function a_i in the CMAC, with generalisation parameter ρ, that has a corner of its support in the centre of a consistency equation of the network with a generalisation parameter $\rho - 1$. The corresponding i^{th} weight is set to ρ with all the other weights being set to zero, and the resulting network output is 1 if the input lies in the support of the ith basis function, and zero otherwise. The desired function therefore has the values:

$$\hat{y}(3) = 1, \qquad\qquad \hat{y}(4) = 0$$
$$\hat{y}(1) = 0, \qquad\qquad \hat{y}(2) = 0$$

which does not satisfy the consistency relationship given in Equation 7.7, and cannot be modelled by the CMAC with a smaller generalisation parameter. $\quad\square$

An example is now given to illustrate this result. Consider two, two-dimensional CMAC networks, with $r_1 = r_2 = 4$, $\mathbf{d} = (1,1)$, and the two generalisation parameters are $\rho = 2$ and $\rho = 3$. The CMAC with $\rho = 3$ is able to model the desired function shown in Figure 7.6, but this mapping violates a consistency equation generated by the network with a generalisation parameter of 2, therefore $\text{CMAC}(3, \mathbf{d}) \not\subset \text{CMAC}(2, \mathbf{d})$.

One corollary of this work which states how the CMAC networks *are* related for different generalisation parameters is Corollary 7.4.

Corollary 7.4 For two well defined multivariate binary CMACs with identical overlay displacement vectors, $\text{CMAC}(i\rho, \mathbf{d}) \subseteq \text{CMAC}(\rho, \mathbf{d})$, for each positive integer i.

Thus for two CMAC networks with generalisation parameters 2 and 4, and identical overlay displacement vectors, $\text{CMAC}(4, \mathbf{d}) \subset \text{CMAC}(2, \mathbf{d})$. From this it follows that if ρ is prime, the modelling capabilities of the CMAC is generally unique, whereas when ρ is even, there exists a network with a generalisation parameter $\rho/2$, which can *exactly* reproduce the first network.

CMACs With Different Overlay Displacement Vectors

The following theorem shows that the basic modelling capabilities of the original Albus CMAC (with a unity overlay displacement vector) and the CMAC with the improved overlay displacements are in general different.

Theorem 7.4 For two different n-dimensional overlay displacement vectors, $\mathbf{d}_1$ and $\mathbf{d}_2$ as defined in Section 6.2.3, the following relationship is true for a well defined multivariate binary CMAC which satisfies $\min_{i=1}^n r_i > \rho$:

$$\text{CMAC}(\rho, \mathbf{d}_1) \neq \text{CMAC}(\rho, \mathbf{d}_2)$$

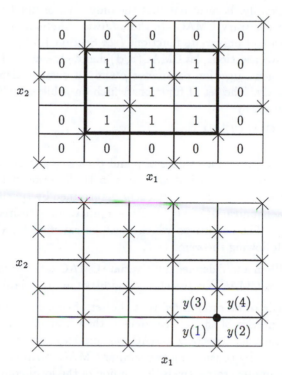

Figure 7.6 Top: desired function and the corners of the CMAC supports (crosses), with $\rho = 3$. The support of the non-zero basis function is shown in bold. Bottom: local desired function and the corners of the CMAC's supports (crosses), $\rho = 2$. The consistency equation (which has its centre denoted by a circle) is not satisfied.

This theorem implicitly assumes that the generalisation parameter is ≥ 3, or else two distinct overlay displacement vectors cannot be generated.

Proof. Again this theorem can easily be proved by construction, as the consistency equations occur, in general, at different positions within the lattice and all that needs to be constructed is an appropriate desired function. This is achieved by setting a single weight on the first overlay (with displacement vector $\mathbf{d}_1$) to be ρ and the remaining weights are all set to zero. A consistency equation for the second network is not satisfied and the result is proved. □

7.3.4 Additive Functions

Given that a CMAC can reproduce a desired function that satisfies all its local consistency equations and that changing the CMAC's structure (generalisation parameter and displacement vector) alters the network's modelling abilities, it is worthwhile investigating the type of mappings which *every* network has the ability to model exactly. It turns out that the only class of functions which can be reproduced by an arbitrary CMAC are the *additive functions*, those formed from a linear combination of univariate functions (see Section 7.3.1).

As in the previous sections, let $\text{CMAC}(\rho, \mathbf{d})$ denote the class of functions which a CMAC, with a generalisation parameter ρ and an overlay displacement vector $\mathbf{d}$, can model exactly. The set of functions which an arbitrary CMAC can model without error is given by:

$$\text{CMAC}() = \bigcap_{\rho, \mathbf{d}} \text{CMAC}(\rho, \mathbf{d}) \tag{7.15}$$

for all valid ρ and $\mathbf{d}$. If various values of ρ and $\mathbf{d}$ can be generated, such that the consistency equations must hold at all points in the interior of the lattice, every value of the desired function can be *uniquely* determined by specifying the desired function along each univariate axis, and the multivariate desired function is an additive mapping. The conditions which must be satisfied for this to occur are contained in the following theorem.

Theorem 7.5 When a well defined multivariate CMAC satisfies $\max_i (r_i + 1) \geq 3$, the only member of CMAC() is the class of additive multivariate look-up tables.

Proof. It only needs to be established that the set of consistency equations covers the whole of the interior of the lattice, since it then immediately follows that the only functions which an arbitrary CMAC can model exactly are additive ones. This can be achieved by considering two similar CMACs with an odd and an even generalisation parameter, respectively. The union of the local consistency equation locations fills the first hypercube of ρ^n occurring on the lattice. In a similar manner, it can also be shown that the local consistency equations must hold at every point in every hypercube and on the boundaries of the hypercubes. Therefore, the locations of the consistency equations completely covers the interior of the input lattice, the desired function is an additive mapping and, from Theorem 7.2, any CMAC can model this function exactly.   □

The proof of the above theorem relied on the premise that different overlay displacement vectors can be considered for the same value of ρ. The original Albus CMAC used only unity displacement vectors, producing overlays which were distributed along the main diagonal. In this case, the above theorem obviously does not apply, as the consistency equations are *not* required to hold along the main diagonal. This does not mean that the original network has improved modelling capabilities, rather that an arbitrary member of a subset of CMAC() can model a slightly larger class of data sets, although the attributes of such an approach are questionable.

7.3.5 Univariate Look-Up Tables

As a special case of the constructive proof of Theorem 7.2, the following corollary can be established.

Corollary 7.5 If the desired function is a bounded univariate look-up table, for any value of ρ which generates a binary well defined univariate CMAC, there exists a network which is output-equivalent to the desired function.

Proof. This follows directly from Algorithm 7.1 in the proof of Theorem 7.2 with $n = 1$. □

This could have been established simply by noting that the rank of the solution matrix of a univariate CMAC is $(r_1 + 1)$, which is equal to the rank of the augmented solution matrix. It is also interesting to note that this construction is similar to the one used by Ellison [1988], where it was first proved that an optimal weight vector for the univariate CMAC always existed. In that paper, the last $(\rho - 1)$ weights were arbitrarily set to zero and backsubstitution was used to initialise the weight vector. In the above approach, the first $(\rho - 1)$ weights of a univariate CMAC are set to zero and the remainder of the weight vector is initialised using *forward* substitution.

Summary

This section has considered the modelling abilities of various multivariate CMAC. Generally, changing the overlay displacement vector or the generalisation parameter *alters* the structure of the underlying model, and it cannot be simply stated that increasing or decreasing the generalisation parameter improves or degrades the model. The results in Section 7.2 indicate that the model's flexibility increases as ρ decreases, but the theorems in this section show that the modelling abilities of various networks are not the same and only in special cases can a CMAC reproduce another network with a different structure.

A complete set of local consistency equations has been derived, which must be satisfied by the training data for the CMAC to be able to store this information exactly. They show how the CMAC produces local additive models when the network generalises locally and also illustrate why the improved overlay displacement vectors, proposed in Section 6.2.3, generally produce better models. Using a large generalisation parameter, together with the original Albus overlay strategy, results in additive models being formed across large diagonal areas in the input space, as illustrated in Figures 6.5 and 7.3.

These results are important if the CMAC is to be used in more complex learning schemes. For instance, Kavli's ASMOD algorithm (see Section 8.5) learns to model a complex function by modifying the *structure* of the network according to the observed data. The network's structure is changed by introducing new inputs into the modelling scheme, combining inputs and introducing new knots on the univariate axes. B-spline AMNs are chosen because, whenever the flexibility of the network is increased, the new one can reproduce the old model *exactly*. If a CMAC network is used instead, new variables can be introduced and the inputs can be combined to form a multivariate network without the modelling capabilities of the network being reduced. However, when a new knot is placed on an axis, the relative positions of the consistency equations change and the new network is unable to reproduce the old network exactly. If instead of adding a new knot, a new set of knots (in the form of an overlay) is introduced [An, 1991], the new network's modelling capabilities would be increased, although the effect of adding a new overlay is *global*.

The CMAC's modelling capabilities have been shown to depend on the generalisation parameter and the overlay displacement vector, both of which must be selected by the designer. These parameters influence the basic modelling capabilities, the computational cost and the learning ability of the network, and so their selection is a design compromise. The reader should not be too alarmed, as experimental evidence indicates that most reasonable choices give acceptable results.

7.4   ORTHOGONAL FUNCTIONS

The previous section investigated the functions which a binary CMAC can model by examining its set of local consistency equations and showing that they are directly related to form of the desired function. In contrast, this section considers the *dual* of these consistency equations: the *orthogonal functions*, by analysing what type of mappings a CMAC is completely unable to model in a least-squares sense. These orthogonal functions are equally as useful as the consistency equations because they allow the ability of the CMAC to model *multiplicative* functions to be investigated, showing that the locally additive models produced by the CMAC may not always be sufficient. The set of local orthogonal functions is also shown to be complete, in the sense that any orthogonal mapping can be expressed as a linear combination of these basis functions.

7.4.1   Matrix Interpretation

It is well known in numerical analysis [Burden and Faires, 1993] that if a model is formed from a linear combination of basis functions and the desired function is orthogonal to *every* basis function, the model is completely unable to reproduce the desired function in the least-square sense. Equivalently, the part of a function that the optimal network (with respect to the MSE) is unable to learn is an orthogonal mapping, and the original function can be uniquely decomposed into two parts [Cotter and Guillerm, 1992]. Trying to model an orthogonal function produces a network with an identically zero output and weight vector. A desired function is said to be orthogonal to every basis function if $\forall\, i = 1, 2, \ldots, p$:

$$\int_{L} \widehat{y}(\mathbf{x})\, a_i(\mathbf{x})\, d\mathbf{x} = 0 \tag{7.16}$$

where $\mathbf{L}$ is the n-dimensional lattice. When Equation 7.16 holds, the CMAC is completely unable to model the desired function, as the optimal network (in a least-squares sense) has an identically zero weight vector.

For a uniform distribution of the inputs across a binary CMAC's lattice (the input has the same probability of occurring in each cell), the above set of expressions reduces to:

$$\sum_{\mathbf{x} \in \mathbf{L}_i} a_i(\mathbf{x})\,\hat{y}(\mathbf{x}) = 0 \tag{7.17}$$

where $\mathbf{L}_i$ is the intersection between the lattice and the support of the ith basis function. The set of constraints that a desired function must satisfy if it is orthogonal to the binary CMAC can then be expressed as:

$$\mathbf{C}\,\hat{\mathbf{y}} = \mathbf{0} \tag{7.18}$$

where $\mathbf{C}$ is a matrix of size $(p \times p')$ with its ij^{th} element, $c_{ij} = a_i(\mathbf{x}(j))$ and $\hat{\mathbf{y}}$ is a vector of length p' composed of the orthogonal function's values.

It should be immediately noticed that this constraint matrix is simply the CMAC's solution matrix *transposed*, i.e.:

$$\mathbf{C} = \mathbf{A}^T \tag{7.19}$$

hence, from Corollary 7.2:

$$\text{rank}(\mathbf{C}) = p - (\rho - 1)$$

This result is not surprising, since the generalised Moore-Penrose pseudo-inverse (see Section 4.3.2) simply multiplies the desired function vector by $\mathbf{A}^T$ and projects the part of the mapping that the network is able to model, filtering out its orthogonal component.

Thus there are $(p - (\rho - 1))$ linearly independent constraints that must be satisfied for a function to be orthogonal to the binary CMAC. Since the look-up table has p' unrelated outputs and, in general, $p \ll p'$, the size of the space of orthogonal functions is *much larger* than the range of mappings that the CMAC is able to model exactly. Hence, the CMAC's coarse coding of the input space means that it is unable to reproduce most of the corresponding look-up tables. Again, the reader should not be alarmed at this statement, as the binary CMAC generalises smoothly, producing locally additive models, and many of the orthogonal functions are distinctly non-smooth. For instance, consider a function where the output takes random values except at one corner of each basis function. Here the network's output is set equal to $-(\rho^n - 1)$ times its average output across the corresponding support. This is an orthogonal function and is certainly not smooth.

7.4.2   Local Orthogonal Functions

A local orthogonal function is defined to be an orthogonal function which has a
non-zero output only on a (2×2) slab, as described in Section 7.3.1, or equivalently, the supports of the local orthogonal functions are the (2×2) slabs. These
local orthogonal functions are important, as they allow the modelling abilities of
the CMAC to be investigated locally, and are complete in the sense that any orthogonal function can be expressed as a linear combination of these "orthogonal
basis functions".

Suppose that an n-dimensional (2×2) slab lies completely within the lattice
and that the two $(n-1)$-dimensional hyperplanes which intersect with the slab lie
on different overlays. When the CMAC is able to model the data exactly, the local
consistency equations must hold. If this does not occur, it does not mean that the
CMAC is unable to model the function, rather that the network cannot model it
exactly. Constructing a desired function which has a zero value for all the inputs
lying outside the slab, and for the inputs lying within the slab, the desired output
of the network is given by:

$$\hat{y}(3) = -1, \qquad \hat{y}(4) = 1$$
$$\hat{y}(1) = 1, \qquad \hat{y}(2) = -1 \qquad\qquad (7.20)$$

Then by Equation 7.7, the CMAC is unable to reproduce this desired function
exactly and it is shown that the desired function is orthogonal to every basis
function. Thus the CMAC is unable to reproduce the desired function in a least-
squares sense.

All but four of the basis function's supports defined on the lattice either completely contain the slab or have a null intersection with the slab; in either case the
desired function is orthogonal to each of these basis functions. The four supports
that cut the slab contain inputs which correspond to two non-zero desired outputs, where the desired outputs are equal in magnitude and have opposite signs.
Therefore the desired function is also orthogonal to these four basis functions and
Equation 7.16 holds for the desired function specified by Equation 7.20.

It is important to realise that the local consistency equations and the local
orthogonal functions occur at the same positions in the lattice. They *locally* represent the desired function, as a part which the CMAC can model and a part which
it is completely unable to model in a least-squares sense, and this is considered
further in Section 7.5.

Example

Figure 7.7 shows a two-dimensional CMAC's lattice and an orthogonal desired
function when the network's parameters are: $r_1 = r_2 = 4$, $\mathbf{d} = (1,1)$ and $\rho = 3$.
The crosses represent the corners of the supports and the solid circles denote the

legal positions for the centres of the two-dimensional slabs. As described previously, there exist 10 (2×2) slabs on which the orthogonal functions (consistency equations) are defined.

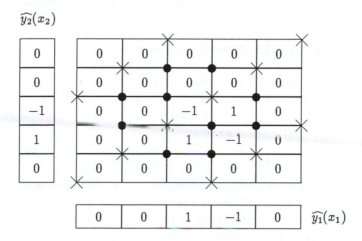

Figure 7.7 A two-dimensional multiplicative function (formed from the product of the two univariate functions shown) which constitutes a local *orthogonal* function that the CMAC is unable to model. The crosses denote the positions of corners of the basis functions and the circles represent the centres of the two-dimensional slabs.

The orthogonal desired function given in Equation 7.20 is generated whenever a slab is completely contained within the lattice, such that the two hyperplanes which cut the slab do not lie on the same overlay. Any linear combination of two or more desired functions which are orthogonal to all the basis functions is also orthogonal to all the basis functions, and so the desired functions specified in Equation 7.20 form a (redundant) set of orthogonal desired functions. They are also *complete*, in the sense that any orthogonal function can be expressed as a linear combination of these local mappings, and this is proved in Section 7.4.3.

Orthogonal Multiplicative Functions

It is possible to generate the local orthogonal functions from simple *multiplicative* mappings which have the form:

$$\hat{y}(\mathbf{x}) = \prod_{i=1}^{n} \hat{y}_i(x_i) \tag{7.21}$$

where $\hat{y}_i(.)$ is the ith *univariate* desired function. It has been shown that the CMAC can model additive functions exactly and it generalises as a locally additive model, but this result illustrates in the inability of the CMAC to model multiplicative functions.

Consider the two-dimensional multiplicative function shown in Figure 7.7. It is obviously an orthogonal function, so by construction, simple multiplicative functions exist which the binary CMAC cannot model in a least squares sense.

7.4.3 Complete Set of Local Orthogonal Functions

The set of local orthogonal mappings is *complete*, meaning that any orthogonal function can be expressed as a linear combination of these orthogonal basis functions. This is an important result, because it means that any look-up table can be uniquely written in the following form:

$$\widehat{y}(\mathbf{x}) = y^{ce}(\mathbf{x}) + y^{of}(\mathbf{x}) \tag{7.22}$$

where $y^{ce}(.)$ is a function which satisfies all the CMAC's consistency equations and can be modelled exactly by the network, and $y^{of}(.)$ is a mapping that can be expressed as a linear combination of the local orthogonal functions. Thus, any look-up table can be expressed as:

$$\widehat{y}(\mathbf{x}) = \sum_{i=1}^{p} w_i a_i(\mathbf{x}) + \sum_{j=1}^{q} w_j^o b_j(\mathbf{x}) \tag{7.23}$$

where $b_j(.)$ is the jth local orthogonal basis function ($j = 1, \ldots, q$) and $\mathbf{w}^o$ is the corresponding weight vector. The local orthogonal functions completely characterise the type of functions which the CMAC is unable to model.

Theorem 7.6 Any orthogonal function $y^o(\mathbf{x})$ can be expressed as a linear combination of the local orthogonal basis functions, $b_j(\mathbf{x})$.

Proof. Each local orthogonal basis function is, by construction, an orthogonal function, so any linear combination is also an orthogonal function and satisfies the set of constraints given in Equation 7.18. An arbitrary orthogonal function $y^o(\mathbf{x})$ also satisfies the equivalent set of constraints given in Equation 7.17, which can be re-expressed as requiring the mapping to satisfy:

$$y^o(\mathbf{x}_i^{\mathbf{max}}) = - \sum_{\substack{\mathbf{x}\in L_i, \\ \mathbf{x}\neq \mathbf{x}_i^{\mathbf{max}}}} y^o(\mathbf{x}) \tag{7.24}$$

for each $i = 1, \ldots, p$, where $\mathbf{x}_i^{\mathbf{max}}$ is the largest input lying in support of the ith basis function, L_i. This can be interpreted as requiring the network's output to be equal to the negative sum of its output at all the remaining points in each basis function's support. Therefore any linear combination of the local basis functions satisfies these orthogonality conditions, and it remains to show that a weight vector, $\mathbf{w}^o$, can be found such that:

$$y^o(\mathbf{x}) = \sum_{j=1}^{q} w_j^o b_j(\mathbf{x}) \tag{7.25}$$

for every remaining input $(\mathbf{x} \neq \{\mathbf{x}_i^{\max}\}_{i=1}^{p})$. The right-hand side of this expression is referred to as an *orthogonal network* for the remainder of this section. When Equation 7.25 is satisfied, the theorem is proved, as the orthogonality constraints mean that the network generalises correctly for the inputs lying in $\{\mathbf{x}_i^{\max}\}_{i=1}^{p}$.

Therefore, consider a weight initialisation procedure similar to Algorithm 7.1. Each weight is initialised, so that orthogonal network and the orthogonal function have the same outputs at the training points. Then it only remains to show that the set of training points and the restricted input space $\{\mathbf{x} : \mathbf{x} \in \mathbf{L}, \mathbf{x} \notin \{\mathbf{x}_i^{\max}\}_{i=1}^{p}\}$ are the same, as the orthogonal network and the orthogonal function will be output equivalent.

Therefore, consider using Algorithm 7.2 to initialise the orthogonal network's weight vector, $\mathbf{w}^o$.

Algorithm 7.2

1. set $x_i = 1$, $\qquad \forall\, i = 1, 2, \ldots, n$
2. if $\mathbf{x}$ corresponds to the first input (1) on the ith orthogonal function slab, set:

$$w_i^o = y^o(\mathbf{x}) - \sum_{\substack{j=1, \\ j \neq i}}^{q} w_j^o b_j(\mathbf{x})$$

   When two or more orthogonal functions have their first input at this point, set all of the corresponding weights to zero, then initialise only one using the above procedure.

3. find the minimum k such that $x_k \neq r_k + 1$:

   increment x_k by one
   set $x_i = 1$, $\qquad \forall\, 0 < i < k$

4. if $(x_i = r_i + 1, \forall\, i = 1, 2, \ldots, n)$ stop
   else goto 2

The initialisation procedure is well defined in the sense that, for each iteration, at most one of the weights that contribute to the output is uninitialised and this is chosen, such that the orthogonal network and the orthogonal function are output-equivalent for this training point.

It can now be argued by contradiction that the set of training inputs and the restricted input space are equivalent. Consider a point in the restricted input space which is not a training input. This input does not lie in the set $\{\mathbf{x}_i^{\max}\}_{i=1}^{p}$, so there exist two unique hyperplanes touching the outer face of the input cell which lie on distinct overlays. But this has defined an orthogonal basis function slab where the input is the first cell, contradicting the original assumption.

Therefore, the orthogonal network and the orthogonal function are output-equivalent on the restricted input space, and the orthogonality conditions mean that they are the same over the whole input space. □

7.5   BOUNDING THE MODELLING ERROR

It is possible to use both the consistency equations and the orthogonal functions to calculate a *cheap* lower bound for the CMAC's modelling error. This process does not involve any matrix inversions, and gives an insight into the type of functions which the network cannot model sufficiently accurately.

The consistency equations and the orthogonal functions occur at the same positions in the lattice, enabling a set of five local equations to be constructed and solved, which produces a local lower bound of the CMAC's modelling error. Then the maximum of the absolute values of these quantities gives a lower bound for the modelling error in the whole network. Consider an n-dimensional slab as described in Section 7.3.1, where each of the hyperplanes which intersect it lie on different overlays. The network's outputs must satisfy the consistency equation given in Equation 7.6, and a local orthogonal function may exist which the network is completely unable to model. This is illustrated in Figure 7.8 which also gives the desired values to be stored. The desired function can be linearly decomposed

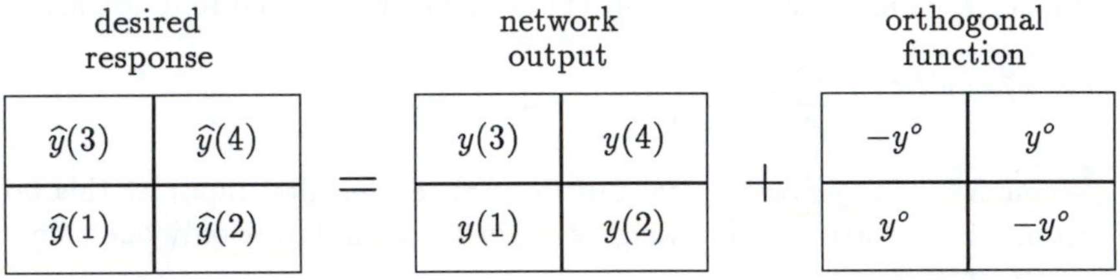

<table>
<tr><td align="center">desired
response</td><td></td><td align="center">network
output</td><td></td><td align="center">orthogonal
function</td></tr>
</table>

$\widehat{y}(3)$	$\widehat{y}(4)$
$\widehat{y}(1)$	$\widehat{y}(2)$

$=$

$y(3)$	$y(4)$
$y(1)$	$y(2)$

$+$

$-y^{\circ}$	y°
y°	$-y^{\circ}$

Figure 7.8   A decomposition of the desired function on a slab into a component which the network can model (its response) and one which it is unable to model, the orthogonal function.

into two parts: that which the CMAC can model and an orthogonal function. This gives five linearly independent equations for the five unknowns $y(1), y(2), y(3), y(4)$ and y°:

$$
\begin{aligned}
\widehat{y}(1) &= y(1) + y^{\circ} \\
\widehat{y}(2) &= y(2) - y^{\circ} \\
\widehat{y}(3) &= y(3) - y^{\circ} \\
\widehat{y}(4) &= y(4) + y^{\circ} \\
0 &= y(1) + y(4) - y(2) - y(3)
\end{aligned}
\tag{7.26}
$$

and solving for the size of the orthogonal function y° gives:

$$
y^{\circ} = \frac{\widehat{y}(1) + \widehat{y}(4) - \widehat{y}(2) - \widehat{y}(3)}{4}
\tag{7.27}
$$

The form of the numerator reflects how much the desired data differ from the consistency equations. When the data satisfy:

$$\hat{y}(1) + \hat{y}(4) - \hat{y}(2) - \hat{y}(3) = 0$$

the local orthogonal function is identically zero, and this is just a rearrangement of the consistency equation, Equation 7.7, which gives necessary conditions for the CMAC to be able to model the data. This relationship between the size of the local orthogonal function and the degree with which the consistency equation is not satisfied *links* both concepts and provides a *simple* and *cheap* algorithm for measuring how well a CMAC can locally reproduce a particular data set.

It has been assumed that the network is free to take the values $y(1), y(2), y(3)$ and $y(4)$, although due to neighbouring orthogonal functions this is not always possible. When the effects of the network not achieving these values are considered, it can easily be shown that y° forms a lower bound for the actual modelling error at these points. For instance, consider when $y(1)$ is less than the value generated in Equations 7.26, the modelling error at the point corresponding to the first input is greater than $|y^{\circ}|$. Similarly, when $y(1)$ is greater than its value in given in Equations 7.26, the modelling error at the fourth input is greater than $|y^{\circ}|$. A similar argument can be developed for the remaining outputs $y(2)$ and $y(3)$, which shows that $|y^{\circ}|$ forms a *lower bound* for the absolute value of the modelling error on this slab.

7.5.1 Multiplicative Functions

It has been shown that any well defined, multivariate, binary CMAC can reproduce any additive look-up table, and also that this is the only type of function which an arbitrary CMAC can model. However, an additive function can also be modelled exactly by summing n univariate CMACs, where each network reproduces the desired univariate function. The memory requirements of an n-dimensional CMAC are substantially greater than that required by n univariate networks. Therefore, it is pertinent to investigate how well the multivariate CMAC is able to model other classes of simple functions.

Local Error Estimate

In Section 7.4.2, it was shown that there exist certain multiplicative functions which the CMAC cannot model in a least-squares sense. This result is now extended by deriving a lower bound of the modelling error for an arbitrary multiplicative function, and showing that for any non-trivial mapping this error is non-zero. Thus the multivariate CMAC cannot model exactly *any* non-trivial multiplicative function.

Consider a (2×2) consistency equation slab in a multivariate CMAC, where (without loss of generality) the desired multiplicative function takes the values given in Figure 7.9. Substituting this into Equation 7.27 gives a lower bound for

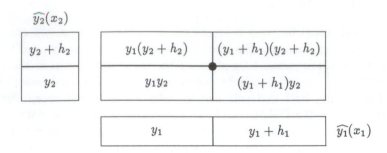

Figure 7.9 The desired multiplicative function defined on a two-dimensional slab.

the output modelling error of the multiplicative function on the slab:

$$y^o = \frac{h_1 h_2}{4} \tag{7.28}$$

The modelling error is *directly* related to the change in values (derivative) of each of the univariate functions (h_1 and h_2), and unless either of these quantities is zero, in which case the multiplicative function reduces locally to a univariate or additive mapping, the CMAC cannot model this function exactly.

It may have been thought that the extra flexibility associated with a multivariate network would enable it to model a larger class of functions, but in general this is not the case. The class of additive functions is the *only* one which an arbitrary multivariate CMAC can model exactly, and the modelling capabilities associated with a *particular* multivariate CMAC depend on its generalisation parameter and displacement vector, although it is readily acknowledged that most sensibly defined CMACs successfully *approximate* most smoothly varying nonlinear functions.

These modelling error results are important for two reasons:

1. The CMAC is typically trained using an instantaneous LMS learning rule (see Chapter 5), and any modelling error results in the parameters converging to a minimal capture zone rather than to a fixed point. The size and shape of these minimal capture zones (which need to be estimated if these networks are adapted on-line) depend on the modelling error and the relative orientations of the solution hyperplanes.

2. When the desired function forms a multiplicative look-up table which varies rapidly (large h_1 and h_2), there is a large modelling error because of the inability of the CMAC to store this type of mapping, and also because of the error introduced when the original (continuous) function was represented as a look-up table. Hence, a satisfactory approximation to this class of functions requires that the network's input intervals (distance between consecutive knots) must be *small*, to reduce both the original sampling error and the network's approximation (modelling) error (smaller h_1 and h_2). However, if the generalisation parameter is doubled as the intervals are halved

(to maintain the physical width of the basis function's support), the modelling error would remain approximately constant. This is because the input regions where the CMAC generalises as a locally additive function would remain the same and the network would still be trying to approximate a locally multiplicative mapping with a locally additive relationship.

This modelling error estimate is poor when ρ is large, as it substantially underestimates the true value. Future work will investigate the possibility of developing more realistic lower and upper bounds, although this measure does give important insights into the type of functions that the multivariate CMAC cannot model.

Example: Modelling Error

These modelling results are now illustrated by using a two-dimensional CMAC to approximate the multiplicative function shown in Figure 7.10. The network

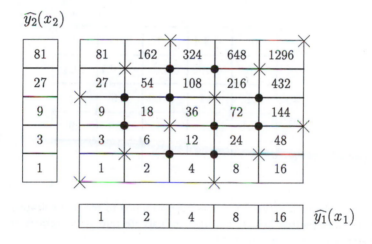

Figure 7.10 A two-dimensional multiplicative function (formed from the product of the two univariate functions shown). The circles represent the centres of the two-dimensional slabs (consistency equations and orthogonal functions) and the crosses denote the positions of corners of the basis functions.

has a generalisation parameter $\rho = 3$, an overlay displacement vector $\mathbf{d} = (1,1)$ and the size of the input space of interior is given by the number of interior knots: $r_1 = r_2 = 4$. The number of weights used by the network is $p = 17$, and the number of training samples is $p' = 25$. Therefore there exist ten consistency equations (and orthogonal functions) and their positions are also shown in Figure 7.10.

The optimal weight vector can be calculated using singular-valued decomposition matrix inversion techniques (nullity($\mathbf{A}$) = 2), and the optimal network approximation is shown in Figure 7.11 along with its residual. It is interesting to note that the modelling error is significant where the desired function changes rapidly (top right of the lattice) as predicted, and the absolute value of the maximum error is 110.7. It is also worth noting that the desired mapping is monotonically increasing with respect to both x_1 and x_2, but the optimal CMAC network is not.

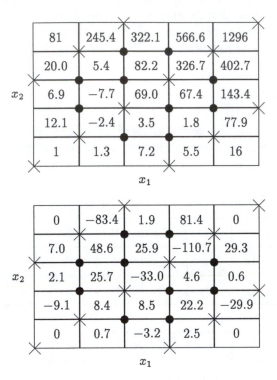

Figure 7.11   The network's least-squares approximation to the multiplicative function (top) and the residuals (bottom). Note that the residuals are zero at the corners of the lattice, as no consistency equations are defined there.

The ten local orthogonal functions (basis functions) were then used to model the residual data, and as predicted by Theorem 7.6 they modelled them exactly. The size of the weights in the orthogonal network (the linear parameters) are shown in Figure 7.12, along with their estimates formed using Equation 7.27. The local estimates are of the same magnitude and the lower bound for the modelling error is 54, which is approximately half the true value.

As stated previously, the calculated lower bound is relatively insensitive to the choice of the generalisation parameter and overlay displacement vector, although this means that for many networks the value may significantly underestimate the true modelling error. This is especially true when the generalisation parameter is large and the overlays are distributed unevenly.

	83.4	81.4	
−7.0	27.8		29.3
−9.3		5.3	29.9
	0.7	−2.5	

	27	54	
4.5	9		36
1.5		6	12
	1	2	

Figure 7.12   The true (left) and the estimated (right) values of the size of the local orthogonal functions. The input lattice is (5×5), therefore there are at most (4×4) orthogonal functions centred on the filled boxes in Figure 7.11. A blank space means that an orthogonal function is not defined at this place in the lattice.

7.6   INVESTIGATING THE CMAC'S COARSE CODING MAP

One of the most interesting (and computationally important) aspects of the CMAC algorithm is the initial nonlinear mapping. This transforms the network's input into a higher dimensional sparse space in which at most ρ of the variables have a non-zero output. The most significant part of this mapping is the arrangement of the basis functions, which ensures that only ρ contribute to the network's output, and this number does *not* depend on the position of the input in the lattice and the *dimension* of the input space. This last result is very surprising when it is considered that the size of the supports ρ^n depends *exponentially* on n, but the number of basis functions used to encode the network input is constant.

In this section, it is shown that the CMAC's encoding mapping is invertible: the original quantised input can be reconstructed from the position of the non-zero elements in the transformed input vector. Then an expression for the accuracy of a coarse coding map is interpreted in terms of the CMAC, which shows that the accuracy of the encoding is approximately *constant* and does not depend on the size (and number) of the supports. For supports of size ρ^n, the CMAC encoding is *minimalist*, using the smallest number of basis functions necessary to represent uniquely the network's input in a distributed fashion.

7.6.1   Inverting the CMAC Coarse Coding

It is now established that the coarse coding mapping used in the CMAC algorithm is one to one, so that each network input has a *unique* signature in the transformed input space.

Theorem 7.7 For any well defined binary CMAC, the coarse coding of the input space is a one-to-one mapping.

Proof. Consider two distinct network inputs, $x(1), x(2)$, where $x(1)$ and $x(2)$ are coarsely coded as $a(1)$ and $a(2)$, respectively. The network inputs lie in distinct cells in a finite, n-dimensional lattice and so a path from $x(1)$ to $x(2)$ exists which is completely contained within the lattice, which always moves parallel to one of the axes and always gets closer to $x(2)$, moving one cell at a time (with respect to $\|.\|_1$). Without loss of generality, the first step moves one cell parallel to the first axis and, due to the uniform projection principle, one bit of $a(1)$ is turned off and one new bit turned on. The bit turned off corresponds to a basis function whose support does not contain the new input. The supports are hypercubes, and so this basis function is never turned back on as the input moves along the path parallel to each of the axes. Therefore, the two transformed input vectors are different. □

The mapping is invertible if it is also onto, and for the remainder of this chapter it is assumed that the range of the coarse coding map is the set of those values generated by the mapping. Thus the coarse coding map has been defined to be invertible.

7.6.2 Accuracy of the Coarse Coding Mapping

The accuracy with which an input is represented by its coarse coding was investigated by Rumelhart and McClelland [1986]. It was assumed that the basis functions are uniformly distributed throughout the input space, that p basis functions are used to cover the whole of the input space, and that each basis function has a (spherical) support of radius ρ. In this case the accuracy of the approximation a (for an n-dimensional space) is:

$$a \propto p\rho^{n-1} \tag{7.29}$$

This measure of accuracy of the approximation is defined to be:

> *proportional to the number of different encodings that are generated as an input moves along a straight path from one side of the space to the other*

and this is illustrated in Figure 7.13.

The distribution of the CMAC's basis functions *approximately* satisfies these conditions; the supports are hypercubes rather than circles, although the basis functions *are* evenly distributed over the input space and exactly ρ basis functions are used to encode each input. Relationship 7.29 can then be used to express the accuracy of the CMAC coarse coding map.

Note: this relationship gives *no* information about how well the CMAC models a particular function. It is only concerned with the accuracy of the *internal* distributed representation.

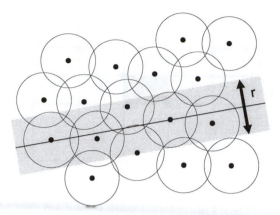

Figure 7.13   The number of different encodings which are generated as an input moves along a straight line in input space is proportional to the number of support centres lying in the shaded area.

Multivariate CMAC

For an n-dimensional multivariate input, the memory requirements for the CMAC is approximately given by:

$$p \propto \frac{1}{\rho^{n-1}}$$

Therefore the accuracy of the coarse coding map is approximately *constant*, i.e.:

$$a \approx c$$

for some constant c. This does *not* depend on ρ, which should not be too surprising since damaging *any* information held in the CMAC's internal structure may cause the encoding to be non-invertible. Thus ρ is the *minimum* number of active basis functions necessary to encode a multivariate input space with supports of size ρ^n evenly and sparsely.

This can be illustrated with an example. Consider a simple two-dimensional CMAC network with a generalisation parameter, $\rho = 3$, four interior knots on each axis and a unity overlay displacement vector. The corresponding input lattice is shown in Figure 7.14. In this figure, the solid box represents the network's input. The input is coarsely coded as a vector $\mathbf{a}$ that has three non-zero elements, which correspond to the basis functions whose supports contain the input. When the information contained in the first overlay is lost (where the overlay displacement vector is $\mathbf{d} = (1,1)$), the mapping is still invertible, although this bit of information is redundant only for this input. If the bit of information on the second overlay is destroyed (displacement vector $\mathbf{d} = (2,2)$), the inverse map gives *four* possible

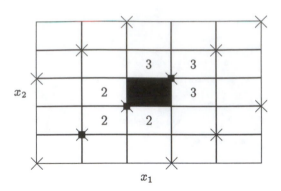

Figure 7.14   An illustration of the inconsistency which arises if any information is damaged in the CMAC coarse coding map. The solid box represents the network input, and the cells containing numbers shows how the network is unable to distinguish between the true input and this cell when information is lost for the corresponding overlay.

answers, the true input and the three cells which contain a 2. A similar situation also occurs on the third overlay.

When the generalisation parameter satisfies, $\rho \leq 2n$, there always exist inputs which lie on the boundary of the support of every non-zero basis function. This is because the $(n-1)$-dimensional hyperplanes which generate the lattice lie on one and only one overlay. Thus every bit of the distributed representation is important. If $\rho > 2n$, there exists at least one basis function, such that the input does not lie on the boundary of its support. This can easily be proved by noting that any n-dimensional cell can be constructed from $2n$ $(n-1)$-dimensional hyperplanes and these hyperplanes can lie on at most $2n$ different overlays. Therefore there exists at least one overlay in which the input does not lie on the boundary of the support of the non-zero basis function. If this bit of information is lost, the coarse coding map can still be successfully inverted, for a *particular* input.

Even if the generalisation parameter satisfies this lower bound, the whole coarse coding map is not truly distributed. Dropping *any* basis function from the output calculation results in the coarse coding map being non-invertible for at least one input. This occurs because each overlay is composed of a unique set of hyperplanes, which form part of the lattice. Losing information about one basis function results in the inverse map being unable to distinguish between inputs lying on either side of the relevant hyperplanes which form its support.

7.7 CONCLUSION

The ability to understand the modelling and generalisation abilities of an artificial neural network is important, especially when these algorithms are applied in on-line learning control loops. Modelling error plays a major part in determining the stability of such systems and, unless it can be verified that the structures of both the network and the unknown plant are appropriate, the system will perform poorly. Usually the structure of the plant is unknown, or only partially known, so the modelling abilities of the network must be analysed to see what type of functions can be stored, and which may be inappropriate for these learning algorithms.

This chapter has analysed the modelling capabilities of the binary CMAC from several viewpoints: its flexibility, the local consistency equations and orthogonal functions, classifying its generalisation ability in terms of local and global additive functions and deriving a lower bound for the network's modelling error. It is shown that increasing the generalisation parameter results in a less flexible model, and the only class of mappings which all the networks can store exactly are additive functions. The local consistency equations provide an insight into how the CMAC generalises, showing that the network forms locally additive models across any monotonic region generated by consistency equation slabs. They are also complete, as any function satisfying all of the consistency equations can be modelled exactly using a CMAC.

The local orthogonal functions completely determine the type of functions which the binary CMAC is unable to model in a least-squares sense. They can be formed from simple multiplicative functions, so while the CMAC generalises as a locally additive model, the desired function may contain multiplicative components which produce modelling error. It was also shown that it is possible to combine these two concepts (local consistency equations and local orthogonal functions) to derive a lower bound for the network's modelling error.

The modelling abilities of a multivariate network are completely determined by its local consistency equations, and the local orthogonal functions represent the component of the desired function which the CMAC is completely unable to model. Apart from providing a complete understanding of how the CMAC operates, these results can also be used to motivate new areas of research. For instance, Figure 6.5 shows the undesirable effects of using the original Albus overlay displacement strategy, and this is due to the large additive model produced when the network generalises. Using an improved overlay displacement strategy, such as those described in Appendix B, means that the network generalises locally, the additive models formed are much smaller and this results in a much better model, as shown in Figure 6.6.

Adaptive B-spline Networks

8.1 INTRODUCTION

The ability to learn highly nonlinear functional relationships, using only observed plant input/output data, is a very appealing concept. Humans perform this type of task every day, learning to stabilise and control complex systems in a robust fashion. We use little *a priori* knowledge about the process, and build up models which can generalise and extrapolate locally to unforeseen states. Learning is both *hierarchical* and *local*. Currently researchers are trying to endow artificial neural networks with these behaviours, although there is still much research which needs to be performed before computers can even come close to rivalling the skills of humans in these areas. It is the authors' opinion that truly autonomous behaviour (adaptive modelling and control strategies) requires the combination of many different learning skills: one shot remembering of facts, iteratively constructing complex nonlinear relationships, etc. Each level of intelligence requires appropriate representations and training mechanisms, and this chapter investigates the use of B-spline networks for servo level adaptive, nonlinear, on-line modelling and control, and in Chapter 9 for guidance level, static, off-line design applications.

B-splines have been employed as surface-fitting algorithms within the graphical visualisation community for the past twenty years. A major landmark in the use of B-splines occurred in 1972, when Cox [1972] and DeBoor [1972] independently derived a stable and efficient recurrence relationship for evaluating the basis functions. A B-spline function is simply a piecewise polynomial mapping, which is formed from a *linear* combination of basis functions, and the multivariate basis functions are defined on a lattice (or in the most general form, $K - d$ trees [Omohundro, 1987]). Thus a B-spline network can be considered as belonging to the class of lattice AMNs, as described in Chapter 3. The only reason for classifying a B-spline function as an AMN rather than as a surface-fitting algorithm, is the way in which the linear coefficients (weights) are generated. The *on-line* B-spline AMN adjusts its weights *iteratively* in an attempt to reproduce a particular function, whereas an *off-line* or *batch* B-spline algorithm typically generates the coefficients by matrix inversion or using conjugate gradient. These two applications are historically very different, although their aims are very similar: to reconstruct a particular nonlinear mapping given data which describe the input/output relationship.

B-spline AMNs adjust their (linear) weight vector, generally using instantaneous LMS-type algorithms, in order to realise a particular mapping, modifying the strength with which a particular basis function contributes to the network output. The network's sparse internal representation simplifies the learning process as only a small percentage of the total weights contribute to the output and only these parameters are modified by the LMS rules. Therefore this network has the potential to be used for on-line learning, as the adaptive algorithm operates in real-time and the adaptive rules are linear. For more complex learning tasks the model structure (definition of the basis function, inputs, etc.) must also be learned from the training data as *a priori* knowledge about its form is not generally available. It is possible to derive B-spline network initialisation algorithms [Kavli, 1993, 1994], which exploit the fact that, when the network's flexibility is increased, the new model can still reproduce the old one exactly. New variables can be introduced to provide extra information, cross-coupling dependencies modelled and the sensitivity of the network with respect to an input variable can be increased. All of these techniques make the network more *intelligent*, but its behaviour becomes more unpredictable and the computational burden is significant. Hence these algorithms should only be applied to off-line (not real-time) problems where data are plentiful.

One of the main reasons for investigating this AMN is because it provides a *direct* link between artificial neural networks and the fuzzy systems described in Chapter 10. From a fuzzy viewpoint, the univariate B-spline basis functions represent fuzzy linguistic statements, such as *the error is positive small*, and multivariate fuzzy sets are formed using the *product* operator to represent fuzzy conjunction. This link enables the B-spline networks to be interpreted as a set of fuzzy rules and allows modelling and convergence results to be derived for the fuzzy networks. These networks therefore embody both a qualitative and a quantitative approach, enabling heuristic information to be incorporated and inferred from neural nets, and allowing fuzzy learning rules to be derived, for which convergence results can be proved. This view allows the above network design algorithm to be interpreted as an automatic method for fuzzy knowledge elicitation, deciding on which input variables are important and the number of rules and concepts necessary to represent the unknown function adequately.

A standard B-spline AMN, designed using common fuzzy heuristics (seven triangular basis functions defined on each axis), is then applied to the time series prediction problem described in Section 6.6. It is shown that the B-spline network can model the data reasonably well, although its performance is worse than the CMACs and the Multi-Layer Perceptron (MLP) and Radial Basis Function (RBF) networks described by An *et al.* [1993c]. The local generalisation associated with each basis function means that the network has difficulty in significantly interpolating and extrapolating information and the internal representation could be considered inappropriate for this problem. The ASMOD algorithm is then used for the same prediction problem, and it produces a much simpler network, which correctly identifies the linear relationships contained in the noisy data. The network's performance is improved, and it is similar to the CMAC's, except that the B-spline system uses only 6 weights instead of 60. This improves its noise-filtering abilities, reduces its computational cost and makes the knowledge stored in the network easier to validate. The complex B-spline network (49 parameters) is more flexible, but is not appropriate for modelling this particular data set, and the ASMOD algorithm correctly parameterises a B-spline system. One of the advantages in using B-splines is that *linear* relationships can be encoded within the network's structure, and when this type of knowledge is unavailable, algorithms such as ASMOD may be used to discover it, thus providing the designer with valuable knowledge about the unknown process.

8.2   BASIC ALGORITHM

Like all AMNs, the output of the B-spline network is formed from a linear combination of a set of basis functions which are defined on the n-dimensional input space. By definition, the B-spline network is a lattice AMN, and so the basis functions are defined on a lattice which is positioned in the input space. Since the support of the basis functions is *bounded*, only a small number of weights are involved in the network output calculation, and as with the CMAC, the B-spline network stores and learns information locally.

When the B-spline network is initially designed, it is necessary to specify the shape (order) of each of the univariate basis functions, and this implicitly determines the number of basis functions mapped to for a particular network input. If the univariate B-splines are all of width k, k^n basis functions contribute to the network output. Unlike the CMAC, the size of the network's active internal representation is *exponentially* dependent on n; whereas for the CMAC it is a user-defined parameter which does *not* depend on the input space dimension. Thus the computational cost and the memory requirements of the B-spline network are exponentially dependent on n. The coarse coding strategy ensures that the computational cost of the CMAC network is only linearly dependent on n, although the underlying modelling capabilities are not as flexible as those of an order 1 B-spline network (look-up table).

Thus B-spline networks should only be used when the number of *relevant* inputs is small and the desired function is nonlinear, or when the desired mapping can be additively decomposed into a number of simple relationships. Algorithms can be used to find these dependencies in the data automatically and to structure the network accordingly, or else *a priori* knowledge about the partially known process can be incorporated into the network's design. However, when the number of inputs is large and simple additive relationships are not present in the data, either the cost of implementing the B-spline network is too great, or its performance is poor because the model is over-parameterised. Even with these reservations, this AMN can be applied to a wide range of tasks which exploit its desirable features.

8.2.1　Notation

The number of basis functions that contribute to the B-spline output is a constant which is determined by the network's other parameters, and in keeping with the notation already developed this is called the *generalisation parameter*, ρ $(= k^n)$. The supports of the basis functions can be divided into ρ sets, or overlays, where, on each overlay, only one basis function is active at any one time.

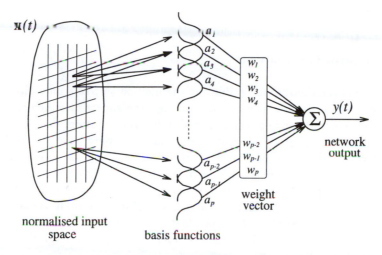

Figure 8.1　A schematic illustration of the B-spline network, showing the basis functions defined on an n-dimensional lattice.

The n-dimensional input to the network is denoted by **x**, and the network's (scalar) output by y. For the reasons given in Section 3.2.1, only scalar output networks are considered. As shown in Figure 8.1, the network can be decomposed into two parts: a static, nonlinear, topology conserving map and an adaptive linear mapping. The output of the ith basis function is denoted by a_i and the output of all the basis functions at time t is contained in the p-dimensional transformed input vector $\mathbf{a}(t)$. Hence the output of the B-spline network is given by:

$$y(t) \;=\; \sum_{i=1}^{p} a_i(t)\, w_i(t-1) \tag{8.1}$$

$$=\; \sum_{i=1}^{\rho} a_{ad(i)}(t)\, w_{ad(i)}(t-1) \tag{8.2}$$

where w_i is the *weight* corresponding to the ith basis function and $ad(i)$ is the address of the ith non-zero basis function $(i = 1, \ldots, \rho)$. The address calculation algorithm ensures that the output of only ρ basis functions needs to be calculated, rather than p, which results in a large reduction in the computational cost.

Simple LMS learning rules which exploit this sparse representation can be used to train the B-spline networks. These training algorithms were extensively discussed in Chapter 5, and therefore they are not considered in this chapter. The shape of the basis functions and the modelling capabilities of the network are determined by the orders of the univariate basis splines, and multivariate basis functions are placed at every available point within the n-dimensional lattice, so (unlike the CMAC) the overlay distribution does *not* influence the modelling capabilities of the network (see Figure 3.13). Hence, the remainder of this section concentrates on the generation of the basis functions and the evaluation of their integrals and differentials.

8.2.2   Univariate Basis Functions

The recurrence relationship for evaluating the membership of a univariate B-spline basis function of order k is now given, although the subscript which refers to the input's index i is temporarily omitted to simplify the notation.

Originally the univariate B-spline basis functions were calculated using a divided difference formula [DeBoor, 1978]. However, when the order of the splines is high or the placement of the knots is strongly non-uniform, this scheme is unstable. A breakthrough occurred in 1972, when a recurrence relationship was derived [Cox, 1972, DeBoor, 1972] which is numerically stable, computationally efficient and can deal with any knot distribution strategy. Denoting the jth univariate basis function of order k by $N_k^j(.)$, the basis functions are defined through the three term recurrence relationship:

$$N_k^j(x) = \left(\frac{x - \lambda_{j-k}}{\lambda_{j-1} - \lambda_{j-k}}\right) N_{k-1}^{j-1}(x) + \left(\frac{\lambda_j - x}{\lambda_j - \lambda_{j-k+1}}\right) N_{k-1}^j(x)$$

$$N_1^j(x) = \begin{cases} 1 & \text{if } x \in I_j \\ 0 & \text{otherwise} \end{cases} \tag{8.3}$$

where λ_j is the jth knot and I_j $(= [\lambda_{j-1}, \lambda_j))$ is the jth interval, as defined in Section 3.2.3.

Univariate basis functions of orders $k = 1, \ldots, 4$ are shown in Figure 8.2 and it is easy to see that the basis functions and also the network's output become *smoother* as the order increases. The relationship between the basis function shape and the order of the polynomial is not obvious from this figure. On each interval the basis function is a polynomial of order k, and at the knots the polynomials are joined smoothly. This is illustrated in Figure 8.3, where the polynomial representations of the quadratic and the cubic basis functions are shown, and each basis function is composed of k polynomial segments. It should therefore be emphasised that the recurrence relationship given in Equation 8.3 simply generates polynomial segments, as can be seen from closer inspection. On each interval, k coefficients are

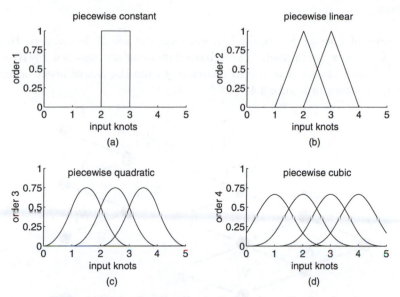

Figure 8.2 Univariate B-spline basis functions of orders $1 - 4$.

necessary to represent a polynomial of order k, and k B-spline weights (basis functions) contribute to the output. There is a *direct* correspondence between the two forms.

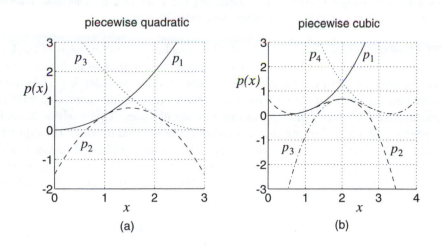

Figure 8.3   Polynomial representations of (a) quadratic and (b) cubic B-spline basis functions. The knots are placed at the integers and on the ith interval $[i - 1, i)$ the basis function is simply the polynomial $p_i(x)$.

Evaluation

The number of operations required to evaluate the set of k non-zero B-splines of order k are now calculated. These basis functions are denoted by N_k^j, N_k^{j+1}, ..., N_k^{j+k-1}, and can be determined recursively from the active lower order basis functions, as shown in Figure 8.4.

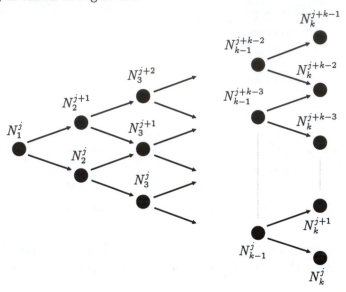

Figure 8.4   The triangular array used for evaluating the k non-zero B-spline basis functions of order k. All the basis functions not shown are identically zero.

The basis functions of order k are calculated as linear combinations of the basis functions of order $k-1$. The evaluation of the basis function of order k involves two multiplications and two divisions for each basis function of order $k-1$. However the divisions are identical, so each order $k-1$ basis function has associated with it a computational cost of two multiplications and one division. This analysis holds for any order $k \geq 2$, and assuming that the index of the interval in which the input lies is available, all that needs to be determined is the number of basis functions in the triangular array (Figure 8.4) up to and including the $(k-1)^{th}$ layer. The number of basis functions can easily be shown to be $k(k-1)/2$ and the total number of arithmetic operations is:

$$k(k-1) \qquad \text{floating point multiplications}$$
$$\frac{k(k-1)}{2} \qquad \text{floating point divisions} \qquad\qquad (8.4)$$

The number of floating point additions and subtractions have not been considered, because they are of the same order and their computational cost is negligible in comparison.

Properties

The B-spline basis functions which are generated by the recurrence relationship given in Equation 8.3 have many desirable properties:

1. The basis functions are defined on a *bounded* support and the output of the basis function is *positive* on its support, i.e. $N_k^j(x) = 0$, $x \notin [\lambda_{j-k}, \lambda_j]$, and $N_k^j(x) > 0$, $x \in (\lambda_{j-k}, \lambda_j)$.

2. The basis functions form a partition of unity. For any network input, the sum of the outputs of the basis functions is always one, i.e. $\sum_j N_k^j(x) \equiv 1$, $x \in [x^{\min}, x^{\max}]$.

3. The basis functions (and hence the output of the network) are members of the continuity class $C^{k-2}(x^{\min}, x^{\max})$, for simple knots. Thus $N_k^j(x)$ and its derivatives up to the $(k-2)^{nd}$ are continuous on $(x^{\min}, x^{\max})$.

4. The output of the B-spline network is bounded above and below by the values of the weights which correspond to the non-zero basis functions. This can be expressed as $\min\{w_j, w_{j+1}, \dots, w_{j+k-1}\} \leq y(x) \leq \max\{w_j, w_{j+1}, \dots, w_{j+k-1}\}$ for an input lying in the jth interval.

The first property shows that the univariate basis functions have a bounded support, which means that only a small number of basis functions ($\rho = k$) contribute to the network's output. Information is stored and learnt *locally* across these k basis functions. Also, it is generally required for the basis function to form a partition of unity, as undesirable oscillatory behaviour can be induced in the network's output if there are variations in the sum of output of the basis functions (see the CMAC examples in Section 6.2). The third property is required when the network is defined as a *spline* function [Cox, 1972]. If there are r coincident knots, the basis functions and the network output have $(k - (r+1))$ discontinuous derivatives at this point. This enables discontinuities to be hardwired into the network. For instance, when basis functions of order 2 are defined on an interval which contains two coincident knots, the network output is discontinuous at this point. Finally, the last property shows how qualitative knowledge about the network's output can be obtained from examining the local values of the weights. The upper and lower bounds are always achieved for networks of orders 1 or 2, although they are generally not sharp for higher order networks.

8.2.3 Differentiation and Integration

The derivative of a univariate B-spline network is simply a linear combination of the derivatives of each basis function, and the derivative of a network of order k is a piecewise polynomial of order $(k-1)$. In fact, it is shown that the derivative is *another* B-spline of order $(k-1)$, whose weights can be calculated using a *simple* recurrence relationship. It follows that the indefinite integral of a B-spline of order

k is another B-spline of order $(k+1)$, whose coefficients can also be found from a similar expression.

These relationships enable the network, and its derivatives, to be constrained in real-time; a property which is used extensively in Chapter 9 for on-line constrained trajectory generation.

Differentiation

For a B-spline network whose output is given in Equation 8.1, the derivative can be evaluated by [Cox, 1982]:

$$\frac{dy(x)}{dx} = \sum_{j=1}^{p} w_j \frac{d}{dx} N_k^j(x) = \sum_{j=1}^{p}(k-1)w_j \left(\frac{N_{k-1}^{j-1}(x)}{\lambda_{j-1} - \lambda_{j-n}} - \frac{N_{k-1}^{j}(x)}{\lambda_j - \lambda_{j-k+1}} \right) \quad (8.5)$$

where Equation 8.3 has been used to evaluate the derivative of the univariate basis functions. This can be rearranged by noticing that $N_{k-1}^0(x)$ and $N_{k-1}^p(x)$ are identically zero on the domain of interest, so the above expression becomes:

$$\frac{dy(x)}{dx} = \sum_{j=1}^{p-1}(k-1)\left(\frac{w_{j+1} - w_j}{\lambda_j - \lambda_{j-k+1}} \right) N_{k-1}^j(x) \quad (8.6)$$

The derivative of a B-spline of order k is therefore a B-spline of order $(k-1)$, whose weights are generated using a simple expression. This differentiation process can be repeated up to $(k-1)$ times. The rth derivative of a B-spline of order k, $y^{(r)}(x)$ is given by:

$$y^{(r)}(x) = \sum_{j=1}^{p-r} w_j^r N_{k-r}^j(x) \quad (8.7)$$

where the weights w_j^r are calculated using the following recurrence relationship:

$$w_j^r = \begin{cases} w_j & \text{if } r = 0 \\ (k-r)\left(\frac{w_{j+1}^{r-1} - w_j^{r-1}}{\lambda_j - \lambda_{j-k+r}} \right) & \text{for } r = 1, \dots, k-1 \end{cases} \quad (8.8)$$

Therefore up to the $(k-1)^{st}$ derivative of a B-spline network can be evaluated using the above simple recurrence relationship. Increasing the order of the basis function, and hence the smoothness of the network's output, also increases the number of derivatives which can be evaluated, and requiring a network to satisfy a given number of differential constraints immediately places a lower bound on the order of the basis functions.

Integration

The expression for the indefinite integral of a B-spline network follows immediately from the recurrence relationships for the network's derivative [Cox, 1982].

The rth indefinite integral of a B-spline network, $y^{(-r)}(x)$, of order k is given by:

$$y^{(-r)}(x) = \sum_{j=1}^{p+r} w_j^{-r} N_{k+r}^j(x) + f(r) \tag{8.9}$$

where the weights w_j^{-r} are recursively defined as:

$$w_i^{-r} = \begin{cases} 0 & \text{if } j \leq r \\ w_{j-1}^{-r} + \left(\frac{\lambda_{i-1}-\lambda_{i-k-r}}{k+r-1}\right) w_{j-1}^{1-r} & \text{for } j > r \end{cases} \tag{8.10}$$

and $f(r)$ is a function which represents the r degrees of freedom introduced by the integration process. This function, $f(r)$, is simply a polynomial of order r, which can be incorporated (in a stable fashion) into the network's weight set $\{w_j^{-r}\}$ [Cox, 1978].

8.2.4 Multivariate Basis Functions

Multivariate B-spline basis functions are formed by taking the tensor product of n univariate basis functions, where one and only one univariate basis function is defined on each input axis. Thus the desirable properties of the univariate B-spline basis functions (bounded support, uniform field strength, piecewise polynomial modelling) are all extended in a natural way to the multivariate basis functions.

The jth multivariate B-spline basis function $N_{\mathbf{k}}^j(\mathbf{x})$ is generated by multiplying n univariate basis functions $N_{k_i,i}^j(x_i)$, $(i = 1, \ldots, n)$:

$$N_{\mathbf{k}}^j(\mathbf{x}) = \prod_{i=1}^{n} N_{k_i,i}^j(x_i) \tag{8.11}$$

where each multivariate basis function is calculated using a different set of univariate basis functions (and conversely every set of univariate basis functions generates a multivariate basis function), and $\mathbf{k}$ is an n-dimensional integer vector composed of the orders of the univariate basis functions, k_i.

The order of the univariate basis functions determines the smoothness of the multivariate basis functions and the network output. Multivariate basis functions of different shapes and sizes are generated when different univariate basis functions are combined, and some low-order two-dimensional basis functions are shown in Figures 8.5-8.9. The first three figures show two-dimensional multivariate basis functions formed using two univariate B-splines of orders $k_1 = k_2 = 1, 2, 3$, respectively, whereas the multivariate basis functions in the last two figures are formed

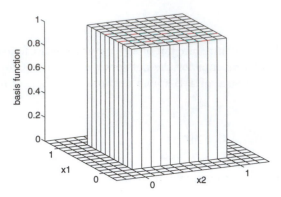

Figure 8.5   Two-dimensional multivariate basis function formed from two, order 1, univariate basis functions.

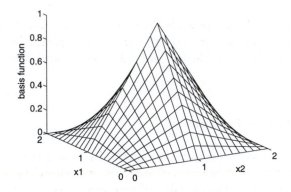

Figure 8.6   Two-dimensional multivariate basis function formed from two, order 2, univariate basis functions.

from univariate basis functions whose orders are different on each axis. Allowing basis functions of different orders to be defined on different input axes means that *a priori* information can be reflected in the structure of the network, and this is illustrated in Section 9.2 and An *et al.* [1993c].

Memory Requirements

The number of basis functions of order k_i defined on an axis with r_i interior knots is $r_i + k_i$. Therefore the total memory requirements for a multivariate B-spline network is given by:

$$p = \prod_{i=1}^{n} (r_i + k_i) \tag{8.12}$$

because every possible combination of univariate basis functions is taken. This number depends *exponentially* on the size of the input space and, unless some

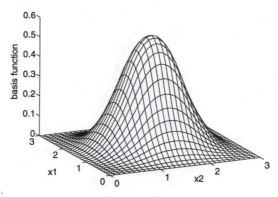

Figure 8.7 Two-dimensional multivariate basis function formed from two, order 3, univariate basis functions.

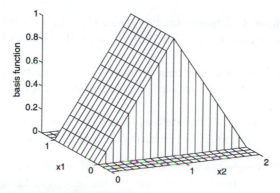

Figure 8.8 Two-dimensional multivariate basis function formed from an order 1 (x_1) and an order 2 (x_2) univariate basis function.

technique is used to reduce the size of the input space, these algorithms are only suitable for applications which have a small number of inputs (typically ≤ 5). There are several ways in which these networks can be applied in higher dimensional spaces. One technique which is suitable for on-line learning is to use memory hash-coding, where the large weight vector is randomly mapped onto a smaller physically realisable memory (see Section 6.2.5). Another more effective approach would be to try and exploit any redundancy in the structure of the desired function and construct a model composed of several smaller subnetworks, and this is the philosophy behind Kavli's ASMOD algorithm [1994] described in Section 8.5.

Properties

The multivariate basis functions share the properties of the univariate basis functions from which they are formed:

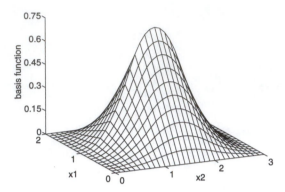

Figure 8.9 Two-dimensional multivariate basis function formed from an order 2 (x_1) and an order 3 (x_2) univariate basis function.

- They are defined on hyperrectangles of size $(k_1 \times k_2 \times \cdots \times k_n)$ and therefore possess a bounded support. The output is positive inside this domain and zero outside.
- The multivariate basis functions form partition of unity $(\sum_{j=1}^{p} N_{\mathbf{k}}^{j}(\mathbf{x}) \equiv 1,$ $\forall \, \mathbf{x})$. Thus the network's surface is not influenced by a non-uniform internal representation.
- The outputs of the multivariate basis functions are piecewise polynomials of order k_i and are members of the class $C^{k_i-2}(x_i^{\min}, x_i^{\max})$.

Therefore the multivariate basis functions retain *all* of the desirable properties of the univariate basis functions. Unfortunately this results in the memory requirements being exponentially dependent on n, and so these networks are best suited to smooth, highly nonlinear, low-dimensional modelling tasks.

Contour Plots

The contour plots generated by multivariate basis functions of different orders are now investigated. It is demonstrated that as the orders of the basis functions increase, the contours become circular close to the centre of the multivariate basis function and so resemble the radial basis functions described in Section 3.3.4.

    Figure 8.10 shows the contour plots of two-dimensional basis functions of orders 2, 3 and 4 (piecewise linear, quadratic and cubic). The contour plot of the basis function of order 2, whose output shape is plotted in Figure 8.6, shows that for inputs close to the centre of the basis function, $\mathbf{c}_j$, the contours are diamond shaped and the output of the basis function can be approximated by:

$$N_2^j(\mathbf{x}) = 1 - \|\mathbf{c}_j - \mathbf{x}\|_1$$

When the input is close to the edge of the support, the output of the basis function

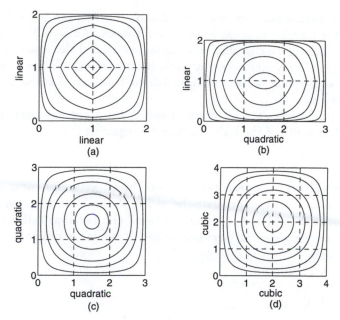

Figure 8.10 Contour plots of two-dimensional linear, quadratic and cubic B-spline basis functions ((a), (c), and (d), respectively) and a linear/quadratic combination (b) on unity-spaced, simple knots.

is approximately proportional to $\|\mathbf{c}_j - \mathbf{x}\|_\infty$, and in between, the higher order terms provide a continuous transformation between the two distance measures.

Higher order basis functions ((c) and (d)) have smoother contour plots as they possess a continuous first derivative at the knots. When the order of the basis function increases, the distance measure appears to approximate the Euclidean norm for inputs close to the centres of the basis functions. If this is true, the network is very similar to the RBF network (under certain assumptions) and the basis functions resemble scaled versions of the superspheres and the compact Gaussian functions (see Section 6.4.1), which were originally proposed for the CMAC algorithm.

8.2.5 Dilated Basis Functions

The basic B-spline network is quite flexible in that it allows the designer to model individual inputs with polynomials of any order. However, there is a direct relationship between the width of the univariate basis functions and its order, the basic algorithm sets them equal; a basis function of order k_i is defined over k_i intervals. It may be necessary to have wide basis functions, for instance, to increase the initial rate of convergence, but this results in a set of high-order splines being generated.

High-order basis functions are generally too flexible, causing the network to *overfit* the data, and the learning problem is badly conditioned. An algorithm is required which can decouple the relationship between the width and the shape of the basis functions and generate wide, low-order basis functions with all the properties of the conventional B-spline basis functions; dilated B-splines [Lane *et al.*, 1992] is one such technique.

Dilated B-splines relax the direct relationship which generally exists between the order and the width of the basis function by requiring that:

$$l_i = m_i k_i$$

where l_i is the width of the basis functions on the ith axis, and m_i is a positive integer. Thus dilated B-splines allow the width of the basis function to be an integer multiple of the order of the basis functions, and two illustrations are given in Figure 6.11, showing conventional and dilated B-splines of orders 2 and 3 with $m_i = 2$.

Evaluation and Properties

These basis functions are worthy of study because there exists a recurrence relationship, equivalent to Equation 8.3, which can be used to calculate their output. Suppressing the dependency on the ith axis, the output of the jth basis function of order k and width l, and constant m is given by:

$$N_{m,k}^j(x) = \left(\frac{x - \lambda_{j-mk}}{\lambda_{j-m} - \lambda_{j-mk}} \right) N_{m,k-1}^{j-m}(x) + \left(\frac{\lambda_j - x}{\lambda_j - \lambda_{j-(k-1)m}} \right) N_{m,k-1}^j(x)$$

$$N_{m,1}^j(x) = \begin{cases} 1 & \text{if } x \in \bigcup_{i=j-m+1}^{j} I_i \\ 0 & \text{otherwise} \end{cases} \qquad (8.13)$$

This recurrence relationship ensures that all of the properties of the B-spline basis functions also hold for the dilated B-splines; they are zero outside their support and positive in the interior, they are piecewise polynomials of order k, belong to the class of functions C^{k-2} and their field strength is constant. The sum of the dilated basis functions' outputs is always m, and they form a partition of unity if, after evaluation, the output of each basis function is divided by m. Relationships which can calculate the derivative and integral of a dilated B-spline are also simple generalisations of the conventional formulae [Lane *et al.*, 1992].

A Conventional Interpretation

The dilated B-splines have an interesting relationship with conventional B-splines, in that they can be regarded as a set of m conventional B-spline basis functions of order k, defined on m modified knot sets. This interpretation explains why the recurrence relationships for evaluating, differentiating and integrating the dilated basis functions are trivial generalisations of the conventional formulae, and demonstrates why the sum over all the basis functions is always m.

Define the modified knot sets, κ_j, $j = 1, \ldots, m$, by:

$$\kappa_{j,i} = \lambda_{j+m(i-k)} \qquad \text{for } i = 1, \ldots \qquad (8.14)$$

and on each of these modified knot sets define a set of conventional univariate B-spline basis functions of order k. The union of these basis functions is *equivalent* to the dilated B-spline basis functions, because the modified knot sets have a null intersection and their union generates the original knot set. The output of the dilated B-splines is therefore calculated using the conventional recurrence relationship on a modified knot set and the field strength is always the sum of m partitions of unity. This interpretation is illustrated in Figure 8.11 for order 2 basis functions with $m = 2$.

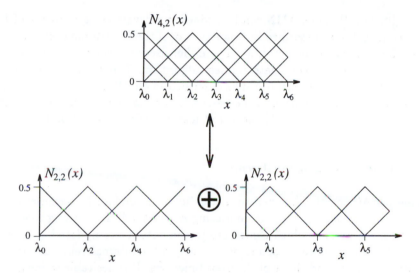

Figure 8.11 The relationship between dilated and conventional B-spline basis functions for order 2 network with basis functions of width 4.

Implementation Cost

The memory requirements for the dilated B-splines is slightly greater than that of the conventional B-spline network, although the number of arithmetic operations is considerably greater and this is due to the increased number of basis functions which contribute to the output.

For a univariate dilated B-spline network, there are always m times as many basis functions involved in calculating the output and the arithmetic cost is m

times that of a conventional network (see the previous network comparison). The multivariate dilated B-spline network has m^n times as many basis functions contributing to the output, but careful programming can ensure that the arithmetic cost of the multivariate dilated B-spline network is only m times as great as the conventional network.

When these dilated basis functions are to be used in other AMNs (such as the CMAC which is described in Section 6.4.2 and Lane *et al.* [1992]), the computational cost of the algorithm can be reduced dramatically, although the modelling capabilities of the network are also substantially decreased, as illustrated in Chapter 7.

8.3   B-SPLINE LEARNING RULES

The weights in a B-spline AMN can be trained iteratively using a variety of LMS-type learning rules, as described in Chapter 5, or if the full training set is available off-line, the weights can be determined by either pseudo-inverting the autocorrelation matrix or using the conjugate gradient algorithm. The iterative learning rules can be used to train any of the AMNs described in this book, and the only practical difference is the *condition* of the basis functions which determines the rate of parameter convergence. It is this aspect of learning that this section focuses on.

8.3.1   Univariate Learning Interference

Both the CMAC and B-spline networks store information locally, because the basis functions have a bounded support, and using *any* of the LMS learning rules means that only those weights which contribute to the output are updated. The extent to which learning is local depends on the network's internal representation and therefore on the size and shape of the basis functions. It is desirable to measure the amount of learning interference for different basis functions to compare their storage ability and provide an indirect measure of the rate of parameter convergence. In Section 5.5, a definition of learning interference was given, and it is shown that such a measure can be generated by calculating the absolute value of the largest root of:

$$\lambda^{\rho-1} + \delta \sum_{i=1}^{\rho-1} \lambda^{\rho-1-i} \left(\sum_{j=1}^{\rho-i} a_j a_{j+i} \right) = 0$$

For univariate B-spline networks, the number of non-zero basis functions, ρ, is equal to the width (and order) of each basis function l.

The output of each basis function depends on the point at which it is evaluated, and so two sets of results for the learning interference measures are presented: the first gives the interval in which the learning measure lies and the second is the expected value (for a uniform probability density function). This is not necessary for the CMAC, because the basis functions are binary and their value does not depend on the evaluation point.

k	B-spline basis functions min, max	$E(.)$	Dilated B-spline basis functions min, max	$E(.)$
1	[0.0, 0.0]	0.0	[0.5, 0.5]	0.5
2	[0.0, 0.5]	0.285	(0.370, 0.409)	0.389
3	(0.162, 0.5]	0.338	(0.390, 0.418)	0.403
4	(0.230, 0.451)	0.343	(0.423, 0.431)	0.427
5	(0.260, 0.375)	0.309	(0.440, 0.450)	0.450

Table 8.1 Learning interference measures for B-spline basis functions of orders $1, \ldots, 5$, and dilated B-splines with $m = 2$.

The learning interference intervals and expected values are shown in Table 8.1 for conventional B-spline basis functions of orders 1 to 5. As the order of the basis function increases, their supports become correspondingly wider, although the learning interference measure increases only slightly. This is because the shape of the basis function becomes more localised, unlike the binary basis functions which were originally proposed for the CMAC. The corresponding learning interference measures have also been generated for the dilated B-spline basis functions, which are twice the width of the conventional basis functions, as shown in Table 8.1. They are greater than the corresponding univariate B-spline basis function learning measures, but less than those generated by the binary basis functions. Generally, learning interference is greater for wide, binary basis functions.

The learning interference measures which have been generated in this book cannot all be directly compared. For instance, it is wrong to directly compare all of the learning interference measures in Table 8.1 because they are generated by assuming that the model and the network have the same basic structure. Changing the order of the basis functions means that the networks are modelling *different* functions. While this gives an *indication* of the amount of learning interference, meaningful comparisons can only be made when the different basis functions belong to the same class, as occurs with the binary CMAC or for the dilated B-splines of the same order and different widths.

8.3.2 Condition of the B-spline Basis

In Chapter 4, it was shown that the rate of convergence of gradient descent learning laws depends directly on the *condition* of the autocorrelation matrix, $C(\mathbf{R})$, and that weight convergence can only be inferred from output convergence if the basis is *well-conditioned*. Thus the success of any learning algorithm depends on the condition number of the nonlinear transformation which occurs in the hidden layer. This is a *much* neglected aspect of research in the neural network community, whereas in the numerical analysis/approximation fields, a large body of relevant theory has been developed, and some of these results are now reviewed.

Condition of an Arbitrary Basis

Consider an AMN whose output is given by $y(\mathbf{x}) = \mathbf{a}^T(\mathbf{x})\mathbf{w}$. The *condition* of the set of basis functions, $\{a_i(.)\}_{i=1}^p$ can be used to infer information about the convergence of the weight vector, given measurements which describe the accuracy of the network's output. As has been stated previously, parameter convergence is essential if the AMNs are expected to generalise correctly outside their training domain, and the condition of the basis functions provides an important measure of how well this is achieved.

For a univariate AMN with a weight vector $\mathbf{w}$, the output of the network defined on the interval $\left[x^{\min}, x^{\max}\right]$ is $y(x) = \mathbf{a}^T(x)\mathbf{w}$. By defining two appropriate vector and function measures $\|.\|$, and two positive numbers m and M as:

$$m = \min_{\mathbf{w}} \frac{\|y(x)\|}{\|\mathbf{w}\|} \tag{8.15}$$

$$M = \max_{\mathbf{w}} \frac{\|y(x)\|}{\|\mathbf{w}\|} \tag{8.16}$$

then:

$$m\|\mathbf{w}\| \le \|y(x)\| \le M\|\mathbf{w}\| \tag{8.17}$$

and the condition number of the basis functions is defined as:

$$C(\mathbf{a}) = \frac{M}{m} \tag{8.18}$$

From this definition, it can easily be shown that the normalised error in the output corresponding to a normalised error in the weight vector of the network can be bounded by:

$$\frac{\|\epsilon_{\mathbf{w}}\|}{\|\mathbf{w}\|} \le C(\mathbf{a})\frac{\|\epsilon_y(x)\|}{\|y(x)\|} \tag{8.19}$$

where $y(x) + \epsilon_y(x) = \mathbf{a}^T(x)(\mathbf{w} + \epsilon_{\mathbf{w}})$. When the condition number of the set of basis functions is large, a small normalised output error does *not* necessarily imply that the normalised error in the weight vector is small.

Univariate B-spline Basis Functions

When the distance measure is taken to be the infinity norm, i.e.:

$$\|\mathbf{w}\|_\infty = \max_{i=1}^{p} |w_i|$$

$$\|y(x)\|_\infty = \max_{x\in[x^{\min},x^{\max}]} |y(x)|$$

then the condition number $C_{k,\infty}(\mathbf{a})$ of the set of B-spline basis functions of order k exists and does not depend on the knot sequence:

$$\|\mathbf{w}\|_\infty \leq C_{k,\infty}(\mathbf{a})\|y(x)\|_\infty$$

The smallest possible values for $C_{k,\infty}(\mathbf{a})$ are given in Table 8.2 [DeBoor, 1976].

k	2	3	4	5	6	7	8	9	10
$C_{k,\infty}(\mathbf{a})$	1	3	5	11.7	21	46.2	85.8	183.9	347.3

Table 8.2 Condition of the univariate B-spline basis functions.

The following relationship also approximately holds for the condition of the B-spline basis functions:

$$C_{k,\infty}(\mathbf{a}) \approx 2^{k-1.5} \tag{8.20}$$

which shows that as the spline order is incremented by one, the corresponding condition of the set of basis functions increases by a factor of two.

The order of the basis functions should therefore be sufficiently high that the desired function can be modelled adequately, but it should also be as small as possible to keep the basis well conditioned. Using basis functions of as low an order as possible also reduces the computational cost of the algorithm and lowers the possibility of overfitting the data.

8.4   B-spline Time Series Modelling

An adaptive B-spline AMN is now applied to the time series prediction problem considered in Section 6.6. The network is not "optimal" and a more appropriate structure is described by An *et al.* [1993c], although the aim of the two time series prediction problems considered in this chapter is to illustrate that:

- **parsimonious** networks have improved generalisation characteristics; and
- **automatic** knowledge extraction algorithms are available for structuring B-spline (fuzzy) networks.

Therefore the B-spline network used in this section has seven univariate order 2 basis functions defined on each axis, representing the seven triangular fuzzy sets commonly used in modelling and control applications. Little prior knowledge is assumed about the form of the prediction surface, even less than the CMAC network, which had an extremely large generalisation parameter on the presumption that the desired surface was very smooth. Hence, the system would be expected to have a large variance, as no information has been incorporated into the network's structure in order to bias its design.

The time series prediction problem reported in Section 6.6, where $\eta(.)$ is a zero mean white Gaussian noise sequence with a variance of 0.01 is again used. This allows a direct comparison to be made between the CMAC and B-spline networks for this prediction problem. The training set of 300 examples is shown in Figure 6.18, and the additive Gaussian noise provides a large set of points around the globally attracting limit cycle, although in its interior and around the exterior the training data are sparsely distributed, thus testing the network's ability to interpolate and extrapolate.

8.4.1   B-spline Network Design

The network's two-dimensional input domain is $[-1.5, 1.5] \times [-1.5, 1.5]$, with seven, order 2 basis univariate basis functions used to represent both $y(t-2)$ and $y(t-1)$. Their respective domains are divided into six equal intervals, each of width 0.5, so the interior knot sets is given by $\{-1.0, -0.5, 0.0, 0.5, 1.0\}$. The multivariate B-spline basis functions are therefore similar to those shown in Figure 8.6; piecewise linear with respect to both variables $y(t-2)$, and the network uses 49 multivariate basis functions (and hence weights) in total, with ($\rho =$) four weights contributing to the output at each time instant. An SALMS learning rule is used to train the weight vector with a learning rate of the form:

$$\delta_i(t) = \frac{1}{1 + t_i/50.0} \tag{8.21}$$

where t_i is the number of times that the ith weight has been updated.

8.4.2   Simulation Results

The performance of the B-spline network is now examined, using a suite of tests similar to those employed in Section 6.6. These tests are designed to evaluate the ability of the network to generalise and extrapolate locally, to assess the smoothness of the network's output and to see how well the network has learnt to approximate the underlying dynamics of the recurrence relationship.

The network's one-step-ahead prediction surface is shown in Figure 8.12, and the basis functions' shape (piecewise bivariate) determines its form. Its ability to

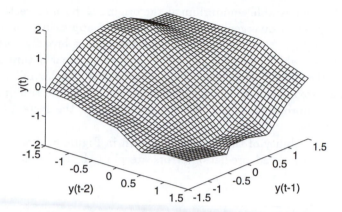

Figure 8.12 The B-spline network approximation to the time series surface where the network's output lies in the range $[-1.35, 1.35]$.

interpolate and extrapolate is also reflected in this figure; for inputs lying within the limit cycle, derivative discontinuities (especially with respect to $y(t-2)$) can be seen that are not present in the original function. Similarly, network's ability to extrapolate the learnt information is quite limited, and both of these effects are due to the localised definitions of the basis functions coupled with the distribution of the training data.

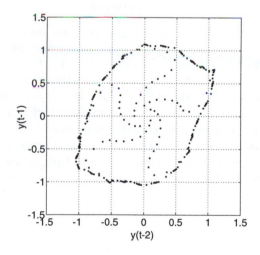

Figure 8.13 Iterated dynamics of a B-spline network from the initial condition $x(1) = (0.1, 0.1)^T$. The jagged nature of the limit cycle and the limited generalisation ability in the interior of the limit cycle is due to the model being too flexible.

The network's limited ability to interpolate is also observed in the dynamic behaviour of the iterated system, as shown in Figure 8.13. Within the interior of the limit cycle, the unstable equilibrium at the origin has been correctly modelled, although the system's divergent behaviour is a poor approximation of the true one (see Figure 6.16). Also, the shape of the limit cycle is broadly correct, but it has many "kinks", which again are due to the piecewise linear nature of the basis functions and the data distribution. This network has the flexibility to provide an excellent approximation of the true function, but when the training data distribution is not appropriate (a network specific concept), the system is unable to generalise correctly.

The prediction ability of the network is shown in Figure 8.14 where the mean square k-step-ahead ahead prediction errors are plotted. It shows that good es-

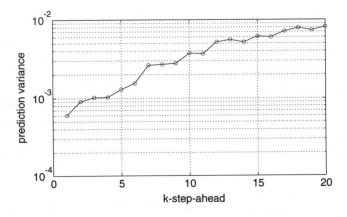

Figure 8.14 Variance of the k-step-ahead prediction errors.

timates can be obtained by iterating the network up to 20 steps into the future, although the variances are significantly higher than those of the CMAC. This is confirmed in Figure 8.15, where the network and the noiseless time series are run in parallel for 50 iterations. As can be seen from these two figures, there is good agreement between the true and the estimated time history, but the network and the true iterated mapping are starting to diverge as t approaches 50. The B-spline network produced an adequate model of the time series, and this is also confirmed in Figure 8.16, where the normalised residual autocorrelations, after 20 batch training cycles, are shown.

8.4.3 Discussion

These results may seem discouraging, as they indicate that the B-spline network performs significantly worse than the CMAC. However, this would be misleading and there is a good reason for including this section. Many fuzzy systems are initialised with seven sets on each axis, solely because this means that the set of linguistic labels (negative big, negative medium, negative small, almost zero,

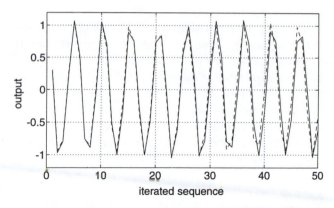

Figure 8.15 The noiseless time series and the B-spline network iterated mappings from an initial condition $y(-1) = 0.5$, $y(0) = 1.0$. The dashed line represents the true time series and the solid line shows the B-spline iterated mapping.

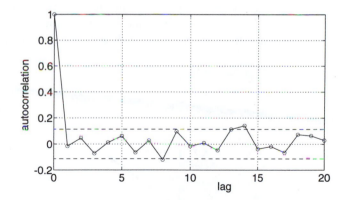

Figure 8.16 The autocorrelation of the prediction errors after training has ceased. The 95% confidence band is represented by the dashed lines.

positive small, positive medium, positive large) can be used to reference the basis functions defined on each axis. This provides a reasonably flexible system which is *not* appropriate for some adaptive problems, either because of the form of the underlying function or the data distribution. Both could result in the network's variance being high, as its structure was chosen without considering the complexity of the learning problem. Despite this, the network's performance is acceptable and in Section 8.6 a B-spline network is developed (automatically) which is appropriate for this prediction problem.

8.5 MODEL ADAPTATION RULES

In this book it has been assumed that the structure of the model remains *constant*, the network's inputs (number and type) do not change and the number and position of the basis functions are not adapted. This implicitly assumes that the network's internal representation is adequate, and that for efficient storage, learning and computation, it is sufficiently flexible. The number of inputs, their internal dependencies and the number of basis functions determine the network's structure, and when these parameters cannot be adequately determined prior to the network being trained, algorithms for automatically generating them must exist. Two such procedures are now discussed.

8.5.1 Network Representation

B-spline networks provide a very flexible modelling algorithm, although, for many practical medium and high-dimensional problems, the representation is *too* flexible as the memory and computational requirements are not feasible. Using such an algorithm would not be in keeping with the heuristic proposed in Chapter 4, where it was postulated that the *simplest* acceptable adaptive system performs the best. As an example, consider the five input, B-spline network shown in Figure 8.17. A conventional multivariate B-spline network with seven univariate sets defined

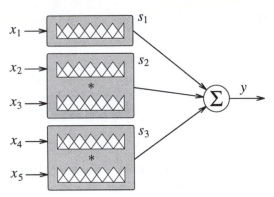

Figure 8.17   The basic ASMOD algorithm, where the B-spline network is implemented as a set of low-dimensional submodels.

on each axis would require 7^5 $(= 16,807)$ storage locations and each input would activate 2^5 $(= 32)$ basis functions, producing a model of the form:

$$y = f(x_1, x_2, x_3, x_4, x_5)$$

If it is supposed that the desired function can be *additively* decomposed, such that it can be successfully modelled from a linear combination of two, two-dimensional subnetworks and a univariate subnetwork:

$$y = s_1(x_1) + s_2(x_2, x_3) + s_3(x_4, x_5)$$

the memory requirements reduce to 105 and only 10 basis functions are activated for each input. The new network requires fewer than 1% of the original storage locations, and the required amount of training data is reduced by a similar factor. This reduction can be even more spectacular if the possibility of redundant input variables is considered. However, the internal representations used in these simpler networks are not always obvious to a designer presented with a data set. This provided the motivation for the development of the B-spline Adaptive Spline Modelling of Observation Data (ASMOD) and Adaptive B-spline Basis function Modelling of Observation Data (ABBMOD) algorithms [Kavli, 1992, 1994].

These algorithms attempt to overcome the *curse of dimensionality* associated with medium or large input space modelling tasks by reproducing the *internal structure* and *dependencies* contained in the training data.

Network Structure

Denoting a B-spline submodel by $s_u(.)$ and its associated input vector $\mathbf{x}_u$, $u = 1, \ldots, U$, the overall network output is given by:

$$y(\mathbf{x}) = \sum_{u=1}^{U} s_u(\mathbf{x}_u) \tag{8.22}$$

as shown in Figure 8.17. The sets of input vectors satisfies:

$$\bigcup_{u=1}^{U} \mathbf{x}_u \subseteq \mathbf{x}$$

If the weight coefficient vector for each submodel is denoted by $\mathbf{w}_u$, $\mathbf{w} = \bigcup_{u=1}^{U} \mathbf{w}_u$, and similarly for the basis function output vectors $\mathbf{a} = \bigcup_{u=1}^{U} \mathbf{a}_u$, the network output is then given by:

$$y(\mathbf{x}) = \sum_{i=1}^{p} a_i(\mathbf{x}) w_i \tag{8.23}$$

Despite this expression being equivalent to the output of a conventional B-spline network, the number of active basis functions and the total number of weights required are generally much less for this representation. It allows B-splines to be applied to higher dimensional, redundant, nonlinear, adaptive modelling and control problems.

Fuzzy Interpretation

As described in Chapter 10, the univariate B-spline basis functions can be used to represent the fuzzy *membership functions* which implement the fuzzy linguistic

terms such as *the error is small*. The product operator which combines the univariate basis functions represents a fuzzy conjunction and addition is used as the
fuzzy disjunction operator. Therefore, the network shown in Figure 8.17 could be
expressed as a set of fuzzy production rules (a fuzzy *algorithm*):

$r_{5,6}$:   IF (x_1 *is positive small*)

          THEN (y *is positive medium*)   ($c_{5,6}$)

$\vdots$

$r_{27,4}$:   OR   IF (x_2 *is almost zero* AND x_3 *is negative small*)

          THEN (y *is negative small*)   ($c_{27,4}$)

$r_{51,6}$:   OR   IF (x_2 *is positive large* AND x_3 *is positive medium*)

          THEN (y *is positive medium*)   ($c_{51,6}$)   (8.24)

$\vdots$

$r_{60,2}$:   OR   IF (x_4 *is negative large* AND x_5 *is positive small*)

          THEN (y *is negative medium*)   ($c_{60,2}$)

$r_{86,3}$:   OR   IF (x_4 *is positive medium* AND x_5 *is negative large*)

          THEN (y *is negative small*)   ($c_{86,3}$)

where $r_{i,j}$ is used to refer to the ij^{th} fuzzy rule and c_{ij} is the associated fuzzy rule
confidence.

The fuzzy algorithm which describes a normal B-spline network has an antecedent which forms the logical conjunction of *all n* input variables. By decomposing the input space in the manner described above, fewer rules are required in
the fuzzy algorithm, their form is much simpler and the condition of the learning
problem is improved as the network's internal representation is appropriate for
that particular data set.

Generalisation and Training Data

Many theoretical results related to the modelling abilities of learning algorithms
assume that there are sufficient training data to populate the relevant input space
adequately (this assumption was made in Chapter 7). However, for most practical
modelling problems, the availability of such data is a very serious issue, due to
the amount required to populate a high-dimensional input space evenly, the well-
known curse of dimensionality [Bellman, 1961], and this is even worse for dynamical
systems whose data trajectories generally lie in restricted parts of lower dimensional
subspaces. The modelling strategy described above can partially overcome these
difficulties (for suitably structured functions) as it no longer requires the data to
be evenly distributed across the whole input space. Only enough data are needed
to identify the structure, and this parsimonious (or minimalist) representation
improves the network's generalisation. This can be illustrated by considering the

example shown in Figure 8.18. The training data lie along the main diagonal *line* in

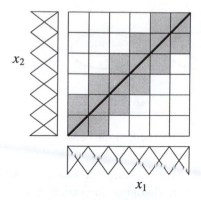

Figure 8.18   Generalisation in two-dimensional weight space for additive and tensor product B-spline networks. When the training data lie along the bold diagonal line, the tensor product network can only generalise this information locally (shaded cells), whereas an additive network learns the correct relationship across the whole domain.

a two-dimensional space, and the structure of the desired function is such that it can be adequately modelled using two univariate subnetworks composed of piecewise linear basis functions. When the two-dimensional, tensor product B-spline network is used to identify this function, it only generalises to the shaded area formed from the union of the supports of the activated basis functions. Alternatively, when two univariate subnetworks are used to model the unknown function, their stored knowledge is distributed across the whole of the input space, and because the model's structure is correct, it generalises appropriately.

8.5.2   Iterative Model Construction

The ASMOD and ABBMOD B-spline construction algorithms are both based on *growing* networks by iteratively refining a very simple model. They start from an initially empty network or one which includes a small number of relevant subnetworks, and gradually new inputs are included, cross-product terms are identified and a better representation of each input is formed. At each iteration step, a number of possible ways in which the network can be made more flexible is identified, and the performance of each is calculated. The optimal refinement step is selected and included in the current model. This process is repeated until the current model's performance is acceptable, as illustrated in Figure 8.19.

It should be emphasised that when B-splines are used to form the above subnetworks, *any* of the above enhancements generates a more complex model, which is capable of *exactly* reproducing the previous model. This is not the case for such networks as the CMAC because, as described in Chapter 7, adding a new knot changes

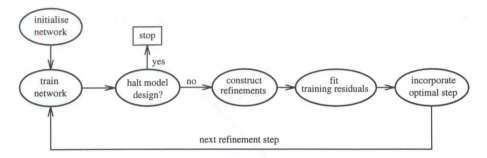

Figure 8.19 The B-spline network design cycle.

its modelling capabilities. One possible refinement that the ASMOD/ABBMOD algorithms do not consider is changing the order of basis functions, which is because the new model is unable to reproduce the old network exactly. Therefore the order of the splines which represent each univariate input has to be determined *before* learning commences.

The main part of this algorithm is the determination of the optimal model refinement which can be one of three possible actions. Once the current optimal model has been determined, a *candidate evaluation model* is constructed for each possible refinement. It can be shown that introducing a new input variable or forming the tensor product of two submodels is *orthogonal* to the current model, and the introduction of a new basis function is orthogonal to the other subnetworks. Therefore all the possible candidate evaluation models are formed and the one which best fits the current *residual* data is incorporated into the model.

The difference between these two algorithms is the representation of the basis functions. ASMOD stores the B-spline subnetworks shown in Figure 8.17, and when a knot is inserted into a multivariate subnetwork, or a multivariate subnetwork is formed from the tensor product of two univariate subnetworks, the refinement is *global* across the appropriate subnetwork. ABBMOD stores each basis function *separately* and, instead of forming the tensor product of univariate subnetworks, it multiplies *single* multivariate basis functions. Similarly, instead of inserting a knot in a multivariate subnetwork, it inserts one in a multivariate basis function, which splits it in two. Thus the ABBMOD refinements are completely *local* and the structure is more general than ASMOD, although computationally its internal structure is more complex, as a knot vector has to be stored for each basis function.

The ASMOD algorithm has been applied to a wide variety of modelling problems: robot actuator modelling, metallurgic process modelling, water content estimation and redundant functional approximation [Kavli, 1992]. It has been benchmarked against multi-layer perceptron, radial basis function and partial least-squares networks, and appears to perform at least as well as the best of the other models. However, it can also be expressed as a set of fuzzy rules which provide insights into both the structure of the learnt mapping and knowledge about the unknown function.

8.5.3   B-spline Submodel Enhancement

There are three basic ways in which the B-spline model can be refined using the ASMOD/ABBMOD algorithm: introducing a new input variable, modelling input dependencies, forming multivariate submodels or basis functions, and refining a particular input's representation by introducing a new basis function (equivalently adding a new knot). These refinements iteratively increase the network's complexity, although in practice it is often necessary to *prune* the model (remove knots, split multivariate subnetworks) and search for a simpler structure, as mistakes can be made in the iterative refinement process. All of these operations are now described.

Single Variable Dependencies

When an input is introduced to the network, it is represented as a single univariate submodel of order k, generally with k basis functions defined on this input axis. The overall model then has the capability of *additively* modelling this input variable.

Cross-Product Dependencies

Internal input data dependencies in the ASMOD algorithm are found by combining the univariate and multivariate submodels to form new tensor product multivariate submodels, whereas the ABBMOD algorithm combines individual basis functions with a univariate model. The former technique generally uses far more basis functions for each tensor product refinement, as the latter increases the number of basis functions by only two or three at each refinement step. The overall aim of these algorithms is to generate a set of *simple*, additive subnetworks, and in this respect the latter may be preferred. However, it is possible to use performance measures which automatically penalise the number of basis functions in a model, thus generating a simple representation. These tensor product refinement steps increase the model's flexibility, but care must be taken to ensure that the algorithm first tries to approximate the data, using low-dimensional submodels.

Addition of New Basis Functions

Finally, the representation of any input variable can be refined by introducing a new univariate basis function and this occurs when a new knot is defined on the appropriate input axis. When a knot is introduced into an ASMOD subnetwork, many new basis functions are formed, due to the tensor product terms. However, only one new basis function is introduced into the ABBMOD model, because all the knot vectors are stored locally and only one basis function is split. It is possible to introduce a new knot *anywhere* between the extrema of each axis, although these algorithms only consider new positions which lie halfway between two previous knots. This approach generally works well and is computationally efficient.

Network Pruning

Including the best refinement step at each iteration means that a near-optimal network is constructed, although it can be too complex, either due to an excess number of knots being inserted or because multiplicative terms are mistakenly introduced instead of univariate knots. These problems can be overcome by *pruning* the network after a certain number of refinement steps, and trying to fit simpler subnetworks to the data stored in each of the current submodels. Knots can be deleted or the tensor product networks can be split, leaving a *set* of additive subsubmodels with which to fit the data modelled by the original submodel. Deleting knots allows them to be positioned 1/4 or 3/4 of the way between two adjacent knots [Lyche and Mörken, 1988], even when new knots are only introduced at midpoints.

8.5.4   Network Training and Validation

Adjusting the weights in each candidate refinement model and the current network is an *off-line* linear optimisation problem, and there are many standard training algorithms that can be used. Kavli [1994] used the stochastic approximation version of the LMS rule (see Section 5.3.3) where the learning rate is slowly reduced through time, although more complex training algorithms, such as conjugate gradient, converge to an *exact* solution in a *finite* number of steps. Alternatively, the weight vector could be found using a numerically stable, singular-valued decomposition algorithm [Haykin, 1991]. LMS learning algorithms are slow to store ill-conditioned data (see Section 5.3.4), although sometimes this information is unimportant and the calculated solution is sufficient. However, as the model becomes more complex, learning is harder as the *condition* number increases. Hence, the authors postulate that the conjugate gradient method would provide the best overall performance, especially when the algorithm is started from a good initial value (such as the previous optimal model), as the matrices are generally sparse.

Network complexity measures can be useful both to assess the relative contributions of different refinements and to decide when to terminate training. A straight MSE performance measure cannot be used to compare two different refinements (such as knot insertion and forming tensor products) directly as a different number of parameters are introduced into the model. The performance measure should balance the complexity of the proposed refinement with the amount of training data and the reduction in the MSE [Pottmann and Seborg, 1992], and there exist several well-known formulae:

Bayesian information criterion: $\quad K = L\ln(J) + p\ln(L)$

Akaike's information criterion: $\quad K(\phi) = L\ln(J) + p\phi, \qquad \phi > 0$

Final prediction error: $\quad K = L\ln(J) + L\ln\left(\frac{L+p}{L-p}\right)$

where K is the performance measure, p is the size of the current model, J is the MSE, and L is the number of data pairs used to train the network. By fitting potential refinement models to the residual data, the iterative regression procedure can proceed in a near-optimal fashion. These tests can also be used to assess the performance of the overall network, as they balance modelling error against network complexity. A simple termination procedure for the refinement process would check if the current performance measure K is greater than the previous, and if so, the current refinement is rejected and the stepwise regression procedure terminated.

8.5.5 Other Algorithms

The ASMOD and ABBMOD B-spline initialisation algorithms are not the only ones which can iteratively construct nonlinear representations of the training data, and recently this has been a very fruitful area of research. Probably the most famous example of this is Friedman's Multivariate Adaptive Regression Splines (MARS) algorithm [Friedman, 1991a, b], which recursively partitions the input space and generates a piecewise polynomial output from a set of *truncated polynomial* basis functions [Sekulic and Kowalski, 1992]. This network can be regarded as an extension of the Classification And Regression Trees (CART) algorithm [Breiman *et al.*, 1984], which is one of the most widely used machine learning procedures.

The output of a MARS network is formed from the linear combination of multivariate basis functions, which in turn are produced by multiplying together a number of univariate truncated polynomial basis functions. A truncated polynomial is a function which is defined as zero when its argument is negative and is a polynomial (constant, linear, quadratic, etc.) mapping when it is positive. There is also a strong relationship between truncated polynomial and B-spline basis functions, as one can be represented as the other and vice versa [Cox, 1982]. The expression $(x - \lambda)_+^k$ is generally used to define a truncated polynomial of order k which is non-zero to the right of λ, and $(x - \lambda)_-^k$ is commonly used instead of $(\lambda - x)_+^k$. Two linear ($k = 2$) truncated polynomial basis functions are shown in Figure 8.20, as well as a typical network output. During the refinement procedure MARS uses only piecewise linear truncated polynomials, although during the final optimisation step it can generate a set of cubic basis functions, and this may be desirable if a network with a continuous first derivative is required.

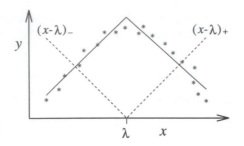

Figure 8.20   A piecewise linear univariate function (solid line) formed using the truncated polynomial (linear) basis functions (dashed lines) used in the MARS algorithm.

Briefly, the MARS algorithm constructs a *tree* by iteratively refining a very simple model (a constant function), which is the root. New variables are included, tensor product terms are modelled and knots are inserted using an *identical* refinement step. Each node in the tree can be considered as a possible *parent* and two *children* are formed which are univariate truncated linear basis functions, as shown in Figure 8.20. The new basis functions are calculated by taking the product of all the nodes in the relevant branch, so additive relationships can be modelled when the parent is a constant function and knots can be inserted along the univariate axis. Tensor product terms can then be formed when the parents are truncated linear basis functions. The number of splits in each branch is at most n, as MARS does not allow the same input variable to appear twice in a branch; a restriction not present in the CART algorithm. Also, each refinement increases the number of basis functions in the network by two, so any refinement validation measure which combines the output MSE with the size of the network and the amount of training data only compares the reduction in the MSE, as the other two terms are constant. One of the main features of the MARS algorithm is that *every* node contributes to the output, not just the leaves, and this procedure is illustrated in Figure 8.21 for a two-dimensional input. After growing a tree, MARS prunes away unnecessary basis function in order to produce a parsimonious final model.

The MARS algorithm requires two parameters from the user: the maximum number of nodes and the maximum degree of interaction (length of each branch). Both parameters affect the refinement procedure and, to obtain a satisfactory model, it is necessary to understand their roles. The refinement procedure keeps operating until it has constructed a tree that has the user-specified number of nodes, then it tries to delete any unnecessary terms and fit a cubic surface to the network. If the term is too large, MARS begins to model any noise contained in the data and the refinement procedure may take a long time, although if it is too small, a poor representation will be formed. Friedman recommends that it should be set to be twice the number of terms used in the final algorithm. Similarly the maximum degree of interaction specifies the maximum length of each branch, as it is the maximum number of inputs involved in any basis function. When it is set equal to 1, MARS is restricted to *additive* (nonlinear univariate functions)

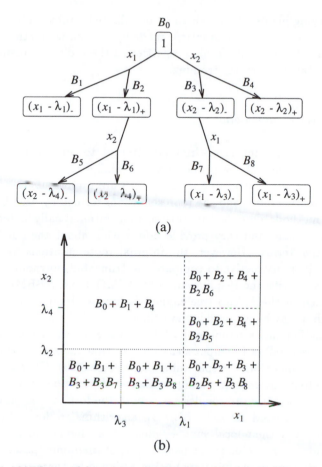

(a)

(b)

Figure 8.21   A MARS tree (a) and its associated partitioning of a two-dimensional input space (b). The dashed partition corresponds to the left-hand side of the MARS tree, whereas the dotted one corresponds to the right, and in the interior of each region, the basis functions contributing to the output are shown. As the output only depends on two inputs, each branch can have (at most) two splits.

modelling tasks and as the parameter is increased, the range of functions which the network can model is larger, but so is the potential to make mistakes and the length of the refinement procedure. If both parameters are properly set, MARS generally finds a good model in a reasonable length of time, although bad choices can lead to poor final networks, which require an excessive computational load.

Other network construction algorithms have been developed that are based on choosing which polynomial or sinusoidal terms are relevant in a functional link network [Ivakhnenko, 1971, Linkens, 1993, Rogers, 1992, Sanger, 1991]. However, the ASMOD and ABBMOD have been described in detail, because of their potential real-time learning. The model selection is performed *off-line*, but in order to deal with time-varying plants, the weights can be adapted on-line. The model's structure remains fixed, but the contributions made by each basis function varies. These algorithms have the same desirable properties as the ordinary B-spline networks, so they are suitable for on-line training.

Local and Global Representations

These network structuring algorithms all attempt to provide the designer with information about the unknown process after training has ceased. Knowledge can be provided in various forms: the inputs which are important, the size and form of the multiplicative terms and the representations formed by each basis function. Each of the methods described in this section automatically determines which inputs are important, and they provide information about the number and type of multiplicative terms. However, the internal representations formed are very different, and it is instructive to compare the truncated polynomials used in the MARS algorithm with the B-splines used by ASMOD and ABBMOD.

Each branch in the MARS tree cannot include an input more than once, so several branches may be involved in modelling a bounded region of the input space where the function varies significantly. The global support of the truncated basis functions (at least half of the input space) means that their effect is *greatest* when the input is *far* away from the split value (see Figure 8.20). Therefore terms must be introduced to counteract this, and the associated partitioning of the input space provides little information to the designer about the form of the desired function.

In contrast, the ASMOD and ABBMOD algorithms are formed using B-spline basis functions and their local support means that significant variations in the desired function are reflected in the partitioning of the input space, with a greater density of knots (and basis functions) being assigned to these regions. The value of the function can also be locally bounded by examining the weights' value, so the internal structure of these networks provides valuable information to the designer, especially because of the relationship between these networks and fuzzy systems.

8.5.6 Summary

The need for an *intelligent* way of initialising a B-spline network is obvious; the number of basis functions (rules) can be automatically determined from a data set, and the number of relevant inputs can also be predicted. The ASMOD algorithm [Kavli, 1994] has been shown to perform well in a variety of nonlinear modelling tasks. However, in order to work well, a sufficient number of representative training examples must be collected, and this generally limits its operation to off-line tasks, although an on-line version of the algorithm has been proposed. These network initialisation problems also occur in fuzzy systems, and the ASMOD algorithm can be used for calculating the structure of the fuzzy rules, due to the invertible relationship between fuzzy and B-spline networks (see Chapter 10).

8.6 ASMOD Time Series Modelling

The ASMOD structuring algorithm is now used to construct a suitable B-spline model for predicting the two-input, single-output nonlinear mapping described in Sections 6.6 and 8.4. The desired recurrence relationship can be expressed as:

$$y(t) = f(y(t-1)) + g(y(t-1))y(t-2) + \eta(t) \qquad (8.25)$$

for two appropriate nonlinear functions $f(.)$ and $g(.)$, and an additive disturbance $\eta(t)$. A CMAC and a poorly initialised B-spline network have already been used to model this function, and the aim of this section is to illustrate how the AS-MOD procedure works, using the same training data set, and to demonstrate that incorporating *a priori* knowledge into the B-spline network design improves its generalisation abilities. Neural networks are often cited as being model free estimators, but this is generally not true. The network's structure is generally decided *a priori* by the designer and a set of parameters (weights) are adapted using an appropriate learning algorithm. Due to the underlying flexibility of many of the networks considered in this book, the authors have termed these algorithms *weak* or *soft* modelling algorithms. One of the main advantages of using B-spline networks is that they allow limited *a priori* knowledge to be incorporated into their structure, although it should be remembered that if the underlying structure of the network is not flexible enough (or even too flexible) to represent the training data adequately (a biased system), it performs poorly.

8.6.1 ASMOD Refinements

Equation 8.25 shows how the discrete recurrence relationship can be expressed as a sum and product of two nonlinear *univariate* functions $f(.)$ and $g(.)$. The important point to notice is that the desired function is a linear function of $y(t-2)$, when $y(t-1)$ is held constant, whereas it is a strongly nonlinear function of $y(t-1)$. If this type of knowledge is available during the network design process, it should be incorporated into the basic network structure, as it simplifies the learning process. Otherwise any algorithm that adapts the network's structure should be able to automatically generate such information. This kind of *a priori* knowledge can easily be incorporated into the B-spline network design by defining univariate basis functions of different shapes and sizes (orders) on each (univariate) axis, as the multivariate basis functions are formed by taking the tensor product.

To begin the ASMOD procedure, two, order 2 univariate basis functions were used to represent both $y(t-1)$ and $y(t-2)$ on the input domain $[-1.5, 1.5]$. The initial optimal ASMOD network was empty, so one of these univariate submodels must be included from the store to start the refinement algorithm. The most statistically significant subnetwork s_2 to include was the one representing $y(t-2)$ which produced the model output shown in Figure 8.22. After this refinement,

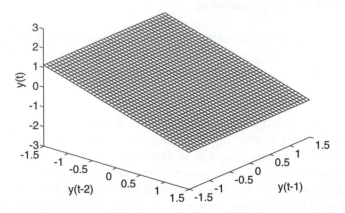

Figure 8.22   The first ASMOD refinement which introduces a univariate subnetwork on $y(t-2)$. The network's output lies in the range $[-1.12, 1.12]$.

the network was able to model the variable $y(t-2)$ additively. Three possible candidate refinement steps could then be performed: inserting a knot at zero in s_2, introducing the univariate function $s_1(y(t-1))$ as another subnetwork or forming a tensor product between s_2 and s_1 resulting in a new B-spline subnetwork $s_3(y(t-1), y(t-2))$. The tensor product refinement was estimated to be the best, and the new output surface is shown in Figure 8.23. Only two possible actions could

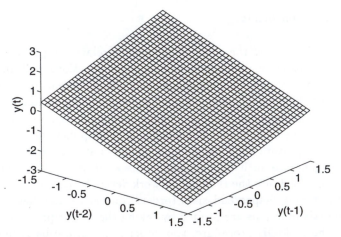

Figure 8.23   The second ASMOD refinement step which forms a tensor produce model. The network's output lies in the range $[-2.45, 2.51]$.

now be taken: a knot insertion at zero on either axis. Inserting a knot at zero on $y(t-1)$ was statistically the most significant and any further refinements reduced the amount of information stored in the network, according to the complexity measures described in Section 8.5.4. Therefore the final model was composed of just one subnetwork $s_3(.)$, which has two piecewise linear basis functions defined on $y(t-2)$ and three on $y(t-1)$, resulting in the prediction surface shown in Figure 8.24.

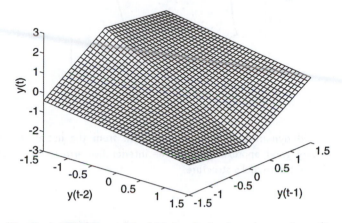

Figure 8.24 The final ASMOD model which has had a knot inserted at $y(t-1) = 0$. The network's output lies in the range $[-1.90, 1.94]$.

It is important to notice that the ASMOD algorithm has discovered the correct representation for $y(t-2)$ and models it using a globally linear mapping.

8.6.2 Model Evaluation

The network's output is plotted in Figure 8.24, and while this may seem a gross simplification of the original surface, it is sufficient for representing the dynamical behaviour. This is confirmed in Figure 8.25 where the iterated dynamics of the B-spline network is displayed. It is a very good approximation to the true plot shown in Figure 6.16, as the knowledge that has been incorporated into the network's structure improves its ability to generalise in the interior of the limit cycle. It also improves the network's ability to extrapolate information, with the minimum and maximum network outputs being -1.83 and 1.89, respectively, which are very close to the true values.

The network's prediction ability is shown in Figure 8.26 where the mean square k-step-ahead prediction errors are plotted. It shows that good estimates can be obtained by iterating the network up to (at least) 20 steps into the future and this is confirmed in Figure 8.27, where the network and the noiseless time series are run in parallel for 50 iterations. As can be seen from these two figures, there is very

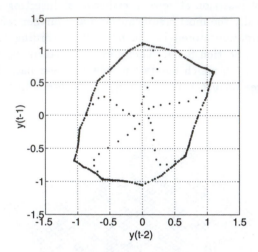

Figure 8.25   Iterated dynamics of a B-spline network from the initial condition $x(1) = (0.1, 0.1)^T$. The very good approximation of the interior five arm spiral is because of the knowledge encoded in the model's structure.

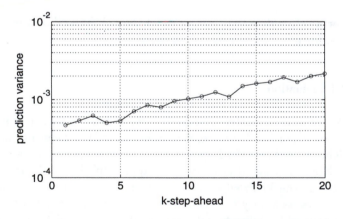

Figure 8.26 Variance of the k-step-ahead prediction errors.

good agreement between the true and the estimated time history, and the final model produced has a satisfactory autocorrelation plot as shown in Figure 8.28.

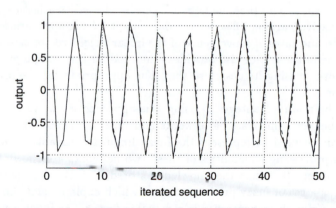

Figure 8.27   The noiseless time series and the B-spline network iterated mappings from an initial condition $y(-1) = 0.5$, $y(0) = 1.0$. The dashed line represents the true time series and the solid line shows the B-spline iterated mapping.

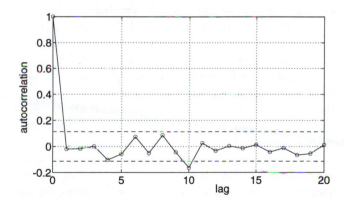

Figure 8.28   The autocorrelation of the prediction errors after training has ceased. The 95% confidence band is represented by the dashed lines.

8.7 DISCUSSION

B-spline networks can be used for many tasks within intelligent control applications; their desirable attributes such as real-time operation and predictable learning behaviours, coupled with the fuzzy interpretation, produce a powerful technique for dealing with ill-defined, nonlinear systems. This chapter has provided a thorough description of the basic algorithm, focusing on the method used to evaluate basis functions and the network's properties. The B-spline basis functions are just convenient forms for representing piecewise polynomial mappings (truncated polynomials are another), so they are universal approximators, being able to approximate any continuous nonlinear function defined on a bounded domain to any degree of accuracy. The B-spline network can be classed as a lattice AMN, hence all of the iterative learning rules described in Chapter 5 are appropriate for training these models, and rates of convergence of the linear weight vector can be estimated. The definition of the lattice also provides a method for imposing *a priori* structure in the network, biasing its design so that it is appropriate for the problem under consideration.

One of the main advantages in using this approach is due to the relationship that exists between this class of networks and fuzzy systems. The B-spline basis functions can be used to represent the fuzzy linguistic variables, and the overall information processing capabilities of these two networks are shown to be equivalent in Chapter 10. Thus the knowledge stored in a B-spline network can be interpreted as a set of fuzzy production rules which explain *what* the network has learnt. This network *transparency* is a valuable feature, as it allows the designer to initialise the network using natural, vague terms that are commonly used by experts to explain their actions. Similarly the network can be interrogated and examined after training to determine what information has been learnt. The autonomous docking example described in Chapter 9 illustrates the usefulness of this feature, as new rules can be initialised, old rules edited and the performance of the rule base improved after training has ceased. It has been argued that the main difference between a B-spline type network and a certain class of fuzzy systems is the basis functions' fuzzy linguistic interpretation [Wang, 1994]. This division can be artificial however, as some fuzzy networks do not use natural language terms to describe their basis (membership) functions or else the membership functions used to represent the vague fuzzy terms are inappropriate for use in learning systems, as was shown in Section 8.4. The authors take the view that the fuzzy interpretation is useful *when* it is appropriate.

The suitability of these networks for both on-line and off-line modelling and control tasks depends on the structure of both the desired function and the B-spline AMN. The B-spline representation is very flexible, although for certain functions a simpler model may suffice. In these cases, the condition of the adaptive B-spline network is poor and its generalisation abilities are limited, as the model is over-parameterised. One possible solution is to use algorithms that automatically determine the network's structure (the input variables, the number of knots and the degree of multiplicative interaction) from a set of training data. These algorithms try to exploit both global and local redundancy in the training data, producing a parsimonious network while fitting the data in some best sense. Simple networks are usually better conditioned and generalise appropriately across a larger portion of the input space. Therefore, the ASMOD and ABBMOD algorithms described in this book are recommended when there are sufficient training data and the prior knowledge about the process is limited. Information about the function can be incorporated in the network before learning commences (such as linear relationships), so the structures can be biased to reduce the variance of the on-line training procedures.

All of these comments were illustrated when the two B-spline networks were applied to the time series prediction problem. The first B-spline network was initialised in a similar way to a typical fuzzy system with seven piecewise linear basis functions on each axis, but the results were only adequate. The network was too flexible for this particular data set. An ASMOD learning procedure was then applied to design an "optimal" B-spline model and this resulted in a network approximately 90% less complex. Its form was much simpler, its performance more reliable and it generalised better. A set of fuzzy linguistic labels could also be applied to the basis functions, but more importantly, the ASMOD algorithm correctly determined a linear relationship contained in the data. Thus these algorithms can be used to determine automatically fuzzy-type rule bases that provide information to the designer about numeric relationships, as well as fuzzy sets.

B-spline Guidance Algorithms

9.1 INTRODUCTION

The B-spline network was described in considerable detail in Chapter 8, and applied to the benchmark time series modelling task. Time series prediction (or plant modelling) is just one of the many areas where this and related algorithms can be successfully used. Chapter 11 discusses how to derive a controller based on a B-spline/fuzzy model of the plant and this chapter describes the development of a static B-spline/fuzzy rule base, as well as a real-time method for incorporating constraints into an Autonomously Guided Vehicle's (AGV) or robot's desired trajectory. This work has been included in order to emphasise the range of potential application areas for this algorithm and to highlight its desirable numerical properties.

One of the main themes of this book is that there is an equivalence between B-spline, CMAC and fuzzy networks, and this relationship is fully explained in Chapter 10. The main difference between these techniques is the *level of abstraction* at which they are interpreted. B-spline and CMAC networks are viewed as numerical processing or computational systems, whereas fuzzy networks can be given a linguistic interpretation as a fuzzy algorithm, using terms such as *small* or *large* to label the basis functions. For practical applications, there is little to distinguish the algorithms, although the equivalence between the networks is useful for initialising or for validating their rules. Whilst the local support property of the B-spline basis functions produces a well-conditioned "hidden layer", it can limit the generalisation abilities of the network, especially when the training data are badly distributed. Hence the fuzzy interpretation of a B-spline network can be invaluable, as shown in the next section, when this system is taught to reverse park a vehicle. The neurofuzzy interpretation is also useful as fuzzy "techniques" such as scaling factors (see Section 11.2.5) can be used to design networks that are more robust and can work in a wider range of conditions.

B-splines can also be used to produce constrained, desired trajectories, which is reminiscent of their more traditional use as data interpolation/approximation schemes [DeBoor, 1978]. The algorithm generates a spline which fits a discrete set of subgoals (such as cartesian position or velocity/curvature) to produce a feasible, continuous desired path. If the only requirement is to fit the subgoal information exactly, it is a standard data interpolation problem and the trajectory generation task is introduced as such. However, when the trajectory is also required to be *feasible*, it is necessary to ensure that none of the derivatives exceed certain limits. These limits are regarded as *hard* constraints, whereas the discrete desired behaviours are *soft* subgoals that should be satisfied, provided that none of the dynamic constraints are violated. A real-time algorithm for satisfying dynamic constraints is proposed and the approach is illustrated by producing a quadratic B-spline velocity/curvature constrained desired trajectory.

9.2 AUTONOMOUS DOCKING

The problem of a vehicle parking autonomously has aroused a considerable amount of interest [Dorst *et al.*, 1991, Sugeno and Nishida, 1985, Sugeno and Murakami, 1985, Sugeno *et al.*, 1989], and this chapter describes the development of a B-spline neurofuzzy network for implementing this task. The B-spline network has a natural interpretation as a set of fuzzy production rules, so it may be referred to as a fuzzy rule base. This adaptive network learns by example, observing a human park a vehicle several times, then the trained rules are validated and new rules are initialised to increase its robustness with respect to different initial conditions. The fuzzy rule base is *static*; it does not learn from its interaction with the environment once the design phase has ceased. Training only occurs during the design cycle, after which the rule set must be verified and validated and new rules can be initialised and old rules edited, and this design process is illustrated in Figure 9.1.

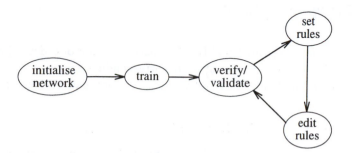

Figure 9.1 A typical B-spline network design cycle.

The success of any application is determined by selecting appropriate input and output variables, as well as choosing the reasonable values for the order of the B-spline basis functions and the overall network's structure. The input variables should uniquely determine the state of the vehicle; the order of the basis functions must be chosen according to the desired continuity constraints and any representational requirements; and the network should be parsimonious, exploiting any redundancy or structure in the training data.

9.2.1   Fuzzy B-spline Rules

Storing the parking routines as a set of B-spline/fuzzy rules is proposed, and before the problem is posed and a rule base constructed, the link between these two algorithms must be described. This explanation is necessarily brief, as the relationship between the B-spline and fuzzy networks forms the basis for Chapter 10. Interpreting the B-spline network as a set of fuzzy production rules allows the adaptive B-spline AMN to be endowed with a linguistic, fuzzy representation, which can be used for initialisation or validation purposes.

The (multivariate) B-spline basis functions may be viewed as defining pattern matching functions, which represent the antecedents of the fuzzy production rules. Each basis function in the B-spline network can then be interpreted as a linguistic rule of the form:

IF $(\mathbf{x}\ is\ \mathbf{A}^i)$ THEN $(y\ is\ w_i)$

where $\mathbf{A}^i$ is a linguistic description of the ith multivariate B-spline basis function. Each multivariate basis function is formed from n univariate basis functions, and each of these represents a vague statement such as x_i *is positive small*. The tensor product of the univariate basis functions is computed, which is equivalent to taking the logical intersection (AND) of each of the univariate linguistic statements.

In addition, there exists an *invertible* map between each weight, w_i, and a set of rule confidences $\{c_{ij}\}_{j=1}^{q}$ such that:

$$w_i = \sum_{j=1}^{q} y_j^c\, N_k^j(w_i) = \sum_{j=1}^{q} c_{ij}\, y_j^c$$

where $N_k^j(.)$ is the jth symmetrical B-spline basis function of order $k\ (\geq 2)$, which is defined on the output domain, y_j^c is its corresponding centre and $c_{ij} = N_k^j(w_i)$. Therefore, the ij^{th} multivariate B-spline basis function can be represented as the linguistic rule:

$r_{ij}:$ IF $(\mathbf{x}\ is\ \mathbf{A}^i)$ THEN $(y\ is\ B^j)$ (c_{ij})

where B^j is the linguistic description of the jth fuzzy output set and c_{ij} is the confidence in the ij^{th} fuzzy rule being true. Instead of storing a p-dimensional weight vector, the algorithm uses a rule confidence matrix of size $(p \times q)$. For this application, however, the fuzzy representation has the advantage that the information learnt from the expert can be displayed in a transparent form as a set of fuzzy production rules, enabling the network to be interrogated, old rules edited and new rules initialised.

This relationship which exists between fuzzy systems and B-spline AMNs gives a *transparent*, fuzzy representation of the network and also means that the convergence and stability results which can be derived for AMNs can also be applied to their equivalent fuzzy systems (see Chapter 11).

9.2.2 Autonomous Docking Problem

In order to implement a fuzzy docking rule base, the problem must be posed in an unambiguous manner. There are many different ways of parking a car in a slot, and those depend on the slot's length, width and orientation, as shown in Figure 9.2. A driver formulates a different docking strategy for each of these three

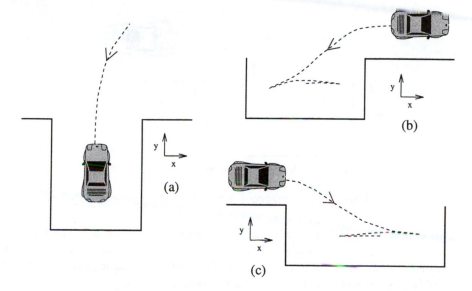

Figure 9.2 Three different parking scenarios: (a) reverse garage, (b) reverse slot and (c) forward slot.

scenarios, and so to implement a fuzzy rule base which simulates a driver, a rule base must be designed for each case. The section only describes the development of a reverse parking rule base, although the other two can be developed in an analogous manner.

Reverse Docking into a Slot

The reverse parking scenario can be stated as:

> *From an initial vehicle position close to a slot, generate a path which safely guides the vehicle to slot's centre.*

From this task definition, certain environmental features are assumed. First, it has been assumed that it is possible to extract a parking slot from the environment. This could be performed by driving past the slot, sensing the size and the shape of the parking area and fitting a rectangular template to this sensor map. Second, it has also been assumed that the desired parking position is in the centre of the slot. If the slot is very large, this may produce a cautious trajectory, and it would be the task of the higher level local navigation system to artificially create a slot of a reasonable size, or initialise a different guidance mode. Finally, it has been assumed that the path is obstacle free, and that any collision detection systems are running in parallel with the parking function. If an obstacle is detected, the system must either stop until the object has moved past it, or else a different obstacle avoidance guidance mode is initiated.

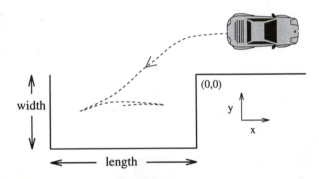

Figure 9.3 The envisaged reverse docking scenario.

This is illustrated in Figure 9.3, where the vehicle has driven past the slot and is in a good position to start the manoeuvre. The docking problem is essentially *kinematic*, it is performed at low speeds (to allow sufficient sensor processing time) and starting and stopping is almost instantaneous.

The state of the vehicle is given by its current cartesian coordinates and its heading relative to some local origin, and the trajectory is characterised by the vehicle's demanded speed and curvature. The origin is the nearest slot corner when the vehicle begins to reverse and the heading is measured with respect to the nearest plane/wall, as illustrated in Figure 9.3. These variables form the inputs and outputs of the B-spline/fuzzy rule base.

Fuzzy Guidance Rules

The inputs to the fuzzy network originally consisted of the states of the vehicle (cartesian x, y coordinates and heading h relative to the local origin) and previous demanded speed of the vehicle s, and the outputs are the demanded speed and curvature. Each of the input variables is covered by a set of univariate basis functions, with their size and positioning proving to be important factors in determining the success of this application.

The ability of the network to cope with slots of varying lengths and widths must be *implicitly* encoded into the B-spline/fuzzy rule base. This is achieved by *normalising*, or scaling, the cartesian coordinates with respect to the width and length of the slot when they are negative:

$$x = \frac{x}{\text{length}} \qquad \text{if } x < 0.$$

$$y = \frac{y}{\text{width}} \qquad \text{if } y < 0.$$

Thus the intervals on which x and y are defined are given by $[-1, x^{\max}] \times [-1, y^{\max}]$, and for this rule base $x^{\max}$ and $y^{\max}$ are chosen to be 3 and 2 metres, respectively. The rule base's cartesian coordinate inputs are now normalised, although this can cause problems, as the length and width of the car do not change in the same way. It is possible to incorporate such information into the definition of the rule base by modifying the *knots* inside the slot, such that the first and last occur at about half a car's length (width) from the front and back (sides), respectively. As the simulated vehicle is 2 metres long and the slot is 4, the knots occur at the relative positions 0.25 and 0.75 and similarly for the knots describing the y position. The heading lies within the range $[-0.3, 1.1]$ rad., and each variable is covered with 7 univariate B-splines of order 2, as shown in Figure 9.4.

This number of basis functions was arrived at both from *practical* considerations, where a minimum number is required, and from *representational* considerations, which dictate that there should be enough basis functions to model the desired function sufficiently well. Using piecewise linear basis functions generates a continuous output (speed and curvature) and while smoother network outputs may be required for some applications, the corresponding learning problem becomes more difficult and the network less transparent as the fuzzy membership functions become increasingly similar. In most fuzzy applications, seven fuzzy sets are used to represent each variable, although this is only a heuristic, as each application has different modelling and computational requirements which decide the structure of the fuzzy rule base. This network design embodies the principle which was proposed in Section 3.4.2; *a parsimonious adaptive system performs the best.*

The fourth input s (the previous desired speed) is necessary, as the other three states all use fuzzy representations which generalise locally and this makes it very difficult for the network to distinguish between the two states shown in Figure 9.5. Two similar inputs demand very different network outputs, and the rule base is unable to model this phenomenon accurately using only fuzzy state (x, y, h) information.

Humans do not park vehicles in an optimal fashion; they generally adopt the strategy of reversing into the slot *as far as possible* and then moving forwards and backwards until the car is parked satisfactorily. Using the previous desired speed as an extra input resolves this possible inconsistency, because the previous

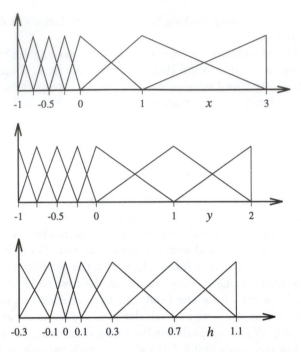

Figure 9.4   Univariate B-spline basis functions which represent the vehicles cartesian x, y coordinates and heading h.

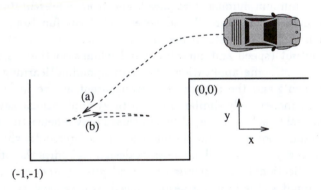

Figure 9.5 Two similar vehicle states, (a) and (b), which require very different outputs.

demanded speed is either negative or positive and this results in the appropriate course of action being generated. Linguistically it only needs to be known if the previous demanded speed is either *positive* or *negative*, so the input variable can be represented by two *crisp* fuzzy sets which are modelled using two B-splines of order 1.

This produced a four input, two output rule base with $7 \times 7 \times 7 \times 2$ ($= 686$) multivariate basis functions. This is a fairly large rule base and all of the relevant (physically realisable) rules must be initialised if it is designed to cope with every possible initial condition.

9.2.3 Rule Base Training and Initialisation

The rule base's structure has now been specified and all that remains is to encode the desired behaviour within a set of rules. It is proposed to perform this task in two ways: first, the rule base is trained from a data set obtained by observing an "expert" park the vehicle, and second, the rule base's performance is evaluated, new rules initialised and old rules edited using a suite of software tools [An *et al.*, 1994a].

Training

As happens in many applications, the designer had little control over the generation of the training data (see Figure 9.6) although there were several problems

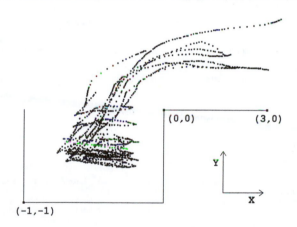

Figure 9.6   The training data for the parking rule base. This data was collected by restarting the parking manoeuvre from approximately the same position several times.

with using this training set. First, the data are very noisy and the final position of the vehicle is different for each trial; this is typical behaviour for a human, but undesirable for an automated docking system. Second, each trial is started from approximately the same position, which results in new rules having to be initialised by the designer so that the rule base can deal with a wide range of initial conditions. The final problem with the data set is that because a similar manoeuvre is performed on each trial, there are parts of the network's input space which are never visited and these rules also have to be initialised by the designer. This initialisation/editing procedure would not be possible if a neural network, such as a multi-layer perceptron, was used to model the data, as the weights (rules) could not be locally modified.

The B-spline network was trained using a stochastic LMS rule and its performance afterwards was reasonable. From some initial conditions it was able to reverse into the slot, although its behaviour in the interior of the slot was not satisfactory and it did not usually tend to the centre of the slot.

Editing and Initialisation

The rule base editing and initialisation was performed with the aid of the software Graphical User Interface (GUI), described in An *et al.* [1994a]. This software toolset allows the designer to select the appropriate multivariate fuzzy input set and edit the corresponding weight vector (two-dimensional in this example). As shown in Figure 9.7, the number of uninitialised rules is also displayed and the designer can automatically select the next empty rule at the touch of a button.

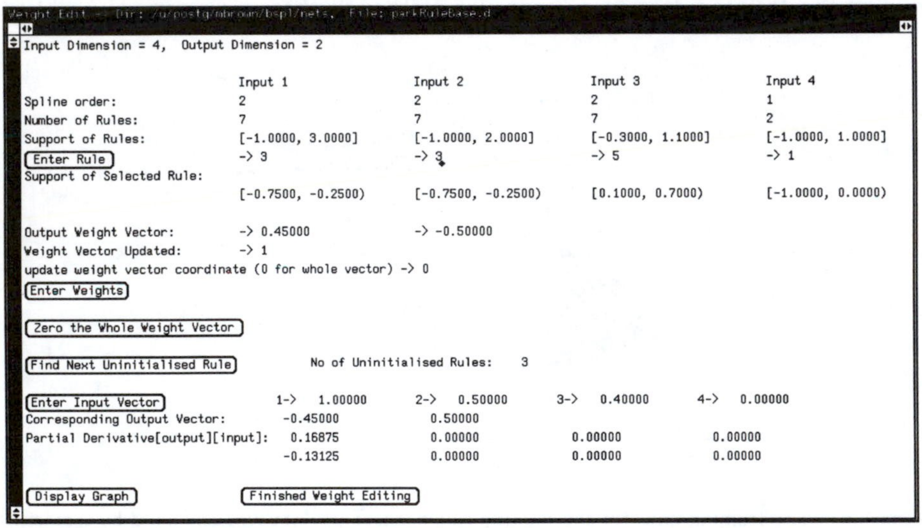

```
Weight Edit -- Dir: /u:postg/mbroun/bspl/nets,  File: parkRuleBase.d

Input Dimension = 4,   Output Dimension = 2

                             Input 1          Input 2          Input 3          Input 4
Spline order:                2                2                2                1
Number of Rules:             7                7                7                2
Support of Rules:            [-1.0000, 3.0000]  [-1.0000, 2.0000]  [-0.3000, 1.1000]  [-1.0000, 1.0000]
(Enter Rule)                 -> 3             -> 3             -> 5             -> 1
Support of Selected Rule:
                             [-0.7500, -0.2500)  [-0.7500, -0.2500)  [0.1000, 0.7000)  [-1.0000, 0.0000)

Output Weight Vector:        -> 0.45000       -> -0.50000
Weight Vector Updated:       -> 1
update weight vector coordinate (0 for whole vector) -> 0
(Enter Weights)

(Zero the Whole Weight Vector)

(Find Next Uninitialised Rule)     No of Uninitialised Rules:    3

(Enter Input Vector)         1->  1.00000     2->  0.50000     3->  0.40000     4->  0.00000
Corresponding Output Vector:      -0.45000         0.50000
Partial Derivative[output][input]: 0.16875         0.00000          0.00000          0.00000
                                  -0.13125         0.00000          0.00000          0.00000

(Display Graph)          (Finished Weight Editing)
```

Figure 9.7   The rule/weight editing panel used to initialise and modify the fuzzy/B-spline rules. Also displayed is the support of the selected rule, number of uninitialised rules and information about the output of the network.

The performance of the resulting rule base depends completely on the information which is encoded by the designer, and after each rule modification session the network's performance must be re-evaluated. There are several ways in which this can be achieved, by inspecting the network's output *surface* and simulating the parking manoeuvre for instance. The former gives the designer information about the smoothness of the resulting rule base and any gross errors which occur when the information is entered are immediately obvious. The latter evaluation method is generally more instructive, as it gives an indication of the true performance of the fuzzy rule set.

It took approximately three days to initialise and edit the rule base until it performed with the desired degree of accuracy. After the first initialisation/editing session, the rule base performed satisfactorily and much of the remaining time was spent ensuring that it could perform well from a wide range of initial conditions.

9.2.4 Simulation Results

The performance of the fuzzy parking rule base was assessed by simulating a vehicle reverse parking into a slot, where the simulation employs kinematic curvature and rate of curvature constraints, typical of those found in real vehicles. A slot sized 4 metres long and 2 metres deep was chosen and the rule base was attempting to park a vehicle of size (2.1×1.1) metres. The vehicle's trajectory was obtained by numerical integration along the desired path, with curvature and rate of change of curvature constraints of ± 0.45rad. m^{-1} and ± 0.1rad. $m^{-1}s^{-1}$, respectively, and the final goal tolerances were for the x, y and h variables ± 0.3m., ± 0.15m., and ± 0.15rad., respectively.

The results of four simulations are shown in Figure 9.8 for a variety of initial conditions. As can be seen from these figures, the performance of the fuzzy rule set is good for a wide variety of initial conditions. This robust behaviour had to be incorporated into the rule base by the designer, and it is expected that the development of each *static* fuzzy rule set will require a similar learning, evaluation, initialisation/editing cycle.

This application of a fuzzy rule base worked well as the following conditions were satisfied:

- **clearly** defined input variables;
- **small** input space (4 variables);
- **local** generalisation was desirable; and
- **flexible** fuzzy sets that were able to represent each input in an appropriate manner.

It is expected that any successful application of static B-spline/fuzzy techniques will at least partially satisfy these first three conditions and the fourth is a desirable property of this fuzzy implementation methodology.

9.3 CONSTRAINED TRAJECTORY GENERATION

While this book has concentrated on the application of adaptive neurofuzzy algorithms for low-level modelling and control, it is not the only area in which they

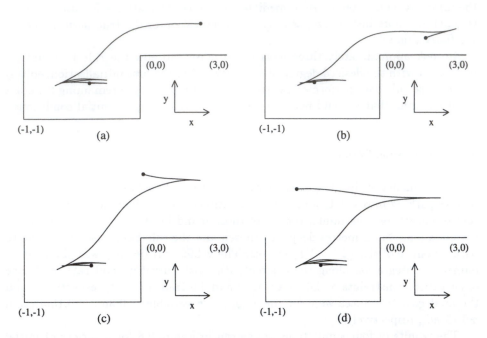

Figure 9.8   Reverse parking trajectory when the vehicle has an initial condition (a) $(x, y, h) =$ $(2.0, 1.0, 0.0)$, (b) $(x, y, h) = (1.5, 0.5, 0.4)$, (c) $(x, y, h) = (0.0, 2.0, -0.3)$ and (d) $(x, y, h) =$ $(-3.0, 1.5, 0.1)$.

can be used. A static fuzzy rule base has been constructed which is to park a vehicle autonomously. This rule base is implemented at the *guidance* level in a model reference control architecture (see Section 1.2.1), receiving commands from the navigation systems and sending ones (desired speed and curvature) to the low-level servo control routines. The autonomous docking routine is only one of several guidance modes which may be used in an AGVs control system. Other possible guidance modes include contour following, road following, manual piloting, fine positioning, convoy following, etc. These modes are distinguished not only by the task that they perform, but also by their interface with the other modules in the vehicle's architecture. While the contribution of this chapter is specifically aimed at a piloting guidance model, the approach is quite general and may be applied to other multi-degree of freedom tasks.

This chapter also describes algorithms which are used for constrained trajectory generation. Information is passed down from the navigation system, which commands the vehicle to achieve a certain velocity, v_i, and curvature, h_i, at specific times t_i:

$$\text{velTraj}(t_1, v_1, h_1, t_2, v_2, h_2, \ldots, t_L, v_L, h_L) \tag{9.1}$$

Alternatively, a set of cartesian (x, y) coordinates could be given:

$$\text{posTraj}(t_1, x_1, y_1, t_2, x_2, y_2, \ldots, t_L, x_L, y_L) \qquad (9.2)$$

where the vehicle is commanded to achieve a relative cartesian position (x_i, y_i) at time t_i, with the vehicle's position at t_1 implicitly defining the local origin. The autonomous piloting guidance mode attempts to fit a *smooth* and *feasible* desired trajectory to these points which is sent to the servo level control routines. The calculated trajectory is not feasible when any of the vehicle's *kinematic constraints* are violated, and the trajectory should be smooth enough to calculate these quantities. This autonomous piloting task therefore reduces to a *constrained spline-fitting* problem, where constrained curves are used to approximate the discrete subgoal information received from the navigation system. The type of spline-fitting algorithm used is determined by the particular requirements of the task: the ability to smoothly join together two consecutive trajectories and real-time operation. Similar algorithms have been developed for robotic arm trajectory generation [Thompson and Patel, 1987, Wang and Horng, 1990], and from this work an approach based on on-line, constrained *B-spline* curve fitting is described.

The most important contribution of this work is the proposed method for constraining the calculated trajectory. The order of the B-spline function depends on the number of the derivatives that need to be constrained. If it is necessary to constrain a cartesian (x, y) trajectory which produces a continuous velocity and acceleration, an order 4 (cubic) B-spline polynomial is the lowest order B-spline curve which can produce this information. Subgoals passed down in the motion commands 9.1 and 9.2 are regarded as *soft*, whereas the vehicle's kinematic constraints are regarded as *hard*. Thus the spline generation algorithm attempts to interpolate the soft subgoals, and if any hard kinematic constraints are violated, the trajectory is replanned locally to ensure that the hard constraints are always satisfied and the soft constraints are *nearly* satisfied. The problem of local replanning in the case of an unforeseen obstruction is not considered, as this task should be handled higher up in the control hierarchy (where the planning horizons are longer).

To ensure that none of the hard constraints are violated, there must exist an efficient algorithm for differentiating and integrating the spline locally. With conventional spline-fitting techniques this is not feasible, as a completely new trajectory would have to be formed. However, this is not the case for the B-spline representation as the algorithm possesses the following properties:

- **compact support** means that the trajectory can be locally modified;
- **piecewise polynomial** interpolants of order k on the relevant domain;
- **Gauss decomposition** without pivoting on a banded matrix of bandwidth k can be used to find the coefficients; and
- **simple** recurrence relationships exist for evaluating a B-spline's derivatives and integrals.

The first three properties are desirable for any interpolation scheme, and the last property is used extensively in this section to incorporate constraints into the vehicle's desired trajectory.

This section begins by reviewing the properties of a conventional B-spline interpolant. The theory is not new; it has been developed over the last twenty years by a number of researchers, [Cox, 1972, 1982, 1990, DeBoor, 1972, 1978, DeBoor and DeVore 1985, Kozak 1980] and the relevant results are simply stated. Then the proposed algorithms for incorporating boundary conditions which ensure that two consecutive trajectories can be joined smoothly, and for modifying the trajectory locally subject to the hard kinematic constraints are described. Both of these algorithms only affect the trajectory *locally*, the calculations are *non-iterative* and can be performed in real-time.

9.4   B-SPLINE INTERPOLANTS

B-spline interpolants are formed from a *linear* combination of a set of univariate B-spline basis functions. The output calculations of a B-spline interpolant and a univariate B-spline AMN (see Section 8.2) are equivalent, as it is only the algorithms used to generate the set of linear coefficients which are different. A B-spline AMN typically trains its weight vector using an instantaneous gradient descent rule, whereas the B-spline interpolant coefficient vector is calculated using a matrix inversion routine. To reflect these similarities and differences, the same notation is used for the B-spline basis functions, whereas the *coefficient* vector of a B-spline interpolant is denoted by $\mathbf{c}$, rather than the equivalent *weight* vector represented by $\mathbf{w}$.

The definition of a univariate B-spline basis function is now reviewed briefly. Each trajectory is now a function of time and so the input to a B-spline interpolant is t, as opposed to x which represented the input to the B-spline AMN. Similarly, only univariate spline curves are considered, so the ith knot and the jth interval are denoted by λ_i and I_j, respectively; the extra subscript used to denote the input axis is not necessary.

9.4.1   B-spline Interpolation

The B-spline basis functions are generated using the numerically stable recurrence relation defined in Section 8.2 and are defined on a compact interval I. For the trajectory generation application, the relevant domain is the time spanned by the motion command, $I = [t_1, t_L]$. For interpolation tasks, the exterior knots are generally set equal to the appropriate extrema (t_1 or t_L), as this improves the condition of the matrix [Kozak, 1980], and facilitates the incorporation of derivative boundary conditions (see Cox [1977] and Section 9.5.1).

B-spline Interpolants

Let $s(t)$ denote a univariate polynomial spline curve of order k with r interior knots, which is defined on the interval $[t_1, t_L]$. The interior knots are simple, so $s(t)$ consists of a set of $(r + 1)$ polynomial pieces of order k. The spline curve is a polynomial of order k in the interior of the jth subinterval I_j, and at λ_j the jth and $(j + 1)^{th}$ polynomial pieces are equal in value and for all derivatives up to and including the $(k - 2)^{nd}$. The B-spline basis functions which have just been defined form a *basis* for polynomial splines of order k on the interval $[t_1, t_L]$ with an interior knot set $\{\lambda_j\}_{j=1}^r$. Hence if $s(t)$ is such a spline, it has a unique B-spline representation given by:

$$s(t) = \sum_{j=1}^{p} c_j N_k^j(t) \qquad (9.3)$$

for a p-dimensional coefficient vector $\mathbf{c}$. The value of the univariate polynomial function $s(t)$ at time t depends on at most k of the coefficients c_j, so Equation 9.3 can be rewritten as:

$$s(t) = \sum_{j=i}^{i+k-1} c_j N_k^j(t), \qquad \text{when } t \in I_i \qquad (9.4)$$

Thus the computational cost of evaluating the k non-zero basis functions for a particular input t is approximately $1.5k^2$ floating point multiplications and divisions, once the subinterval index i has been determined (see Section 8.2.2).

9.4.2 Polynomial Interpolation

A univariate spline interpolation problem [Cox, 1990] may be stated as:

> Given L data points $\{t_i, f_i\}_{i=1}^L$, where $t_1 < t_2 < \cdots < t_L$, calculate a spline function $s(t)$ which satisfies the following condition for each i:

$$s(t_i) = f_i \qquad (9.5)$$

For a B-spline interpolant there are three parameters to be chosen: the order of the spline, the number of the knots and the knot set. Choosing $k = 1$ gives a piecewise constant spline function, $k = 2$ gives a piecewise linear spline function and $k = 3, 4$ give piecewise quadratic and cubic spline interpolants, respectively. The number of interior knots follows directly from the number of data points and the order of the spline ($r = L - k$) as the number of basis functions ($p = r + k$) must equal the number of data points.

Interior Knot Choice

The existence and shape of the spline $s(t)$ depend on the *position* of the interior knots. It has been established [Schoenberg and Whitney, 1953] that necessary and sufficient conditions for the existence of a unique spline are given by:

$$t_i < \lambda_i < t_{i+k} \tag{9.6}$$

for each interior knot, $i = 1, 2, \ldots, r$. There must exist at least one data point lying in the support of each basis function.

Whilst this condition ensures the existence and uniqueness of $s(.)$, it is necessary for these conditions to be *well satisfied*, otherwise the polynomial interpolant may have oscillatory behaviour (demonstrated in Section 9.6.1). Generally, the interior knot placement strategy given by:

$$\lambda_i = \begin{cases} t_{i+j} & \text{if } k = 2j \ (k \text{ even}) \\ (t_{i+j-1} + t_{i+j})/2 & \text{if } k = 2j - 1 \ (k \text{ odd}) \end{cases} \tag{9.7}$$

for $i = 1, 2, \ldots, r \ (= L - k)$, produces a well-behaved spline interpolant. For an order 2 piecewise linear spline, the knots are placed at each value of t_i, and for an order 4 piecewise cubic polynomial, the knots also are placed at each value of t_i apart from t_2 and t_{L-1}. When k is odd, the knots are placed between two adjacent data points, ignoring $((k-1)/2)$ data points at each end.

Matrix Interpretation

Combining Equations 9.3 and 9.5 generates the following set of equations:

$$\sum_{j=1}^{p} c_j N_k^j(t_i) = f_i$$

for $i = 1, 2, \ldots, p \ (= L)$, which can be rewritten as:

$$\mathbf{Ac} = \mathbf{f} \tag{9.8}$$

where $\mathbf{A}$ is a $(p \times p)$ matrix with $a_{i,j} = N_k^j(t_i)$, $\mathbf{c}$ is the p-dimensional coefficient vector and $\mathbf{f}$ is the p-dimensional vector composed of the desired points.

The matrix $\mathbf{A}$ is *banded* of bandwidth k, as each row contains at most k non-zero entries, due to the compact support property of the B-spline basis functions. It is also totally positive [DeBoor and DeVore, 1985] hence Equation 9.8 can be solved safely using Gauss elimination *without* pivoting with a computational cost $O(pk^2)$, requiring only pk memory locations. Also, the interior knot placement strategy proposed in Expression 9.7 ensures that the existence and uniqueness of the coefficient vector $\mathbf{c}$.

9.5 BOUNDARY AND KINEMATIC CONSTRAINTS

The proposed application of the B-spline interpolants requires that two consecutive splines should be smoothly joined and that the derivatives of the spline should be constrained. This requires two modifications to the B-spline interpolant algorithm just described: the incorporation of boundary conditions, and a real-time derivative constraining method.

9.5.1 Incorporation of Boundary Conditions

During the operation of an AGV, it is necessary for the navigation system to issue motion commands which must be smoothly joined, as a complete trajectory cannot be determined before a mission starts, and local trajectory replanning is required in order to compensate for incomplete or inaccurate information.

If the desired trajectory is a B-spline of order k, the output and all its derivatives up to the $(k-2)^{nd}$ are continuous. However, it is not possible to specify the initial derivative values using the previously described algorithm, and when a cartesian coordinate motion command is issued, different initial derivative values result in a discontinuous change in the desired velocity and acceleration. Generally, the order of the spline is the *minimum* which satisfies the smoothness conditions of the required trajectory, and these conditions apply for the overall trajectory, at the joining points as well as for each interval. For a B-spline of order k, $(k-2)$ initial boundary conditions must therefore be specified and the modified spline interpolation problem can be stated as:

> *Given L data points (t_i, f_i) $(i = 1, 2, \ldots, L)$, with $t_1 < t_2 < \cdots < t_L$, and a total of $(k-2)$ derivative boundary conditions at $t = t_1$. The problem is to determine a spline of order k which interpolates the given data points and satisfies the specified boundary conditions.*

Matrix Formulation

For a B-spline interpolant $s(t)$ of order k with $(k-2)$ derivative boundary constraints of the form:

$$
\begin{aligned}
s^{(1)}(t_1) &= f^{(1)}(t_1) \\
s^{(2)}(t_1) &= f^{(2)}(t_1) \\
&\vdots \\
s^{(k-2)}(t_1) &= f^{(k-2)}(t_1)
\end{aligned}
\tag{9.9}
$$

the structure of $\mathbf{A}$ is very simple *if* the first set of exterior knots are all chosen to be equal to t_1. At the point $t = t_1$, there are the $(k-2)$ derivative boundary conditions as well as the data point, so the top $(k-1) \times (k-1)$ submatrix of $\mathbf{A}$ has the form:

$$
\begin{bmatrix}
1 & & & & \\
* & * & & & \\
* & * & * & & \\
\vdots & & \vdots & & \\
* & * & * & \cdots & *
\end{bmatrix}
\tag{9.10}
$$

where $*$ denotes a non-zero basis function output and the tth row corresponds to the derivative condition $s^{(t-1)}(t_1) = f^{(t-1)}(t_1)$. The top $(k-1) \times (k-1)$ submatrix is *lower triangular* and the first $(k-1)$ coefficients can be determined *independently* of the remaining coefficients.

An extra $(k-2)$ *distinct* basis functions and interior knots must be introduced to account for the extra $(k-2)$ degrees of freedom introduced by the boundary conditions, and the positioning of these knots is discussed in Section 9.6.1.

9.5.2   The Incorporation of Hard Derivative Constraints

The incorporation of kinematic constraints in the B-spline trajectory requires that the spline's output and its derivatives be constrained, and it is assumed that these constraints are available for use in the trajectory planning routines. Up to and including the $(k-1)^{st}$ derivative can be constrained, even though the spline has only $(k-2)$ continuous derivatives.

Due to the initial boundary constraints, the coefficients $c_1, c_2, \ldots, c_{k-1}$ can be determined immediately, as the top-left submatrix of size $(k-1) \times (k-1)$ is lower triangular. Also the rest of matrix $\mathbf{A}$ may be reduced to a lower triangular form by matrix row addition in a stable manner without pivoting. The remaining coefficients $c_k, c_{k+1}, \ldots, c_{L+k-2}$ can then be found using feedforward substitution. It is worth emphasising that this reduction of $\mathbf{A}$ to lower triangular form only needs to be performed once for each motion command. For example, the matrix associated with the velTraj(.) motion command for the velocity interpolant is *identical* to the matrix associated with the curvature subgoals, hence the computation only needs to be performed for one of the interpolants. In the following sections, only the coefficient vectors are distinguished by subscripts as the same matrix is used to generate both spline interpolants.

The value of the spline and its derivatives on an interval I_i depend on (at most) $c_i, c_{i+1}, \ldots, c_{i+k-1}$. By induction, every term except c_{i+k-1} has already been determined, so the values of the preceding coefficients and the hard constraint can be combined to form an *interval*, in which c_{i+k-1} must lie if the hard constraints are to be satisfied. If the desired value of c_{i+k-1} (obtained using feedforward substitution) lies within this interval, the spline interpolates this data point exactly. Otherwise the desired value lies outside the interval, and c_{i+k-1} is set equal to the appropriate endpoint. This ensures that the spline passes close to the desired value while still limiting the derivatives. Therefore the hard derivative constraints are always satisfied and if possible the soft data constraints as well.

This algorithm for incorporating derivative constraints is not optimal, as it does not attempt to fit the data points in any best sense, although it is simple, operates in real-time and is well behaved if the constraints are not greatly violated. This technique is now applied to a quadratic spline trajectory fitted to a velocity/curvature motion command.

9.6   EXAMPLE: A QUADRATIC VELOCITY INTERPOLANT

A quadratic B-spline is used to interpolate the velocity and curvature velTraj(.) motion command passed down from the navigation system. The spline generates a *continuous* demanded acceleration, and this is expected to be the minimum continuity requirement for any practical system.

9.6.1   The Problem Description

The piecewise quadratic velocity/curvature interpolants provide continuous velocity/acceleration and curvature/rate of curvature trajectories that can be constrained using the previously described algorithm. Continuous acceleration is required if the velocity trajectory is to be physically realisable, and a lot of control effort can be wasted in trying to achieve a trajectory which is not. As stated previously, the algorithms for forming a velocity and a curvature trajectory are equivalent, so only the former is described.

The initial boundary conditions require the velocity and the acceleration to be specified at $t = t_1$ for two consecutive spline interpolants to be joined smoothly. These boundary conditions are expressed as:

$$
\begin{aligned}
s_v(t_1) &= v(t_1) \\
s_v^{(1)}(t_1) &= v^{(1)}(t_1)
\end{aligned}
\qquad (9.11)
$$

The velocity trajectory is again limited by the vehicle's velocity and acceleration kinematic constraints. Hence it is required that the following expressions be satisfied:

$$
\begin{aligned}
v_{\min} &\leq s_v(t) \leq v_{\max} \\
v_{\min}^{(1)} &\leq s_v^{(1)}(t) \leq v_{\max}^{(1)}
\end{aligned}
\qquad (9.12)
$$

for $t \in [t_1, t_L]$.

The interior and exterior knots are chosen according to Expression 9.7:

Exterior $\lambda_{-2} = \lambda_{-1} = \lambda_0 = t_1,$ $\lambda_{L-1} = \lambda_L = \lambda_{L+1} = t_L$

Interior $\lambda_i = \dfrac{t_i + t_{i+1}}{2}$

for $i = 1, 2, \ldots, L - 2$. The derivative boundary condition again requires that an extra distance interior knot be defined, and for the reasons explained below this is placed at $(t_1 + t_2)/2$. Thus $\mathbf{A}$ is again a banded matrix of bandwidth 3, and the constraining algorithm proposed in Section 9.5.2 can be applied; reduce the matrix to lower triangular form, evaluate the unconstrained coefficients using feedforward substitution, and limit these values using the kinematic constraints.

A Bad Knot Choice

Two consistent places for this new knot appear to be either $(t_1 + t_2)/2$ or $(t_{L-1} + t_L)/2$, and these arrangements are now discussed.

Choosing to place the extra knot at $(t_{L-1} + t_L)/2$ and renumbering the subsequent knots resulted in the matrix $\mathbf{A}$ being in lower triangular form. Each row contains at most three non-zero elements and its structure is shown below:

$$
\begin{bmatrix}
1 & & & & & & & & \\
* & * & & & & & & & \\
* & * & * & & & & & & \\
 & * & * & * & & & & & \\
 & & * & * & * & & & & \\
 & & & & \ddots & & & & \\
 & & & & * & * & * & & \\
 & & & & & * & * & * & \\
 & & & & & & & & 1 \\
\end{bmatrix}
$$

where $*$ denotes a non-zero basis function output.

As the first two coefficients for the position spline are determined uniquely by the initial boundary conditions, the remaining coefficients, $c_{v,j}$ $(j = 3, 4, \ldots, L+1)$, can be found using feedforward substitution. This is a very desirable feature as it simplifies the incorporation of the hard kinematic and safety constraints, although the interpolant generated by this knot choice, as shown in Figure 9.9, is not satisfactory. The spline is a piecewise quadratic interpolant which passes through the subgoals, but its behaviour is far from desirable. The structure of $\mathbf{A}$ gives some indication of this, as the three non-zero elements in each row are typically $\{0.125, 0.75, 0.125\}$, with the third element occurring on the main diagonal. Thus the remaining $(L - 1)$ coefficients are strongly influenced by the first two coefficients and the initial boundary conditions; an undesirable property. The condition of the matrix (for a set of 21 equispaced subgoals) is $C(\mathbf{A}) = 4.8 \times 10^{14}$, hence this knot distribution produces an extremely ill-conditioned problem.

Although this velocity interpolant is clearly unsatisfactory, the motivation for reducing the matrix to lower triangular form and performing feedforward substitution, while constraining each coefficient, came from this initial work.

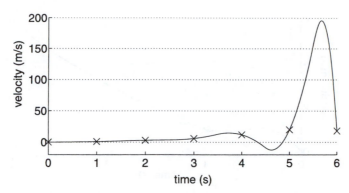

Figure 9.9   A piecewise quadratic velocity interpolant produced by a bad interior knot placement. The spline interpolates the subgoals (crosses) although its behaviour between these points is unsatisfactory.

A Good Knot Placement Strategy

Choosing to place the extra knot at $(t_1 + t_2)/2$, and reindexing the subsequent knots resulted in $\mathbf{A}$ being a banded matrix of bandwidth 3, as shown below:

$$
\begin{bmatrix}
1 & & & & & & & \\
* & * & & & & & & \\
& * & * & * & & & & \\
& & * & * & * & & & \\
& & & * & * & * & & \\
& & & & \ddots & & & \\
& & & & & * & * & * \\
& & & & & & * & * & * \\
& & & & & & & & 1 \\
\end{bmatrix}
$$

This can be reduced to a lower triangular form (as the $(L-1) \times (L-1)$ lower submatrix is totally positive) using matrix row addition without pivoting. The unconstrained coefficients $c_{v,j}\ j = 3, 4, \ldots, L+1$, can then be determined using feedforward substitution. The three non-zero elements in each row of $\mathbf{A}$ are still generally $\{0.125, 0.75, 0.125\}$, but the 0.75 values now occur on the main diagonal and so the effect of any change in the boundary conditions decreases very quickly. For this knot placement strategy, the condition number of the matrix is $C(\mathbf{A}) = 8.1$, so the matrix inversion routines are numerically stable.

The insensitivity of the global spline to changes in the left-hand derivative constraints is illustrated in Figure 9.10. Changing the boundary conditions only affects the spline *locally*, which is a very desirable property, and this interior knot placement strategy is adopted in the remainder of this section.

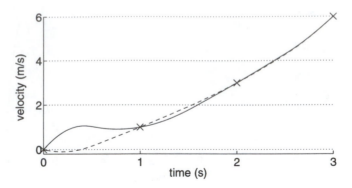

Figure 9.10   Two piecewise quadratic velocity interpolants produced by a good interior knot placement. The interpolants are insensitive to bad initial boundary conditions, as the splines incorporating two very different derivative boundary conditions $\left(v^{(1)}(t_1) = -1, 5\text{ms}^{-1}\right)$ are nearly the same after just two intervals.

9.6.2   Incorporation of Hard Constraints

Consider constraining the quadratic velocity trajectory using the velocity and acceleration vehicle constraints. The hard velocity constraint is incorporated into the trajectory using the B-spline's local bound property. Thus it is required that:

$$v_{\min} \leq c_{v,j} \leq v_{\max}$$

for $j = 1, 2, \ldots, L + 1$. This expression is sufficient to guarantee that the velocity spline interpolant does not exceed the kinematic bounds, but it is not necessary. It is possible for a small number of coefficients to exceed the local bounds and for the interpolant still to satisfy the kinematic constraints. The constrained trajectory will not be optimal, although it is sufficient for most reasonable subgoal specifications. The hard velocity constraints are taken into account by ensuring that on the jth interval:

$$c_{v,j+2} \in [v_{\min}, v_{\max}] \tag{9.13}$$

as by induction, the previous two coefficients also satisfy this relationship.

On same interval I_j, the vehicle's demanded acceleration is a piecewise linear function. By induction, $s_v^{(1)}(\lambda_{j-1})$ satisfies the hard acceleration constraint, and so it remains to prove that $s_v^{(1)}(\lambda_j)$ also satisfies these bounds. This can be achieved by ensuring that:

$$v_{\min}^{(1)} \leq 2\frac{c_{v,j+2} - c_{v,j+1}}{\lambda_{j+1} - \lambda_{j-1}} \leq v_{\max}^{(1)}$$

which can be rewritten as:

$$c_{v,j+2} \in \left[c_{v,j+1} + v_{\min}^{(1)} \frac{\lambda_{j+1} - \lambda_{j-1}}{2}, c_{v,j+1} + v_{\max}^{(1)} \frac{\lambda_{j+1} - \lambda_{j-1}}{2} \right] \qquad (9.14)$$

Hence the hard velocity and acceleration kinematic constraints can be incorporated into the desired velocity trajectory by taking the intersection of Expressions 9.13 and 9.14, and using this interval to limit the values of the coefficients obtained from the feedforward substitution process. The second derivative (jerk) can also be constrained using an analogous expression.

9.6.3 Results

This is illustrated in Figure 9.11 where a piecewise quadratic velocity desired trajec-

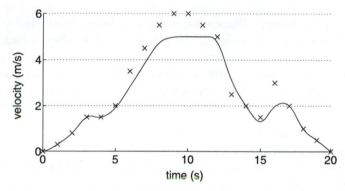

Figure 9.11 A constrained piecewise quadratic velocity interpolant. The crosses show the desired subgoals and the solid line is the desired trajectory which incorporates both velocity and acceleration constraints.

tory is constrained as previously described. The ease with which these constraints can be incorporated into this trajectory supports the proposal that the velTraj(.) motion command should be used as the main communication interface between the navigation system and the guidance unit. The posTraj(.) motion command can then be used to describe tasks which operate at low speeds and are mainly servoing in nature.

9.7 DISCUSSION

The aim of this chapter has been to show both the flexibility of the B-spline networks and their application to a wide range of apparently different tasks. Their range of potential applications include nonlinear adaptive modelling (time series prediction, etc.) and control, providing a framework for static fuzzy rule sets and constrained surface fitting algorithms. All of these tasks rely on the excellent numerical properties of the basis functions (compact support, partition of unity, real-time computations, well-conditioned bases, etc.), so numerical analysis techniques and tools are very important in the study of learning systems.

The development of the static rule base illustrated many of the potential problems with using these techniques. The limited amount of badly distributed training data causes these networks only to generalise locally, and unless a more complex algorithm is used to determine the rule base's structure automatically, many rules must be set and edited by the designer. For this task, the *interpretation* of the B-spline network as a set of fuzzy rules is invaluable. Also the ability to incorporate *a priori* knowledge in the network (in the form of the shape of the basis functions) simplifies its structure and provides a useful design tool. Another "fuzzy technique" which proved useful in the design of the B-spline network was the basis function scaling factors (making the width and position of the membership functions depend on the length and width of the slot). They allowed the *same* rule base to be used with many different slots and vehicles, without explicitly using the slot's length and width as inputs to the rule base. This latter option would have made the size of the rule base unmanageable.

This chapter has also described an algorithm for on-line, real-time constrained trajectory generation, based on univariate B-splines, which has many desirable features:

- **well-conditioned** basis functions;
- **simple** incorporation of boundary conditions;
- **non-iterative** spline constraining algorithms and so does not require a search procedure; and
- **smooth** splines of any order can be used.

The bounded support of the B-spline basis functions also make on-line, local trajectory replanning feasible, since changing a small number of coefficients modifies only the trajectory in that region [Thompson and Patel, 1987]. Thus piloting algorithms can be developed which modify a precalculated trajectory locally, such that unexpected/unforeseen obstacles are avoided.

10

The Representation of Fuzzy Algorithms

10.1   INTRODUCTION: HOW FUZZY IS A FUZZY MODEL?

Research into fuzzy systems is currently receiving a lot of attention, thus mimicking the revival of interest in artificial neural networks which occurred in the late eighties. Both fields are aimed at developing systems, motivated by current biological and physiological understanding of the brain, which possess the ability to learn, reason and generalise using noisy, redundant information. There has been a wide variety of alternative systems proposed for implementing different tasks (supervised, unsupervised and reinforcement learning, feedforward and recurrent networks), although some neural and fuzzy networks are similar, and often it is only the implementation method which differs. This chapter compares several neural and fuzzy networks, using the notation developed for representing fuzzy systems, and shows that the two apparently different fuzzy implementation methods, *continuous* and *discrete*, are equivalent under certain conditions and the relationship between these fuzzy networks and radial basis function and B-spline neural networks is explained.

The term "fuzzy" is widely (ab)used in the literature. It has been said that fuzzy systems are inherently robust with respect to both plant parametric variations and measurement uncertainty. Similarly, it has been claimed that fuzzy systems are easily understood by the non-specialist because they use vague terms to explain their control actions, and that fuzzy systems can be verified and validated quicker than conventional models and controllers. These remarks are often based strictly on simulation results and have little theoretical foundations, and recent survey articles [Cox, 1993, Schwartz and Klir, 1992, Self, 1990] often add to the mystique that surrounds the subject rather than trying to clarify it. Fuzzy practitioners often have to defend their research both against hyped comments and ill-informed criticism. Fuzzy algorithms which are composed of rules such as:

IF (*I am late for a meeting*) THEN (*I should hurry up*)

are (necessarily) imprecise, and are frequently subjective and context dependent. For instance, if a person was explaining their actions using the above rule, the relative weighting placed on how much they speed up would depend on the importance of the meeting. If the meeting was a job interview, there would be a greater urgency as opposed to if the meeting was merely to decide who was organising a dinner. Often these details are *implicitly* assumed to be known. However, if the rules are to be implemented, they have to be explicitly stated or else different fuzzy algorithms constructed for different situations. The imprecision associated with a set of fuzzy rules is generally completely resolved once they have been implemented on a computer, as specific meanings have been assigned to the vague statements [Pedrycz, 1993].

The power of a fuzzy approach lies in the way these vague, imprecise rules can be given a specific meaning and the manner in which the fuzzy system produces outputs for inputs which only partially match the rules created during initialisation and training. This partial matching ensures that similar inputs produce similar outputs whereas dissimilar inputs produce independent outputs, so the network *generalises locally.* The smooth interpolation between rules sometimes results in a smaller rule base (compared to an equivalent expert system), although many of the fuzzy systems developed do not consider output smoothness (minimum curvature) as being a prime consideration. Representing concepts locally also ensures that only a small part of the system is involved in the output calculation, and their effect can easily be interpreted, such a network is said to be *transparent.* The robustness of a fuzzy system depends on its structure and on the design procedure. Fuzzy networks are not inherently robust; they can process uncertain (noisy) information, although the success of such an operation depends on the *a priori* information given to the system, and to claim that a fuzzy design is robust simply because it is nonlinear is not justified. It is also shown in this chapter that similar techniques for dealing with noisy inputs can be used with neural models, illustrating the close relationship between these two techniques.

Traditionally fuzzy modelling and control systems have been implemented using a discrete approach where the fuzzy sets and relations are represented by the fuzzy membership value at a set of discrete points [Kouatli and Jones, 1991, Linkens, 1993, Sutton and Towill, 1985]. This representation is very flexible in that an explicit membership function does not need to be stored, although generally the number of sample points required to represent the functions is large and depends on the method used to fuzzify the information presented to the network. Each fuzzy membership function and each fuzzy relation (rule) is stored at several sample points, which implies that the parameters may be *redundant* and that to modify a fuzzy expression or rule, a subset of the discrete parameters must be updated. The discrete approach is arguably more flexible than the continuous implementation scheme as all of the common fuzzy operators can easily be used within this framework, although it is argued that the type of operators which are inappropriate for use within continuous fuzzy systems reduce the quality of the decision surfaces and would not be used because of this factor alone.

Continuous fuzzy systems have gained in popularity in recent years, partly due to their link with certain neural networks. In a continuous fuzzy network, the fuzzy membership functions are stored as continuous mappings which depend on a set of parameters, and a rule confidence is used to denote the strength with which a particular rule fires. The memory requirements are generally much less than that required for a discrete fuzzy implementation, although the computational cost is sometimes slightly greater. One significant advantage with this approach is that it allows concepts such as normalised basis functions, rule confidences, least-mean square learning, universal approximation and learning interference to be investigated and analysed within a consistent framework. Previously, fuzzy systems have been shown to outperform standard nonlinear modelling and control designs on certain problems, although there has been a lack of theoretical rigour associated with this approach. The results derived in this chapter relating to continuous fuzzy systems, allow new insight to be gained into how vague linguistic knowledge is implemented in a rule base and how approximate reasoning algorithms *generalise*.

This chapter describes a common framework for the investigation of discrete and continuous fuzzy systems and outlines the similarities that exist between the two representations. The role of the fuzzy membership functions and operators are discussed and arguments that support the adoption of *algebraic* rather than *truncation* fuzzy operators are expounded. It is shown that under certain conditions the continuous and discrete fuzzy implementations are *equivalent* and that a continuous approach which uses algebraic fuzzy operators generally requires a smaller number of parameters as well as reducing the computational cost. These results also illustrate the strengths and weaknesses of using fuzzy networks within modelling and control systems, and in particular they show that a fuzzy system can reproduce exactly any bounded linear mapping. The role of the rule confidence and relational matrices are also discussed and potential problems with certain implementations are highlighted (such as using *binary* rule confidence matrices). Several fuzzy training algorithms are proposed in Chapter 11 for which convergence to a minimum Mean Square Error (MSE) can be proved. Training can take place in either *weight* or *rule confidence* space, and it is shown that the outputs are equivalent in either case. However, it is computationally cheaper to train the fuzzy network in weight space and then to generate fuzzy algorithms (with their associated rule confidences) for network validation and verification purposes.

10.1.1 Fuzzy Modelling and Control

Fuzzy networks can be used for many different tasks within intelligent control systems because they represent general, nonlinear, relationships which can be initialised using expert knowledge [Berenji, 1992, Harris *et al.*, 1992, Lee, 1990]. The networks have been used as plant models in predictive control algorithms, as nonlinear estimators [Moore *et al.*, 1993, Pacini and Kosko, 1992], to form nonlinear controllers [Moore and Harris, 1994, Procyk and Mamdani, 1979], to represent expert knowledge in PID autotuning algorithms and also at a supervisory level for plant monitoring, fault detection and diagnosis [Linkens and Abbod, 1993].

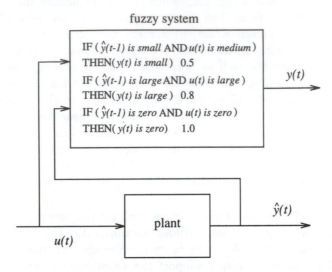

Figure 10.1   A fuzzy plant model. The inputs are time-delayed plant outputs and control
actions and the output is an estimate of the current plant output.

For a basic plant model or controller, the inputs to a fuzzy system generally
contain discrete measurements of the system's past and present state, and the
output is some nonlinear function of these values (see Figure 10.1). The fuzzy rules
contain local knowledge about the relationship between these quantities, and it is
the interaction between these rules which determines the type of model produced by
the fuzzy system. Many different fuzzy modelling, control and estimation systems
have been proposed where self-organising fuzzy networks are used to learn arbitrary
relationships, although very little work has been focused on *how* these modules
adapt. This chapter investigates the type of mappings formed by a fuzzy network
and relates them to more conventional nonlinear systems, which allows claims made
about fuzzy systems to be investigated (their robustness qualities and their ability
to process uncertain information), as well as describing possible limitations of some
implementation schemes. The training algorithms described all assume that the
desired response of the network is available, hence the results derived in this chapter
apply directly to adaptive plant models, and the problems of inferring the control
action errors which cause undesirable behaviour is considered in Chapter 11.

10.2   FUZZY ALGORITHMS

Fuzzy logic is widely used in intelligent control to reason about vague rules which
describe the relationship between imprecise, qualitative, linguistic assessments of

the system's input and output states. These production rules are generally natural language representations of a human's (or expert's) knowledge, and provide an easily understood knowledge representation scheme for explaining information learnt by a computer or for initialising a particular system. Most conventional expert systems can be described using a set of binary production rules [Quinlan, 1993], but the discontinuities that occur in the system's output, as different rules become active, do not resemble a human's actions, where there is generally a *smooth* cause and effect relationship. The vague manner in which fuzzy information is presented naturally means that similar inputs map to similar concepts with similar degrees of membership. Hence fuzzy systems inherently possess this important property. Fuzzy production rules are described using vague terms such as *small* or *medium* to categorise the input and output variables, and the set of all these rules forms a fuzzy algorithm.

A *fuzzy algorithm* is defined to be the set of these rules which describes a mapping between the system's input and output states. If the fuzzy algorithm has n input $(x_1, \ldots, x_n)^T$ and m output states $(y_1, \ldots, y_m)^T$, it can equivalently be written as m fuzzy algorithms each with n inputs and one output. This argument is identical to the one expounded in Section 3.2.1, where it was stated that only single output networks were considered in this book as an m output network could be decomposed into m single output ones. These fuzzy algorithms are composed of fuzzy rules of the form:

$$\text{IF } (x_1 \text{ is } A_1^i) \text{ AND } \cdots \text{ AND } (x_n \text{ is } A_n^i) \text{ THEN } (y \text{ is } B^j) \qquad (c_{ij})$$

where A_k^i, $k = 1, \ldots, n$ and B^j are linguistic variables which represent vague terms such as *small, medium* or *large* defined on the input and output variables, respectively. A fuzzy rule maps the antecedent, formed by the intersection (AND) of n univariate linguistic statements $(x_k \text{ is } A_k^i)$, to the consequent formed by a single univariate linguistic statement $(y \text{ is } B^j)$. It can therefore be rewritten as:

$$r_{ij} : \quad \text{IF } (\mathbf{x} \text{ is } \mathbf{A}^i) \text{ THEN } (y \text{ is } B^j) \qquad (c_{ij})$$

where $\mathbf{A}^i$ is the ith multivariate fuzzy set formed from the fuzzy intersection of the individual univariate fuzzy sets and r_{ij} represents the ij^{th} fuzzy rule. Associated with each rule is a variable $c_{ij} \in [0, 1]$ that denotes the *confidence* in the rule being true. A confidence of $c_{ij} = 0$ means the rule never fires whereas if $c_{ij} > 0$, the rule influences the output whenever the input is a partial member of the antecedent. A more detailed explanation of these rule confidences is provided in Section 10.6.6, although it should be noted that they are similar to Wang's rule degrees [1994], and the *prior* state probabilities used in Bayesian systems (see Section 10.6.6).

In general, the fuzzy algorithm consists of a set of these fuzzy rules which are connected together (using a union operator or the fuzzy OR) to form the rule base. For example, the rules of a fuzzy algorithm might have the form:

$$r_{1,1} : \qquad \text{IF } (\mathbf{x} \text{ is } \mathbf{A}^1) \text{ THEN } (y \text{ is } B^1) \qquad (c_{1,1})$$

$$r_{1,2}: \quad \text{OR} \quad \text{IF } (\text{x } is \text{ } \mathbf{A}^1) \text{ THEN } (y \text{ } is \text{ } B^2) \quad (c_{1,2})$$
$$r_{2,1}: \quad \text{OR} \quad \text{IF } (\text{x } is \text{ } \mathbf{A}^2) \text{ THEN } (y \text{ } is \text{ } B^1) \quad (c_{2,1})$$
$$\vdots \qquad\qquad\qquad \vdots$$
$$r_{pq}: \quad \text{OR} \quad \text{IF } (\text{x } is \text{ } \mathbf{A}^p) \text{ THEN } (y \text{ } is \text{ } B^q) \quad (c_{pq})$$

A fuzzy algorithm that maps p input sets to q output sets has pq rules and associated confidences.

To implement a fuzzy algorithm, the fuzzy sets that represent the linguistic statements such as $(x_k$ is $A_k^i)$ need to be defined and the functions used to implement the fuzzy operators, AND, IF(.) THEN(.) (implication) and OR, chosen. Various different schemes are examined in this chapter and claims about *robustness* and *insensitivity* to the fuzzy set shapes are discussed. Once the set shapes and the operators have been chosen, the relationship between the input and output variables is no longer uncertain or vague; it is a *deterministic*, nonlinear multivariable function. Therefore a distinction is made in this book between a *fuzzy algorithm* which contains a set of imprecise, qualitative, linguistic rules, and a *fuzzy system* which refers to a *specific*, context dependent implementation of the rules, and this is illustrated in Figure 10.2. The distinction between a fuzzy algorithm and a fuzzy

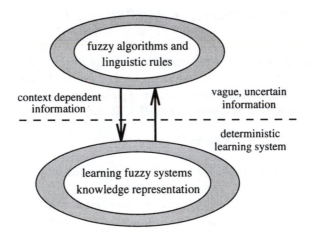

Figure 10.2 The distinction between a fuzzy algorithm and a fuzzy system.

system was originally made by Procyk and Mamdani [1979], who wrote "the (fuzzy) theory acts as an interface between the real imprecise world outside expressed linguistically and mathematical representation inside the controller". Recently this subtle but important point has been forgotten when unsubstantiated claims have been made about the subject.

10.3   FUZZY SETS

To represent linguistic statements such as (x *is small*), Zadeh [1965,1973] introduced the concept of a fuzzy set. A fuzzy set A is a collection of elements defined in a *universe of discourse* labelled X when the universe is continuous and X^d when it is discrete. It generalises the concept of a classical set by allowing its elements to have *partial* membership ($\in [0,1]$), and the degree to which the generic element x belongs to A is characterised by a *membership function* $\mu_A(x)$. Associated with a classical *binary* or *crisp* set is a *characteristic function* which returns 1 if the element is a member of that set and 0 otherwise. The fuzzy membership function generalises this concept by allowing elements to be partial members of a set, reflecting degrees of uncertainty about the information.

Each linguistic term is represented by a membership function and the set of all of these terms determines how an input variable is represented within the fuzzy system. Typically, about seven membership functions are used to represent each input, and they define fuzzy terms such as x *is positive small*, as shown in Figure 10.3.

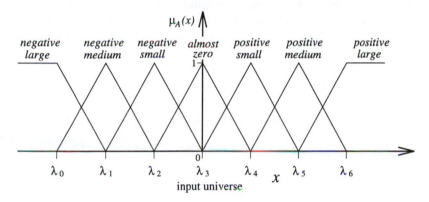

Figure 10.3 A typical set of seven triangular fuzzy membership functions.

The *support* of a fuzzy set A is the set of inputs that have a non-zero membership function value, ie. $\{x : \mu_A(x) > 0\}$. The support is said to be *compact* if it is a (strict) subset of the universe of discourse. For on-line, adaptive modelling and control applications it is often important to use fuzzy sets which have a compact support as this means only a small proportion of all the rules contribute to the output (a localised response region), and only these rules need to be modified when the network is trained using instantaneous gradient descent algorithms. If the fuzzy sets are also evenly distributed across the input space, this reduces the computation time required for calculating the system's response as efficient algorithms can be developed to produce the addresses of the relevant fuzzy sets (non-zero memberships). This does not require the output of *every* membership function to be calculated, as would generally occur if the membership functions did not have a compact support or were unevenly placed in the input space. Also fuzzy sets which do not have a compact support can be artificially endowed with this property by setting the membership function value to zero if it is less than some predefined value α (performing an α-cut).

If the universe of discourse is bounded, an input may lie outside the permitted values. This problem can easily be overcome by simply mapping any such input to the closest point which lies in the permitted set of input points. Thus the network's output is extrapolated as a constant function, and this process is equivalent to extending both the original universe of discourse and the membership function at each end (see Figure 10.3). Similarly, if (normalised) Gaussian membership functions are used, which have an infinite support, the network response for inputs that lie outside the relevant domain tends to a constant value. Therefore the fuzzy sets can be artificially extended to the complete real line, although it is only on some *domain of interest* $((\lambda_0, \lambda_6)$ in the above diagram) that any significant nonlinear behaviour is exhibited.

Scaling factors are often used to scale the real data so that the relevant information always lies within the domain of interest of the fuzzy sets. This also provides a convenient parameter with which to modify the performance of the rule base by adjusting the *sensitivity* of the network with respect to a particular input or output [Linkens and Abbod, 1993] and this is discussed further in Section 11.2.5.

10.3.1   Continuous Fuzzy Sets

A continuous fuzzy set A is defined on a *continuous* universe of discourse X, and the membership function $\mu_A : X \to [0,1]$ assigns to each element $x \in X$ a real number $\mu_A(x)$ in the interval $[0,1]$. It may then be represented by the set of ordered pairs:

$$A = \{(x, \mu_A(x)) \mid x \in X\}$$

As $\mu_A(.)$ is continuous, it must be stored as a functional relationship that is defined by a finite set of parameters. These parameters are different for each fuzzy membership function, and several popular representations are described below.

B-splines

Consider using B-spline basis functions to define the hth univariate fuzzy set membership function $\mu_{A_h}(.)$. B-splines have proved very popular in the numerical approximation literature due to their excellent numerical properties (low computational cost, numerically stable evaluation algorithm, local representations, partition of unity) and also because they form piecewise polynomial mappings of a desired smoothness (see Chapter 8). The recursive evaluation algorithm for basis functions of order k is given by:

$$N_h^k(x) = \left(\frac{x - \lambda_{h-k}}{\lambda_{h-1} - \lambda_{h-k}}\right) N_{h-1}^{k-1}(x) + \left(\frac{\lambda_h - x}{\lambda_h - \lambda_{h-k+1}}\right) N_h^{k-1}(x)$$

$$N_h^1(x) = \begin{cases} 1 & \text{if } x \in I_h \\ 0 & \text{otherwise} \end{cases} \tag{10.1}$$

where $\mu_{A_h}(.) = N_h^k(.)$ is the hth basis function of order k, and I_h is the hth interval $[\lambda_{h-1}, \lambda_h)$. The parameters associated with each basis (fuzzy membership) function are therefore its order and the knot vector on which it is defined. The *knots* $\{\lambda_h\}$ determine the size of the intervals and hence define the *width* of each fuzzy set on the original input space. For order $k = 2$, the basis function $N_h^2(.)$ has the popular triangular shape used in many fuzzy logic implementations and the parameter that needs defining is the knot vector $[\lambda_{h-2}, \lambda_{h-1}, \lambda_h]$. Order $k = 3$ basis functions have a quadratic shape and are defined by the knot vector $[\lambda_{h-3}, \lambda_{h-2}, \lambda_{h-1}, \lambda_h]$. In general, an order k B-spline basis function has a knot vector of length $k + 1$, and only has a non-zero response over the k intervals that form its compact support. It is also possible to have low-order basis functions defined over wide intervals if *dilated* B-splines are used to represent the fuzzy sets (see Section 8.2.5), and these two schemes are shown in Figure 10.4.

Gaussian Functions

Another popular choice is the Gaussian membership function:

$$\mu_{A_h}(x) = \exp\left(-\frac{(x - c_h)^2}{2\sigma_h^2}\right)$$

where the parameters defining the hth basis function are the centre c_h and the variance σ_h. Gaussian fuzzy sets do not have a compact support as they always have a positive response, but they can be modified (using the α-cut) so that this property is incorporated into the fuzzy sets. Alternatively, a Gaussian-type fuzzy set [Werntges, 1993] can be used:

$$\mu_{A_h}(x) = \begin{cases} \exp^{-1}(-1) * \exp\left(-\frac{(\lambda_{h,2} - \lambda_{h,1})^2/4}{(\lambda_{h,2} - x)(x - \lambda_{h,1})}\right) & \text{if } x \in (\lambda_{h,1}, \lambda_{h,2}) \\ 0 & \text{otherwise} \end{cases}$$

which is defined over a compact support and is *infinitely* differentiable. These two Gaussian membership functions both have an input in their supports for which $\mu_A(x) = 1$ (see Figure 10.4), and although this is generally required for a fuzzy set to be a fuzzy number [Pedrycz, 1993], it is probably more important for the input sets to satisfy:

$$\sum_h \mu_{A_h}(x) \equiv 1 \qquad \forall x \tag{10.2}$$

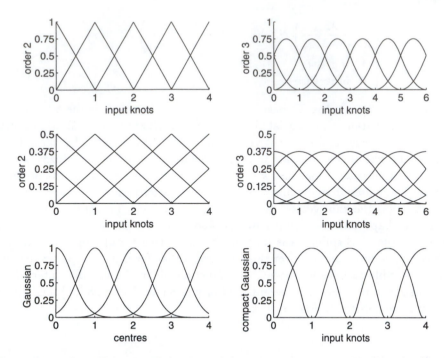

Figure 10.4 Six different possible types of continuous fuzzy sets: B-splines (top), dilated B-splines (middle), Gaussian and compact support Gaussian (bottom).

Such fuzzy sets are said to form a *partition of unity*, which generally results in smoother output surfaces that are partially invariant to shift and linear deformation of the data set. Fuzzy sets which do not form a partition of unity can be endowed with this property by dividing them by the sum over all the fuzzy sets at each point, although this can produce unexpected results. The normalised defuzzification algorithms (centre of gravity, etc.) generally produce fuzzy systems with this property, even if the original fuzzy sets do not as explained in Section 10.6.1.

10.3.2   Discrete Fuzzy Sets

When A is defined on a *discrete* universe of discourse, the fuzzy membership function maps a finite set of input values to the unit interval, i.e. $\mu_A : X^d \to [0,1]$, and A may be represented by the set of ordered pairs:

$$A = \left\{ (x^d, \mu_A(x^d)) \mid x^d \in X^d \right\}.$$

The discrete input space X^d is generally finite, and the value of the membership function, $\mu_A(.)$, at each point in X^d is stored instead of a functional relationship.

Its membership can then be calculated very quickly by simply evaluating which x^d is the closest and looking up its corresponding discrete membership value, and as the intervals between the discrete values tend to zero, the discrete fuzzy sets resemble the original continuous fuzzy sets more closely. This process of finding the closest discrete value is analogous to finding in which interval the input lies when the continuous membership functions are evaluated. Therefore the number of arithmetic operations used to evaluate a discrete membership function is small compared to the number used to evaluate a continuous one, as the extra step of calculating the value of the membership does not need to be performed. However, the memory requirements for the continuous representations are generally much smaller as the knot values (centres) produce a *compact* representation of these functions compared to (over)sampling the function at each of the discrete input points.

A possible advantage of using discrete instead of continuous fuzzy sets is the freedom they allow the designer in choosing set shapes that are not described by simple functions. Despite this, many discrete fuzzy systems are implemented using simple set shapes such as the discrete triangular or quadratic. Also to obtain the membership of elements that lie between discrete values, some form of interpolation is required. If the closest match is calculated, a reconstructed continuous fuzzy set is piecewise constant, and the piecewise constant areas are centred on the discrete input values. Alternatively, if a piecewise linear interpolation technique is used (calculate the discrete output for each of the surrounding points and then fit a linear interpolant to their values), the reconstruction of a triangular fuzzy set is exact if the knot set is contained in the set of discrete input values, as shown in Figure 10.5.

Therefore a discrete representation of a triangular fuzzy set can be used to reconstruct the continuous set function exactly if the knot values form a subset of the discrete points. However, for smoother fuzzy sets such as the quadratic (order 3) B-splines or the Gaussian fuzzy sets, more discrete values are needed to adequately *approximate* the set shape and the storage cost greatly exceeds that required for the parameters of the continuous membership function.

10.4 LOGICAL OPERATORS

To implement the fuzzy algorithms described in Section 10.2, a set of operators is required to manipulate the fuzzy quantities described by the membership functions. A fuzzy production rule is formed using three operations: intersection (AND), implication (IF(.) THEN(.)) and union (OR), and these all need to be appropriately represented within a fuzzy system. These operators determine how the univariate fuzzy sets interact and therefore influence the smoothness of the output surface, and this is a prime consideration in the following discussions.

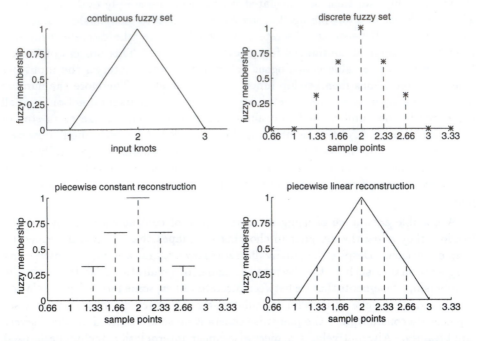

Figure 10.5   A continuous triangular fuzzy set (top left) and a discrete representation when the sample points are 0.33 units apart (top right). The reconstructed fuzzy sets are shown when the technique is based on nearest sample point (bottom left) and linear interpolation of the neighbouring values (bottom right).

10.4.1   Fuzzy Intersection

The fuzzy intersection (AND) of n univariate linguistic statements:

$$\left(x_1 \ is \ \mu_{A_1^i}\right) \ \text{AND} \ \cdots \ \text{AND} \ \left(x_n \ is \ \mu_{A_n^i}\right)$$

is represented by the membership function $\mu_{A_1^i \cap \cdots \cap A_n^i}(x_1, \ldots, x_n)$ or $\mu_{A^i}(\mathbf{x})$ which is defined in the product space $A_1 \times \cdots \times A_n$ by:

$$\mu_{A^i}(\mathbf{x}) = t\left(\mu_{A_1^i}(x_1), \ldots, \mu_{A_n^i}(x_n)\right)$$

where t is a class of functions called *triangular norms*. Triangular norms provide a wide range of functions to implement intersection of which the two most popular are the *min* and the *product* operators. The shape of the multivariate fuzzy membership function $\mu_{A^i}(.)$ is influenced both by the shapes of the univariate membership functions and by the operator used to represent the triangular norm. This is illustrated in Figure 10.6 where the two-dimensional membership functions are shown when two univariate triangular membership functions have been combined using the *min* and *product* operators. The multivariate membership functions formed

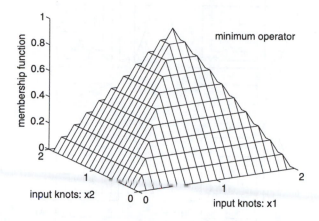

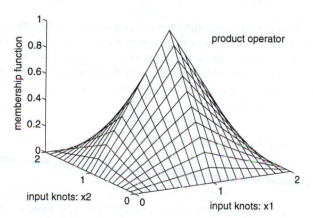

Figure 10.6 Two-dimensional fuzzy membership functions formed from the intersection of two univariate triangular fuzzy membership functions using the *min* operator (top) and the *product* operator (bottom).

using the *product* operator retain more information than when the *min* operator is used to implement the fuzzy AND because the latter scheme only retains one piece of information whereas the *product* operator combines n-pieces. This fact has led some people to claim that the *min* operator is more robust than the *product*, as noise in any of the $(n-1)$ inputs which do not affect the output is not propagated through the network. However, this does not happen in practice because when the fuzzy sets are distributed evenly throughout the input space, at least one fuzzy multivariate membership function depends on each input variable and so the information propagated to the next layer of the fuzzy system is sensitive to any measurement noise, as shown in Figure 10.7. Using the *product* operator generally gives a *smoother* output surface, a desirable attribute in modelling and control sys-

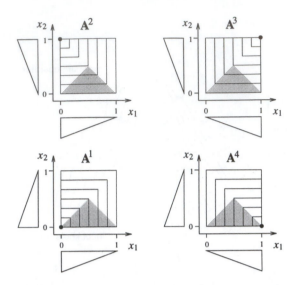

Figure 10.7   An illustration of how information is propagated through the system when a *min* intersection operator is used. For a two-dimensional input with triangular membership functions, each two-dimensional fuzzy input set is sensitive to only one input, although two depend on x_1 (horizontal contours that intersect with the shaded areas) and two depend on x_2 (vertical contours intersecting with the shaded area). Thus *any* measurement noise is propagated through the network.

tems. Similarly, it may be supposed from Figure 10.6 that the *min* operator gives greater weight to the knowledge closer to the set centre than the *product* operator, although the output of the fuzzy system is generally formed using a *normalised* calculation and a fuzzy system which uses the *min* fuzzy intersection operator actually places *less* confidence on the rules nearby than the *product* operator (see Appendix D). This can be illustrated by considering the output surfaces generated in Figure 10.8. Two univariate triangular fuzzy membership functions are defined on each axis, and the system's output is formed from a *normalised*, linear combination of these sets, as is shown in Section 10.6.1. The output of the fuzzy system employing the *min* operator has derivative discontinuities lying along the diagonals and the surface changes rapidly near the point $\mathbf{x} = (1, 1)^T$, which is due to the basis function *furthest* away. In contrast, the system that uses the *product* operator has a smooth output surface which is not affected by the (constant) normalisation factor.

Continuous and Discrete Fuzzy Intersection

If the univariate fuzzy membership functions $\mu_{A_h^i}(x_h)$ are expressed as functions, the multivariate membership function generated by the fuzzy intersection, $\mu_{\mathbf{A}^i}(\mathbf{x})$,

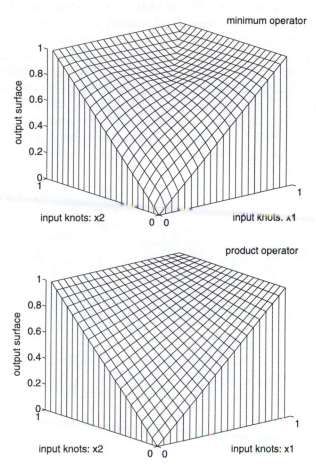

Figure 10.8   Two-dimensional fuzzy output surfaces formed using four two-dimensional fuzzy triangular membership functions with the *min* operator (top) and the *product* operator (bottom).

requires the storage of n parameter sets, one for each dimension. For example, the B-splines discussed in Section 10.3.1 must store n knot vectors, each of length $k+1$. The result of a fuzzy intersection of two continuous membership functions is a surface in the two-dimensional space whose support has a length and width equal to the supports of the two univariate membership functions, as shown in Figure 10.6. This naturally generalises to taking the logical intersection of n univariate fuzzy sets.

The fuzzy intersection of the discrete univariate fuzzy membership functions, $\mu_{A_h^i}(x_h^d)$, can be calculated by either storing the discrete univariate functions and forming the fuzzy intersection on-line, or by storing a discrete representation of the multivariate membership function $\mu_{A^i}(\mathbf{x}^d)$. For the former method, n discrete fuzzy sets must be stored, whereas the storage of the discrete multivariate membership function would generate an n-dimensional matrix, where the number of discrete values used to store the univariate fuzzy membership functions determines the size of the matrix. It is important to realise that this discrete fuzzy set only forms a relationship between discrete points and does *not* contain information about how the system interpolates between these values. This depends *totally* on the representation of the (measured) input variable, and this is discussed further in Section 10.6.2. Two discrete two-dimensional fuzzy membership functions are shown in Figure 10.9, which are formed from the fuzzy intersection of two sampled triangular membership functions, similar to those reproduced in Figure 10.5.

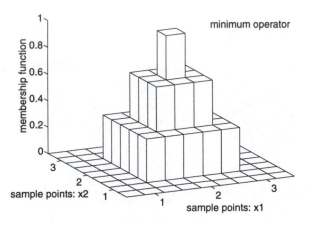

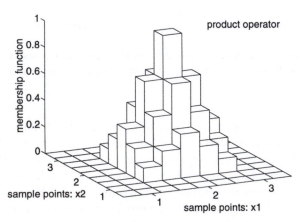

Figure 10.9 Two discrete two-dimensional fuzzy sets formed from the intersection of two discrete univariate fuzzy sets using the *min* (top) and *product* (bottom) operators. The *min* operator induces an infinity norm on the discrete input space when the two univariate sets are equivalent.

10.4.2 Fuzzy Implication

To represent a relation (IF *antecedent* THEN *consequent*), let the rule that maps the ith multivariate fuzzy input set $\mathbf{A}^i$ to the jth univariate output set with confidence c_{ij} be labelled by r_{ij}, i.e.:

$$r_{ij}: \qquad \text{IF } (\mathbf{x} \text{ } is \text{ } \mathbf{A}^i) \text{ THEN } (y \text{ } is \text{ } B^j) \qquad (c_{ij})$$

Then the degree to which element $\mathbf{x}$ is *related* to element y is represented by the membership function $\mu_{r_{ij}}(\mathbf{x}, y)$ that is defined in the product space $A_1 \times \cdots \times A_n \times B$ by:

$$\mu_{r_{ij}}(\mathbf{x}, y) = t\left(\mu_{\mathbf{A}^i}(\mathbf{x}), c_{ij}, \mu_{B^j}(y)\right) \qquad (10.3)$$

where again t is a triangular norm usually chosen to be the *min* or the *product* operator, although a much larger selection can be found in Kiszka *et al.* [1985]. The fuzzy set $\mu_{r_{ij}}(\mathbf{x}, y)$ represents the confidence in the output being y given that the input is $\mathbf{x}$ for the ij^{th} fuzzy rule.

The above operation is the same as forming the fuzzy intersection of $(n + 1)$ univariate membership functions as again a t-norm is employed to combine the fuzzy knowledge, and from this viewpoint, c_{ij} represents the confidence in the following statements being true:

$$(x_1 \text{ } is \text{ } A_1^i) \text{ AND } (x_2 \text{ } is \text{ } A_2^i) \text{ AND } \cdots \text{ AND } (x_n \text{ } is \text{ } A_n^i) \text{ AND } (y \text{ } is \text{ } B^j)$$

or

$$\text{IF } (x_1 \text{ } is \text{ } A_1^i) \text{ THEN IF } (x_2 \text{ } is \text{ } A_2^i) \cdots \text{ THEN IF } (x_n \text{ } is \text{ } A_n^i) \text{ THEN } (y \text{ } is \text{ } B^j)$$

The order in which these statements are evaluated and the position of the unknown quantity (y in this case) are immaterial. Hence for a known output and one unknown input, the knowledge stored in the above rules can be *inverted* to infer the value of the unknown variable. This is analogous to forming a fuzzy plant model (mapping present state and control to next state) and inverting the rules to infer the control necessary to move the plant from its present to the desired state [Moore and Harris, 1994].

Continuous and Discrete Fuzzy Implication

To explain the differences between continuous and discrete fuzzy implication, the same arguments used in Section 10.4.1 apply. This is because a fuzzy relation is formed by applying fuzzy intersection. Hence, for univariate fuzzy sets, μ_{A^i} and μ_{B^j} expressed as functions, the fuzzy relation $\mu_{r_{ij}}(\mathbf{x}, y)$ requires the storage of $(n + 1)$ parameter sets. The fuzzy relation for the discrete univariate fuzzy sets can either be computed on-line or it can be stored using an $(n + 1)$-dimensional matrix. In Figure 10.10, the *min* and *product* operators are used to implement

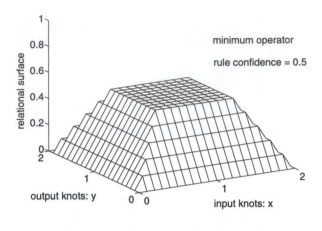

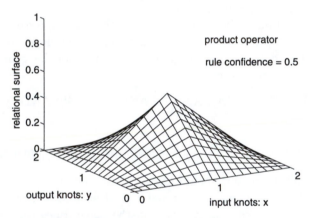

Figure 10.10   Using the *min* and the *product* operators to implement fuzzy implication. The smoother effect of the latter operator is obvious.

the *t*-norm for a single input, single output fuzzy rule with continuous triangular fuzzy membership functions defined on each axis. The *min* operator thresholds the two-dimensional fuzzy input/output set at the value of the rule confidence which produces a sharp nonlinearity in the fuzzy reasoning algorithm, whereas the *product* operator simply scales the set. This also occurs for the discrete systems as shown in Figure 10.11. Given the fuzzy relational membership function $\mu_{r_{ij}}(\mathbf{x}, y)$ and a crisp input $\mathbf{x}$, the corresponding fuzzy output membership function is obtained by taking a *slice* through $\mu_{r_{ij}}(\mathbf{x}, y)$ for the input $\mathbf{x}$. This induced membership function represents the fuzzy output distribution of the ij^{th} rule.

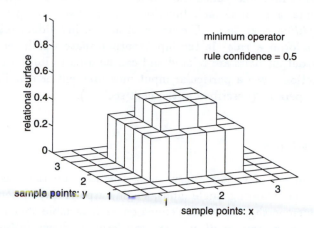

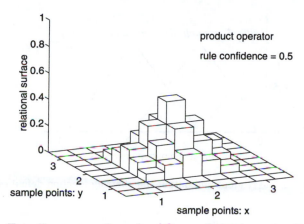

Figure 10.11 Two discrete two-dimensional fuzzy relational membership functions formed using the *min* (top) and *product* (bottom) operators. The *min* operator again induces an infinity norm on the discrete input space.

10.4.3 Fuzzy Union

If p multivariate fuzzy input sets $\mathbf{A}^i$ map to q univariate fuzzy output sets B^j, there are pq, $(n+1)$-dimensional membership functions, one for each relation. The pq relations may then be connected to form a fuzzy rule base $\mathbf{R}$ by taking the union (OR) of the individual membership functions:

$$\mu_{\mathbf{R}}(\mathbf{x}, y) = \bigcup_{ij} \mu_{r_{ij}}(\mathbf{x}, y).$$

This operation is defined by:

$$\mu_{\mathbf{R}}(\mathbf{x}, y) = s\left(\mu_{r_{11}}(\mathbf{x}, y), \ldots, \mu_{r_{1q}}(\mathbf{x}, y), \ldots, \mu_{r_{p1}}(\mathbf{x}, y), \ldots, \mu_{r_{pq}}(\mathbf{x}, y)\right) \qquad (10.4)$$

where s is a class of functions called the *triangular co-norm*. Triangular co-norms also provide a wide range of suitable functions but the two most popular are the *max* and the *addition* operators. The union of all the individual relation membership functions forms a *ridge* in the input/output space which represents how individual input/output pairs are related and can be used to infer a fuzzy output membership function given a particular input measurement or distribution using the *composition* operator (described in the next section).

Continuous Fuzzy Union

To illustrate the effect of various fuzzy union operators on the fuzzy relational surface, consider the following single input, single output fuzzy system which has four triangular membership functions defined on each variable that represent the linguistic terms *almost zero*, *small*, *medium* and *large*, as shown in Figure 10.12.

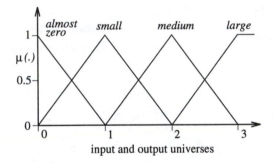

Figure 10.12 Fuzzy input and output membership function definitions.

Binary Fuzzy Algorithms

The fuzzy algorithm for this system is given by:

$$
\begin{array}{lll}
& \text{IF } (x \text{ is almost zero}) \text{ THEN } (y \text{ is almost zero}) & (c_{11} = 1.0) \\
\text{OR} & \text{IF } (x \text{ is small}) \text{ THEN } (y \text{ is small}) & (c_{22} = 1.0) \\
\text{OR} & \text{IF } (x \text{ is medium}) \text{ THEN } (y \text{ is medium}) & (c_{33} = 1.0) \\
\text{OR} & \text{IF } (x \text{ is large}) \text{ THEN } (y \text{ is large}) & (c_{44} = 1.0)
\end{array}
$$

and all the fuzzy production rules not shown have a zero rule confidence and do not therefore influence the output of the system. The set of all the rule confidences forms the rule confidence matrix which relates each fuzzy input set to every possible fuzzy output set. For the above fuzzy algorithm, the rule confidence vector associated with each input set is *binary* (only one element is non-zero and this has a value of unity), which means that each row in the rule confidence matrix is normalised (sum to unity). If the *product* operator is used as the t-norm, the fuzzy system will be normalised, justifying the use of the *addition* operator to implement the fuzzy union, and the fuzzy relational surface is shown in Figure 10.13.

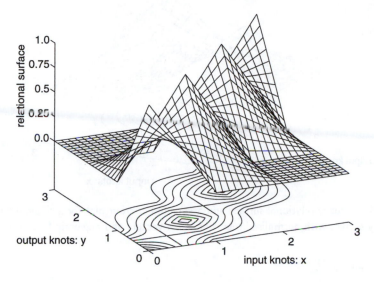

Figure 10.13 A fuzzy relational surface and associated contour plot for a single input, single output fuzzy system with binary rule confidences and algebraic fuzzy operators. Each peak corresponds to a particular fuzzy rule.

Continuous Fuzzy Algorithms

Consider the non-binary fuzzy algorithm given by:

$$
\begin{array}{lll}
& \text{IF } (x \text{ is almost zero}) \text{ THEN } (y \text{ is almost zero}) & (c_{11} = 1.0) \\
\text{OR} & \text{IF } (x \text{ is small}) \text{ THEN } (y \text{ is almost zero}) & (c_{21} = 0.4) \\
\text{OR} & \text{IF } (x \text{ is small}) \text{ THEN } (y \text{ is small}) & (c_{22} = 0.6) \\
\text{OR} & \text{IF } (x \text{ is medium}) \text{ THEN } (y \text{ is small}) & (c_{32} = 0.2) \\
\text{OR} & \text{IF } (x \text{ is medium}) \text{ THEN } (y \text{ is medium}) & (c_{33} = 0.8) \\
\text{OR} & \text{IF } (x \text{ is large}) \text{ THEN } (y \text{ is large}) & (c_{44} = 1.0)
\end{array}
$$

which has normalised rule confidence vectors associated with each input membership function. The *product* and the *addition* operators again represent the t and s-norms, respectively, and the fuzzy relational surface produced is shown in Figure 10.14. Continuous rule confidence vectors have a greater modelling flexibility than binary ones, and this is discussed further in Section 10.6.3.

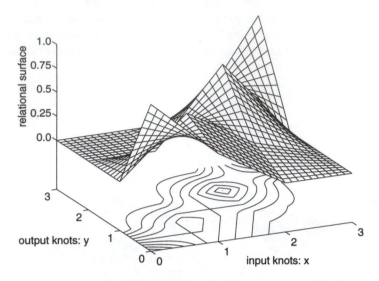

Figure 10.14   A fuzzy relational surface and associated contour plot for a single input, single output fuzzy system with normalised rule confidence vectors and algebraic fuzzy operators.

Truncation Fuzzy Operators

Finally, the relational surface shown in Figure 10.15 represents the continuous fuzzy algorithm except that now *all* of the fuzzy operators are implemented using truncation operators (*min* and *max*). The surface is very regular since for each input/output pair only one rule is active. However, this also means that the surface has severe derivative discontinuities both along lines parallel to the input axes (a consequence of using triangular input/output membership functions) and along the main and minor diagonals (which is due to using the *min* intersection operator). The relational surface also has large areas where the system is not sensitive to any change in x (around 0.8 for instance), and the output of the fuzzy system is *constant* in these regions. This implies that the overall network is insensitive to measurement noise, but in the regions where one rule is dropped from the calculation and another is substituted, the output is extremely sensitive to the input. Therefore fuzzy systems which use truncation operators are not inherently robust, rather the information lost when they are used produces a very regular but undesirable fuzzy relational surface.

Discrete Fuzzy Relational Matrices

A continuous fuzzy system represents the fuzzy algorithm by choosing the membership functions, which implement the fuzzy linguistic statements, and the fuzzy

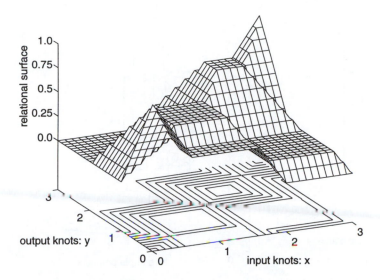

relational surface

1.0
0.75
0.5
0.25
0.0

output knots: y

input knots: x

Figure 10.15 A fuzzy relational surface and associated contour plot for a single input, single output fuzzy system with normalised rule confidence vectors where truncation operators are used to implement all the fuzzy operators.

operators. The rule confidence matrix is stored *separately*. Discrete fuzzy systems however store the input/output information in the form of a *fuzzy relational matrix* which is simply the continuous relational surface *sampled* at the appropriate points. The relational matrix is therefore an $(n+1)$-dimensional matrix where the number of discrete points used to represent each variable determines the size of each axis. It contains information about the shape of the fuzzy input and output sets, the operators used to implement the underlying fuzzy logic *and* the fuzzy algorithm, and its ij^{th} element d_{ij} is given by:

$$
\begin{aligned}
d_{ij} &= \bigcup_{kl} \mu_{r_{kl}}(\mathbf{x}_i^d, y_j^d) \\
&= \bigcup_{kl} t\left(\mu_{\mathbf{A}^k}(\mathbf{x}_i^d), c_{kl}, \mu_{B^l}(y_j^d)\right)
\end{aligned}
\tag{10.5}
$$

for discrete input/output points $\mathbf{x}_i^d, y_j^d$, respectively. The relational matrix attempts to approximate the form of the relational surface by sampling it at a large number of discrete points and manipulating this information directly rather than the rule confidence matrix. This can be extremely costly in terms of the storage requirements, as can be seen by considering the relational matrix shown in Figure 10.16 that has been generated by the single input, single output fuzzy system represented by Figure 10.14. Each variable is sampled at the points:

$$\{0, 0.33, 0.67, 1.0, 1.33, 1.67, 2.0, 2.33, 2.67, 3.0\}$$

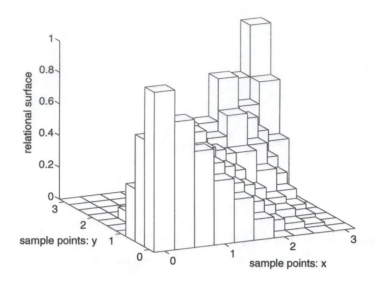

y	0.0				1.0				2.0	3.0
3.0	0.0	0.0	0.0	0.0	0.0	0.0	0.0	0.33	0.67	1.0
	0.0	0.0	0.0	0.0	0.09	0.18	0.27	0.4	0.53	0.67
	0.0	0.0	0.0	0.0	0.18	0.36	0.53	0.47	0.4	0.33
2.0	0.0	0.0	0.0	0.0	0.27	0.53	0.8	0.53	0.27	0.0
	0.0	0.07	0.13	0.2	0.33	0.47	0.6	0.4	0.2	0.0
	0.0	0.13	0.27	0.4	0.4	0.4	0.4	0.27	0.13	0.0
1.0	0.0	0.2	0.4	0.6	0.47	0.33	0.2	0.13	0.07	0.0
	0.33	0.4	0.47	0.53	0.40	0.27	0.13	0.09	0.04	0.0
	0.67	0.6	0.53	0.47	0.33	0.2	0.07	0.04	0.02	0.0
0.0	1.0	0.8	0.6	0.4	0.27	0.13	0.0	0.0	0.0	0.0

$$0.0 \qquad\qquad 1.0 \qquad\qquad 2.0 \qquad\qquad 3.0$$
$$x$$

Figure 10.16   The sampled fuzzy relational surface (top) and the corresponding relational matrix (bottom). The relational elements at the knot points correspond to the rule confidences as the input/output sets have a unity membership at these values.

which produces the above relational matrix of size (10×10). This is substantially larger than the rule confidence matrix (size (4×4)) as the relational matrix also contains information about the discrete representation of the fuzzy input and output sets which define the linguistic variables.

The relational matrix contains the linguistic rules used either to initialise the system or to explain its actions after training has ceased. Equation 10.5 shows how the elements of the relational matrix are generated from the rule confidences and to be *consistent*, the knowledge stored in the relational matrix should be able to be interpreted as a set of linguistic rules. This immediately imposes constraints on the method updating neighbouring elements of the relational matrix, and this is discussed further in Chapter 11.

10.5 COMPOSITIONAL RULE OF INFERENCE

Once a fuzzy rule base has been formed, it can then be used to generate a fuzzy output distribution for a given input set using a procedure known as the *compositional rule of inference*. Let $\mu_R(\mathbf{x}, y)$ be a fuzzy rule base, then the fuzzy output set $\mu_B(y)$ induced by the fuzzy input set $\mu_A(\mathbf{x})$ is given by:

$$\mu_B(y) = \mu_A(\mathbf{x}) \circ \mu_R(\mathbf{x}, y)$$

where $\circ$ is the composition operator. It is important to note that these fuzzy distributions are generally *different* from those used to implement the linguistic terms. Different forms of uncertainty are being modelled, and there is no reason why the fuzzy set representing the system's input should be the same as that used to implement the rules (unless of course the input is a linguistic term).

Composition is the procedure that allows a fuzzy model to produce *sensible* outputs for previously unseen inputs. It is equivalent to generalisation in neural networks or interpolation and local extrapolation in approximation theory, and the output is acceptable provided that the fuzzy rule base and the fuzzy input sets are appropriate. Before composition can be applied, the input to the fuzzy model must first be represented as the fuzzy set $\mu_A(\mathbf{x})$. In most fuzzy modelling and control applications, the input is a *crisp* or non-fuzzy measurement that may be corrupted by noise, although the input could equally be a linguistic statement such as *x is large*. This process is called *fuzzification*, and there are a variety of different techniques that are used to represent the uncertainty associated with each input.

When the model input is a measurement, the most commonly used fuzzifier is the *singleton* that maps the input x^s to a binary or *crisp* univariate fuzzy set A with membership:

$$\mu_A(x) = \begin{cases} 1 & \text{if } x = x^s \\ 0 & \text{otherwise} \end{cases}$$

For inputs that are corrupted by noise, the shape of the fuzzy set can reflect the uncertainty associated with the measurement process. For example, a triangular fuzzy set can be used where the vertex corresponds to the mean of some measurement data and the base width is a function of the standard deviation. If the model input is a linguistic statement, a fuzzy set must be found that adequately represents the statement. Unless the input is a linguistic statement, there is *no* justification for fuzzifying the input using the same membership functions used to represent the linguistic statements such as *x is small*. The latter membership functions are chosen to represent *vague* linguistic statements whereas the input fuzzy sets reflect the uncertainty associated with the *imprecise* measurement process, and these two quantities are generally distinct. The effect of different input set shapes on the network output is shown in Figure 10.19, and it can be clearly seen that as the width of the input set grows (increasingly imprecise measurements), a greater emphasis is placed on neighbouring output values and the system becomes more conservative in its recommendations.

The composition operator can now be defined. It depends on whether the fuzzy sets are defined on a continuous or a discrete universe and on the choice of triangular norm and co-norm operators chosen to implement the fuzzy logic.

10.5.1 Continuous Fuzzy Composition

The composition operator for fuzzy sets defined on a continuous universe is defined by:

$$\mu_B(y) = s\left(t\left(\mu_A(\mathbf{x}), \mu_R(\mathbf{x}, y)\right)\right) \tag{10.6}$$

where the *s*-norm is taken over *all* possible values of $\mathbf{x}$ in the domain $\mathbf{X}$, and the *t*-norm computes a match between two membership functions for each value of $\mathbf{x}$. When *s* and *t* are chosen to be the *integration* and the *product* operators, respectively, then:

$$\mu_B(y) = \int_{\mathbf{X}} \mu_A(\mathbf{x})\,\mu_R(\mathbf{x}, y)\,d\mathbf{x} \tag{10.7}$$

which for an arbitrary fuzzy input set requires an *n*-dimensional integral to be evaluated. *Discrete* numerical integration routines can be used to approximate the integrand $\mu_A(\mathbf{x})\,\mu_R(\mathbf{x}, y)$, and in this case the continuous and the discrete fuzzy composition algorithms are equivalent (if the sample/evaluation points are the same).

If *s* and *t* are chosen to be the *max* and *min* operators, the expression becomes:

$$\mu_B(y) = \max_{\mathbf{X}}\left(\min\left(\mu_A(\mathbf{x}), \mu_R(\mathbf{x}, y)\right)\right)$$

which also introduces numerical difficulties since a global, nonlinear optimisation problem in *n* dimensions must be solved to obtain the maximum over $\mathbf{X}$.

The numerical difficulties of performing a continuous fuzzy composition can be avoided if a *singleton* fuzzifier is used. Equation 10.6 then reduces to $\mu_B(y) = \mu_R(\mathbf{x}^s, y)$ which involves neither an *n*-dimensional integration nor an optimisation problem, and is simply a *slice* through the relational surface at the point $\mathbf{x}^s$. It should also be noted that both implementation algorithms require knowledge of the fuzzy rule base $\mu_R(\mathbf{x}, y)$ which is stored as a set of $(n+1)$ parameter sets containing information about the fuzzy input and output sets, and also an $(n+1)$-dimensional rule confidence matrix.

10.5.2 Discrete Fuzzy Composition

When the fuzzy sets defined on a discrete universe, the composition operator is defined by:

$$\mu_B(y^d) = s\left(t\left(\mu_A(\mathbf{x}^d), \mu_R(\mathbf{x}^d, y^d)\right)\right)$$

where the s-norm is taken over all possible discrete values of $\mathbf{X}^d$, and the t-norm performs a comparison of the two sampled membership functions at those points. If s and t are chosen to be the *addition* and *product* operators, the above reduces to:

$$\mu_B(y^d) = \sum_{\mathbf{X}^d} \mu_A(\mathbf{x}^d)\,\mu_R(\mathbf{x}^d, y^d) \tag{10.8}$$

and when s and t are chosen to be the *max* and *min* operators then:

$$\mu_B(y^d) = \max_{\mathbf{X}^d}\left(\min\left(\mu_A(\mathbf{x}^d), \mu_R(\mathbf{x}^d, y^d)\right)\right)$$

Because the input domain $\mathbf{X}^d$ is discrete, both are easier to calculate than the continuous case, but they still require knowledge of the discrete relational matrix. Also for a fuzzy singleton input which is represented using a crisp discrete fuzzy set, the discrete fuzzy composition reduces to:

$$\mu_B(y^d) = \mu_R(\mathbf{x}^{dc}, y^d)$$

where $\mathbf{x}^{dc}$ is the discrete value closest to the input $\mathbf{x}^s$. The fuzzy output distribution is therefore given by the appropriate column of the discrete relational matrix.

10.6   DEFUZZIFICATION

If a crisp output is required from the fuzzy rule base rather than the fuzzy output set, $\mu_B(y)$, a process known as *defuzzification* is used to *compress* this information. The crisp output is generally obtained using a *mean of maxima* or a *centre of gravity* defuzzification strategy. It may seem that a lot of information about the uncertainty associated with the input measurements is lost when the fuzzy output set is compressed to a single value, but the *distributed* fuzzy representations (such as storing a *single* value as the membership of *several* fuzzy sets) do not always contain any *extra* information. The shape of the fuzzy output distribution depends on the fuzzy output sets used indirectly in the system through the knowledge stored in the rule base, although the *output* sets are generally designed to represent vague linguistic values and not to reflect any uncertainty associated with the *input* measurements. It is possible to use the fuzzy information contained in the output distribution, although the system must be specifically designed to incorporate this feature [Moore *et al.*, 1993].

The two most popular defuzzification algorithms are the mean of maxima and the centre of gravity (sometimes called the first moment of area). The former only uses part of the information contained in the fuzzy output distribution unlike the latter method which uses all of this information. In the mean of maxima defuzzification algorithm, the output which has the highest membership of the fuzzy output distribution is selected, and if this value is not unique an average is taken. The centre of gravity defuzzification algorithm weights each possible output value by its corresponding membership in the output distribution and then calculates a normalised average. This enables a discrete fuzzy system to generate any possible output value, and for both continuous and discrete fuzzy systems it generally produces a *smoother* output than the mean of maxima method.

Both defuzzification algorithms are easier to calculate if a discrete fuzzy system is used, although an extremely important insight is obtained when the centre of gravity method is applied to certain continuous fuzzy systems since it:

- **allows** a direct link to be made with artificial neural networks;
- **provides** an implementation method which greatly reduces both the computational cost and the storage requirements of the algorithm; and
- **explains** how fuzzy systems *generalise* and the role of the fuzzification algorithm.

The results and insights do not apply directly to *all* fuzzy systems, although it has been the authors' experience that adopting any of the proposed implementation methods (continuous representations, algebraic fuzzy operators, centre of gravity defuzzification) generally results in a smoother output surface and improved system performance.

10.6.1 Continuous Centre of Gravity Defuzzification

The centre of gravity defuzzification algorithm for continuous fuzzy systems is given by:

$$y(\mathbf{x}) = \frac{\int_Y \mu_B(y)\, y\, dy}{\int_Y \mu_B(y)\, dy}. \qquad (10.9)$$

Let s and t be the *addition* and *product* operators, respectively, then substituting the equations for continuous fuzzy composition 10.7, union 10.4 and implication 10.3 into Equation 10.9 gives:

$$y(\mathbf{x}) = \frac{\int_Y \int_X \mu_A(\mathbf{x}) \sum_{ij} \mu_{A^i}(\mathbf{x}) \mu_{B^j}(y) c_{ij} \, y \, d\mathbf{x} \, dy}{\int_Y \int_X \mu_A(\mathbf{x}) \sum_{ij} \mu_{A^i}(\mathbf{x}) \mu_{B^j}(y) c_{ij} \, d\mathbf{x} \, dy}. \tag{10.10}$$

But for bounded and symmetric fuzzy output sets, the integrals $\int_Y \mu_{B^j}(y) \, dy$, for all j, are equal and so the following relationship holds:

$$\frac{\int_Y \mu_{B^j}(y) \, y \, dy}{\int_Y \mu_{B^j}(y) \, dy} = y_j^c$$

where y_j^c is the centre of the jth output set. Equation 10.9 therefore reduces to:

$$y(\mathbf{x}) = \frac{\int_X \mu_A(\mathbf{x}) \sum_i \mu_{A^i}(\mathbf{x}) \sum_j c_{ij} \, y_j^c \, d\mathbf{x}}{\int_X \mu_A(\mathbf{x}) \sum_i \mu_{A^i}(\mathbf{x}) \sum_j c_{ij} \, d\mathbf{x}} \tag{10.11}$$

Suppose that the multivariate fuzzy input sets form a partition of unity, i.e. $\sum_i \mu_{A^i} = 1$ and that the ith rule confidence vector $\mathbf{c}_i = (c_{i1}, \ldots, c_{iq})^T$ is normalised, i.e. $\sum_j c_{ij} = 1$, then the defuzzified output becomes:

$$y(\mathbf{x}) = \frac{\int_X \mu_A(\mathbf{x}) \sum_i \mu_{A^i}(\mathbf{x}) w_i \, d\mathbf{x}}{\int_X \mu_A(\mathbf{x}) \, d\mathbf{x}} \tag{10.12}$$

where $w_i = \sum_j c_{ij} \, y_j^c$ is the *weight* associated with the ith fuzzy membership function. The transformation from the weight w_i to the vector of rule confidences $\mathbf{c}_i$ is a many-to-one mapping, although for fuzzy sets defined by symmetric B-splines of order $r \geq 2$, it can be inverted in the sense that for a given w_i there exists a *unique* $\mathbf{c}_i$ that generates the desired output. This is explained further in Section 10.6.4 and Appendix E. It should also be emphasised that using weights in place of rule confidence vectors provides a considerable reduction in both the storage requirements and the computational cost, and is also relevant to the discussion on training given in Chapter 11.

Singleton Input

When the fuzzy input set $\mu_A(\mathbf{x})$ is a singleton, the numerator and denominator integrals in Equation 10.12 cancel to give:

$$y^s(\mathbf{x}) = \sum_i \mu_{A^i}(\mathbf{x}) \, w_i \tag{10.13}$$

where $y^s(\mathbf{x})$ is called the fuzzy singleton output. This is an important and surprising result since $y^s(\mathbf{x})$ is a *linear* combination of the fuzzy input sets and does *not* depend on the choice of fuzzy output sets. It also provides a useful link between fuzzy and neural networks and allows both approaches to be treated within a unified framework, and this is discussed further in Section 10.6.5. The reduction in the computational cost of implementing a fuzzy system in this manner and the overall algorithmic simplification is illustrated in Figure 10.17.

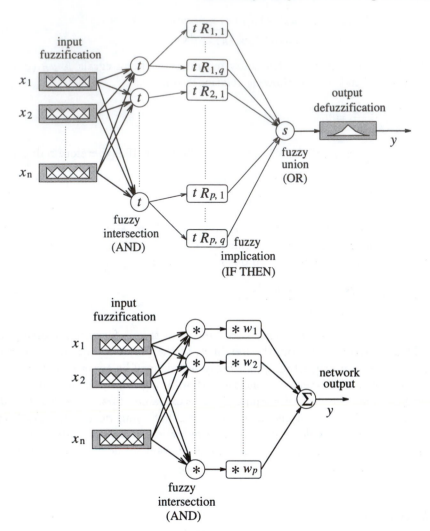

Figure 10.17   An illustration of the information flow through a continuous fuzzy system (top) and the resulting simplification (bottom) when algebraic operators are used in conjunction with a centre of gravity defuzzification algorithm. The input is represented as a fuzzy singleton and there exist p multivariate fuzzy input sets and q univariate fuzzy output sets.

Normalised Membership Functions

The analysis also illustrates how the centre of gravity defuzzification procedure *implicitly* imposes a partition of unity on the fuzzy input membership functions. Consider the above system when the fuzzy input sets do not sum to unity, which could be due to either their univariate shape or the operator used to represent

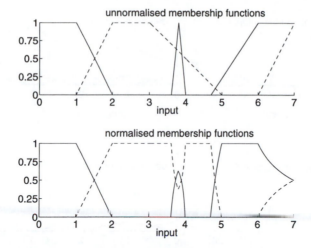

Figure 10.18   Piecewise linear fuzzy membership functions (top) and the resulting normalised sets (bottom). The normalisation process can make previously convex sets, non-convex and can vastly alter their shape.

fuzzy intersection. The output is then given by:

$$
\begin{aligned}
y^s(\mathbf{x}) &= \frac{\sum_i \mu_{\mathbf{A}^i}(\mathbf{x})\,w_i}{\sum_j \mu_{\mathbf{A}^j}(\mathbf{x})} \\
&= \sum_i \mu^*_{\mathbf{A}^i}(\mathbf{x})\,w_i
\end{aligned}
\tag{10.14}
$$

where the modified fuzzy input membership functions $\mu^*_{\mathbf{A}^i}(\mathbf{x})\left(= \mu_{\mathbf{A}^i}(\mathbf{x})\big/\sum_j \mu_{\mathbf{A}^j}(\mathbf{x})\right)$ form a partition of unity. This normalisation step is very important because it determines the *actual* influence of the fuzzy set on the system's output and can make previously convex sets, non-convex, as shown in Figure 10.18.

Fuzzy Input Distributions

If the input to the fuzzy system is a fuzzy distribution rather than a singleton, it is possible to substitute Equation 10.13 into 10.12 giving:

$$
y(\mathbf{x}) = \frac{\int_X \mu_{\mathbf{A}}(\mathbf{x})\,y^s(\mathbf{x})\,d\mathbf{x}}{\int_X \mu_{\mathbf{A}}(\mathbf{x})\,d\mathbf{x}}.
\tag{10.15}
$$

The defuzzified output is a weighted average of the fuzzy singleton outputs over the support of the fuzzy input set $\mu_{\mathbf{A}}(\mathbf{x})$, and the effect is to *smooth* or *low pass filter* the system's output y. This is illustrated in Figure 10.19. It can be seen that as the width of the fuzzy input set increases, the overall output of the system becomes *less sensitive* to the shape of either the input set or the sets used to represent the

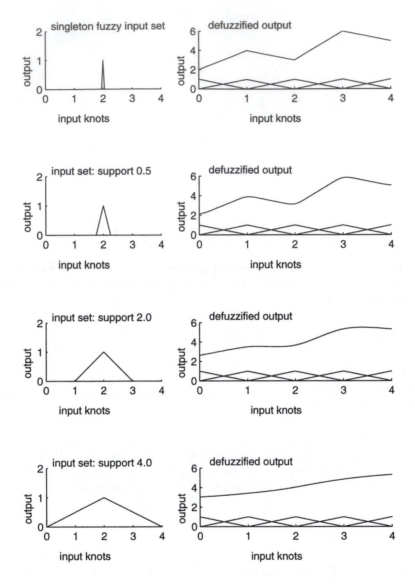

Figure 10.19   Four fuzzy input sets and their corresponding defuzzified outputs, when the fuzzy rule base consists of triangular membership functions. The original triangular membership functions used to represent the linguistic terms are shown on the bottom of the graphs on the right, and it can clearly be seen that as the width of the input set increases, the system becomes less sensitive to the input variable and the set shapes.

linguistic terms. However, this is not always desirable as the output also becomes less sensitive to individual rules and input variations. Indeed, when the input set shape has an arbitrarily large width which represents complete uncertainty about the measurement, the system's output is constant everywhere.

An important consequence of the above analysis is that using centre of gravity defuzzification in conjunction with the *addition* and *product* operators has reduced fuzzy composition and defuzzification to a single operation. It is no longer necessary to calculate and store $\mu_R(\mathbf{x}, y)$.

Example

In Section 10.4.3, two fuzzy relational surfaces were generated and the effects of using truncation and algebraic fuzzy operators were compared. When a centre of gravity defuzzification algorithm is employed, it is also possible to compare the outputs of the single input, single output fuzzy systems and to investigate how different operators affect the *generalisation* properties of the network. These two nonlinear functions are plotted in Figure 10.20 when the input is represented by

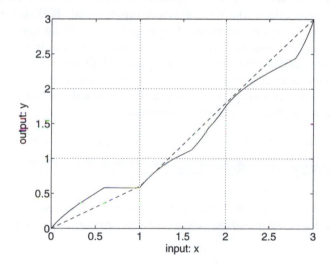

Figure 10.20 The output of a fuzzy system when algebraic (dashed line) and truncation (solid line) operators are used to implement the underlying fuzzy logic.

a singleton fuzzy input membership function and the fuzzy output sets have been extended by one interval at either end (see Figure 10.12) so that the *almost zero* and the *large* membership functions are implemented using symmetrical triangles.

The output surface is *significantly* affected by the choice of operators used to implement the underlying fuzzy logic. Using truncation operators, the fuzzy system has a constant region (zero gain) at around $x = 0.8$ whereas the algebraic fuzzy system produces a conventional piecewise linear surface that is representative of the fuzzy input membership functions. The *form* of the output of the truncation fuzzy system strongly depends on the value of rule confidences and the fuzzy output set definitions whereas in an algebraic fuzzy network these quantities only influence the output's *value*. Hence, it is extremely difficult to design a truncation fuzzy system with well-understood interpolation properties, and possibly the only relationship that can be proved is that the value of the fuzzy system in the interior of an interval lies between its output at either end (it may also be possible to establish monotonic behaviour).

Polynomial Models

Before describing the discrete defuzzification algorithm, it is worthwhile considering the implications of the results derived in this section on the modelling abilities of the fuzzy system. It was shown in Equation 10.13 that the output of the fuzzy system (for a singleton input and algebraic operators) is simply a linear combination of the multivariate fuzzy input membership functions. Therefore, the modelling abilities of the fuzzy system and its ability to generalise depend on the size, shape and distribution of these sets in the n-dimensional input space. If B-splines are used to represent the membership functions, the output of the fuzzy system is a (low-order) piecewise polynomial with appropriate continuity constraints as fuzzy rules are dropped from, and introduced to, the output calculation. Polynomials have been used for nonlinear system modelling for many years, and the advantages of using local, low order piecewise polynomial models are well known to the surface fitting community. The properties which they possess such as:

- **network stability** where the rules are modified only in response to *local* variations in the training data;
- **linear optimisation** where the unknown parameters appear linearly in the optimisation calculation; and
- **sparse calculations** where only a small number of parameters influence the output and the optimisation calculations are performed in a sparse, linear space

are *all* desirable for fuzzy systems, and are achieved by using the B-spline membership functions.

A trivial consequence of this work is that the class of functions which a fuzzy system can reproduce includes the set of polynomial models (controllers) defined on a compact space, and by the Stone-Weierstrass theorem, the set of fuzzy systems can approximate any continuous nonlinear mapping defined on a compact domain arbitrarily closely[1]. Therefore fuzzy controllers can always reproduce *exactly* any linear controller defined on a bounded input space. It may be slightly surprising that the defuzzified output of a fuzzy system is simply a piecewise polynomial mapping, but fuzzy rules simply *interpolate* the information stored at their set

---

[1]Fuzzy systems are therefore *universal approximators*.

centres, and this is directly influenced by the type of membership function used. Many fuzzy systems use simple triangular membership functions which results in a piecewise linear output if the algorithm is implemented as described above. Other representations produce an almost piecewise linear output, although the surface will be corrupted by small nonlinearities which are due solely to the fuzzy operators used. For most systems which require a smooth, nonlinear model or control surface, the former is desirable [Farrall and Jones, 1993].

Although it has been argued that fuzzy controllers can reproduce exactly any linear controller defined on a bounded input space, their relative computational cost has not been compared. If two piecewise linear fuzzy sets are used to represent each input variable (the smallest number required), the fuzzy system requires 2^n weights whereas the conventional linear model requires only n parameters, and this is also reflected in the relative computational costs. Alternatively, if the fuzzy system's rule base is constructed so as to exploit the structural information in the training data (no cross-coupling terms between inputs), the storage requirements can be reduced to $2n$, although advanced tree building algorithms (see Section 8.5) are required. For the more complex fuzzy model (controller), the condition of the learning problem is worse, as is the network's ability to generalise, and it is always desirable to use the *simplest* possible model which can achieve the desired results.

If the input fuzzy set is not a singleton but rather a fuzzy distribution, the output surface is limited by the size and shape of the membership functions which are used in the rule base and by the set which represents the input distribution. The mapping no longer depends directly on the membership functions defined in the rule base as this information is *smoothed*, and the type of nonlinearities that can be represented by systems presented with singleton and fuzzy distribution inputs are therefore different.

10.6.2 Discrete Centre of Gravity Defuzzification

For fuzzy sets defined on a discrete universe, the centre of gravity defuzzification algorithm is given by:

$$y(\mathbf{x}^d) = \frac{\sum_{Y^d} \mu_B(y^d)\, y^d}{\sum_{Y^d} \mu_B(y^d)}$$

and a similar analysis to that applied in Section 10.6.1 can be followed. Using algebraic fuzzy operators and Equation 10.8 gives:

$$y(\mathbf{x}^d) = \frac{\sum_{Y^d} \sum_{X^d} \mu_A(\mathbf{x}^d)\mu_R(\mathbf{x}^d, y^d)\, y^d}{\sum_{Y^d} \sum_{X^d} \mu_A(\mathbf{x}^d)\mu_R(\mathbf{x}^d, y^d)}$$

where $\mu_R(\mathbf{x}^d, y^d)$ is an element in the *relational matrix*. If an arbitrary set of fuzzy logic operators were used to generate the relational matrix, this would be the most computationally efficient way to retrieve the desired information despite the size of the relational matrix (this is still true if truncation operators are used in the retrieval process except that the above sum and products are replaced by max and min operators, respectively). However, if algebraic operators are used, a considerable reduction in the computational cost can be achieved by exploiting the structure of the relational matrix. Substituting from Equation 10.5 produces:

$$y(\mathbf{x}^d) = \frac{\sum_{Y^d} \sum_{X^d} \mu_A(\mathbf{x}^d) \sum_{kl} \mu_{A^k}(\mathbf{x}^d) c_{kl} \mu_{B_l}(y^d) y^d}{\sum_{Y^d} \sum_{X^d} \sum_{kl} \mu_{A^k}(\mathbf{x}^d) c_{kl} \mu_{B_l}(y^d)}.$$

If the discrete fuzzy output sets are equivalent (same shape, size and sample points), their set centres are given by $\frac{\sum_{Y^d} \mu_{B_l}(y^d) y^d}{\sum_{Y^d} \mu_{B_l}(y^d)}$ which is equivalent to the continuous representation as the number of discrete sample points increases. Also substituting for the *weights*, the defuzzified output becomes:

$$y(\mathbf{x}^d) = \frac{\sum_{X^d} \mu_A(\mathbf{x}^d) \sum_i \mu_{A^i}(\mathbf{x}^d) w_i}{\sum_{X^d} \mu_A(\mathbf{x}^d)} \qquad (10.16)$$

and for a singleton input which is represented by a crisp fuzzy membership function:

$$y^s(\mathbf{x}^d) = \sum_i \mu_{A^i}(\mathbf{x}^d) w_i. \qquad (10.17)$$

This is the discrete analogue of Equation 10.13 since again, $y^s(\mathbf{x}^d)$ is a weighted linear sum of the fuzzy input sets. Substituting Equation 10.17 into 10.16 then gives:

$$y(\mathbf{x}^d) = \frac{\sum_{X^d} \mu_A(\mathbf{x}^d) y^s(\mathbf{x}^d)}{\sum_{X^d} \mu_A(\mathbf{x}^d)} \qquad (10.18)$$

which provides a weighted average of the fuzzy singleton outputs over the support of the fuzzy input set $\mu_A(\mathbf{x}^d)$. As the number of points in $\mathbf{X}^d$ increases, this quantity becomes closer to the output of the equivalent continuous fuzzy system. It also allows the effect of different fuzzy input sets, $\mu_A(\mathbf{x}^d)$, to be investigated, as shown in Figure 10.21. If the crisp input is represented using a *singleton* membership function which has a value of one at the closest discrete sample point (top), the system's output is piecewise constant and the fuzzy output distribution is the appropriate column of the relational matrix. Alternatively, if a piecewise linear fuzzy input set is used which provides a linear combination of the system's output at the nearest discrete input points (for each variable), the output of the discrete system is piecewise linear (continuous). This input set can be represented either by a triangular membership function of width twice the sampling interval or by a simple linear weighting scheme [Sutton and Jess, 1991], as shown second from top. The discrete system is then equivalent to a continuous system *if* the set of discrete sample points contains the knot set when B-splines of order 2 (triangular) are used to represent the membership functions. Using fuzzy input sets which are similar to those used to represent the linguistic terms (bottom and second from bottom) produces a smoothed network output, just as occurs in continuous systems. If the

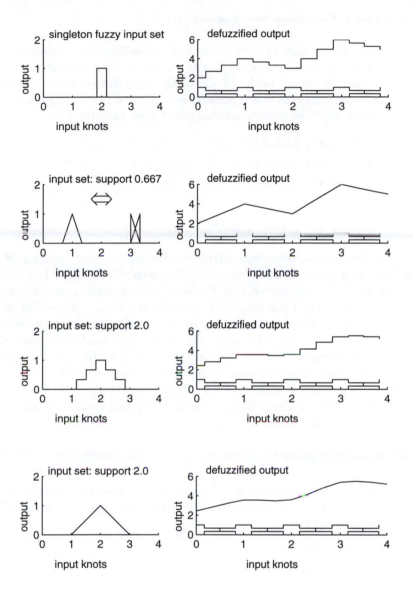

Figure 10.21 Four different methods of accessing the information held in a discrete fuzzy system. The first and third input membership functions result in a piecewise constant output whereas the second and fourth generate a continuous output.

fuzzy input set is always centred on the closest sample point (second from bottom) the overall system produces a piecewise constant output, whereas if the fuzzy set is centred on the actual crisp input (bottom), the output is a smoothed, continuous function.

10.6.3   Binary or Continuous Rule Confidences?

Many applications of fuzzy logic to problems in modelling and control use binary rule confidences, i.e. $c_{ij} \in \{0, 1\}$, so that the belief in the rule that maps the ith multivariate fuzzy input set to the jth univariate fuzzy output set is either one or zero. Using binary rule confidences means that only one fuzzy output set fires for each fuzzy input set. The fuzzy algorithm might then have the form:

$$
\begin{array}{lll}
\text{IF } (\mathbf{x} \text{ is } \mathbf{A}^1) \text{ THEN } (y \text{ is } B^1) & (c_{11} = 1) \\
\text{OR} \quad \text{IF } (\mathbf{x} \text{ is } \mathbf{A}^2) \text{ THEN } (y \text{ is } B^3) & (c_{23} = 1) \\
\text{OR} \quad \text{IF } (\mathbf{x} \text{ is } \mathbf{A}^3) \text{ THEN } (y \text{ is } B^1) & (c_{31} = 1) \\
\quad \vdots \qquad\qquad\qquad \vdots \\
\text{OR} \quad \text{IF } (\mathbf{x} \text{ is } \mathbf{A}^p) \text{ THEN } (y \text{ is } B^5) & (c_{p5} = 1)
\end{array}
$$

The aim of this section is to demonstrate the limitations that binary rule confidences impose on the defuzzified output and the restrictions on the functions that can be represented. It is shown that the binary fuzzy system cannot reproduce certain functions, even if the network's structure is correct. The effect of using fewer output sets than input sets is then considered and the restrictions which this imposes on the design of the rule base is described. These limitations are overcome by using rule confidences that lie in the continuous interval $[0, 1]$, and a consistent rule confidence generation algorithm is described in Section 10.6.4.

Therefore, consider a singleton input to a fuzzy system $\mu_{\mathbf{A}}(\mathbf{x})$ that uses a centre of gravity defuzzification algorithm and has a binary rule confidence matrix. The defuzzified output is given by:

$$
y^s(\mathbf{x}) = \sum_i \mu_{\mathbf{A}^i}(\mathbf{x}) \sum_j c_{ij}\, y_j^c. \tag{10.19}
$$

Associated with each fuzzy input set is a *single* fuzzy output set and this rule has a confidence of 1, all other rule confidences relating to this input set are zero. Hence, for the ith fuzzy input set:

$$
c_{ij} = \begin{cases} 1 & \text{if } j = b(i) \\ 0 & \text{otherwise} \end{cases}
$$

where $b(i)$ is a function which returns the index of the single active fuzzy output set corresponding to the ith fuzzy input set. Equation 10.19 now becomes:

$$
y^s(\mathbf{x}) = \sum_i \mu_{\mathbf{A}^i}(\mathbf{x})\, y_{b(i)}^c. \tag{10.20}
$$

The restrictions that this equality places on the defuzzified output are more easily understood by letting the fuzzy input sets be univariate and defined by B-splines of order $r = 2$ (triangular set shapes). Figure 10.22 shows that the fuzzy singleton input corresponding to the centre of the ith B-spline has $\mu_{\mathbf{A}^i}(\mathbf{x}) = 1$ and from

Equation 10.20, the defuzzified output is one of q possible values, i.e. $y = y_{b(i)}^c$, $b(i) = 1, \ldots, q$. The value obtained depends only on the centres of the fuzzy output sets, and when the system is required to output a value which does not correspond to one of these values, it is unable to reproduce this function exactly. For fuzzy singletons lying between the centres of the input membership functions, the fuzzy rule base provides linear interpolation, although the slope can again be only one of a finite number of possible values [Jager, 1992].

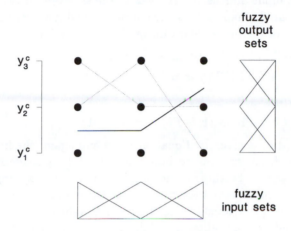

Figure 10.22 The restrictions placed on the output of a fuzzy system when binary rule confidences are used. The network output (dotted line) may only pass through the solid dots at the centres of the input sets, and hence is unable to reproduce the solid line.

This restriction is overcome by letting $c_{ij} \in [0,1]$ and allowing each input set to activate more than one output set. The fuzzy algorithm that maps the ith multivariate fuzzy input set to the q fuzzy sets defined on the univariate output universe might then have the form:

$$\text{IF } (\mathbf{x} \textit{ is } \mathbf{A}^1) \text{ THEN } (y \textit{ is } B^1) \qquad (c_{11} = 1.0)$$

$$\vdots \qquad\qquad\qquad \vdots$$

$$\text{OR } \text{IF } (\mathbf{x} \textit{ is } \mathbf{A}^i) \text{ THEN } (y \textit{ is } B^1) \qquad (c_{i1} = 0.0)$$
$$\text{OR } \text{IF } (\mathbf{x} \textit{ is } \mathbf{A}^i) \text{ THEN } (y \textit{ is } B^2) \qquad (c_{i2} = 0.7)$$
$$\text{OR } \text{IF } (\mathbf{x} \textit{ is } \mathbf{A}^i) \text{ THEN } (y \textit{ is } B^3) \qquad (c_{i3} = 0.3)$$
$$\text{OR } \text{IF } (\mathbf{x} \textit{ is } \mathbf{A}^i) \text{ THEN } (y \textit{ is } B^4) \qquad (c_{i4} = 0.0)$$

$$\vdots \qquad\qquad\qquad \vdots$$

$$\text{OR } \text{IF } (\mathbf{x} \textit{ is } \mathbf{A}^p) \text{ THEN } (y \textit{ is } B^q) \qquad (c_{pq} = 1.0)$$

Notice that the ith rule confidence vector has been normalised, i.e. $\sum_j c_{ij} = 1$, which implies that there exists total knowledge of the output corresponding to this fuzzy input set. Consider again the fuzzy input membership functions defined by order $k = 2$ B-splines, when $\mu_{A^i}(\mathbf{x}) = 1$. Allowing c_{ij} to take any value in the interval $[0, 1]$ means that any defuzzified output lying in the support of the output sets can be obtained.

The second scenario occurs when there are more fuzzy input sets than output sets $(p > q)$. This generally occurs for multivariable input since the basis functions are defined on a lattice across the input space, and their number increases exponentially with the input dimension. For $p > q$, the situation arises where several input sets could possibly map to the *same* output set and if binary rule confidences are employed, part of the fuzzy algorithm would be of the form:

$$\text{IF } (\mathbf{x} \ is \ \mathbf{A}^1) \text{ THEN } (y \ is \ B^j) \qquad (c_{1j} = 1)$$
$$\text{OR } \ \text{IF } (\mathbf{x} \ is \ \mathbf{A}^2) \text{ THEN } (y \ is \ B^j) \qquad (c_{2j} = 1)$$
$$\vdots \qquad\qquad\qquad \vdots$$
$$\text{OR } \ \text{IF } (\mathbf{x} \ is \ \mathbf{A}^i) \text{ THEN } (y \ is \ B^j) \qquad (c_{ij} = 1)$$

The defuzzified output is given by Equation 10.20 and again, the limitations that this places on the fuzzy system are best understood by considering input sets defined by triangular membership functions. For this case, a fuzzy singleton applied to the centre of all input sets that map to the same output set has the *same* defuzzified output $y^c_{b(i)}$ restricted to one of q possible values.

If the rule confidences are allowed to take any value in the interval $[0, 1]$, this restriction will be overcome because it is no longer required that an input set maps to just one output set. The fuzzy system then has the flexibility to allow the defuzzified output to take *any* desired value in the support of the output set at the centre of any fuzzy input set.

10.6.4   The Rule Confidence Distribution

Using rule confidences that lie in the interval $[0, 1]$ increases the flexibility of the fuzzy system by relaxing its dependency on the output sets' definitions. The main problem is how to generate the rule confidences so that the desired behaviour can be incorporated into a fuzzy system. It was shown in Equation 10.13 that the output of a continuous fuzzy system that uses algebraic operators and receives a crisp fuzzy input set is given by:

$$y^s(\mathbf{x}) = \sum_{i=1}^{p} \mu_{A^i}(\mathbf{x}) \, w_i$$

where $w_i = \sum_{j=1}^{q} c_{ij} y_j^c$. There should be no restrictions on the weights and it is desirable for there to be no interdependency between the different input sets. This generates one constraint for each of the rule confidence vectors:

$$w_i = \sum_{j=1}^{q} c_{ij} \, y_j^c \qquad\qquad\qquad (10.21)$$

where w_i is the *desired* weight associated with each fuzzy input set. The weight can be considered as the possible *output* associated with the fuzzy input set, and it is therefore consistent to evaluate their grade of memberships in the fuzzy output sets. This equation ensures that the data stored and retrieved from the fuzzy system is *equivalent*, and unless relationships such as these are embodied in the learning rules, convergence, rates of convergence and stability results are very difficult to establish. The rule confidences are still not uniquely determined as Equation 10.21 gives only one equation in q unknowns, therefore other constraining equations must be generated. A second constraint which has been assumed to exist is that the rule confidence vectors are *normalised*, and so the following expression must hold:

$$\sum_{j=1}^{q} c_{ij} \equiv 1 \qquad \text{for each } i \tag{10.22}$$

It is also generally required that the rule base should be as *sparse* as possible, to minimise the computational cost and storage requirements, and to keep the rule base *simple* for user verification and validation. This can be achieved by restricting the non-zero rule confidences to coincide with the fuzzy output sets for which the desired weight has a non-zero degree of membership. The rule confidence vector should also be *convex* (a single maximum value) as this ensures that the fuzzy output distribution is also convex.

Triangular Fuzzy Output Sets

Consider the commonly occurring case when there are two overlapping sets on the output universe and each desired weight has a non-zero degree of membership in two output sets. For convenience assume that these are the first two output sets, then Equations 10.21 and 10.22 give:

$$w_i = c_{i1}y_1^c + c_{i2}y_2^c$$
$$1 = c_{i1} + c_{i2}$$

There exist two independent equations in two variables, and solving for the rule confidences gives:

$$c_{i1} = \frac{y_2^c - w_i}{y_2^c - y_1^c}$$
$$c_{i2} = \frac{w_i - y_1^c}{y_2^c - y_1^c} \tag{10.23}$$

These functions are simply B-splines of order 2 or (equivalently) the commonly used triangular fuzzy membership functions. Therefore if the fuzzy output sets are B-splines of order 2, the rule confidences can be calculated from:

$$c_{ij} = \mu_{B^j}(w_i) \tag{10.24}$$

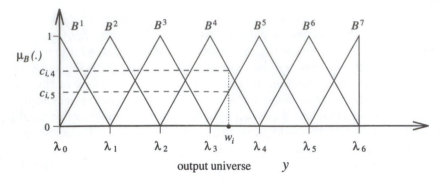

Figure 10.23   The rule confidence vector $\mathbf{c}_i = (0,0,0,0.6,0.4,0,0)^T$ which corresponds to the weight w_i is found by evaluating the membership of the weight in the fuzzy output sets.

as illustrated in Figure 10.23. This provides a *direct, invertible* relationship between the weights, rule confidences and fuzzy output sets. Given a particular weight to store, a rule confidence vector can be evaluated using the fuzzy output sets which produces the original weight when the fuzzy information is defuzzified. It is possible to use other fuzzy output membership functions which have different shapes, although a simple and consistent method must be found for generating the rule confidences, and to the authors' knowledge none have been proposed apart from this algorithm.

Higher Order Fuzzy Output Sets

For higher order fuzzy output sets, it might be hoped that the above relationship can be generalised, and indeed if the fuzzy output membership functions are *symmetric* B-splines of order ≥ 2, the following relationship is satisfied:

$$w_i = \sum_{j=1}^{q} c_{ij}\, y_j^c$$

where

$$c_{ij} = \mu_{B^j}(w_i) \tag{10.25}$$

and this is proved in Appendix E. The B-spline membership functions are also normalised and produce convex membership values, thus the rule confidences produced satisfy all of the constraints.

The *invertible* relationship which exists between weights and rule confidences is very important as it makes it possible to implement and train the network in weight space (with a considerable reduction in the computational cost), while explaining its responses with linguistic rules and the associated rule confidences. These systems are therefore *transparent* to the designer, and the network's performance (fuzzy rules) can be modified after training and reimplemented in weight space.

10.6.5   A Comparison With Neural Networks

In the simplest form, fuzzy systems calculate their response by taking a *linear* combination of the input membership functions, which is a strategy adopted by several neural networks commonly used within learning modelling and control architectures. The input membership functions represent *pattern matching* mappings which calculate how close the input is to the centre of the basis functions [Linkens and Nie, 1994]. These networks differ according to the size, shape and placement of the basis functions in the input space, although this book has described the three most common systems: radial basis functions, the CMAC and B-splines. A modified type of fuzzy reasoning system has also been proposed in the literature where instead of associating a weight with each fuzzy input membership function, a *functional mapping* is stored instead. This section describes the relationship between these two types of fuzzy system as well as describing their equivalent neural algorithms.

Radial Basis Function Networks

This class of networks was first proposed by Broomhead and Lowe [1988], and its output is given by:

$$y(\mathbf{x}) = \sum_{i=1}^{p} f_i(\|\mathbf{x} - \mathbf{c}_i\|_2)\, w_i$$

where $\mathbf{c}_i$ is the centre of the ith radial basis function f_i, and $\|.\|_2$ denotes the common Euclidean norm. Different univariate functions f_i, produce networks which generalise differently [Chen *et al.*, 1992], although one of the most popular is the standard multivariate Gaussian function which can be obtained by *multiplying* together n univariate Gaussian functions. As noted by Wang [1992], this is equivalent to forming the fuzzy AND of n univariate Gaussian membership functions. Therefore there is a direct equivalence between radial basis function networks and several Gaussian fuzzy systems which have recently been proposed [Bruske, 1993, Hollantz and Tresp, 1992, Tresp *et al.*, 1993].

Lattice Networks

The Cerebellar Model Articulation Controller (CMAC) (see Chapter 6), was originally proposed by Albus in the mid-seventies, and in its most basic form, is a distributed look-up table which generalises locally. The network's response is:

$$y(\mathbf{x}) = \sum_{i=1}^{p} f_i(\mathbf{x})\, w_i$$

where the nonlinear function $f_i(.)$ can take various forms. It can be expressed as $f_i(\|\mathbf{x} - \mathbf{c}_i\|_\infty)$ about a centre $\mathbf{c}_i$, or it can be formed by combining n univariate basis functions (fuzzy membership functions) using a *product* or *min* operator. Again there is a direct relationship between the CMAC and fuzzy systems. The CMAC possesses a special feature which ensures that the number of basis functions (fuzzy rules) which contribute to the output is *not* a function of the input space dimension, so the cost of calculating the network's response depends *linearly* on the input space dimension. This method of distributing the basis functions can be used as an even, sparse coarse coding algorithm for fuzzy systems. The basic B-spline algorithm can also be expressed as a linear combination of piecewise polynomial basis functions which are formed by multiplying together n univariate basis functions (see Chapter 8).

Functional Fuzzy Rules

This relationship between neural networks and fuzzy systems also unifies two apparently different schemes developed in each field. Sugeno and Nishida [1985] proposed a type of fuzzy system where instead of the consequent of the production rule being a linguistic term, a *numerical function* is stored. These production rules have the general form:

$$y(\mathbf{x}) = \sum_i \mu_{A^i}(\mathbf{x})\, w_i(\mathbf{x}) \tag{10.26}$$

where the weights are no longer just numbers but are functions. A simple and commonly used example is when $w_i(\mathbf{x})$ represents a linear mapping:

IF x_1 is A_1^i AND $\cdots$ AND x_n is A_n^i
THEN $y = w_0^i + w_1^i x_1 + \cdots + w_n^i x_n$

where $\mathbf{w}^i$ is the *linear* parameter (weight) vector associated with i^{th} multivariate fuzzy input set. The fuzzy input sets provide a *nonlinear* scheduling of the linear parameters, while the output is smooth due to the linear functions which only have a local influence. It should be noted that the conventional fuzzy systems that have been extensively investigated in this chapter (algebraic operators, normalised fuzzy membership functions and centre of gravity defuzzification) are a *special* case of these rules, as the output associated with each antecedent is simply a single weight, w_0^i, rather than a linear function. A similar representation has also been proposed for normalised Gaussian Radial Basis Functions [Shao *et al.*, 1993], where the system's output is given by:

$$y(\mathbf{x}) = \frac{\sum_{i=1}^{p}(w_0^i + w_1^i x_1 + \cdots + w_n^i x_n) f_i(\|\mathbf{x} - \mathbf{c}_i\|_2)}{\sum_{j=1}^{p} f_j(\|\mathbf{x} - \mathbf{c}_j\|_2)}.$$

Whilst there is no first-order term in the Taylor series expansion of a Gaussian function, this modification to the standard radial basis function representation enables it to model locally linear functions using only a small number of terms. This is similar to Sugeno's fuzzy system where trapezoidal membership functions are generally employed. These membership functions are constant near their centre, so the linear function means that the overall system can model locally linear functions using a small number of rules. A multi-layer network architecture that incorporates this principle has also been developed recently [Jacobs and Jordan, 1993], allowing a hierarchical, tree-like decomposition of the input space to be formed using the available training data.

The relationship between neural and fuzzy systems is not solely restricted to singleton inputs. As shown in Equations 10.15 and 10.18, when the input is represented by a fuzzy distribution, the defuzzified network output is calculated by taking a locally weighted average of the outputs which are generated by singleton inputs. It is irrelevant whether a neural or a fuzzy system is used to store the desired function; rather research should be concentrating on uncertainty representations (membership function shapes) and knowledge processing operations (truncation and algebraic operators).

10.6.6 A Probabilistic Interpretation of a Fuzzy System

It is possible to give a probabilistic interpretation of the continuous fuzzy systems described by Equation 10.26, where probability is interpreted as a subjective degree of belief in an event [Tresp *et al.*, 1993]. As well as explaining the close relationship between Bayesian approximation theory and the centre of gravity defuzzification algorithm, this relationship is useful for initialising a fuzzy system when an expert provides knowledge in terms of probability distributions (such as the probability of a rule being active, etc.).

In order to describe this equivalence, consider a system which has a number of unobservable states s_{ij} that correspond to the ij^{th} fuzzy rule r_{ij}. Let $P(s_{ij})$ denote the *prior* probability that the system is in state s_{ij} and $P(\mathbf{x}, y|s_{ij})$ is the probability distribution for $\mathbf{x}$ and y given that the system is in state s_{ij}. Then the probability of the input, output pair $(\mathbf{x}, y)$ occurring is given by:

$$
\begin{aligned}
P(\mathbf{x}, y) &= \sum_{ij} P(\mathbf{x}, y|s_{ij})P(s_{ij}) \\
&= \sum_{ij} P(y|\mathbf{x}, s_{ij})P(\mathbf{x}|s_{ij})P(s_{ij})
\end{aligned}
\tag{10.27}
$$

The *a posteriori* probability of the input belonging to the ij^{th} class is:

$$
P(s_{ij}|\mathbf{x}) = \frac{P(\mathbf{x}|s_{ij})P(s_{ij})}{\sum_{kl} P(\mathbf{x}|s_{kl})P(s_{kl})}
\tag{10.28}
$$

and Bayesian theory enables the optimal output to be predicted such that the expected error is minimised by:

$$E\left(y|\mathbf{x}\right) = \frac{\sum_{ij} \int y P(y|\mathbf{x}, s_{ij}) \, dy \, P(\mathbf{x}|s_{ij}) P(s_{ij})}{\sum_{kl} P(\mathbf{x}|s_{kl}) P(s_{kl})}$$

where $P(\mathbf{x}|s_{ij}) = \int P(\mathbf{x}, y|s_{ij}) \, dy$. Substituting $a_{ij}(\mathbf{x}) = \frac{P(\mathbf{x}|s_{ij})P(s_{ij})}{\sum_{kl} P(\mathbf{x}|s_{kl})P(s_{kl})}$ and also $w_{ij}(\mathbf{x}) = \int y P(y|\mathbf{x}, s_{ij}) \, dy$, the above expression can be written as:

$$E\left(y|\mathbf{x}\right) = \sum_{ij} a_{ij}(\mathbf{x}) \, w_{ij}(\mathbf{x}) \tag{10.29}$$

This is quite similar to Equation 10.26, in that the output can be expressed as a fuzzy functional rule base, although the mappings now depend on the index of the input/output fuzzy rule rather than just on the index of the input set. If certain assumptions are made about the *prior* state probabilities $P(s_{ij})$ (similar to the restrictions placed on the rule confidences), this dependence on the fuzzy output sets can be relaxed and the above sum will just be taken over i.

10.7 CONCLUSIONS

This chapter has considered the underlying modelling capabilities of a set of fuzzy rules. The approach taken has been to investigate the internal structure of the fuzzy network, and to propose modifications which result in a smoother, basic model. This was achieved by using B-spline basis functions to represent the fuzzy membership functions, and implementing the fuzzy logic operators using *algebraic* operators rather than the more common thresholding operators. The use of algebraic fuzzy operators is becoming increasingly popular [Chiu et al., 1991, Mizumoto, 1992, Pacini and Kosko, 1992, Wang and Mendel, 1992], due to the fact that they generate a *smoother* output surface and because their use ensures that the defuzzification algorithm is much simpler (see Section 10.6.1). Choosing the fuzzy sets to be B-splines basis functions may seem an unnecessary restriction, although many of the popular fuzzy set shapes (triangular, trapezoidal) can be reproduced using B-splines, and these basis functions provide one way of choosing higher order fuzzy sets. If the underlying fuzzy operators could not produce a smooth output surface, there would be no reason for choosing higher order fuzzy sets. However, it has been shown that the output is linearly dependent on the fuzzy input sets and so there is a requirement for a general method for the design of the fuzzy sets; the B-spline basis functions is one such technique which has many desirable numerical properties.

When the modelling capabilities of the network were being analysed, it was proved that the defuzzified output of the fuzzy network was *linearly dependent* on the set of rule confidences. The rule confidences denote the strength with which each rule fires and the rule confidences are the adjustable parameters of the fuzzy system. It was also shown that there exists a many-to-one, invertible mapping between the rule confidence vector and a weight associated with each multivariate fuzzy input set, hence the two representations are equivalent in that the input/output mapping is identical. The internal representations of these networks are different, although they have an *identical* black box behaviour even when the parameters are trained (see Chapter 11). Hence it possible to apply *all* of the theory developed in Chapters 4 and 5 to understand the dynamical behaviour of an adaptive fuzzy network. In particular, the effect of the size and shape of the fuzzy input sets on the rate of convergence (both output and parameter) can be estimated. This is useful because heuristics [Kosko, 1992a] have often been proposed in an attempt to describe the relationship which exists between the fuzzy set shapes, the modelling capabilities of the fuzzy system and the amount of learning interference. The results derived in Chapter 5 give *specific* measures of the amount of gradient noise and learning interference, and a geometrical insight into the learning laws for different fuzzy set shapes and sizes.

Probably the most important result of this work has been in using B-splines to provide a common framework with which to study the application of self-organising fuzzy control and neurocontrol in the intelligent control field. This work shows that they should not be considered as two separate approaches; they are equivalent algorithms, differing only in their information representation schemes. The fact that the input/output behaviour is identical means that the B-spline neural network representation is computationally the most efficient, although when knowledge is stored in terms of rule confidences it allows (in certain cases) the rule base to be linguistically inverted. This motivated the development of the *indirect* fuzzy adaptive control scheme, where a fuzzy plant model is constructed from plant input/output data and a fuzzy inversion is performed on-line in order to realise the control signal [Harris *et al.*, 1993]. The close relationship which exists between the two techniques also allows Kavli's network construction algorithm ASMOD (see Section 8.5) to be viewed as a method for automatically generating fuzzy rules and determining the width of the supports of the basis functions. This has traditionally been an ill-defined problem, although it can now be posed as an iterative, maximum error reduction strategy. Also the ASMOD decomposition of the overall map into a set of linearly combined submodels allows these fuzzy techniques to be applied in higher dimensional spaces and partially overcomes the curse of dimensionality.

Currently, there are many different methods for implementing fuzzy algorithms and although this chapter has compared two representation schemes (continuous and discrete) and two sets of operators (algebraic and truncation) in detail, there are still many different fuzzy operators which have not been considered. This is especially true for the current generation of fuzzy software where each package has different fuzzy membership functions and operators and these are just for *static* fuzzy rule bases. It is not surprising therefore that general theories about the quality and type of fuzzy output surfaces are rare, and those that exist are so general that their usefulness is debatable. The approach described in this book (continuous fuzzy systems coupled with algebraic operators) is limited, but it is precisely for this reason that it is useful. Its modelling capabilities can be determined, the relationship between the system's parameters and the output is easily understood, learning rules can be derived for which standard theories can be applied and the basic representation can be extended (using an ASMOD approach) to cope with many inputs. It is for these reasons that the authors believe that the work described in this chapter represents an important contribution to the field of fuzzy systems.

11

Adaptive Fuzzy Modelling and Control

11.1 INTRODUCTION

Static fuzzy systems have been used for the past twenty years to control a wide range of poorly understood plants. Their success was attributed to the fact that inherently nonlinear control strategies, expressed in a (restricted) natural language framework, could be obtained from human operator and then implemented as a fuzzy controller. The strength and weakness of this approach is the assumption that a fuzzy algorithm, which describes the input, output behaviour of the controller, is available. Fuzzy logic provides a framework for interpreting the vague, linguistic rules, although the process of obtaining such rules from an expert, known as *knowledge elicitation*, is one of the major *bottlenecks* in the development of reliable expert systems. Rules are required to be *correct*, *relevant* and *complete*. The former ensures that the controller operates satisfactorily while the latter two mean that only useful knowledge is stored. Frequently, the fuzzy algorithms provided by experts do not satisfy these requirements; the rules are vague and can be misinterpreted, the rule base is incomplete (due to its size or because the expert simply forgets or is unable to articulate certain situations/action pairs) and often irrelevant features, or inputs, are used to describe the input, output mapping. These problems can be overcome using adaptive (or learning/self-organising) fuzzy systems which automatically find an appropriate set of rules. In addition, fuzzy systems which adapt their rules *on-line* have the potential to cope with time-varying plants.

The first adaptive fuzzy controllers were developed at Queen Mary College by Procyk and Mamdani [1979]. It was called a self-organising controller as *its control policy is one that can change with respect to the process it is controlling and the environment it is operating in.* However, it is also worth noting another comment that was made in this paper:

> *It is impossible to design a controller which need not assume anything about its environment. One can only strive to lessen its dependency and sensitivity to it.*

Learning fuzzy systems offer a potential solution to the knowledge elicitation problem, although (just like any learning algorithm) they are not a universal panacea.

The basic representations should be flexible enough to model a wide variety of plants, but they should also be well conditioned.

11.1.1 Fuzzy Representation

Chapter 10 described, in considerable detail, how fuzzy algorithms may be represented. The designer of a static rule base has to select the input variables, the corresponding fuzzy membership functions (their shape and size), the logical operators and the rules contained in the fuzzy algorithm. While this uniquely determines a fuzzy system, there are a wide variety of parameters that can be potentially adjusted when the models and controllers are trained on-line. Traditionally, only the relational elements were changed when new knowledge was stored in a discrete fuzzy system, and this is equivalent to altering the rule confidences in a continuous fuzzy network. The basic structure of the rule base (the size and shape of the input and output membership functions) remains static, only the strengths with which the rules fire are altered.

It has been shown that fuzzy systems which use algebraic logical operators are *linearly dependent* on the rule confidences. Therefore, it is possible to train the rule confidence matrix using modified versions of the instantaneous learning rules described in Chapter 5. However, a relationship between a weight and a rule confidence vector was also derived, and again the network depends *linearly* on these weights and the behaviour of the adaptive system can be predicted using the theory described in Chapter 5. Section 11.2 compares and contrasts these two training procedures and demonstrates that there exists a learning equivalence between the neural and fuzzy representations, although it is shown that it is computationally simpler to implement the learning rules in weight space, and the rule confidences should only be used for initialisation, verification and validation purposes.

These comments apply to the ASMOD-type fuzzy systems as well as to the more conventional rule structures. The ASMOD procedure, which automatically determines the structure of a B-spline network (see Section 8.5), can also be used to initialise fuzzy systems due to the equivalence that exists between these two forms, and reasons why it may be desirable to use such an initialisation algorithm include:

- **transparency**, to gain insight into the relationships contained in the data;
- **parsimony**, to produce a complete and compact set of fuzzy rules; and
- **conditioning**, which simplifies the on-line training procedure.

These techniques generally require a large amount of relevant data, hence it is infeasible to adjust the network's structure on-line. Section 11.2.5 describes how these algorithms compare with other fuzzy learning algorithms that adjust the parameters associated with their internal structure (membership functions, number of rules, relevant input variables, etc.).

11.1.2 Fuzzy Modelling and Control

Once a set of supervised learning rules has been derived for the adaptive fuzzy systems, it is relatively straightforward to use these networks to construct plant models. This book has focused on learning rules for feedforward neurofuzzy models rather than recurrent mappings, and this approach is continued in this chapter, although recurrent networks and training rules are briefly considered in Section 11.3.2.

A common problem in control engineering is to synthesise an appropriate controller from a learnt plant model, and as most of the work in this book has focused on adaptive plant modelling, it is natural to develop this approach. This is called *indirect* adaptive control [Åström and Wittenmark, 1989], and it is slightly unusual to describe learning fuzzy controllers within this context as the vast majority of self-organising fuzzy controllers considered in the literature are *direct*. Direct adaptive fuzzy control does not update a process model, rather an explicit fuzzy controller is used to store the control actions and this is trained directly. This chapter focuses on the structure of these two control schemes and provides illustrative, simple examples.

The indirect fuzzy control approach has the advantage that the control design is separated from the adaptive mechanism, which enables standard controller designs (such as k-step ahead) to be incorporated into the overall architecture. Stability results can be derived for such systems, thus *convergence* and *stability* theorems do exist for these neurofuzzy control algorithms, although now the challenge is to derive new theories which exploit specific features of these networks.

Direct fuzzy controllers generally use a static incremental process model to relate the error in the calculated control action to the deviation in the desired behaviour of the plant. This is necessary because the desired control signal is never known, and it needs to be inferred from knowledge about the plant, and the current plant's performance. Therefore, it is incorrect to say that direct fuzzy controllers do not require a prior plant model, as the learning mechanisms must incorporate knowledge about its Jacobian and also some type of reference model. This information is usually contained in the performance index.

11.2 LEARNING ALGORITHMS

There are many ways in which the rules stored in a fuzzy system can be altered and they depend on the particular knowledge representation scheme. Conventional self-organising fuzzy controllers have three different sets of parameters which can be altered:

- **scaling factors** for the input and output variables;
- **relational elements** contained in the relational matrix; and

- **membership functions** associated with the input and output variables,

although generally, only the relational elements are adapted on-line.

The most useful network parameters are those:

- which are **sensitive** to changes in the input, output relationship; and
- when there exists **simple** and **direct** relationships between the network's performance and the direction and magnitude of the parametric change.

This latter point is important as the knowledge stored in a trained fuzzy system is used to provide insight into the unknown mapping, and when the cause/effect relationships of the parameters that are adapted are not well understood, the transparency of the fuzzy system is lost. The network operates simply as a black-box mapping and the learning procedure is generally badly conditioned. This section discusses how the relational elements, the rule confidences and the equivalent weights can be adapted and proves that an input, output equivalence exists between the different representations.

11.2.1 Weight Training Rules

Due to the invertible relationship between a weight and a rule confidence vector, it is possible to calculate the output of the fuzzy system using only a weight vector. It has also been shown that the network depends linearly on these weights, hence they can be adapted using *any* of the instantaneous learning rules described in Chapter 5 or the weight vector can be either calculated directly using a singular-valued decomposition algorithm, or determined using the iterative conjugate gradient and gradient descent rules [Brown *et al.*, 1994]. Once it has been established that the networks are linear in their adjustable parameters, there is a large body of theoretical work which can be applied to analyse a learning system's behaviour.

Fuzzy Interpretation

When the weight vector has been trained using any of the previously mentioned learning rules, the fuzzy rule confidence matrix can be inferred using the algorithm described in Section 10.6.4. This may be an *indirect* way to update the rule confidences, but when the output membership functions are represented by B-splines of order k it has some important properties:

- each rule confidence lies in the interval $[0, 1]$ and each rule confidence vector is **normalised**, as the fuzzy output membership functions form a partition of unity; and
- the rule confidence vectors (matrix) are **sparse**, as only k elements are non-zero.

The transformation of the weight vector into a rule confidence matrix can be a computationally costly process if it is performed each time the weights are updated, as the fuzzy output membership functions need to be evaluated ρ times (once for each updated weight). However, it is more important to have a learning algorithm which has low on-line computational requirements, and calculating the rule confidence vectors is only necessary when the knowledge stored in the network is being verified. Hence, these fuzzy systems should be implemented and trained using the weight space representation described in Chapter 10.

11.2.2 Rule Confidence Training Rules

For a variety of reasons, it may be desirable to store and train the rule confidences directly, and although this is possible, it is also recommended that fuzzy learning algorithms should update the weights rather than the rule confidences. The fuzzy network's output is linearly dependent on the rule confidences, and the weight associated with the ith fuzzy input membership function is generated by taking the dot product of the rule confidence vector with the vector of output set centres (see Equation 10.12). Therefore, in order to be consistent, any change in the rule confidences should be the same as the *equivalent* weight update once the information is defuzzified.

Instantaneous Gradient Descent

As an example, consider deriving an instantaneous gradient descent learning algorithm which can be used to train the rule confidences directly. Computing the partial derivatives of the instantaneous output error $\epsilon_y(t)$ with respect to the adjustable rule confidences gives the following gradient descent learning rule:

$$\Delta \mathbf{c}_i(t-1) = \delta\, \epsilon_y(t)\, \mu_{\mathbf{A}^i}(\mathbf{x}(t))\, \mathbf{y}^c \tag{11.1}$$

To evaluate the equivalent change in the weights, it is necessary to "defuzzify" Equation 11.1 by taking the dot product of the change in the rule confidence vector with $\mathbf{y}^c$, producing:

$$\Delta \mathbf{c}_i^T(t-1)\mathbf{y}^c = \delta\, \epsilon_y(t)\, \mu_{\mathbf{A}^i}(\mathbf{x}(t))\, \|\mathbf{y}^c\|_2^2$$

This is $\|\mathbf{y}^c\|_2^2$ times as great as the weight change recommended by the simple LMS rule. This factor could easily be incorporated into the learning rate (as it is a constant), although it is worthwhile considering the form of this training algorithm in greater detail. From Equation 11.1, the rule confidences are updated in proportion to the value of the output membership function's centre. They do not generally lie in the unit interval, all the rule confidences are non-zero and the rule confidence vectors are not normalised, so there is no logical (transparent) interpretation of the trained system. Hence, it is not possible to interpret its internal rule base, even though this is the main reason for training the rule confidences directly.

These problems occur because of the *redundant* manner in which information is stored using rule confidences. There are an infinite number of different rule confidence vectors which defuzzify to the same weight, and unless the learning rules are structured to resolve this linear dependence, unexpected, opaque results may be obtained.

Modified Instantaneous Gradient Descent

The instantaneous rule confidence gradient descent learning rule just derived can easily be modified to resolve some of the redundancy contained in the rule confidence vector by noting that Equation 11.1 can be rewritten as:

$$\Delta \mathbf{c}_i(t-1) = \delta \left(\hat{\mathbf{c}}^T(t)\mathbf{y}^c - \sum_j \mu_{A^j}(\mathbf{x}(t))\mathbf{c}_j^T(t-1)\mathbf{y}^c \right) \mu_{A^i}(\mathbf{x}(t))\,\mathbf{y}^c \qquad (11.2)$$

where $\hat{\mathbf{c}}(t)$ is the *desired* rule confidence vector obtained from "fuzzifying" the desired output by finding its membership of the fuzzy output sets:

$$\hat{c}_i(t) = \mu_{B^i}(\hat{y}(t))$$

Taking the dot product of Equation 11.2 with $\mathbf{y}^c$ is $\|\mathbf{y}^c\|_2^2$ times as great as the weight updates calculated using the LMS algorithm, although it can be seen that the output set centre vector already occurs twice on the right-hand side. Therefore, consider the following training algorithm:

$$\Delta \mathbf{c}_i(t-1) = \delta \left(\hat{\mathbf{c}}(t) - \sum_j \mu_{A^j}(\mathbf{x}(t))\mathbf{c}_j(t-1) \right) \mu_{A^i}(\mathbf{x}(t)) \qquad (11.3)$$

which can be interpreted as updating the rule confidence vector in proportion to *network's current rule confidence vector error*. This learning rule is also generally modified by setting the rule confidence vector equal to the desired one, when it has not been initialised and the corresponding fuzzy input set has a non-zero degree of membership. Taking the dot product with the vector composed of the output set centres, gives the *same* weight change as that recommended by the LMS rule. Hence systems that are based on weights and rule confidences are *learning equivalent* in the sense that the outputs of the two networks are identical even when they are trained; it is only the internal representations which are different.

This rule confidence learning algorithm has the following properties:

- the rule confidence vector is **normalised** (sums to unity) as the fuzzy output sets form a partition of unity;
- the individual rule confidences do **not** necessarily lie in the unit interval, which is due to the error estimate being calculated using a weighted sum;

- in general every rule confidence is non-zero, so the rule confidence matrix is **not sparse**; and
- although only one rule confidence vector has to be calculated for each learning iteration, there is generally more training data than weights, hence training in rule confidence space is computationally **more expensive**.

Learning rules, such as the one described above, *can* be formulated to train the rule confidences directly. However, they are generally more complex than the equivalent weight training algorithms, often the learnt rule confidence matrix is non-sparse, thus limiting its interpretability, and it is possible for the rule confidences to stray outside the unit interval, making it even more difficult to understand the rule base. Training the weights is computationally much simpler, and a sparse, sensible rule confidence matrix can be generated whenever a set of fuzzy rules is required to explain the system's actions.

Rule Confidence Correction Rules

One simple training rule which ensures that the rule confidences lie in the unit interval and produces a reasonably sparse rule confidence matrix is given by:

$$\Delta \mathbf{c}_i(t-1) = \delta \left(\hat{\mathbf{c}}(t) - \mathbf{c}_i(t-1) \right) \mu_{\mathbf{A}^i}(\mathbf{x}(t)) \tag{11.4}$$

This is similar to Equation 11.3, except that the network's rule confidence vector output, which is calculated using a weighted sum, is replaced by each individual rule confidence vector. Hence the networks which are adapted using this training rule no longer minimise the output Mean Squared Error (MSE), but learn to associate input and output fuzzy sets.

11.2.3 Relational Elements Training Rules

Originally, discrete fuzzy models and controllers were adapted by changing the elements in the relational matrix. Section 10.4.3 explains the strong link that exists between the rule confidences and the relational elements (see Equation 10.5), and to be consistent the learning rules for the relational elements should preserve this relationship. It is possible to derive such learning laws (their form is only slightly different from that given above), although the comments about the computational cost and the redundant manner in which information is stored again apply. In fact, the learning algorithms are even more complex as the input, output space has been sampled, meaning that there are even more parameters to adapt.

When the relational elements are trained using a (general) learning algorithm which modifies their values such that they are no longer equivalent to a sampled relational surface, it could be argued that the discrete fuzzy systems are more flexible. The greater flexibility of the discrete relational matrix has the potential to generate a more complex relational surface. However, it would not be possible to explain this knowledge using the original linguistic terms. Therefore the updated relational matrix should be equivalent to a sampled relational surface after adaptation, and the discrete and continuous fuzzy systems are learning equivalent.

11.2.4   Example: Instantaneous Adaptation

This example illustrates the reasons why it is preferable to implement and train a fuzzy network in weight space rather than in rule confidence (relational element) space.

Consider trying to approximate the squared function on the symmetrical input interval $[-1, 1]$. Three triangular fuzzy sets (B-splines of order 2) of width 2 are defined on the input universe, x, and four triangular fuzzy sets of width 1 are used to represent the output variable y, as shown in Figure 11.1. The training input

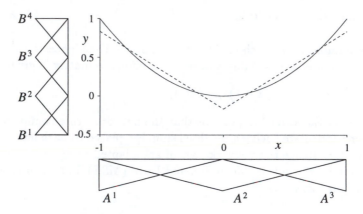

Figure 11.1   The fuzzy input and output membership functions used to model the desired function $\widehat{y}(x) = x^2$ on the interval $[-1, 1]$. The solid line is the desired output and the dashed is the optimal network's output.

is randomly selected from the interval $[-1, 1]$ with a uniform probability density function, and the optimal weight vector, with respect to the MSE, is given by:

$$\widehat{\mathbf{w}} = (0.833, -0.167, 0.833)^T$$

Also the "optimal" rule confidence matrix, which is generated by fuzzifying the optimal weights, is:

$$\widehat{\mathbf{C}} = \begin{bmatrix} 0 & 0 & 0.333 & 0.667 \\ 0.333 & 0.667 & 0 & 0 \\ 0 & 0 & 0.333 & 0.667 \end{bmatrix}$$

The initial (normalised) rule confidence matrix was set equal to

$$\mathbf{C}(0) = \begin{bmatrix} 0 & 0 & 0 & 1 \\ 0 & 1 & 0 & 0 \\ 0 & 0 & 0 & 1 \end{bmatrix}$$

which is a good estimate of the optimal one. The fuzzy system was then trained from 50,000 randomly generated training pairs using the instantaneous rule confidence training algorithm with a learning rate of 0.002. This produced an adaptive fuzzy system which converged slowly but was able to filter out most of the modelling error, producing the following rule confidence matrix:

$$\mathbf{C}(50,000) = \begin{bmatrix} 0 & -0.198 & 0.737 & 0.461 \\ 0 & 1.136 & 0.062 & -0.198 \\ 0 & -0.197 & 0.730 & 0.467 \end{bmatrix}$$

that corresponds to the defuzzified weight vector:

$$\mathbf{w}(50,000) = (0.830, -0.167, 0.832)^T$$

which is close to the optimal weight vector.

The trained rule confidence matrix, $\mathbf{C}(50,000)$, is certainly not sparse and its elements do not lie in the unit interval, although it has some "interesting" features which are worth commenting on. First, the rule confidences which relate to the first output set are all zero, and this occurs because the desired output is never less than zero, hence the first element in the desired rule confidence vector (the fuzzified optimal output) is always zero. This could be interpreted as meaning that the network never produces a negative output, but when the input is zero, the fuzzy system's output is simply $w_2(50,000)$ which equals -0.167. The negative output is generated by having a negative rule confidence which corresponds to the largest, *positive* output set and while this may be numerically correct, it would be more logical if c_{21} was positive. Second, each rule confidence vector is normalised as predicted in Section 11.2.2. Finally, the equivalent relational matrix training rules are similar to the rule confidence learning algorithms except that more parameters must be updated.

It may be that there exists simple rule confidence training algorithms which ensure that the rule confidences lie in the unit interval and that many are zero, but the only one known to the authors is described in Section 11.2.1.

11.2.5 Structure Determination

This section has so far concentrated on supervised learning algorithms which alter the contents of the rule base when the desired response signal is available. There are many other different methods for altering the behaviour of a fuzzy control system, and some of these are now discussed.

Scaling Factors

Scaling factors have been used to map the real inputs to a normalised input space on which the membership functions are defined, and to scale the output of a fuzzy system. The parameter multiplies the appropriate input or output variable, thus altering the domain of interest for the respective variable. For a fixed fuzzy rule base, altering a scaling factor effectively modifies the shape and size of *all* of the membership functions defined on the appropriate domain, thus changing the *gain* of the overall system.

Sutton and Jess [1991], have performed an extensive series of experiments which show how the dynamic response of a process changes when the scaling factors of the fuzzy controller are altered, and their experiences are reproduced in Table 11.2 (page 436). However, it should be re-emphasised that changing the scaling factors simply alters the definition of the membership functions on the real input (output) space, and thus they provide a simple way for changing the system's gain. Generally, scaling factors are used in direct adaptive fuzzy control schemes whose inputs are based on error and error rate measurements which ignore the current operating point of a (nonlinear) plant. In order to take account of process variations due to the operating point, some controller gain scheduling is necessary, and this is achieved using scaling factors. Storing a tabulated scaling factor for each input is equivalent to designing a scheduling strategy for a linear controller.

Changing the definition of the fuzzy input and output membership functions in systems used for *nonlinear* modelling and control can be very important. As explained in Section 10.1, fuzzy membership functions are context dependent although this feature is rarely reflected in their specification. Consider the definition of the fuzzy set *young*, in either a primary school or an old peoples home. Both membership functions are reasonably well defined, but they are both *different* [Wang, 1993]. Similarly, a fuzzy set that represents *smallness* may depend on both the input variable and the plant's current operating point. Altering the definition of the membership functions according to the sensitivity of the model is one way in which the *same* fuzzy algorithm (set of fuzzy rules) may be employed over the whole operating space, and the scaling factors are a convenient set of parameters to alter. In Section 9.2.2, a static fuzzy rule base was developed which could reverse park an autonomous vehicle. The fuzzy sets defined on the two cartesian coordinates which represent the vehicle's position were *scaled* by the slot's length and width so that the same rule base could be applied to many (apparently) different parking manoeuvres.

Membership Functions

As was shown in Section 10.6.1, the type, shape and size of the fuzzy membership functions used to represent the linguistic inputs determines the form of the system's output. Piecewise linear (triangular) membership functions produce a piecewise

linear network output with position of the linear regions being determined by
the boundaries of the triangular sets. Modifying the form of these membership
functions changes the type of mappings which can be approximated by the fuzzy
system, and altering the size and shape of the fuzzy output sets changes the *range*
of possible outputs in a manner analogous to changing the output scaling factor.

Whether a set of fuzzy rules is capable of adequately approximating a particular function depends on the form of the desired mapping. It may require more
(fewer) fuzzy membership functions if the network is not sufficiently flexible (over-
parameterised), and a learning algorithm which aims to recover structure from the
training data should have the ability to alter the number of fuzzy membership
functions, as well as their positions and shapes. The number and positions of
the univariate B-spline fuzzy membership functions is determined by a knot vector and adding (deleting) a knot introduces (removes) a basis function. Similarly,
moving a knot alters the shape of k membership functions and several techniques
for changing the structure of a fuzzy rule base are discussed in Section 8.5. The
centre of a Gaussian fuzzy membership function can be trained in an unsupervised
fashion using the clustering algorithms described in Section 2.4.1, or the centres
and variances can be adapted using a gradient descent rule equivalent to back
propagation.

Input Variables

Finally, any truly intelligent algorithm which can alter a fuzzy rule base should be
able to discover which inputs are important and its internal organisation. Most
fuzzy systems simply assume that the antecedent of the production rules involves
taking a fuzzy intersection which involves *all* the input variables. If some relationships are not necessary, as occurs when the network's output is a linear function
of some of the inputs, the overall fuzzy system is too flexible and may model noise
contained in the training data and generalise poorly. Similarly, when variables are
not important in the reasoning process, their presence increases the complexity
of the network but provides no extra information. An *off-line* learning algorithm
should have the ability to structure the network into smaller fuzzy sub rule bases
(see Section 8.5.1 where the ASMOD procedure is described) and if necessary to
structure the knowledge hierarchically, using deep, nested concepts.

Learning algorithms also exist for structuring Gaussian fuzzy systems automatically, as an input variable becomes redundant when the variance becomes
infinite. Note that for this approach, it is necessary to store a vector of variances for each basis function $\{\sigma_{ij}\}_{j=1}^{n}$. Similarly, a Gaussian basis function can be
completely removed from the output calculation when a weight shrinks to zero or
the centres move far away from the data or the variances become zero or infinite
[Hartman and Keeler, 1991, Tresp *et al.*, 1993].

11.3 PLANT MODELLING

Often a model of the plant needs to be constructed, either to be used in fault detection and diagnosis algorithms or for indirect fuzzy control (see Section 11.4), and similar to the plant being modelled, the fuzzy system relates deterministic input signals (previous states and control actions) to deterministic outputs (current state). Despite the vagueness or uncertainty inherent in the fuzzy algorithm, it was argued in Chapter 10 that a fuzzy system is deterministic and its output depends on the inputs and how they are represented (their membership functions). A fuzzy system forms a nonlinear mapping, but the input signals should be representative and sufficiently rich, otherwise it is not able to store the data and learn the unknown function.

11.3.1 Fuzzy Model Representation

Many fuzzy algorithms are expressed in the continuous time domain, using inputs such as *the rate of change of error*, although most fuzzy modelling and control systems are actually implemented as sampled systems. Once a discrete or a continuous representation has been decided on, the order of the plant (the delays or differentials) must be determined. Over-estimating their values results in poor convergence rates and generalisation as the model is over-parameterised, although choosing too small a value means that there exists unmodelled dynamics which may affect the stability of any adaptive control system. It has been claimed that the adaptive fuzzy systems can be used when the order of the plant is underestimated, and although this can occur for certain controllers, fuzzy mappings are simply nonlinear functions. If the information which guarantees stability is not available in the input vector, the control loop will become unstable.

11.3.2 Recurrent System

For some plants, the output measurements which form both the training signals and the inputs for future predictions may be corrupted by noise. Additive noise can be filtered out and the true underlying function approximated if some information about the structure of the disturbance is available and an appropriate learning algorithm is used (see Sections 5.3.3 and 5.6.4). However, when the disturbance forms part of the input signal, a previously correct network may be adjusted due to output errors caused by the noisy input. This can lead to poor learning convergence rates and biased solutions.

One solution to this problem is to use a *recurrent* network. Recurrent networks have connections between their output and input nodes, using the estimated value of the plant as an input for subsequent predictions. This is based on the assumption that the modelling error is smaller than the measurement noise, so the network's output is a better estimate of the true plant output. These modelling

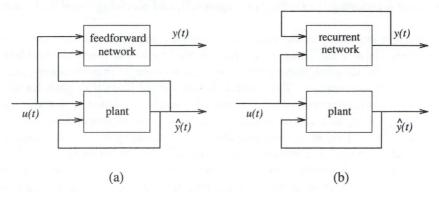

(a)   (b)

Figure 11.2 The difference between a feedforward network (a) and a recurrent system (b).

and control architectures are more advanced than most of those considered in this book, although their training rules are more complex and the stability/convergence analysis of the overall system is extremely difficult.

Recurrent neural networks were first proposed by Robinson and Fallside [1988], and they drew parallels between standard ARMA models and these dynamic systems. A gradient descent learning rule was proposed for training a series of static networks (or a recurrent network whose weights are held constant for a certain time period) and this was subsequently used by Nguyen and Widrow [1990], when they developed a method for training a network to reverse park a truck (see Section 1.4.3). On-line recursive training rules can be developed which iteratively accumulate information [Williams and Zipser, 1989], and although the computational cost of implementing these algorithms is high [Hush and Horne, 1993], their advantages can sometimes outweigh these considerations [Connor *et al.*, 1992].

11.3.3 Fuzzy On-Line Learning

Assuming that the fuzzy model is trying to learn a feedforward mapping rather than a recurrent relationship, the problems associated with on-line learning can be easily identified. Chapter 5 derived and described the LMS instantaneous training rules in considerable detail, and this section discusses those points that are applicable to on-line learning. At the beginning of this chapter, it was argued that while it is computationally cheaper and easier to train the fuzzy system's weights rather than adapting the rule confidences directly, the outputs of both systems are equivalent and it is only the internal representation that differs.

To accurately identify a plant on-line, the model's structure should be as simple as possible, to reduce the computational cost of the algorithm and improve the condition of the learning procedure. Generally, parameters converge faster in a parsimonious learning system and this improves the model's generalisation abilities. However, the two main problems which can occur in on-line modelling are generating a persistently exciting training signal, and receiving correlated training data.

Persistent excitation of the plant is a necessary condition for the fuzzy model to correctly identify the plant, and although this topic has received considerable attention in the adaptive linear control field, new concepts must be developed for adaptive *nonlinear* control. The control signals should be rich in both frequency and magnitude content in order to excite all the plant's states across the whole of the state space, and this is typically achieved using superimposed sinusoidal signals of different frequencies and amplitudes. For linear systems, it is sufficient to excite all of the states, as there are no (theoretical) restrictions on their amplitude. Nonlinear system's however, cannot make assumptions about the *global* behaviour of the plant, and information must be received from all possible operating points in order to correctly identify the plant, although using well-conditioned networks with minimal learning interference can partially relax this restrictive condition.

As described in Section 5.3.5, when the training data is collected on-line, the input vectors are often highly correlated which can slow down the rate of parameter convergence. The training signals may be persistently exciting, but the instantaneous LMS learning algorithms are slow to correctly identify the plant. This problem can be reduced and overcome by introducing orthogonality into the training data. Higher order learning algorithms can be used to search the weight space in different directions, thus minimising the effect of correlated training data (see Section 5.6).

11.4   INDIRECT FUZZY CONTROL

Indirect learning fuzzy controllers have only recently been developed [Harris *et al.*, 1993, Moore, 1991, Wang and Vachtsevanos, 1992, Wang, 1994], although their direct fuzzy counterparts have a long history in the control field. This type of fuzzy controller can be distinguished by the fact that a separate adaptive fuzzy model is constructed of the process, then a design procedure is used to calculate the control signal. This separation of the learning procedure from the control calculation makes it possible to analyse the performance of this type of system, thus yielding convergence and stability theories. The training signal for the adaptive fuzzy model is *directly* available, unlike the direct adaptive fuzzy controller which has to try and infer the control error which caused the plant output error.

11.4.1   Architecture

The basic system architecture of an indirect fuzzy adaptive controller is extremely simple, as shown in Figure 11.3. A fuzzy system is used to model the unknown (or

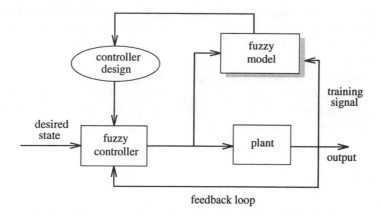

Figure 11.3 An indirect fuzzy adaptive control system.

partially known) plant, learning to predict the current plant state from an input vector that contains information about its past states and control signals.

The type of controller and its associated design process determines what information is passed to the controller. It may be a desired next state, in which case a reference model or desired trajectory must also be included in the architecture, or it could be a step input and the controller's structure must contain information about the desired response of the system. In either case, knowledge about the system's desired dynamic response must be incorporated within the overall architecture. The direct adaptive fuzzy controllers also assume this type of information is available when the performance index (see Section 11.5.3) is specified. It specifies the tolerated dynamic response of the plant, as well as acting as a critic in order to adapt the rules. Therefore both types of adaptive fuzzy controllers incorporate a desired or reference trajectory.

11.4.2 Controller Design

Given a fuzzy system which has learnt to associate past states and control signals with the current state, and a desired current state generated using a reference model, the control design procedure aims to produce a controller that can calculate the control signal necessary for the plant to achieve the desired dynamic response. The fuzzy plant model performs the following map:

$$(\text{past states} \times \text{control signals}) \rightarrow \text{current state}$$

whereas the controller aims to implement the following relationship:

$$(\text{past states} \times \text{desired current state}) \rightarrow \text{control signal}$$

Hence, the controller attempts to perform an inverse plant mapping achieving a unity transfer function across the controller/plant system. This fuzzy plant inversion process can take many forms; it may use the fuzzy rules directly, or it can simply assume that the fuzzy model implements an unknown "black box" function and may only use the input/output data. Both types of systems are described below.

The second type of controller design does not rely on an explicit reference model, rather it is implicitly encoded in the design process. The fuzzy aspects of the plant model are not used in the control design process, and this allows *standard* stability and convergence results to be applied.

Fuzzy Rule Base Inversion

When a fuzzy system is used to model the plant and the desired next state is known, there are two common methods for inverting the stored plant mapping. The first accesses the fuzzy rule base with "inputs" consisting of the desired state and past states (this is possible because the inputs and outputs are stored in an analogous manner, as described in Section 10.4.2), whereas the second method uses the rule base as a one step-ahead predictor, and calculates the control signal which realises the desired state. Both techniques assume that the control signal/predicted state relationship is *monotonic*.

Inverting the fuzzy rule base by treating the desired output as an input, and defuzzifying the resulting fuzzy membership function defined on the control signal, uses the rule confidence/relational matrix fuzzy representation rather than storing a weight vector. This is possibly the only reason why it may be preferable to use this knowledge representation in an on-line adaptive fuzzy model, although the rule base should be sparse to reduce the computational cost which results in complex fuzzy learning rules. Also, this technique performs poorly during the initial stages of learning as the training data may not have accessed the relevant part of the input/output space. Another problem which can arise is solely due to the fuzzy inversion process. The normalised rule confidence vector property, which holds for the forward chaining process, is no longer true for backward chaining systems. Hence, the form of the (inverse) calculated control surface is influenced by the size of the rule confidence vectors, which in turn depends on both the input and output membership functions.

Using a fuzzy plant model as a predictor means that it can be inverted as the *form* of the output surface is known (see Section 10.6.1). When triangular fuzzy input membership functions are used, the fuzzy output is a piecewise linear function which can be evaluated by finding the predicted state at the knots defined on the control variable and fitting linear mappings in between. The control signal necessary to realise a particular plant state can then be simply calculated. This approach generalises to higher order fuzzy systems that use algebraic operators, and a centre of area defuzzification algorithm. For more general fuzzy systems whose interpolation properties are unknown, conventional relationships (such as a globally linear function [Harris *et al.*, 1993]) can be fitted to the fuzzy input/output

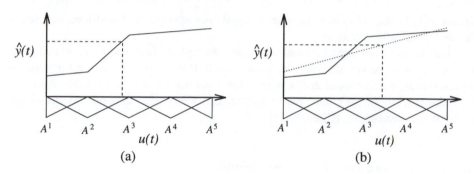

Figure 11.4   Fuzzy model inversion by prediction (a) and by approximating the network output with a known function (b).

data in some optimal sense (for instance, using the mean squared error), which can then be inverted. These techniques partially overcome the problem of poor initial control due to an incomplete rule base, as they can interpolate the stored knowledge globally (in Figure 11.4, the fuzzy rules associated with the input set A^4 are assumed to be unknown), although the computational cost is generally greater as the fuzzy model has to be accessed for a range of different control signals.

There is no reason why fuzzy models should only be used for one step-ahead prediction and several researchers have developed multi-step-ahead optimisation schemes [Montague *et al.*, 1991, Saint-Donat *et al.*, 1994]. This generally improves the performance of the system, although the computational burden can be significant.

Stabilising Learning Control Systems

Using input, output linearisation techniques [Chen, 1994], many nonlinear, single input, single output, minimum phase plants can be expressed as:

$$
\begin{aligned}
\widehat{y}(t) \;=\; & \widehat{f}(\widehat{y}(t-1),\ldots,\widehat{y}(t-n),u(t-1),\ldots,u(t-m)) + \\
& \widehat{g}(\widehat{y}(t-1),\ldots,\widehat{y}(t-n),u(t-1),\ldots,u(t-m))\,u(t) \qquad (11.5)
\end{aligned}
$$

where $\widehat{f}(.)$ and $\widehat{g}(.)$ are unknown nonlinear functions which are learnt by two fuzzy systems $f(.)$ and $g(.)$ (this structure could be identified using the ASMOD and ABBMOD model building algorithms described in Section 8.5). This type of structure has also been studied by [Sanner and Slotine, 1992, Tzirkel-Hancock and Fallside, 1991, Wang, 1994]. When certain assumptions are made about $\widehat{f}(.)$ and $\widehat{g}(.)$ (smoothness and $\widehat{g}(.)$ is bounded away from zero), the fuzzy model is trained using an NLMS instantaneous learning rule and the control signal is calculated by:

$$
u(t) = \frac{r(t) - f(t)}{g(t)}
$$

where $r(t)$ is the reference trajectory signal, the system is closed-loop stable and it asymptotically tracks $r(.)$.

The model tracking proof relies on the assumption that the fuzzy systems can exactly reproduce the desired functions. While this is too restrictive for practical applications, output dead-zones within the learning rules ensure stable operation and the tracking error converges to a small ball whose size depends on the magnitude of the dead-zone, [Chen, 1991].

D Step Ahead and Pole Assignment Control

In Wang *et al.* [1994], a fuzzy system is used to model and control specific plants whose *linear* parameters are unknown nonlinear functions of measurable operating points. Each linear parameter is modelled by a fuzzy subnetwork, as shown in Figure 11.5, and the motivation for using this type of architecture is that the space of

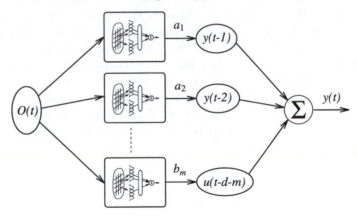

Figure 11.5   A fuzzy system where each subnetwork is used to predict a linear gain of the plant.

measurable operating points is generally considerably smaller than a conventional NARMAX input space. This approach is an automatic method for gain scheduling, which embodies the heuristic proposed in Chapter 4, that the simplest acceptable network performs the best. The operating points can either be independent on the system's states or not, and this characterises the system being considered. For the former case, the system's physical constraints determine the size of the bounded input space on which the fuzzy sets are defined, although it is difficult to determine the boundedness of the input when they include system states. In this situation, it is often necessary to have a *fixed stabilising* controller (sliding mode, for instance) that is activated whenever the operating points lie outside the fuzzy system's compact input domain.

Given that the fuzzy subnetworks can model the nonlinear gains sufficiently accurately, the closed-loop stability of the overall system can be assured when a normalised instantaneous learning rule is used in conjunction with an output dead-zone and a pole placement controller design is used. The use of a conventional control algorithm and a simplified fuzzy system design reduces the computational burden (both training and recall), and means that the performance of the closed-loop system can be analysed using both conventional techniques and fuzzy phase plane analysis tools [Moore and Harris, 1992].

How Fuzzy are These Controllers?

It is worthwhile noting that the fuzzy representation is only explicitly used in one of these control design algorithms, indeed the majority of the control design techniques summarised in this section have also been developed in the neurocontrol field. The reasons it is appropriate to use a fuzzy system as a plant model and why it may be preferable over a conventional neural network (such as a multi-layer perceptron) rely on the analysis described in Chapter 10. It was shown that the output of a fuzzy system is *linearly* dependent on the set of adjustable parameters (either weights or rule confidences), hence a large body of theory can be used to analyse the robustness and stability of these control loops [Sastry and Bodson, 1989]. Also, the compact support of the fuzzy input membership functions (basis functions) ensures real-time operation coupled with fast parameter convergence and minimal learning interference or gradient noise. The linguistic interpretation of the fuzzy system as a set of fuzzy rules does not influence the operation of the learning system; its primary role is for explaining the stored knowledge in a transparent manner.

11.4.3 Example: Ship Heading Control

To illustrate some of the desirable features of an indirect adaptive fuzzy controller, it is now applied to a ship heading control problem which is a standard International Federation of Automatic Control (IFAC) benchmark [Davison, 1990]. These benchmark problems allow meaningful comparisons to be made between various control design algorithms and are useful for developing new techniques, as they include many of the problems and features associated with real-world systems.

The examples which are used to illustrate the potential of both the indirect and the direct adaptive fuzzy controllers have been implemented using a discrete fuzzy system, and while this may seem contrary to the advice given in this chapter and Chapter 10, it should be remembered that the different representations are learning equivalent. The input, output behaviours of the adaptive fuzzy systems are the same and it is only the internal representations which differ. Generally, the discrete representation is computationally more expensive to implement, but its performance is the same as a continuous fuzzy system.

Problem Statement

The controller's objective is to regulate the ship's heading such that it follows a predefined course whilst satisfying a set of performance constraints. A linearised, continuous time model of the ship moving with a constant velocity can be expressed as:

$$\dot{\mathbf{x}} = \mathbf{Ax} + \mathbf{b}u \qquad\qquad (11.6)$$
$$y = \mathbf{cx}$$

where u is the rudder angle (rad.), $\mathbf{x}$ is a three-dimensional state vector composed of the ship's lateral velocity x_1, (ms^{-1}), the turning yaw rate x_2, (rad. s^{-1}), the ship's heading angle x_3, (rad.) and y is also the ship's heading angle.

The structure of $\mathbf{A}$, $\mathbf{b}$ and $\mathbf{c}$ are given by:

$$\mathbf{A} = \begin{bmatrix} -0.895 & -0.286 & 0 \\ -4.367 & -0.918 & 0 \\ 0 & 1 & 0 \end{bmatrix}, \qquad \mathbf{b} = \begin{bmatrix} 0.108 \\ -0.918 \\ 0 \end{bmatrix}, \qquad \mathbf{c} = \begin{bmatrix} 0 & 0 & 1 \end{bmatrix}$$

for a ship travelling at 15.2ms^{-1} and whose length is 160 m (the units have been normalised against the ship's length and the time taken to travel its own length).

The controller must regulate the ship's heading angle to some desired value, in response to a series of step demands, subject to the constraints that negligible overshoot occurs and the rudder's motion u satisfies the following bounds:

$$|u| < 0.7\text{rad.}$$
$$|\dot{u}| < 0.175\text{rad. s}^{-1}$$

To provide a meaningful comparison with the indirect adaptive fuzzy controller, a three-term PID controller was also designed which satisfies the system's performance constraints and produces a satisfactory dynamic response. The PID controller has the following form:

$$\dot{u} = -7.42e + 53.18\dot{e} + 104.3\ddot{e}$$

where the parameters have been selected such that the pair of complex, conjugate closed-loop poles dominates the dynamical response of the plant and the real pole at -2 has little effect. This PID controller is able to regulate the ship's heading when the desired course is expressed as a series of step demands, as is shown in Figure 11.6.

Adaptive Fuzzy Controller

A second-order model was formed of the ship's dynamic response by using a discrete fuzzy system with 21 quantisation levels on each variable. The fuzzy model was

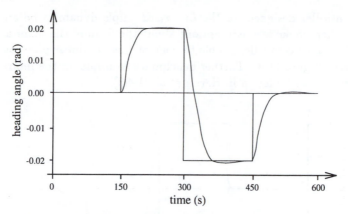

Figure 11.6 A PID controller regulating the ship's heading.

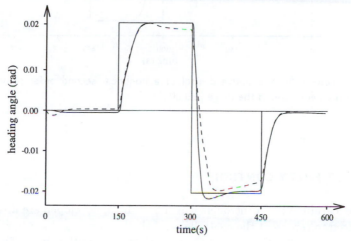

Figure 11.7   An indirect adaptive fuzzy controller with no *a priori* knowledge. The solid line denotes the first training sequence and the dashed line represents the second.

then inverted by prediction and the desired response was obtained from a static fuzzy rule base which acts as a reference model [Harris *et al.*, 1993].

Figure 11.7 shows the closed-loop response of the plant when it is controlled via an indirect adaptive fuzzy system. The fuzzy controller learns quickly and even on the second run, the plant's dynamic response is similar to that controlled using a PID scheme designed using substantial *a priori* knowledge.

This indirect adaptive fuzzy controller operates in a robust fashion (when compared with the PID controller), as the plant's response is noticeably less sensitive to additive measurement noise, variations in its parameters and rejecting external disturbances. In fact, when the following nonlinearity was introduced into the state equations:

$$\dot{x}_1 = -0.895x_1 - 0.286x_2 + 0.7x_1 \left| x_1 \right| + 0.108u$$

the PID controller designed for the *linearised* ship's dynamics became unstable. Fuzzy controllers make few assumptions about the form of the plant and the indirect adaptive fuzzy controller is able to control this nonlinear plant as well as the linear one (see Figure 11.8). Further benchmark examples of the indirect adaptive fuzzy controller can be found in Harris *et al.* [1993].

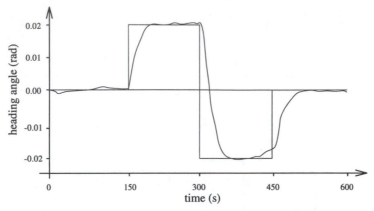

Figure 11.8   Indirect fuzzy adaptive control of a nonlinear second order plant subject to additive measurement noise in the range [±0.003] rad..

11.5 DIRECT FUZZY CONTROL

Most of the learning fuzzy controllers which have been developed are *direct*; a fuzzy system is used to represent the controller and the learning mechanism recommends changes to the control actions based on the performance of the plant. This architecture was popularised by Mamdani and his co-workers in the late seventies and since then, it has been used to control a variety of plants, such as autonomous vehicles [Fairbrother *et al.*, 1991, Harris *et al.*, 1993, Sutton and Jess, 1991], muscle relaxant anaesthesia [Linkens and Hasnain, 1990], and a variety of real and simulated processes [Linkens and Abbod, 1991, Procyk and Mamdani, 1979, Wang, 1994].

11.5.1   Architecture

The basic architecture of a direct fuzzy controller is shown in Figure 11.9. It consists of a normal fuzzy controller which operates on proportional and derivative error input signals, and produces an output which is either the control action or the calculated change in control. Thus it functions as either a nonlinear PD (proportional + derivative) or a PI (proportional + integral) controller. The learning layer which is built around the fuzzy controller acts as a critic, assessing the cur-

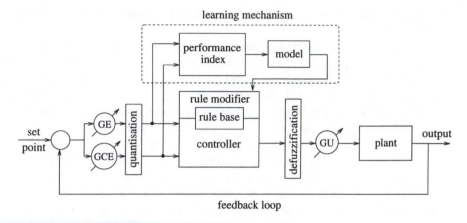

Figure 11.9 A direct, self-organising fuzzy control architecture.

rent state of the plant and recommending changes to the control signal via the performance index and the model.

The direct fuzzy controller operates by measuring the current deviation of the plant from the desired set point, producing an error, and the change in error (derivative) is given by the difference between two successive errors. Other inputs, such as an integral term, can be included as inputs to the rule base, although it is sufficient for this description just to consider the two variables. Both inputs are multiplied by scaling factors (GE and GCE in Figure 11.9) and these can be adapted to alter the gain of the system, as discussed in Section 11.2.5. The input membership functions for direct fuzzy controllers have traditionally been defined on a *discrete* input space, hence a quantisation layer is shown in the above figure which maps each scaled input to an integer that typically lies in the interval $[-5, 5]$. Fuzzy input membership functions are then defined on this space. A control action is calculated using the standard fuzzy reasoning algorithms described in Chapter 10, and after being scaled by the factor GU, the control is applied to the plant. The discrete inputs are also passed to the performance index which assesses the current state of the plant (by means of a look-up table), and any old rules are modified when the plant not operating as required. This process is repeated for a range of different step inputs until the designer is satisfied that the plant is operating satisfactorily.

11.5.2 Control Rule Base

The majority of direct fuzzy controllers are error based; their inputs are functions of the error between a desired set point and the current output. They use the current error, change in error and sometimes the change in change in error as inputs and produce a recommended change in the control signal as output. Thus an error based fuzzy system acts as a nonlinear PI or a PID (proportional + integral + derivative) controller.

For a single input, single output, continuous fuzzy PID controller, the size of the rule confidence matrix is (343×7) when seven membership functions are defined on each axis, and a greater number of relational elements would be required to store the same information in a discrete fuzzy system. Therefore, the memory requirements effectively limits the application of these techniques to low-order, single input, single output plants.

The above error-based structure presupposes that the dynamic response of the plant does *not* depend on the current operating point, which is *not* generally true for nonlinear systems. When the plant's response is a function of the operating point, the controller's rule base also depends on the operating point and convergence about one state does not imply that the rules can be successfully used about another. In addition, if the control actions have been successfully learnt around a particular operating point, and the set point moves away from and then returns to the original location, the first set of control rules will have been overwritten (forgotten) and the controller must relearn them. When a fuzzy PID controller outperforms a linear one, it must be because the desired control surface (about a particular operating point) is *significantly* nonlinear and the PID gains should depend on the current operating state. One way to achieve this is to make the definition of the membership functions depend on the operating point (using scaling factors for instance, see Section 11.2.5). However, when their definitions remain fixed and the inputs to the rule base are error based, the controller must continually adapt as the operating point changes over the process envelope.

11.5.3 Learning Mechanism

A common problem in all direct learning controllers is the derivation of learning rules which can relate errors in the plant's output to errors in the calculated control signal. Many different schemes have been proposed which assume various amounts of *a priori* knowledge about the plant's structure. The indirect adaptive fuzzy controllers assume that a coarse plant model is available in the form of an inverse Jacobian, and that the desired response can be implicitly encoded in the performance index.

Performance Index

The performance index is used to assess the current plant state, storing expert knowledge about the designer's performance requirements. Typically, it maps the current error and change in error (see Figure 11.9) to a scalar value which is an assessment of the recommended change in control action necessary to achieve an acceptable performance. Thus the performance index incorporates *two* features: a desired behaviour in terms of the error and change in error (a phase plane portrait), and the assessment of the eligibility of a particular state. The performance index can be expressed as a static set of fuzzy rules, and because the indirect fuzzy controllers are generally discrete, the performance index is usually stored in a look-up table, as shown in Table 11.1.

Error	Change in error										
	-5	-4	-3	-2	-1	0	1	2	3	4	5
-5	-9.0	-8.0	-7.0	-6.0	-5.0	-4.0	-3.0	-1.5	-0.5	0.0	0.0
-4	-8.0	-7.0	-6.0	-5.0	-4.0	-3.0	-2.0	-0.8	0.0	0.0	0.0
-3	-7.0	-6.0	-5.0	-4.0	-3.0	-2.0	-1.0	0.0	0.0	0.0	0.5
-2	-6.0	-5.0	-4.0	-3.0	-2.0	-1.0	-0.2	0.0	0.0	0.8	1.5
-1	-5.0	-4.0	3.0	2.0	1.0	-0.5	0.0	0.2	1.0	2.0	3.0
0	-4.0	-3.0	-2.0	-1.0	-0.5	0.0	0.5	1.0	2.0	3.0	4.0
1	-3.0	-2.0	-1.0	-0.2	0.0	0.5	1.0	2.0	3.0	4.0	5.0
2	-1.5	-0.8	0.0	0.0	0.2	1.0	2.0	3.0	4.0	5.0	6.0
3	-0.5	0.0	0.0	0.0	1.0	2.0	3.0	4.0	5.0	6.0	7.0
4	0.0	0.0	0.0	0.8	2.0	3.0	4.0	5.0	6.0	7.0	8.0
5	0.0	0.0	0.5	1.5	3.0	4.0	5.0	6.0	7.0	8.0	9.0

Table 11.1 A discrete fuzzy performance index.

As the system's state evolves along the main diagonal, it is deemed to be operating acceptably. Otherwise, the current error and change in error lie outside this band and the system's performance is undesirable, and the rules which caused the plant to move to this operating point must be modified. The element in the performance index which corresponds to the current state (or the output of the performance index rule base for a continuous fuzzy system) gives an indication of the amount that the control should be modified by, although knowledge of the system in the form of a gain or a model also needs to be incorporated into the learning procedure. The form of the performance index shown in Table 11.1 is analogous to a sliding mode controller as the fuzzy rules are modified until the plant's trajectory lies along the main diagonal of a phase plane portrait. The shape of the acceptable band (zero output region) changes the dynamic response of the system and altering the non-zero values in the performance index changes the rule modification algorithm and the rate of parameter convergence.

There is no reason why the performance index has to be a (fuzzy) look-up table, indeed more sophisticated optimisation procedures for process evaluation maybe constructed by utilising the more usual cost functionals. For instance the performance index could be a multi-objective function including control effort, output error and state parameters. Equally, it can be constructed to relate system performance more directly to rule change than the heuristic approach described below.

A Priori System Knowledge

Most adaptive control algorithms assume some structure about the plant that is to be controlled, whether it is the order, or time delay of the system, or the sign of the control gain. Two main facts about a system must be known before an adaptive fuzzy controller can be applied to a plant:

- the **time delay**, to determine which past control actions caused the present undesirable performance; and
- the **system's gain** for single input, single output plants or an estimate of the plant's Jacobian for multi-input, multi-output processes.

This is used to relate the performance measures to the recommended changes in the control signals. Usually an approximate diagonal matrix (a scalar for single input, single output systems) is sufficient, as the controller *iteratively* improves its performance over several training runs.

Control Rule Modification

Having calculated the recommended changes in the control signals, it is relatively simple to update the rule confidence matrix (or relational matrix or weight vector) using one of the instantaneous learning algorithms described in Chapter 5 or Section 11.2. These algorithms update the set of adjustable parameters (fuzzy rule base) using instantaneous information about the output error (control error in this application), although the learning mechanism must take into account the time delay described above.

This procedure for training the fuzzy rule base has certain deficiencies, although the basic learning algorithm can be simply modified. The first improvement was to use a set of fuzzy *meta-rules* or *over-rules* which supervises the learning procedure. These rules represent such facts as the system is symmetric, or the output corresponding to a zero error and a change in error is also zero. They oversee the learning process and improved performance is generally obtained [Sugiyama, 1988].

The second way in which increased convergence rates, and hence improved system performance, can be obtained is to notice that the performance index is essentially *position* based, therefore it is unable to deal with disturbances such as set point changes [Zhang and Edmunds, 1992]. For instance, consider when the control rules are correct (zero error and change in error), but a set point change causes the error to either increase or decrease. Thus, a previously correct rule base is modified as the system moves outside the acceptable zone of behaviour and the performance decreases. This work also proposes to make the rule modification algorithm depend on the *direction of movement of the state* rather than solely on the values of the errors.

11.5.4   Example: Ship's Yaw Control

To complement the indirect adaptive fuzzy controller developed in Section 11.4.3, a direct adaptive fuzzy controller is now applied to the problem of ship yaw control. This problem has been extensively investigated by Sutton and his colleagues for a number of years [Fairbrother *et al.*, 1991, Sutton and Towill, 1985, Sutton and Jess, 1991], and the results which are described in this section are taken from the 1991 paper.

The ship's yaw dynamics are simulated using the following nonlinear, second-order model:

$$\ddot{\psi} + 0.1\dot{\psi} + 0.024\left(\dot{\psi}\right)^3 = 0.012\left(\delta_a + S_d\right) \qquad (11.7)$$

where ψ the yaw angle, δ_a is the rudder angle, S_d represents the yaw disturbance due to the sea's state. Actuator constraints are incorporated by rate limiting the rudder at $6°s^{-1}$ and the rudder's value was not allowed to exceed $35°$ in either direction.

Direct Adaptive Fuzzy Controller

The adaptive fuzzy controller was a two-term, error-based system, which measured the current error and change in error and outputs a signal, which represented the required change in control. However, it was found that a control system using just this information was unable to meet the performance requirements, so the two term fuzzy controller was augmented by a third term so that the change in control signal was calculated by:

$$\Delta u = f(e, \Delta u) + \text{GCCE}\Delta^2 u \qquad (11.8)$$

where $f(.)$ represents the fuzzy mapping and GCCE is a *linear* scaling factor. This extra term adds a derivative action into the control system.

The fuzzy algorithm was implemented using a discrete relational matrix with 13 discrete values (and fuzzy membership functions) defined on each variable. A piecewise linear output was obtained by linear interpolation between the neighbouring inputs and a centre of gravity defuzzification method was used to infer the change in the control signal.

The choice of suitable scaling factors forms an important part of any error-based direct adaptive fuzzy control scheme, and a lot of effort was spent on producing an algorithm which was capable of tuning these parameters. For this particular controller, there are four scaling factors that can be modified, GE, GCE, GCCE and GO (see Figure 11.9 and Equation 11.8) and also the delay in reward training parameter; and after extensive simulations, the guidelines shown in Table 11.2 were produced.

Parameter	Initial selection	Tuning guideline
GE	Map maximum measurable value to maximum quantisation level.	Increasing GE: overshoot increases.
GCE	Map maximum measurable value to lower end of maximum quantisation level.	Increasing GCE: rise time increases marginally, overshoot decreases, steady-state error increases.
GCCE	Map maximum measurable value to lower end of maximum quantisation level.	Increasing GCCE: rise time increases, overshoot increases. Best technique: reduce GCCE until convergence just occurs.
GO	Set GO = 1.0	None established
delay	Set delay = 2	2 < delay < time constant

Table 11.2 Tuning guidelines for an error based direct adaptive fuzzy controller.

Results

Figure 11.10 shows a series of yaw responses at 16 knots as the fuzzy controller adapts from an initially empty rule base. The input and output scaling factors and the delay in the reward signal were chosen using the guidelines contained in Table 11.2. To be effective, the fuzzy controller's performance should be robust

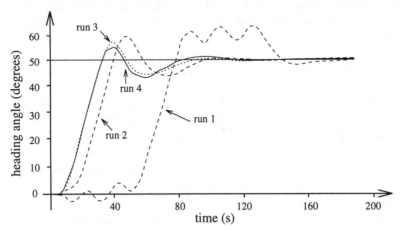

Figure 11.10   A ship's yaw responses for a series of step demands using a direct adaptive fuzzy controller.

with respect to variations in the ship's parameters, course and speed, as well as any external disturbances due to the sea's state. It was found to be necessary to schedule the time delay against the ship's speed (as the yaw's dynamics time constant changes), and the GCCE gain against the error's magnitude. This is because an error-based controller is being used at a variety of operating points for a nonlinear plant, hence it is either necessary to schedule the scaling factors or to develop state based adaptive fuzzy systems. With these modifications, the fuzzy controller produced the step responses shown in Figure 11.11, and it has very

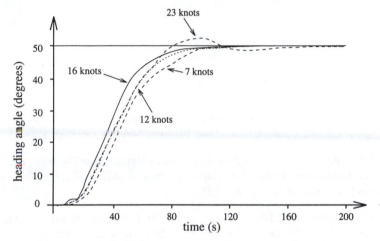

Figure 11.11 A ship's yaw responses for a range of speeds.

similar dynamical responses for all its normal operational conditions.

These observations about robustness and fast adaptation were independently confirmed by Layne and Passion [1993], where a direct adaptive fuzzy controller was compared with a gradient-based and a Lyapunov-based model reference adaptive (linear) controller for a similar ship's yaw control problem. The fuzzy controller learnt quicker than the other two, its disturbance rejection properties were better and it used an order of magnitude less control energy! These improvements in performance are not untypical when an adaptive fuzzy controller is benchmarked against a conventional control design technique.

References

[AGARD, 1991] AGARD Lecture Series 179. 1991. *Artificial Neural Network Approaches in Guidance and Control*, Specialised Printing Services Limited, Essex.

[Albus, 1975a] Albus J.S. 1975. *A New Approach to Manipulator Control: The Cerebellar Model Articulation Controller (CMAC)*, Trans. ASME. Jnl. Dyn. Sys. Meas. and Control, Vol. 63, No. 3, pp. 220-227.

[Albus, 1975b] Albus J.S. 1975. *Data Storage in the Cerebellar Model Articulation Controller (CMAC)*, Trans. ASME. Jnl. Dyn. Sys. Meas. and Control, Vol. 63, No. 3, pp. 228-233.

[Albus, 1979a] Albus J.S. 1979. *Mechanisms of Planning and Problem Solving in the Brain*, Mathematical Biosciences, Vol. 45, pp. 247-293.

[Albus, 1979b] Albus J.S. 1979. *A Model of the Brain for Robot Control, Part 2: A Neurological Model*, BYTE, July, pp. 54-95.

[Albus, 1979c] Albus J.S. 1979. *A Model of the Brain for Robot Control, Part 3: A Comparison of the Brain and Our Model*, BYTE, August, pp. 66-80.

[Albus, 1989] Albus J.S. 1989. *The Marr and Albus theories of the cerebellum. Two early models of associative memory*, Proc. COMPCON, 34th IEEE Comp. Soc. Int. Conf. Washington DC, pp. 577-582.

[Albus, 1990] Albus J.S. 1990. *Robotics: Where Has it Been? Where is it Going?*, Robotics and Autonomous Systems, Vol. 6, pp. 199-219.

[Albus, 1991] Albus J.S. 1991. *Outline of a Theory of Intelligence*, IEEE Trans. Sys. Man and Cyb., Vol. 21, No. 3, pp. 473-509.

[Albus, 1992] Albus J.S. 1992. *A Reference Model Architecture for Intelligent Systems Design*, in "An Introduction to Intelligent and Autonomous Control", Eds Antsaklis P.J., Passino K.M., Kluwer Academic Publishers, Boston, MA, Chapter 2.

[Albus *et al.*, 1990] Albus J.S., Quintero R., Lumia R., Herman M., Kilmer R., Goodwin K. 1990. *Concept for a Reference Model Architecture for Real-Time Intelligent Control Systems (ARTICS)*, NIST Technical Note 1277.

[An, 1991] An P.C. 1991. *An Improved Multi-Dimensional CMAC Neural Network: Receptive Field Function and Placement*, D. Phil Dissertation, University of New Hampshire.

[An et al., 1991] An P.C., Miller W.T., Parks P.C. 1991. *Design Improvements in Associative Memories for Cerebellar Model Articulation Controllers (CMAC)*, Proc. Int. Conf. on Artificial Neural Networks, Helsinki, North-Holland, Vol. 2, pp. 1207-1210.

[An et al., 1993a] An P.C., Brown M., Harris C.J., Lawrence A.J., Moore C.G. 1993. *Comparative Aspects of Associative Memory Networks for Modelling*, Proc. European Control Conference, Groningen, The Netherlands, Vol. 2, pp. 454-459.

[An et al., 1993b] An P.C., Harris C.J., Tribe R., Clarke N. 1993. *Aspects of Neural Networks in Intelligent Collision Avoidance Systems for Prometheus*, JFIT Conference, Keele University, March.

[An et al., 1993c] An P.C., Brown M., Harris C.J., Chen S. 1993. *Comparative Aspects of Neural Network Algorithms for On-line Modelling of Dynamic Processes*, IMechE, Proc, Instn. Mech. Engrs., Part I, J. Sys. Cont. Eng., Vol. 207, pp. 223-241.

[An et al., 1994a] An P.C., Brown M., Harris C.J., Lawrence A.J., Moore C.G. 1994. *Associative Memory Neural Networks: Adaptive Modelling Theory, Software Implementations and Graphical User Interface*, Engr. Applic. Artif. Intell., Vol. 7, No. 1, pp. 1-21.

[An et al., 1994b] An P.C., Brown M., Harris C.J. 1994. *A Glogal Gradient Noise Covariance Expression for Stationary Real Gaussian Inputs*, submitted for publication.

[Anderson, 1989] Anderson C.W. 1989. *Learning to Control an Inverted Pendulum Using Neural Networks*, IEEE Control Sys. Mag., April, pp. 31-37.

[Antsaklis and Passino, 1992] Antsaklis P.J., Passino K.M. (Eds) 1992. *An Introduction to Intelligent and Autonomous Control*, Kluwer Academic Publishers, Boston, MA.

[Åström and Hägglund, 1988] Åström K.J., Hägglund T. 1988. *Automatic Tuning of PID Controllers*, Instrument Society of America.

[Åström and Wittenmark, 1989] Åström K.J., Wittenmark B. 1989. *Adaptive Control*, Addison Wesley, Reading, MA.

[Atkeson, 1991] Atkeson C.G. 1991. *Using Modular Neural Networks with Local Representations to Control Dynamic Systems*, AI Lab, MIT, Cambridge, Tech. Report. AFOSR-TR-91-0452.

[Barnard and Wessels, 1992] Barnard E., Wessels L.F.A. 1992. *Extrapolation and Interpolation in Neural Network Classifiers*, IEEE Control Sys. Mag., October, pp. 50-53.

[Barto and Anandan, 1985] Barto A.G., Anandan P. 1985. *Pattern-Recognizing Stochastic Learning Automata*, IEEE Trans. Sys. Man and Cyb., Vol. 15, No. 3, pp. 360-375.

[Barto *et al.*, 1993] Barto A.G., Bradtke S.J., Singh S.P. 1993. *Learning to Act using Real-Time Dynamic Programming*, Dept. of Computer Science, University of Massachusetts, MA.

[Barto *et al.*, 1983] Barto A.G., Sutton R.S., Anderson C.H. 1983. *Neuronlike Adaptive Elements That Can Solve Difficult Learning Control Problems*, IEEE Trans. Sys. Man and Cyb. Vol. 13, No. 5, pp. 834-846.

[Beale and Jackson, 1990] Beale R., Jackson T. 1990. *Neural Computing: An Introduction*, Adam Hilger, Bristol.

[Bellman, 1961] Bellman R.E. 1961. *Adaptive Control Processes*, Princeton University Press, Princeton, NJ.

[Berenji, 1992] Berenji H.R. 1992. *Fuzzy Logic Controllers*, in "An Introduction to Fuzzy Logic Applications and Intelligent Systems", Eds Yager R.R., Zadeh L.A., Kluwer Academic Publisher, Boston, MA, Chapter 4.

[Berenji and Khedkar, 1992] Berenji H.R., Khedkar P. 1992. *Learning and Tuning Fuzzy Logic Controllers Through Reinforcements*, IEEE Trans. Neural Networks, Vol. 3, No. 5, pp. 724-740.

[Bernard, 1988] Bernard J.A. 1988. *Use of a Rule-Based System for Process Control*, IEEE Control Sys. Mag., October, pp. 3-13.

[Biegler-König and Bärmann, 1993] Biegler-König F., Bärmann F. 1993. *A Learning Algorithm for Multilayered Neural Networks Based on Linear Least Squares Problems*, Neural Networks, Vol. 6, No. 1, pp. 127-132.

[Billings and Chen, 1992] Billings S., Chen S. 1992. *Neural Networks and System Identification*, in "Neural Networks for Control and Systems", Eds Warwick K., Irwin G.W., Hunt K.J., IEE Control Engineering Series 46, Peter Peregrinus, Stevenage, Chapter 9, pp. 181-205.

[Billings and Voon, 1983] Billings S.A., Voon W.S.F. 1983. *Structure Detection and Model Validity Tests in the Identification of Nonlinear Systems*, Proc. IEE Part D, Vol. 130, pp. 193-199.

[Billings and Voon, 1986] Billings S.A., Voon W.S.F. 1986. *Correlation Based Model Validity Tests for Nonlinear Models*, Int. J. Control, Vol. 44, No. 1, pp. 235-244.

[Billings *et al.*, 1989] Billings S.A., Chen S., Backhouse R.J. 1989. *The Identification of Linear and Nonlinear Models of a Turbocharged Automotive Diesel Engine*, Math. Sys. and Sig. Proc., Vol. 3, No. 2, pp. 123-142.

[Billings *et al.*, 1992] Billings S.A., Jamaluddin H.B., Chen S. 1992. *Properties of Neural Networks With Application to Modelling Nonlinear Systems*, Int. J. Control, Vol. 55, No. 1, pp. 193-224.

[Bohlin, 1978] Bohlin T. 1978. *Maximum-power Validation of Models Without Higher-order Fitting*, Automatica, Vol. 4, pp. 321-355.

[Breiman *et al.*, 1984] Breiman L., Friedman J.H., Olshen R.A., Stone C.J. 1984. *Classification and Regression Trees*, Wadsworth, Belmont, CA.

[Brooks, 1986] Brooks R.A. 1986. *A Robust Layered Control System for a Mobile Robot*, IEEE J. Robot. and Auto., Vol. 2, No. 3, pp. 14-23.

[Brooks, 1990] Brooks R.A. 1990. *Elephants Don't Play Chess*, Robotics and Autonomous Systems, Vol. 6, pp. 3-15.

[Broomhead and Lowe, 1988] Broomhead D.S., Lowe D. 1988. *Multivariable Functional Interpolation and Adaptive Networks*, Complex Systems, Vol. 2, pp. 321-355.

[Brown and Harris, 1991] Brown M., Harris C.J. 1991. *A Nonlinear Adaptive Controller: A Comparison Between Fuzzy Logic Control and Neurocontrol*, IMA J. Math. Control and Info., Vol. 8, No. 3, pp. 239-265.

[Brown and Harris, 1993] Brown M., Harris C.J. 1993. *The B-spline Neurocontroller*, in "Parallel Processing in a Control Systems Environment", Eds Rogers E., Li Y., Prentice Hall, Hemel Hempstead, Chapter 5.

[Brown *et al.*, 1993] Brown M., Harris C.J., Parks P.C. 1993. *The Interpolation Capabilities of the Binary CMAC*, Neural Networks, Vol. 6, No. 3, pp. 429-440.

[Brown *et al.*, 1994] Brown M., Mills D.J., Harris C.J. 1994. *The Representation of Fuzzy Algorithms Used in Adaptive Modelling and Control Schemes*, submitted to Intelligent Systems Engineering.

[Bruske, 1993] Bruske J. 1993. *Neural Fuzzy Decision Systems*, Diploma Thesis, Universität Kaiserslautern, Germany.

[Buhmann and Powell, 1990] Buhmann M.D., Powell M.J.D. 1990. *Radial Basis Function Interpolation on an Infinite Grid*, "Algorithms for Approximation II", Eds Mason J.C., Cox M.G., Chapman and Hall, London, pp. 146-169.

[Burden and Faires, 1993] Burden R.L., Faires J.D. 1993. *Numerical Analysis*, 5th Edition, PWS-KENT, Boston, MA.

[Carpenter and Grossberg, 1988] Carpenter G.A., Grossberg S. 1988. *The ART of Adaptive Pattern Recognition by a Self-Organizing Neural Network*, IEEE Computer, March, pp. 77-88.

[Carpenter and Grossberg, 1990] Carpenter G.A., Grossberg S. 1990. *ART 3: Hierarchical Search Using Chemical Transmitters in Self-Organising Pattern Recognition Architectures*, Neural Networks, Vol. 3, No. 2, pp. 129-152.

[Carpenter *et al.*, 1991a] Carpenter G.A., Grossberg S., Rosen D.B. 1991. *ART 2-A: An Adaptive Resonance Algorithm for Rapid Category Learning and Recognition*, Neural Networks, Vol. 4, No. 4, pp. 493-504.

[Carpenter *et al.*, 1991b] Carpenter G.A., Grossberg S., Rosen D.B. 1991. *Fuzzy ART: Fast Stable Learning and Categorization of Analog Patterns by an Adaptive Resonance System*, Neural Networks, Vol. 4, No. 6, pp. 759-772.

[Chen, 1991] Chen F.C. 1991. *A Dead-Zone Approach in Nonlinear Adaptive Control Using Neural Networks*, Proc. 30th IEEE Conf. on Decision and Control, Brighton, Vol. 1, pp. 156-161.

[Chen, 1994] Chen F.C., Khalil H.K. 1994. *Adaptive Control of Nonlinears Systems Using Neural Networks*, in "Advances in Intelligent Control", Ed Harris C.J., Taylor and Francis, London, Chapter 7.

[Chen and Billings, 1994] Chen S., Billings S.A. 1994. *Neural Networks for Nonlinear Dynamic System Modelling and Identification*, in "Advances in Intelligent Control", Ed Harris C.J., Taylor and Francis, London, Chapter 4.

[Chen and Lin, 1993] Chen F.C., Lin M.H. 1993. *On the Learning and Convergence of Radial Basis Networks*, Proc. IEEE Int. Conf. Neural Networks, San Francisco, Vol. 2, pp. 983-988.

[Chen *et al.*, 1990] Chen S., Billings S.A., Cowan C.F.N., Grant P.M. 1990. *Practical Identification of NARMAX Models Using Radial Basis Functions*, Int. J. Control, Vol. 52, No. 6, pp. 1327-1350.

[Chen *et al.*, 1992] Chen S., Billings S.A., Grant P.M. 1992. *Recursive Hybrid Algorithm for Non-linear System Identification Using Radial Basis Function Networks*, Int. J. Control, Vol. 55, No. 5, pp. 1051-1070.

[Chiu *et al.*, 1991] Chiu S., Chand S., Moore D., Chaudhary A. 1991. *Fuzzy Logic for Control of Roll and Moment for a Flexible Wing Aircraft*, IEEE Control Sys. Mag., June, pp. 42-48.

[Connor *et al.*, 1992] Connor J., Atlas L.E., Martin D.R. 1992. *Recurrent Networks and NARMA Modelling*, in "Advances in Neural Information Processing Systems 4", Eds Moody J.E., Hanson S.J., Lippmann R.P., Morgan Kaufmann, San Mateo, CA, pp. 301-308.

[Corfield *et al.*, 1991] Corfield S.J., Fraser R.J.C., Harris C.J. 1991. *Architectures for Real-time Intelligent Control of Autonomous Vehicles*, IEE J. Comput. Control. Eng., Vol. 2, No. 6, pp. 254-262.

[Cotter, 1990] Cotter N.E. 1990. *The Stone-Weierstrass Theorem and Its Application to Neural Networks*, IEEE Trans. on Neural Networks, Vol. 1, No. 4, pp. 290-295.

[Cotter and Guillerm, 1992] Cotter N.E., Guillerm T.J. 1992. *The CMAC and a Theory of Kolmogorov*, Neural Networks, Vol. 5, No. 2, pp. 221-228.

[Cox, 1972] Cox M.G. 1972. *The Numerical Evaluation of B-splines*, J. Inst. Math. Appl., Vol. 10, pp. 134-149.

[Cox, 1977] Cox M.G. 1977. *The Incorporation of Boundary Conditions in Spline Approximation Problems*, National Physical Laboratory NAC Report No. 80, 19 pages.

[Cox, 1978] Cox M.G. 1978. *The Representation of Polynomials in Terms of B-splines*, Proc. 7th Manitoba Conf. on Num. Methods and Comp., Eds McCarthy D. and Williams H.C., Winnipeg, University of Manitoba, pp. 73-105.

[Cox, 1982] Cox M.G. 1982. *Practical Spline Approximation*, NPL Report DITC 1/82, 33 pages.

[Cox, 1990] Cox M.G. 1990. *Algorithms for Spline Curves and Surfaces*, NPL Report DITC 166/90, 36 pages.

[Cox, 1993] Cox E. 1993. *Adaptive Fuzzy Systems*, IEEE Spectrum, February, pp. 27-31.

[Cox and Wilfong, 1990] Cox I.J., Wilfong G.T. (Eds) 1990. *Autonomous Robot Vehicles*, Springer-Verlag, Berlin.

[DARPA, 1988] DARPA. 1988. *DARPA Neural Network Study*, AFCEA Int. Press.

[Dabis and Moir, 1991] Dabis H.S., Moir T.J. 1991. *Least Mean Square as a Control System*, Int. J. Control, Vol. 54, No. 2, pp. 321-335.

[Davison, 1990] Davison E.J. 1990. *Benchmark Examples for Control System Design*, IFAC, Laxenberg, Austria.

[Dayan, 1991] Dayan P. 1991. *Navigating Through Temporal Difference*, in "Advances in Neural Information Processing Systems 3", Eds Lippmann R.P., Moody J.E., Touretzky D.S., Morgan Kaufmann Publishers, San Mateo, CA, pp. 464-470.

[DeBoor, 1972] DeBoor C. 1972. *On Calculating with B-splines*, J. Approx. Theory, Vol. 6, pp. 50-62.

[DeBoor, 1976] DeBoor C. 1976. *On Local Linear Functionals Which Vanish at All B-splines But One*, in "Theory of Approximation with Applications", Eds Law A.G., Sahney B.N., Academic Press, New York, pp. 120-145.

[DeBoor, 1978] DeBoor C. 1978. *A Practical Guide to Splines*, Springer-Verlag, NY.

[DeBoor and DeVore, 1985] DeBoor C., DeVore R. 1985. *A Geometric Proof of Total Positivity for Spline Interpolation*, Math. Comput., Vol. 45, pp. 497-504.

[Dorst et al., 1991] Dorst L., Mandhyan I., Trovato K. 1991. *The Geometrical Representations of Path Planning Problems*, Robotics and Autonomous Systems, North-Holland, Vol. 7, pp. 181-195.

[Ellacott, 1994] Ellacott S.W. 1994. *Aspects of the Numerical Analysis of Neural Networks*, Acta Numerica, pp. 145-202.

[Ellison, 1988] Ellison D. 1988. *On the Convergence of the Albus Perceptron*, IMA J. of Math. Control and Info., Vol. 5, pp. 315-331.

[Ellison, 1991] Ellison D. 1991. *On the Convergence of the Multidimensional Albus Perceptron*, Int. J. of Robotics Research, Vol. 10, No. 4, pp. 338-357.

[El-Zorkany et al., 1985] El-Zorkany H.I., Liscano R., Tondu B. 1985. *A Sensor Based Approach for Robot Programming*, Proc. SPIE Conf. Intel. Robots and Comp. Vision, Vol. 579, pp. 289-297.

[Eng. App. of AI, 1991] 1991. Engineering Applications of Artificial Intelligence, Special Issue on Intelligent Autonomous Vehicles, Vol. 4, No. 4.

[Ersü and Tolle, 1988] Ersü E., Tolle H. 1988. *Hierarchical Learning Control - An Approach With Neuron-like Associative Memories*, in "Neural Information Processing Systems", Ed Anderson D.Z., American Institute of Physics, NY, pp. 249-261.

[Ersü and Wienand, 1986] Ersü E., Wienand S. 1986. *An Associative Memory Based Learning Control Scheme With PI-controller for SISO Processes*, IFAC Microcomputer Appl. Process Control Conf., Istanbul, Turkey, pp. 99-105.

[Fairbrother et al., 1991] Fairbrother H.N., Stacey B.A., Sutton R. 1991. *Fuzzy Self-organising Control of a ROV*, Proc. IEE Control 91, Vol. 1, pp. 499-504.

[Farrall and Jones, 1993] Farrall S.D., Jones R.P. 1993. *Energy Management in an Automotive Electric/Heat Engine Hybrid Powertrain Using Fuzzy Decision Making*, Proc. 8th IEEE Int. Symp. on Intelligent Control, Chicago, IL, pp. 463-468.

[Feldkamp and Puskorius, 1993] Feldkamp L.A., Puskorius G.V. 1993. *Trainable Fuzzy and Neural Fuzzy Systems for Idle Speed Control*, Proc. 2nd IEEE Fuzzy Systems Conf., San Francisco, CA, Vol. 1, pp. 45-51.

[Fraser et al., 1991] Fraser R.J.C., Harris C.J., Mathias L.W., Rayner N.J.W. 1991. *Implementing Task-Level Mission Management for Intelligent Autonomous Vehicles*, Engng. Applic. Artif. Intelli., Vol. 4, No. 4, pp. 257-268.

[Friedman, 1991a] Friedman J.H. 1991. *Multivariate Adaptive Regression Splines*, The Annals of Statistics, Vol. 19, No. 1, pp. 1-141.

[Friedman, 1991b] Friedman J.H. 1991. *Adaptive Spline Networks*, in "Advances in Neural Information Processing Systems 3", Eds. Lippmann R.P., Moody J.E., Touretzky D.S., Morgan Kaufmann, San Mateo, CA, pp. 675-683.

[Fletcher, 1987] Fletcher R. 1987. *Practical Methods for Optimization*, John Wiley & Sons, Chichester, 2nd edition.

[Fu, 1971] Fu K.S. 1971. *Learning Control Systems and Intelligent Control Systems: An Intersection of Artificial Intelligence and Automatic Control*, IEEE Trans. Auto. Control, Vol. 16, pp. 70-72.

[Fukunaga, 1990] Fukunaga K. 1990. *Introduction to Statistical Pattern Recognition*, 2nd Edition, Academic Press, London.

[Gardner, 1987] Gardner W.A. 1987. *Nonstationary Learning Characteristics of the LMS Algorithm*, IEEE Trans. on Circuits and Systems, Vol. 34, No. 10, pp. 1199-1207.

[Geman et al., 1992] Geman S., Bienenstock E., Doursat R. 1992. *Neural Networks and the Bias/Variance Dilemma*, Neural Computation, Vol. 4, No. 1, pp. 1-58.

[Gill et al., 1981] Gill P.E., Murray W., Wright M.H. 1981. *Practical Optimisation*, 2nd Edition, Academic Press Inc., London.

[Girosi and Poggio, 1990] Girosi F., Poggio T. 1990. *Networks and the Best Approximation Property*, Biol. Cybern., Vol. 63, pp. 169-176.

[Goles and Martinez, 1990] Goles E., Martinez S. 1990. *Neural and Automata Networks, Dynamical Behaviour and Applications*, Mathematics and Its Applications, Vol. 58, Kluwer Academic Publishers, Dordrecht.

[Goodwin and Sin, 1984] Goodwin G.C., Sin K.S. 1984. *Adaptive Filtering Prediction and Control*, Prentice Hall, Englewood Cliffs, NJ.

[Grossberg, 1988] Grossberg S. 1988. *Nonlinear Neural Networks: Principles, Mechanisms, and Architectures*, Neural Networks, Vol. 1, No. 1, pp. 17-61.

[Grossberg et al., 1993] Grossberg S., Guenther F., Bullock D., Greve D. 1993. *Neural Representations for Sensory-Motor Control II: Learning a Head-Centred Visuomotor Representation of 3-D Target Position*, Neural Networks, Vol. 6, No. 1, pp. 43-68.

[Guez and Selinsky, 1988] Guez A., Selinsky J. 1988. *A Trainable Neuromorphic Controller*, J. Robotic Systems, Vol. 5, No. 4, pp. 363-388.

[Guo et al., 1993] Guo P., Zi X.Z., Zheng J. 1993. *Polymer Extrusion Production Using Active Recognition and Adaptive Control*, Proc. 2nd IEEE Int. Conf. on Fuzzy Systems, San Francisco, CA, Vol. 2, pp. 779-788.

[Handelman et al., 1990] Handelman D.A., Lane S.H., Gelfand J.J. 1990. *Integrating Neural Networks and Knowledge-Based Systems for Intelligent Robotic Control*, IEEE Control Sys. Mag., April, pp. 77-87.

[Harris and Billings, 1981] Harris C.J., Billings S.A. (Eds) 1981. *Self-Tuning and Adaptive Control*, IEE Control Eng. Ser. 16, Peter Peregrinus, Stevanage.

[Harris, 1994] Harris C.J. (Ed) 1994. *Advances in Intelligent Control*, Taylor and Francis, London.

[Harris and Rayner, 1993] Harris C.J., Rayner N.J.W. 1993. *Fuzzy Planning for Task Level Mission Management of Autonomous Vehicles*, IEE Colloq. on "Two Decades of Fuzzy Control", London.

[Harris et al., 1993] Harris C.J., Moore C.G., Brown M. 1993. *Intelligent Control: Some Aspects of Fuzzy Logic and Neural Networks*, World Scientific Press, London & Singapore.

[Hartman and Keeler, 1991] Hartman E., Keeler J.D. 1991. *Predicting the Future: Advantages of Semilocal Units*, Neural Computation, Vol. 3, No. 4, pp. 566-578.

[Haykin, 1991] Haykin S. 1991. *Adaptive Filter Theory*, 2nd Edition, Prentice Hall, Englewood Cliffs, NJ.

[Hecht-Nielson, 1990] Hecht-Nielson R. 1990. *Neurocomputing*, Addison-Wesley, Reading, MA.

[Hertz *et al.*, 1991] Hertz J., Krogh A., Palmer R.G. 1991. *Introduction to the Theory of Neural Computation*, Addison-Wesley Publishing Company, Redwood City, CA.

[Holden, 1994] Holden S.B. 1994. *On the Theory of Generalisation and Self-Structuring in Linearly Weighted Connectionist Networks*, DPhil dissertation, Department of Engineering, Cambridge University.

[Holland, 1975] Holland J.M. 1975. *Adaptation in Natural and Artificial Systems*, University of Michigan Press, Ann Arbor, MI.

[Hollantz and Tresp, 1992] Hollantz J., Tresp V. 1992. *A Rule-Based Network Architecture*, Proc. Int. Conf. Artificial Neural Networks 2, Eds Aleksander I., Taylor J., Elsevier Science, North-Holland, Vol. 1, pp. 757-761.

[Hopfield, 1984] Hopfield J.J. 1984. *Neurons With Graded Responses Have Collective Computational Properties Like Those of Two-state Neurons*, Proc. Natl. Acad. Sci., May, Vol. 81, pp. 3088-3092.

[Hopfield and Tank, 1986] Hopfield J.J., Tank D.W. 1986. *Computing with Neural Circuits: A Model*, Science, Vol. 233, pp. 625-633.

[Horn and Johnson, 1985] Horn R.A., Johnson C.R. 1985. *Matrix Analysis*, Cambridge University Press.

[Hornik *et al.*, 1989] Hornik K., Stinchcombe M., White H. 1989. *Multilayer Feedforward Networks are Universal Approximators*, Neural Networks, Vol. 2, pp. 551-560.

[Hunt, 1992] Hunt K.J. 1992. *Induction of Decision Trees for Rule-based Modelling and Control*, Proc. IEEE Int. Symp. Intelligent Control, Glasgow, pp. 306-311.

[Hunt and Sbarbaro-Hofer, 1992] Hunt K.J., Sbarbaro-Hofer D. 1992. *Studies in Neural Network Based Control*, in "Neural Networks for Control and Systems", Eds Warwick K., Irwin G.W., Hunt K.J., IEE Cont. Eng. Series 46, Peter Peregrinus Ltd, Stevenage, Chapter 6, pp. 94-122.

[Hunt and Sbarbaro-Hofer, 1991] Hunt K., Sbarbaro-Hofer D. 1991. *Neural Networks for Nonlinear Internal Model Control*, IEE Proc. D, Vol. 138, No. 5, pp. 431-438.

[Hunt *et al.*, 1992] Hunt K.J., Sbarbaro-Hofer D., Zbikowski R., Gawthrop P.J. 1992. *Neural Networks for Control Systems - A Survey*, Automatica, Vol. 28, No. 6, pp. 1083-1112.

[Hush and Horne, 1993] Hush D.R., Horne B.G. 1993. *Progress in Supervised Neural Networks: Whats New Since Lippmann?*, IEEE Sig. Proc. Mag., January, pp. 8-39.

[IFAC IAV, 1993] Proc. 1st IFAC Conf. on Intelligent Autonomous Vehicles, Southampton, 1993.

[IEEE Cont. Sys. Mag., 1988] IEEE Control Sys. Mag., Special Issue on Neural Networks, April, 1988.

[IEEE Cont. Sys. Mag., 1990] IEEE Control Sys. Mag., Special Issue on Neural Networks, April, 1990.

[IEEE Cont. Sys. Mag., 1992] IEEE Control Sys. Mag., Special Issue on Neural Networks, April, 1992.

[IEEE Cont. Sys. Mag., 1993] IEEE Control Sys. Mag., Special Issue on Intelligent Control, June, 1993.

[IEEE Expert, 1991] IEEE Expert, August 1991.

[IEEE Fuzzy Systems, 1992] Proc. 1st IEEE Int. Conf. on Fuzzy Systems, Vol. 1, San Diego, CA, 1992.

[IEEE Fuzzy Systems, 1993] Proc. 2nd IEEE Int. Conf. on Fuzzy Systems, Vols. 1 & 2, San Francisco, CA, 1993.

[IEEE Trans. Neural Nets., 1992] IEEE Trans. Neural Networks, Special Issue on Fuzzy Logic, September 1992.

[Ivakhnenko, 1971] Ivakhnenko A.G. 1971. *Polynomial Theory of Complex Systems*, IEEE Trans. Sys. Man and Cyb., Vol. 1, No. 4,, pp. 364-378.

[Jacobs and Jordan, 1993] Jacobs R.A., Jordan M.I. 1993. *Learning Piecewise Control Strategies in a Modular Neural Network*, IEEE Trans. on Sys. Man and Cyb., Vol. 23, No. 3, pp. 337-345.

[Jager, 1992] Jager R. 1992. *Adaptive Fuzzy Control*, in "Applications of Artificial Intelligence in Process Control", Eds Boullart L., Krijgsman A., Vingerhoeds R.A., Pergamon Press, Oxford, pp. 368-387.

[Johnson, 1988] Johnson C.R. 1988. *Lectures on Adaptive Parameter Estimation*, Prentice Hall, Englewood Cliffs, NJ.

[Jordan and Rumelhart, 1991] Jordan M.I., Rumelhart D.E. 1991. *Forward Models: Supervised Learning With a Distal Teacher*, Occasional paper No. 40, Centre for Cognitive Sciences, MIT, MA.

[Kaczmarz, 1937] Kaczmarz S. 1937. *Angenöherte Auflösung von Systemen Linearer Gleichungen*, Bull. Int. Acad. Pol. Sci. Lett., Cl. Sci. Math. Nat. Ser. A., pp. 355-357 (in German), a translated version appeared in the Int. J. Control (1993), Vol. 57, No. 6, pp. 1269-1271.

[Kalman and Kwasny, 1992] Kalman B.L., Kwasny S.C. 1992. *Why Tanh: Choosing a Sigmoidal Function*, Proc. Int. Joint Conf. Neural Networks, Baltimore, Vol. 4, pp. 578-581.

[Kanerva, 1988] Kanerva P. 1988. *Sparse Distributed Memory*, MIT Press, Cambridge, MA.

[Kanerva, 1992a] Kanerva P. 1992. *Associative-memory Models of the Cerebellum*, Artificial Neural Networks 2, Proc. Int. Conf. on Artificial Neural Networks 2, Eds Aleksander I., Taylor J., Elsevier Science, North-Holland, Vol. 1, pp. 23-34.

[Kanerva, 1992b] Kanerva P. 1992. *Sparse Distributed Memory and Related Models*, NASA Ames RIACS Tech. Report 92.10, 38 pages.

[Kavli, 1992] Kavli T. 1992. *Learning Principles in Dynamic Control*, Dr. Sci. Thesis, Institute for Informatics, University of Oslo, Norway.

[Kavli, 1994] Kavli T. 1994. *ASMOD - an Algorithm for Adaptive Spline Modelling of Observation Data*, in "Advances in Intelligent Control", Ed Harris C.J., Taylor and Francis, London, Chapter 6.

[Kindermann and Linden, 1990] Kindermann J., Linden A. 1990. *Inversion of Neural Networks by Gradient Descent*, Parallel Computing, Vol. 14, pp. 277-286.

[Kiszka *et al.*, 1985] Kiszka J.B., Kochanska M.A., Sliwinska D.S. 1985. *The Influence of some Fuzzy Implication Operators on the Accuracy of a Fuzzy Model - Parts I & II*, Fuzzy Sets and Systems, Vol. 15, pp. 111-128 and 223-240.

[Kohonen, 1988] Kohonen T. 1988. *The "Neural" Phonetic Typewriter*, IEEE Computer, March, pp. 11-22.

[Kohonen, 1990] Kohonen T. 1990. *The Self-Organizing Map*, Proc. IEEE, Vol. 78, No. 9, pp. 1464-1480.

[Kosko, 1992a] Kosko B. 1992. *Neural Networks and Fuzzy Systems*, Prentice-Hall, Englewood Cliffs, NJ.

[Kosko, 1992b] Kosko B. (Ed.) 1992. *Neural Networks for Signal Processing*, Prentice Hall, Englewood Cliffs, NJ.

[Kouatli and Jones, 1991] Kouatli I., Jones B. 1991. *An Improved Design Procedure for Fuzzy Control Systems*, Int. J. Mach. Tools Manufact., Vol. 31, No. 1, pp. 107-122.

[Kozak, 1980] Kozak J. 1980. *On the Choice of Exterior Knots in the B-spline Basis for a Spline Space*, Technical Report 2148, University of Wisconsin, WI.

[Kraft and Campagna, 1990] Kraft L.G., Campagna D.P. 1990. *A Comparison between CMAC Neural Network Control and Two Traditional Control Systems*, IEEE Control Sys. Mag., April, pp. 36-43.

[Kraiss and Küttelwesch, 1990] Kraiss K.F., Küttelwesch H. 1990. *Teaching Neural Networks to Guide a Vehicle Through an Obstacle Course by Emulating a Human Teacher*, IJCNN, Jan. Washington, Vol. 1, pp. 333-337.

[Kröse and Van der Smagt, 1993] Kröse B.J.A., Van der Smagt P.P. 1993. *An Introduction to Neural Networks*, 5th Edition, University of Amsterdam, Fac. Math. and Comp. Sci., Kruislaan, Amsterdam, The Netherlands.

[Kumar and Guez, 1991] Kumar S.S., Guez A. 1991. *ART Based Adaptive Pole Placement for Neurocontrollers*, Neural Networks, Vol. 4, No. 3, pp. 319-335.

[Lane *et al.*, 1992] Lane S.H., Handelman D.A., Gelfand J.J. 1992. *Theory and Development of Higher Order CMAC Neural Networks*, IEEE Control Sys. Mag. April, pp. 23-30.

[Lau and Widrow, 1990] Lau C.G.Y., Widrow B. 1990. *Guest Editorial*, Proc. IEEE, Vol. 78, No. 10, pp. 1547-1549.

[Lawrence and Harris, 1992] Lawrence A., Harris C.J. 1992. *A Label Driven CMAC Intelligent Control Strategy*, in "Application of Neural Networks to Modelling and Control", Eds Page G.F., Gomm J.B., Williams D., Chapman and Hall, Chapter 7.

[Layne and Passino, 1993] Layne J.R., Passino K.M. 1993. *Fuzzy Model Reference Learning Control for Cargo Ship Steering*, IEEE Cont. Sys. Mag., December, pp. 23-34.

[Leontaritis and, Billings, 1987] Leontaritis I.J., Billings S.A. 1987. *Model Selection and Validation Methods for Nonlinear Systems*, Int. J. Control, Vol. 45, No. 1, pp. 311-341.

[Le Cun, 1985] Le Cun Y. 1985. *Une procedure d'apprentissage pour reseau a seuil assymetrique*, Proc. Cognitiva, Vol. 85, pp. 599-604.

[Lee, 1990] Lee C.C. 1990. *Fuzzy Logic in Control Systems: Fuzzy Logic Controller: Parts 1 and 2*, IEEE Trans. Sys. Man and Cyb., Vol. 20, No. 2, pp. 404-435.

[Lin and Kim, 1991] Lin C.S., Kim H. 1991. *CMAC-Based Adaptive Critic Self-Learning Control*, IEEE Trans. Neural Networks, Vol. 2, No. 5, pp. 530-533.

[Linkens, 1993] Linkens D.A. 1993. *Parallel Processing for Self-organizing Control Systems*, in "Parallel Processing in a Control Systems Environment", Eds Rogers E., Li Y., Prentice Hall, Hemel Hempstead, Chapter 6, pp. 168-203.

[Linkens and Abbod, 1991] Linkens D.A., Abbod M.F. 1991. *Self-organising Fuzzy Logic Control for Real-time Processes*, Proc. IEE Conf. Control '91, Edinburgh, pp. 971-976.

[Linkens and Abbod, 1993] Linkens, D.A., Abbod, M.F. 1993. *Supervisory Intelligent Control Using a Fuzzy Logic Hierarchy*, Trans. Inst. Meas. Control, Vol. 15, No. 3, pp. 112-132.

[Linkens and Hasnain, 1990] Linkens D.A., Hasnain S.B. 1990. *Self-organising Fuzzy Logic Control and its Application to Muscle Relaxant Anaesthesia*, Proc. IEE Part D, pp. 274-284.

[Linkens and Nie, 1994] Linkens D.A., Nie J. 1994. *Unified Real Time Approximate Reasoning*, in "Advances in Intelligent Control", Ed Harris C.J., Taylor and Francis, London, Chapter 11.

[Linkens and Shieh, 1992] Linkens D.A., Shieh J.S. 1992. *Self-Organising Modelling for Nonlinear System Control*, Proc. IEEE Int. Symp. on Intelligent Control, Glasgow, pp. 210-215.

[Lippmann, 1987] Lippmann R.P. 1987. *An Introduction to Computing With Neural Nets*, IEEE ASSP Mag., April, pp. 4-22.

[Lippmann, 1989] Lippmann R.P. 1989. *Pattern Classification Using Neural Networks*, IEEE Comm. Mag., November, pp. 47-69.

[Ljung and Soderstrom, 1983] Ljung L., Soderstrom T. 1983. *Theory and Practice of Recusive Identification*, MIT Press, Cambridge, MA.

[Lowe, 1991] Lowe D. 1991. *On the Iterative Inversion of RBF Networks: A Statistical Interpretation*, IEE 2nd Int. Conf. on Artificial Neural Networks, Bournemouth, pp. 29-39.

[Luo, 1991] Luo Z.Q. 1991. *On the Convergence of the LMS Algorithm With Adaptive Learning Rate for Linear Feedforward Networks*, Neural Computation, Vol. 3, No. 2, pp. 226-245.

[Lyche and Mörken, 1988] Lyche T., Mörken K. 1988. *A Data Reduction Strategy for Splines*, IMA J. Numerical Analysis, Vol. 8, pp. 185-208.

[Mamdani, 1974] Mamdani E.H. 1974. *Application of Fuzzy Algorithms for Control of Simple Dynamic Plant*, Proc. IEE, Vol. 121, No. 12, pp. 1585-1588.

[Mason and Parks, 1992] Mason J.C., Parks P.C. 1992. *Selection of Neural Network Structures - Some Approximation Theory Guidelines*, in "Neural Networks for Control Systems", Eds Warwick K., Irwin G.W., Hunt K.J., IEE Control Engineering Series 46, Peter Peregrinus, Stevenage, Chapter 8, pp. 151-180.

[Mathews, 1991] Mathews V.J. 1991. *Adaptive Polynomial Filters*, IEEE Sig. Proc. Mag., July, pp. 10-26.

[Maximov and Meystel, 1992] Maximov Y., Meystel A. 1992. *Optimum Design of Multiresolution Hierarchical Control Systems*, Proc. IEEE Int. Symp. Intelligent Control, Glasgow, pp. 514-520.

[McGregor *et al.*, 1992] McGregor D.R., Odetayo M.O., Dasgupta D. 1992. *Adaptive Control of a Dynamic System using Genetic-Based Methods*, Proc. IEEE Int. Symp. Intellgent Control, Glasgow, pp. 521-525.

[Michie and Chambers, 1968] Michie D., Chambers R.A. 1968. *BOXES: An Experiment in Adaptive Control*, Machine Intelligence 2, Eds Dale E., Michie D., Oliver and Boyd, Edinburgh, pp. 137-52.

[Miller, 1987] Miller W.T. 1987. *Sensor-Based Control of Robotic Manipulators Using a General Learning Algorithm*, IEEE J. Robot. and Auto., Vol. 3, No. 2, pp. 157-165.

[Miller, 1989] Miller W.T. 1989. *Real-Time Application of Neural Networks for Sensor-Based Control of Robots with Vision*, IEEE Trans. on Sys. Man and Cyb., Vol. 19, No. 4, pp. 825-831.

[Miller, 1993] Miller W.T. 1993. *Real-Time Control of a Biped Walking Robot*, Proc. INNS WCNN, Portland OR, Vol. 3, pp. 153-156.

[Miller *et al.*, 1993] Miller W.T., Arehart K.F., Scalera S.M., Gresham H.L. 1993. *On-Line Hand-Printed Character Recognition using CMAC Neural Networks*, Proc. INNS WCNN, Portland OR, Vol. 4, pp. 10-13.

[Miller *et al.*, 1991] Miller W.T., Box B.A., Whitney E.C., Glynn J.M. 1991. *Design and Implementation of a High Speed CMAC Neural Network Using Programmable CMOS Logic Cell Arrays*, in "Advances in Neural Information Processing Systems 3", Eds Lippmann R.P., Moody J.E., Touretzky D.S., Morgan Kaufmann Publishers, San Mateo, CA, pp. 1022-1027.

[Miller *et al.*, 1987] Miller W.T., Glanz F.H., Kraft L.G. 1987. *Application of a General Learning Algorithm to the Control of Robotic Manipulators*, Int. J. Robot. Res., Vol. 6, No. 2, pp. 84-98.

[Miller *et al.*, 1990a] Miller W.T., Glanz F.H., Kraft L.G. 1990. *CMAC: An Associative Neural Network Alternative to Backpropagation*, Proc. IEEE, Vol. 78, No. 10, pp. 1561-1567.

[Miller *et al.*, 1990b] Miller W.T., Glanz F.H., Kraft L.G. 1990. *Real-Time Dynamic Control of an Industrial Manipulator Using a Neural-Network-Based Learning Controller*, IEEE Trans. on Robot. and Auto., Vol. 0, No. 1, pp. 1-9.

[Miller *et al.*, 1990c] Miller W.T., Sutton R.S., Werbos P.J. (Eds) 1990. *Neural Networks for Control*, MIT Press, Cambridge, MA.

[Millington and Baker, 1990] Millington P.J., Baker W.L. 1990. *Associative Reinforcement Learning for Optimal Control*, AIAA Guid. Nav. and Cont. Conf., Portland, OR, Vol. 2, pp. 1120-1128.

[Mills, 1992] Mills D.J. 1992. *The Optimisation of Neural Networks for Approximation*, PhD Thesis, University of Hertfordshire.

[Minsky and Papert, 1969] Minsky M., Papert S. 1969. *Perceptrons*, MIT Press, Cambridge, MA.

[Mischo, 1992] Mischo W.S. 1992. *Receptive Fields for CMAC. An Efficient Approach*, Proc. Int. Conf. on Artificial Neural Networks 2, Eds Aleksander I., TaylorJ., Elsevier Science, North-Holland, Vol. 1, pp. 595-598.

[Miyamoto *et al.*, 1988] Miyamoto H., Kawato M., Setoyama T., Suzuki R. 1988. *Feedback-Error-Learning Neural Network for Trajectory Control of a Robotic Manipulator*, Neural Networks, Vol. 1, pp. 251-265.

[Mizumoto, 1992] Mizumoto M. 1992. *Realization of PID Control by Fuzzy Control Methods*, Proc. IEEE Int. Conf. on Fuzzy Systems, San Diego, CA, pp. 709-715.

[Montague *et al.*, 1991] Montague G.A., Willis M.J., Tham M.T., Morris A.J. 1991. *Artificial Neural Network Based Multivariable Predictive Control*, Proc. IEE 2nd Int. Conf. Artificial Neural Networks, Bournemouth, pp. 119-123.

[Monzingo and Miller, 1980] Monzingo R.A., Miller W.T. 1980. *Introduction to Adaptive Arrays*, John Wiley and Sons, NY.

[Moody, 1989] Moody J. 1989. *Fast Learning in Multi-Resolution Hierarchies*, in "Advances in Neural Information Processing Systems 1", Ed Touretzky D.S., Morgan Kaufmann, San Mateo, CA, pp. 29-39.

[Moody and Darken, 1989] Moody J., Darken C.J. 1989. *Fast Learning in Networks of Locally-Tuned Processing Units*, Neural Computation, Vol. 1, No. 2, pp. 281-294.

[Moore, 1991] Moore C.G. 1991. *Indirect Adaptive Fuzzy Controllers*, PhD. Thesis, Department of Aeronautics and Astronautics, University of Southampton, UK.

[Moore and Harris, 1992] Moore C.G., Harris C.J. 1992. *Phase Plane Analysis Tools for a Class of Fuzzy Control Systems*, Proc. IEEE Int. Conf. on Fuzzy Systems, San Diego, CA, pp. 511-518.

[Moore and Harris, 1994] Moore C.G., Harris C.J. 1994. *Indirect Adaptive Fuzzy Logic Control*, in "Advances in Intelligent Control", Ed Harris C.J., Taylor and Francis, London, Chapter 12.

[Moore et al., 1993] Moore C.G., Harris C.J., Rogers E. 1993. *Utilising Fuzzy Models in the Design of Estimators and Predictors: An Agile Target Tracking Example*, Proc. 2nd IEEE Int. Conf. on Fuzzy Systems, San Francisco, CA, Vol. 2, pp. 679-684.

[Mukhopadhyay and Narendra, 1993] Mukhopadhyay S., Narendra K.S. 1993. *Disturbance Rejection in Nonlinear Systems Using Neural Networks*, IEEE Trans. Neural Networks, Vol. 4, No. 1, pp. 63-72.

[Nagumo and Noda, 1967] Nagumo J.I., Noda A. 1967. *A Learning Method for System Identification*, IEEE Trans. Auto. Control, Vol. 12, No. 3, pp. 282-287.

[Narendra and Annaswamy, 1989] Narendra K.S., Annaswamy A.M. 1989. *Stable Adaptive Systems*, Prentice Hall, Englewood Cliffs, NJ.

[Narendra and Mukhopadhyay, 1992] Narendra K.S., Mukhopadhyay S. 1992. *Intelligent Control Using Neural Networks*, IEEE Cont. Sys. Mag., April, pp. 11-18.

[Narendra and Parthasarathy, 1990] Narendra K.S., Parthasarathy K. 1990. *Identification and Control of Dynamic Systems Using Neural Networks*, IEEE Trans. Neural Networks, Vol. 1, No. 1, pp. 4-27.

[Narendra and Thathachar, 1974] Narendra K.S., Thathachar M.A.L. 1974. *Learning Automata - A Survey*, IEEE Trans. Sys. Man and Cyb., Vol. 4, No. 4, pp. 323-334.

[Nasrabadi and Feng, 1988] Nasrabadi N.M., Feng Y. 1988. *Vector Quantisation of Images based upon the Kohonen Self-Organizing Feature Maps*, Proc. IEEE Int. Conf. Neural Networks, San Diego, CA, Vol. 1, pp. 101-108.

[Nguyen and Widrow, 1990] Nguyen D., Widrow B. 1990. *The Truck Backer-Upper: An Example of Self-Learning in Neural Networks*, in "Neural Networks for Control", Eds Miller W.T., Sutton R.S., Werbos P.J., MIT Press, Cambridge, MA, Chapter 12, pp. 288-299.

[Omohundro, 1987] Omohundro S.M. 1987. *Efficient Algorithms with Neural Network Behaviour*, Complex Systems, Vol. 1, pp. 273-347.

[Ooyen and Nienhuis, 1992] Ooyen A., Nienhuis B. 1992. *Improving the Convergence of the Back Propagation Algorithm*, Neural Networks, Vol. 5, No. 3, pp. 465-471.

[Ozawa and Hayashi, 1992] Ozawa J. Hayashi I., Wakami N. 1992. *Formulation of CMAC-fuzzy system*, Proc. IEEE Int. Conf. Fuzzy Systems, San Diego, pp. 1179-1186.

[Pacini and Kosko, 1992] Pacini P.J., Kosko B. 1992. *Adaptive Fuzzy Systems for Target Tracking*, Intelligent Systems Engineering, Vol. 1, No. 1, pp. 3-21.

[Pao, 1989] Pao Y.H. 1989. *Adaptive Pattern Recognition and Neural Networks*, Addison-Wesley Publishing Company, Reading, MA.

[Pao et al., 1994] Pao Y.H., Phillips S.M., Sobajic D.J. 1994. *Neural-net Computing and Intelligent Control of Systems*, in "Advances in Intelligent Control", Ed Harris C.J., Taylor and Francis, London, Chapter 3.

[Parker, 1985] Parker D. 1985. *Learning-Logic*, Tech. Report. TR-47, Center for Computational Research in Economics and Management Science, MIT, MA.

[Parks and Militzer, 1989] Parks P.C., Militzer J. 1989. *Convergence Properties of Associative Memory Storage for Learning Control Systems*, Automation and Remote Control, Vol. 50, No. 2, Part 2, pp. 254-286.

[Parks and Militzer, 1991a] Parks P.C., Militzer J. 1991. *Improved Allocation of Weights for Associative Memory Storage in Learning Control Systems*, Proc. 1st IFAC Symp. on Design Methods for Control Systems, Zürich, Pergamon Press II, pp. 777-782.

[Parks and Militzer, 1991b] Parks P.C., Militzer J. 1991. Personal Communication.

[Parks and Militzer, 1992] Parks P.C., Militzer J. 1992. *A Comparison of Five Algorithms for the Training of CMAC Memories for Learning Control Systems*, Automatica, Vol. 28, No. 5, pp. 1027-1035.

[Pedrycz, 1993] Pedrycz W. 1993. *Fuzzy Control and Fuzzy Systems*, 2nd Edition Research Studies Press, Taunton, John Wiley and Sons.

[Poggio and Girosi, 1990] Poggio T., Girosi F. 1990. *Networks for Approximation and Learning*, Proc. IEEE, Vol. 78, No. 9, pp. 1481-1497.

[Porcino and Collins, 1990] Porcino D.P., Collins J.S. 1990. *An Application of Neural Networks to the Guidance of Free-Swimming Submersibles*, IJCNN, Washington, Vol. 2, pp. 417-420.

[Pottmann and Seborg, 1992] Pottmann M., Seborg D.E. 1992. *Identification of Nonlinear Processes Using Reciprocal Multiquadratic Functions*, J. Proc. Cont., Vol. 2, No. 4, pp. 189-203.

[Powell, 1987] Powell M.J.D. 1987. *Radial basis functions for multivariable interpolation: A Review*, in "Algorithms for Approximation of Functions and Data", Eds Mason J.C., Cox M.G., Oxford University Press, pp. 143-167.

[Prager and Fallside, 1989] Prager R.W., Fallside F. 1989. *The Modified Kanerva Model for Automatic Speech Recognition*, Computer Speech and Language, Vol. 3, pp. 61-81.

[Prescott and Mayhew, 1992] Prescott T.J., Mayhew J.E.W. 1992. *Obstacle Avoidance Through Reinforcement Learning*, in "Advances in Neural Information Processing Systems 4", Eds Moody J.E., Hanson S.J., Lippmann R.P., Morgan Kaufmann, San Mateo, CA, pp. 523-530.

[Procyk and Mamdani, 1979] Procyk T.J., Mamdani E.H. 1979. *A Linguistic Self-Organising Process Controller*, Automatica, Vol. 15, pp. 15-30.

[Pryce and Parks, 1993] Pryce J.D., Parks P.C. 1993. *Private Communication*, RMCS, Shrivenham.

[Psaltis *et al.*, 1987] Psaltis D., Sideris A., Yamamura A.A. 1987. *Neural Controllers*, IEEE 1st Int. Conf. Neural Networks, San Diego, CA, Vol. 4, pp. 551-558.

[Psaltis *et al.*, 1988] Psaltis D., Sideris A., Yamamura A.A. 1988. *A Multilayered Neural Network Controller*, IEEE Cont. Sys. Mag., April, pp. 17-21.

[Quinlan, 1993] Quinlan J.R. 1993. *C4.5: Programs for Machine Learning*, Morgan Kaufmann Publishers, San Mateo, CA.

[Renders and Hanus, 1992] Renders J.M., Hanus R. 1992. *Biological Learning Metaphors for Adaptive Process Control: A General Strategy*, Proc. IEEE Int. Conf. Intelligent Control, Glasgow, pp. 469-474.

[Reusch, 1993] Reusch B. 1993. *Industrial Application of Fuzzy Logic in North Rhine Westphalia*, Proc. 2nd IEEE Int. Conf. on Fuzzy Systems, San Francisco, CA, Vol. 1, pp. 200-204.

[Rigler *et al.*, 1991] Rigler A.K., Irvine J.M., Vogl T.P. 1991. *Rescaling of Variables in Back Propagation Learning*, Neural Networks, Vol. 4, No. 2, pp. 225-229.

[Ritter *et al.*, 1989] Ritter H.J., Martinetz T.M., Schulten K.J. 1989. *Topology-Conserving Maps for Learning Visuo-Motor-Coordination*, Neural Networks, Vol. 2, No. 2, pp. 159-168.

[Robbins and Monro, 1951] Robbins H., Monro S. 1951. *A Stochastic Approximation Method*, Ann. Math. Stat., Vol. 22, pp. 400-407.

[Robinson and Fallside, 1988] Robinson A.J., Fallside F. 1988. *Static and Dynamic Error Propagation Networks With Application to Speech Coding*, in "Neural Information Processing Systems", Ed Anderson D.Z., American Institute of Physics, NY, pp. 632-641.

[Rogers, 1992] Rogers D. 1992. *Data Analysis using G/Splines*, in "Advances in Neural Information Processing Systems 4", Eds Moody J.E., Hanson S.J., Lippmann R.P., Morgan Kaufmann Publishers, San Mateo, CA, pp. 1088-1095.

[Rosenblatt, 1961] Rosenblatt F. 1961. *Principles of Neurodynamics: Perceptrons and the Theory of Brain Mechanisms*, Spartan Books, Washington DC.

[Ruano *et al.*, 1992] Ruano A.E.B., Flemming P.J., Jones D.I. 1992. *Connectionist Approach to PID Autotuning*, IEE Proc. D, Vol. 139, No. 3, pp. 279-285.

[Rumelhart and McClelland, 1986] Rumelhart D.E., McClelland J.L. (Eds) 1986. *Parallel Distributed Processing: Explorations in the Microstructure of Cognition*, Vol. 1, MIT Press, Cambridge, MA.

[Saint-Donat *et al.*, 1994] Saint-Donat J., Bhat N., McAvoy T.J. 1994. *Neural Net Based Model Predictive Control*, in "Advances in Intelligent Control", Ed Harris C.J., Taylor and Francis, London, Chapter 8.

[Sakaguchi *et al.*, 1993] Sakaguchi S., Sakai I., Hagen T. 1993. *Application of Fuzzy Logic to Shift Scheduling for Automotive Transmission*, Proc. 2nd IEEE Int. Conf. on Fuzzy Systems, San Francisco, CA, Vol. 1, pp. 52-58.

[Sanger, 1991] Sanger T.D. 1991. *A Tree-Structured Adaptive Network for Function Approximation in High-Dimensional Spaces*, IEEE Trans. Neural Networks, Vol. 2, No. 2, pp. 285-293.

[Sanner and Slotine, 1992] Sanner R.M., Slotine J.J.E. 1992. *Gaussian Networks for Direct Adaptive Control*, IEEE Trans. Neural Networks, Vol. 3, No. 6, pp. 837-863.

[Saridis, 1989] Saridis G.N. 1989. *Analytic Formulation of the Principle of Increasing Precision with Decreasing Intelligence for Intelligent Machines*, Automatica, Vol. 25, No. 3, pp. 461-467.

[Saridis and Valavanis, 1988] Saridis G.N., Valavanis K.P. 1988. *Analytical Design of Intelligent Machines*, Automatica, Vol. 24, No. 2, pp. 123-133.

[Sastry and Bodson, 1989] Sastry S., Bodson M. 1989. *Adaptive Control: Stability, Convergence and Robustness*, Prentice Hall, Englewood Cliffs, NJ.

[Sbarbaro-Hofer *et al.*, 1992] Sbarbaro-Hofer D., Neumerkel D., Hunt K. 1992. *Neural Control of a Steel Rolling Mill*, Proc. IEEE Int. Symp. on Intelligent Control, Glasgow, pp. 122-127.

[Sbarbaro-Hofer *et al.*, 1993] Sbarbaro-Hofer D., Neumerkel D., Hunt K. 1993. *Neural Control of a Steel Rolling Mill*, IEEE Cont. Sys. Mag., Vol. 13, No. 3, pp. 70-75.

[Schoenberg and Whitney, 1953] Schoenberg I.J., Whitney A. 1953. *On Pólya Frequency Functions III*, Trans. Am. Math., Vol. 74, pp. 246-259.

[Schwartz, 1990] Schwartz T.J. 1990. *Fuzzy Systems Come to Life in Japan*, IEEE Expert, February, pp. 77-78.

[Schwartz and Klir, 1992] Schwartz D.G., Klir G.J. 1992. *Fuzzy Logic Flowers in Japan*, IEEE Spectrum, July, pp. 32-35.

[Sekulic and Kowalski, 1992] Sekulic S., Kowalski B.R. 1992. *MARS: A Tutorial*, J. Cheometrics, Vol. 6, pp. 199-216.

[Self, 1990] Self K. 1990. *Designing With Fuzzy Logic*, IEEE Spectrum, November, pp. 42-44 & 105.

[Shao *et al.*, 1993] Shao J., Lee Y.C., Jones R. 1993. *Orthogonal Projection Method for Fast On-line Learning Algorithm of Radial Basis Function Neural Networks*, Proc. INNS World Congess on Neural Networks, Portland OR, Vol. 3, pp. 520-535.

[Shelton and Peterson, 1992] Shelton R.O., Peterson J.K. 1992. *Controlling a Truck With an Adaptive Critic CMAC Design*, Simulation, Vol. 58, No. 5, pp. 319-326.

[Shepanski and Macy, 1987] Shepanski J.F., Macy S.A. 1987. *Teaching Artificial Neural Systems to Drive: Manual Training Techniques for Autonomous Systems*, in Proc. 1987 Neural Information Processing Systems Conf., Ed Anderson D.Z., American Institute of Physics, NY, pp. 693-700.

[Slock, 1993] Slock D.T.M. 1993. *On the Convergence Behavior of the LMS and the Normalized LMS Algorithms*, IEEE Trans. Sig. Proc., Vol. 41, No. 9, pp. 2811-2825.

[Soloway and Bialasiewicz, 1992] Soloway D.I., Bialasiewicz J.T. 1992. *Neural Network Modelling of Nonlinear Systems based on Volterra Series Extension of a Linear Model*, Proc. IEEE Int. Symp. Intelligent Control, Glasgow, pp. 7-12.

[Specht, 1991] Specht D.F. 1991. *A General Regression Neural Network*, IEEE Trans. Neural Networks, Vol. 2, No. 6, pp. 568-576.

[Stengel, 1992] Stengel R.F. 1992. *Intelligent Flight Control Systems*, Proc. IMA Conf. Aero. Veh. Dyn. and Control, Cranfield Institute of Technology, UK.

[Sugeno, 1985a] Sugeno M. (Ed.) 1985. *Industrial Applications of Fuzzy Control*, Elsevier Science Publishers BV, North-Holland.

[Sugeno, 1985b] Sugeno M. 1985. *An Introductory Survey of Fuzzy Control*, Information Sciences, Vol. 36, pp. 59-83.

[Sugeno and Murakami, 1985] Sugeno M., Murakami K. 1985. *An Experimental Study of Fuzzy Parking Control Using a Model Car*, Industrial Applications of Fuzzy Control, Ed Sugeno M., Elsevier Science Publishers BV, North-Holland, pp. 125-138.

[Sugeno and Nishida, 1985] Sugeno M., Nishida M. 1985. *Fuzzy Control of Model Car*, Fuzzy Sets and Systems, Vol. 16, pp. 103-113.

[Sugeno *et al.*, 1989] Sugeno M., Murofushi T., Mori T., Tatematsu T., Tanaka J. 1989. *Fuzzy Algorithmic Control of a Model Car by Oral Instructions*, Fuzzy Sets and Systems, Vol. 32, pp. 207-219.

[Sugiyama, 1988] Sugiyama K. 1988. *Rule-based Self-organising Controller*, in "Fuzzy Computing", Eds. Gupta M. and Yamakawa T., Elsevier Science Publishers BV, North-Holland.

[Sutton, 1990] Sutton R.S. 1990. *First Results With Dyna, an Integrated Architecture for Learning, Planning and Reacting*, in "Neural Networks for Control", Eds Miller W.T., Sutton R.S., Werbos P.J., MIT Press, Cambridge, MA.

[Sutton and Jess, 1991] Sutton R., Jess I.M. 1991. *A Design Study of a Self-organizing Fuzzy Autopilot for Ship Control*, IMechE, Proc. Instn. Mech. Engrs., Vol. 205, pp. 35-47.

[Sutton and Towill, 1985] Sutton R., Towill D.R. 1985. *An Introduction to the Use of Fuzzy Sets in the Implementation of Control Algorithms*, J. Inst. Elec. Radio. Eng., Vol. 55, No. 10, pp. 357-367.

[Sutton et al., 1992] Sutton R.S., Barto A.G., Williams R.J. 1992. *Reinforcement Learning is Direct Adaptive Optimal Control*, IEEE Cont. Sys. Mag., April, pp. 19-22.

[Thompson and Patel, 1987] Thompson S.E., Patel R.V. 1987. *Formulation of Joint Trajectories for Industrial Robots Using B-splines*, IEEE Trans. on Ind. Elec., Vol. 34, No. 2, pp. 192-199.

[Tolle and Ersü, 1992] Tolle H., Ersü E. 1992. *Neurocontrol: Learning Control Systems Inspired by Neuronal Architectures and Human Problem Solving*, Lecture Notes in Control and Information Sciences, 172, Springer-Verlag, Berlin.

[Tolle et al., 1994] Tolle H., Parks P.C., Ersü E., Hormel M., Militzer J. 1994. *Learning Control With Interpolating Memories*, in "Advances in Intelligent Control", Ed Harris C.J., Taylor and Francis, London, Chapter 5.

[Tong, 1977] Tong R.M. 1977. *A Control Engineering Review of Fuzzy Systems*, Automatica, Vol. 13, pp. 559-569.

[Tresp et al., 1993] Tresp V., Hollatz J., Ahmad S. 1993. *Network Structuring and Training Using Rule-based Knowledge*, in "Advances in Neural Information Processing Systems 5", Eds Giles C.L., Hanson S.J., Cowan J.D., Morgan Kaufman, San Mateo, CA.

[Tsai et al., 1990] Tsai W.K., Parlos A., Fernandez B. 1990. *A Novel Associative Memory for High Level Control Functions*, Proc. IEEE 29th Conf. Decision and Control, Honolulu, pp. 2374-2379.

[Tzirkel-Hancock and Fallside, 1991] Tzirkel-Hancock E., Fallside F. 1991. *A Direct Control Method for a Class of Nonlinear Systems Using Neural Networks*, Proc. 2nd IEE Int. Conf. on Artificial Neural Networks, Bournemouth, pp. 134-138.

[Walter and Schulten, 1993] Walter J.A., Schulten K.J. 1993. *Implementation of Self-Organizing Neural Networks for Visuo-Motor Control of an Industrial Robot*, IEEE Trans. Neural Networks, Vol. 4, No. 1, pp. 86-95.

[Wang and Vachtsevanos, 1992] Wang B.H., Vachtsevanos G. 1992. *Learning Fuzzy Logic Control: An Indirect Control Approach*, Proc. IEEE Int. Conf. on Fuzzy Systems, San Diego, CA, pp. 297-304.

[Wang and Horng, 1990] Wang C.H., Horng J.G. 1990. *Constrained Minimum-Time Path Planning for Robot Manipulators via Virtual Knots of the Cubic B-spline Functions*, IEEE Trans. Auto. Control, Vol. 35, No. 5, pp. 573-577.

[Wang et al., 1994a] Wang H., Brown M., Harris C.J. 1994. *Neural Network based Modelling of Unknown Nonlinear Systems Subject to Immeasurable Disturbances*, accepted for publication in IEE Part D.

[Wang et al., 1994b] Wang H., Brown M., Harris C.J. 1994. *Modelling and Control of Nonlinear, Operating Point Dependent Systems via Associative Memory Networks*, submitted for publication.

[Wang, 1992] Wang L.X. 1992. *Fuzzy Systems are Universal Approximators*, Proc. IEEE Int. Conf. on Fuzzy Systems, San Diego, CA, pp. 1163-1169.

[Wang, 1994] Wang L.X. 1994. *Adaptive Fuzzy Systems and Control: Design and Stability Analysis*, Prentice Hall, Englewood Cliffs, NJ.

[Wang and Mendel, 1992] Wang L.X., Mendel J.M. 1992. *Fuzzy Basis Functions, Universal Approximation, and Orthogonal Least-Squares Learning*, IEEE Trans. Neural Networks, Vol. 3, No. 5, pp. 807-814.

[Wang, 1993] Wang P. 1993. *The Interpretation of Fuzziness*, Technical Report, Centre for Research on Concepts and Cognition, Indiana University.

[Watkins, 1989] Watkins C.J.C.H. 1989. *Learning From Delayed Rewards*, PhD Thesis, University of Cambridge, UK.

[Weiss and Kulikowski, 1991] Weiss S.M., Kulikowski C.A. 1991. *Computer Systems That Learn: Classification and Prediction Methods From Statistics, Neural Nets, Machine Learning, and Expert Systems*, Morgan Kaufmann Pub. Inc., San Mateo, CA.

[Werbos, 1974] Werbos P.J. 1974., *Beyond regression: New Tools for Prediction and Analysis in the Behavioral Sciences*, Masters Thesis, Harvard University.

[Werbos et al., 1992] Werbos P.J., McAvoy T., Su T. 1992. *Neural Networks, System Identification and Control in the Chemical Process Industries*, in "Handbook of Intelligent Control", Eds White D.A., Sofge D.A., Van Nostrand Reinhold, NY, Chapter 10.

[Werntges, 1993] Werntges H.W. 1993. *Partitions of Unity Improve Neural Function Approximation*, Proc. IEEE Int. Conf. Neural Networks, San Francisco, CA, Vol. 2, pp. 914-918.

[White and Sofge, 1992] White D.A., Sofge D.A. (Eds) 1992. *The Handbook of Intelligent Control*, Van Nostrand Reinhold Inc., NY.

[White et al., 1992] White D., Bowers A., Iliff K., Noffz G., Gonda M., Menousek J. 1992. *Flight, Propulsion and Thermal Control of Advanced Aircraft and Hypersonic Vehicles*, in "Handbook of Intelligent Control", Eds White D.A., Sofge D.A., Van Nostrand Reinhold, NY, Chapter 11.

[Widrow, 1987] Widrow B. 1987. *The Original Adaptive Neural Net Broom-Balancer*, Proc. IEEE Int. Symp. Circuits and Systems, pp. 351-357.

[Widrow and Lehr, 1990] Widrow B., Lehr M.A. 1990. *30 Years of Adaptive Neural Networks: Perceptron, Madaline and Backpropagation*, Proc. IEEE, Vol. 78, No. 9, pp. 1415-1441.

[Widrow and Stearns, 1985] Widrow B., Stearns S.D. 1985. *Adaptive Signal Processing*, Prentice Hall, Englewood Cliffs, NJ.

[Wiener, 1948] Wiener N. 1948. *Cybernetics: or Control and Communication in the Animal and the Machine*, MIT Press, Cambridge, MA.

[Williams, 1990] Williams R.J. 1990. *Adaptive State Representations and Estimation using Recurrent Connectionist Networks*, in "Neural Networks for Control", Eds Miller W.T., Sutton R.S., Werbos P.J., MIT Press, Cambridge, MA, Chapter 4, pp. 97-114.

[Williams and Zipser, 1989] Williams R.J., Zipser D. 1989. *A Learning Algorithm for Continually Running Fully Recurrent Neural Networks*, Neural Computation Vol. 1, No. 2, pp. 270-280.

[Wong and Sideris, 1992] Wong Y., Sideris A. 1992. *Learning Convergence in the Cerebellar Model Articulation Controller*, IEEE Trans. Neural Networks, Vol. 3, No. 1, pp. 115-121.

[Wright, 1991] Wright W. A. 1991. *Neural Networks for Military Robots*, Agard Lecture Series 179, "Artificial Neural Network Approaches in Guidance and Control", Chapter 7.

[Yager and Zadeh, 1992] Yager R.R., Zadeh L.A. (Eds.) 1992. *An Introduction to Fuzzy Logic Appications and Intelligent Systems*, Kluwer Academic Press, Boston, MA.

[Yasunobu and Miyamoto, 1985] Yasunobu S., Miyamoto S. 1985. *Automatic Train Operation by Predictive Fuzzy Control*, in "Industrial Applications of Fuzzy Control", Ed Sugeno M., Elsevier Science Publishers, North-Holland, pp. 1-18.

[Zadeh, 1965] Zadeh L.A. 1965. *Fuzzy Sets*, Information and Control, Vol. 8, pp. 338-353.

[Zadeh, 1973] Zadeh L.A. 1973. *Outline of a New Approach to the Analysis of Complex Systems and Decision Processes*, IEEE Trans. Sys. Man and Cyb., Vol. 3, No. 1, pp. 28-44.

[Zhang and Edmunds, 1992] Zhang B.S., Edmunds J.M. 1992. *Self-organising Fuzzy Logic Controller*, IEE Proc. D, Vol. 139, No. 5, pp. 460-464.

[Zhang and Grant, 1992] Zhang B., Grant E. 1992. *Using Competitive Learning for State-Space Partitioning*, Proc. IEEE Int. Symp. Intelligent Control, Glasgow, pp. 391-396.

Appendix A

Modified Error Correction Rule

It is now shown that when the modified NLMS error correction rule, given in Equation 5.11, is used to train the weight vector of an AMN, the *a posteriori* network output is equal to the desired output with $\delta \equiv 1$. At time t, the unnormalised output of a basis function is denoted by $\tilde{a}_i(t)$, and the *normalised* basis function vector $\bar{a}(t)$ is defined by:

$$\bar{a}_i(t) = \frac{\tilde{a}_i(t)}{\sum_{j=1}^{p} \tilde{a}_j(t)} \qquad \text{for } i = 1, 2, \ldots, p$$

This vector is composed of two parts; the weights which correspond to the basis functions in the first set have been initialised, whereas the basis function in the second set have weights which have not been initialised at time t. Let $\gamma(t)$ be the sum of the normalised basis functions whose weights have been initialised, then the sum of the basis functions whose weights have not been initialised is given by $(1 - \gamma(t))$, and the *initialised* basis functions are defined by:

$$\check{a}_i(t) = \begin{cases} \bar{a}_i(t) & \text{if } w_i(t-1) \text{ has been initialised} \\ 0 & \text{otherwise} \end{cases}$$

for $i = 1, 2, \ldots, p$.

The *a priori* network output is given by:

$$y(t) = \mathbf{a}^T(t)\mathbf{w}(t-1) = \frac{\check{\mathbf{a}}^T(t)\mathbf{w}(t-1)}{\gamma(t)}$$

and the *a posteriori* network output is:

$$\underline{y}(t) = \mathbf{a}^T(t)\mathbf{w}(t) = \bar{\mathbf{a}}^T(t)\mathbf{w}(t)$$

because all the weights have now been initialised. The modified NLMS learning rule is defined as:

$$w_i(t) = \begin{cases} \hat{y}(t) : & \text{if the } i^{th} \text{ weight was not initialised and } \tilde{a}_i(t) > 0 \\ w_i(t-1) + \delta \epsilon_y(t) \gamma(t) \check{a}_i(t) / \|\check{\mathbf{a}}(t)\|_2^2 : & \text{otherwise} \end{cases}$$

and in this appendix, the learning rate δ is assumed to be unity. Then the *a posteriori* output is given by:

$$
\begin{aligned}
\underline{y}(t) &= \bar{\mathbf{a}}^T(t)\mathbf{w}(t) \\
&= \bar{\mathbf{a}}^T(t)\left(\mathbf{w}(t-1) + \Delta\mathbf{w}(t-1)\right) \\
&= y(t)\gamma(t) + \bar{\mathbf{a}}^T(t)\Delta\mathbf{w}(t-1)
\end{aligned}
$$

The term $\Delta\mathbf{w}(t-1)$ can be decomposed into two parts; the first deals with the weights which have been initialised (for the first time) and the second updates the remaining values that are active. In the first case, all the weights are set equal to $\hat{y}(t)$ and in the second case an appropriate multiple of the output error is added to each weight. This gives:

$$
\underline{y}(t) = y(t)\gamma(t) + (1 - \gamma(t))\hat{y}(t) + \bar{\mathbf{a}}^T(t)\left(\frac{\epsilon_y(t)\gamma(t)}{\|\breve{\mathbf{a}}(t)\|_2^2}\breve{\mathbf{a}}(t)\right)
$$

However, from the definition of $\breve{\mathbf{a}}(t)$, $\|\breve{\mathbf{a}}(t)\|_2^2 = \bar{\mathbf{a}}^T(t)\breve{\mathbf{a}}(t)$ and so the above expression reduces to:

$$
\begin{aligned}
\underline{y}(t) &= y(t)\gamma(t) + (1 - \gamma(t))\hat{y}(t) + \epsilon_y(t)\gamma(t) \\
&= y(t)\gamma(t) + (1 - \gamma(t))\hat{y}(t) + (\hat{y}(t) - y(t))\gamma(t) \\
&= \hat{y}(t)
\end{aligned}
$$

Hence when the modified NLMS learning rule is employed with $\delta = 1$, the *a posteriori* output error is zero. This also occurs when the standard NLMS learning rule is used, although the direction of the weight changes are different when $\gamma(t) \neq 1$. Using this modified error correction rule means that the weight change no longer adopts the principle of minimal disturbance at this time-instant, but it can increase the initial rate of convergence especially when the data lie close to the boundaries of the input space.

Appendix B

Improved CMAC Displacement Tables[1]

This section contains the improved overlay displacement for the multivariate CMAC generated by Parks and Militzer [1991a, b]. The tables contain "optimal" displacement vectors for input space dimensions, $n = 2, \ldots, 15$, and generalisation parameters, $\rho = 2, \ldots, 100$. They give the designer the optimal overlay displacements for specific values of n and ρ, and provide information about how well the overlays are distributed for similar values of ρ. Particularly good overlay displacements are achieved when ρ is prime and equal to $2n + 1$, and bad overlay displacements are achieved when it is divisible by 6, since both 2 and 3 (and any multiples thereof) are not co-prime.

These tables use the following notation:

- n is the dimension of the input space ($2 \leq n \leq 15$).
- ρ is the user-defined generalisation parameter ($2 \leq \rho \leq 100$).
- d is the n-dimensional overlay vector.
- D is the minimum distance between any two points in the lattice.
- D_c is given by $\rho^{(n-1)/n}$, and is the length of an edge of an n-dimensional hypercube of volume ρ^{n-1}. This value is supplied as an aid for estimating the quality of the displacement vectors.
- D/D_c gives an indication of the quality of the overlay displacements, and values > 1 may be considered as good.

The symbols ($\dagger$, $\ddagger$) which may follow the displacement vector have the following meaning:

- $\dagger$for this pair of n and ρ, the displacement vectors which contain equal components were not completely evaluated, but the displacement vectors containing different components were.
- $\ddagger$for this pair of n and ρ, the components of the displacement vectors were not completely evaluated.

---

[1]Reprinted from Parks P.C. and Militzer J. 1991. *Improved Allocation of Weights for Associative Memory Storage in Learning Control Systems*, Proc. 1st IFAC Symp. on Design Methods for Control Systems, Zürich, Pergamon Press **II**, pp. 777-782, with kind permission from Elsevier Science Ltd, The Boulevard, Langford Lane, Kidlington, OX5 1GB UK.

The symbols (*,**) which follow the quality index (D/D_c) have the following meanings:

- * This coarse coding has a larger minimum distance between any two points than the perfect cubic lattice, i.e. $D/D_c > 1$.
- ** This coarse coding has a minimum distance between any two points that exceeds the distance in the perfect cubic lattice by a factor of 1.3 or larger, i.e. $D/D_c > 1.3$.

Any value of ρ for which the quality index is greater than 1 can be considered as a good choice for ρ.

ρ	$n=2$			$n=3$			$n=4$		
	d	D	D/D_c	**d**	D	D/D_c	**d**	D	D/D_c
2	(1,1)	2	1.000	(1,1,1)	3	1.091*	(1,1,1,1)	4	1.189*
3	(1,1)	2	0.816	(1,1,1)	3	0.833	(1,1,1,1)	4	0.877
4	(1,1)	2	0.707	(1,1,1)	3	0.687	(1,1,1,1)	4	0.707
5	(1,2)	5	1.000	(1,1,2)	6	0.838	(1,1,2,2)	10	0.946
6	(1,1)	2	0.577	(1,1,1)	3	0.525	(1,1,1,1)	4	0.522
7	(1,2)	5	0.845	(1,2,3)	14	1.023*	(1,1,2,3)	15	0.900
8	(1,3)	8	1.000	(1,1,3)	11	0.829	(1,1,3,3)	16	0.841
9	(1,2)	5	0.745	(1,2,4)	21	1.059*	(1,1,2,4)	22	0.903
10	(1,3)	10	1.000	(1,1,3)	11	0.715	(1,1,3,3)	20	0.795
11	(1,3)	10	0.953	(1,2,4)	21	0.927	(1,1,3,5)	34	0.965
12	(1,5)	8	0.816	(1,1,5)	12	0.661	(1,1,5,5)	16	0.620
13	(1,5)	13	1.000	(1,2,5)	29	0.974	(1,2,3,6)	46	0.991
14	(1,3)	10	0.845	(1,3,5)	35	1.018*	(1,1,3,5)	36	0.829
15	(1,4)	17	1.065*	(1,2,4)	21	0.753	(1,2,4,7)	70	1.098*
16	(1,3)	10	0.791	(1,3,5)	35	0.932	(1,3,5,7)	64	1.000
17	(1,4)	17	1.000	(1,2,6)	41	0.968	(1,2,4,8)	85	1.101*
18	(1,5)	18	1.000	(1,5,7)	27	0.757	(1,1,5,7)	36	0.687
19	(1,4)	17	0.946	(1,3,7)	50	0.993	(1,2,4,8)	85	1.013*
20	(1,3)	10	0.707	(1,3,7)	44	0.900	(1,3,7,9)	80	0.946
21	(1,8)	18	0.926	(1,2,8)	45	0.881	(1,2,5,8)	90	0.967
22	(1,5)	20	0.953	(1,3,7)	56	0.953	(1,3,5,7)	84	0.902
23	(1,5)	25	1.043*	(1,3,8)	65	0.997	(1,2,6,10)	117	1.030*
24	(1,5)	26	1.041*	(1,5,7)	48	0.833	(1,5,7,11)	64	0.738
25	(1,7)	25	1.000	(1,3,8)	74	1.006*	(1,2,7,11)	125	1.000
26	(1,5)	26	1.000	(1,3,9)	91	1.087*	(1,3,5,11)	156	1.085*
27	(1,5)	26	0.981	(1,4,10)	81	1.000	(1,2,5,10)	130	0.963
28	(1,5)	26	0.964	(1,3,9)	91	1.035*	(1,3,5,11)	144	0.986
29	(1,12)	29	1.000	(1,3,9)	91	1.011*	(1,3,7,12)	174	1.056*
30	(1,7)	20	0.816	(1,7,11)	75	0.897	(1,7,11,13)	100	0.780
31	(1,12)	29	0.967	(1,3,11)	94	0.982	(1,3,5,12)	179	1.018*
32	(1,7)	32	1.000	(1,3,9)	91	0.946	(1,3,5,13)	176	0.986
33	(1,7)	29	0.937	(1,4,10)	90	0.922	(1,2,8,14)	181	0.977
34	(1,13)	34	1.000	(1,3,13)	104	0.972	(1,9,13,15)	204	1.014*

ρ	$n=2$			$n=3$			$n=4$		
	d	D	D/D_c	**d**	D	D/D_c	**d**	D	D/D_c
35	(1,6)	37	1.028*	(1,11,16)	133	1.078*	(1,3,8,13)	222	1.035*
36	(1,5)	26	0.850	(1,5,17)	108	0.953	(1,5,7,17)	144	0.816
37	(1,6)	37	1.000	(1,7,17)	133	1.039*	(1,3,8,14)	231	1.013*
38	(1,7)	34	0.946	(1,7,11)	152	1.091*	(1,3,7,13)	228	0.987
39	(1,7)	41	1.025*	(1,4,14)	138	1.021*	(1,4,10,16)	238	0.989
40	(1,7)	40	1.000	(1,3,11)	131	0.979	(1,3,9,13)	260	1.014*
41	(1,9)	41	1.000	(1,4,13)	142	1.002*	(1,3,9,14)	287	1.046*
42	(1,5)	26	0.787	(1,5,13)	147	1.003*	(1,5,11,13)	196	0.849
43	(1,12)	41	0.976	(1,4,15)	157	1.021*	(1,3,13,19)	302	1.035*
44	(1,7)	40	0.953	(1,3,17)	139	0.946	(1,3,7,19)	272	0.965
45	(1,19)	50	1.054*	(1,4,17)	171	1.034*	(1,4,7,16)	307	1.008*
46	(1,7)	50	1.043*	(1,5,13)	171	1.019*	(1,3,7,19)	300	0.981
47	(1,7)	50	1.031*	(1,4,18)	189	1.056*	(1,3,8,17)	354	1.048*
48	(1,7)	50	1.021*	(1,5,17)	171	0.990	(1,5,11,23)	256	0.877
49	(1,9)	41	0.915	(1,4,19)	182	1.007*	(1,3,8,18)	395	1.073*
50	(1,19)	50	1.000	(1,7,11)	171	0.963	(1,3,7,19)	380	1.037*
51	(1,7)	50	0.990	(1,4,19)	189	1.000	(1,4,10,22)	394	1.040*
52	(1,7)	50	0.981	(1,9,23)	195	1.002*	(1,3,9,19)	380	1.007*
53	(1,8)	58	1.046*	(1,4,14)	213	1.034*	(1,3,10,24)	429	1.054*
54	(1,7)	50	0.962	(1,7,25)	216	1.029*	(1,7,13,17)	324	0.904
55	(1,16)	53	0.982	(1,6,23)	225	1.037*	(1,3,8,21)	465	1.068*
56	(1,9)	40	0.845	(1,5,13)	195	0.954	(1,5,13,27)	448	1.034*
57	(1,16)	53	0.964	(1,4,14)	213	0.985	(1,5,8,23)	481	1.057*
58	(1,17)	58	1.000	(1,7,27)	216	0.981	(1,7,11,27)	428	0.984
59	(1,9)	61	1.017*	(1,4,15)	242	1.026*	(1,3,8,21)	515	1.066*
60	(1,13)	50	0.913	(1,7,19)	171	0.853	(1,7,19,29)	400	0.928
61	(1,8)	65	1.032*	(1,4,17)	253	1.026*	(1,3,8,22)	558	1.082*
62	(1,23)	58	0.967	(1,5,25)	248	1.005*	(1,3,9,23)	620	1.127*
63	(1,8)	65	1.016*	(1,4,16)	273	1.044*	(1,4,13,20)	495	0.995
64	(1,19)	58	0.952	(1,5,19)	251	0.990	(1,3,11,29)	560	1.046*
65	(1,8)	65	1.000	(1,4,17)	275	1.026*	(1,4,11,24)	625	1.092*
66	(1,25)	68	1.015*	(1,17,29)	243	0.954	(1,5,7,23)	484	0.950
67	(1,9)	65	0.985	(1,4,18)	278	1.011*	(1,3,9,25)	653	1.091*

ρ	$n=2$			$n=3$			$n=4$		
	d	D	D/D_c	**d**	D	D/D_c	**d**	D	D/D_c
68	(1,9)	74	1.043*	(1,5,25)	283	1.010*	(1,5,13,33)	560	0.999
69	(1,8)	65	0.971	(1,4,19)	297	1.024*	(1,5,8,25)	670	1.081*
70	(1,9)	68	0.986	(1,9,33)	275	0.976	(1,3,9,23)	620	1.029*
71	(1,21)	73	1.014*	(1,7,30)	321	1.045*	(1,3,8,25)	682	1.068*
72	(1,11)	72	1.000	(1,5,19)	243	0.901	(1,5,19,31)	576	0.971
73	(1,16)	74	1.007*	(1,8,20)	341	1.057*	(1,6,9,26)	718	1.073*
74	(1,31)	74	1.000	(1,5,19)	347	1.057*	(1,3,23,33)	740	1.078*
75	(1,8)	65	0.931	(1,4,17)	306	0.984	(1,4,11,31)	721	1.054*
76	(1,9)	80	1.026*	(1,5,21)	332	1.016*	(1,3,9,25)	716	1.040*
77	(1,34)	85	1.051*	(1,5,30)	338	1.016*	(1,12,15,20)	770	1.068*
78	(1,17)	74	0.974	(1,5,19)	324	0.986	(1,5,23,35)	676	0.991
79	(1,9)	82	1.019*	(1,5,19)	374	1.050*	(1,3,9,28)	844	1.096*
80	(1,9)	82	1.012*	(1,7,19)	304	0.939	(1,9,13,37)	820	1.071*
81	(1,31)	74	0.956	(1,4,32)	357	1.009*	(1,5,8,29)	846	1.077*
82	(1,9)	82	1.000	(1,5,23)	376	1.027*	(1,7,19,23)	844	1.066*
83	(1,11)	85	1.012*	(1,5,32)	398	1.048*	(1,3,10,29)	871	1.073*
84	(1,25)	90	1.035*	(1,5,19)	387	1.026*	(1,5,19,31)	784	1.009*
85	(1,38)	85	1.000	(1,22,26)	386	1.016*	(1,3,27,38)	947	1.099*
86	(1,9)	82	0.976	(1,5,33)	403	1.030*	(1,7,11,27)	900	1.062*
87	(1,10)	90	1.017*	(1,4,19)	378	0.990	(1,4,11,28)	895	1.050*
88	(1,9)	82	0.965	(1,5,23)	419	1.035*	(1,9,13,41)	844	1.011*
89	(1,34)	89	1.000	(1,12,20)	437	1.049*	(1,12,34,37)	979	1.080*
90	(1,11)	68	0.869	(1,7,17)	339	0.917	(1,7,11,31)	900	1.027*
91	(1,27)	98	1.038*	(1,22,29)	481	1.084*	(1,3,10,32)	1015	1.081*
92	(1,9)	82	0.944	(1,9,39)	464	1.057*	(1,3,11,29)	972	1.050*
93	(1,34)	90	0.984	(1,8,20)	465	1.051*	(1,7,10,32)	1066	1.090*
94	(1,11)	100	1.031*	(1,5,23)	420	0.991	(1,5,21,29)	1000	1.048*
95	(1,11)	97	1.010*	(1,23,33)	481	1.053*	(1,6,9,33)	1078	1.079*
96	(1,11)	90	0.968	(1,7,31)	432	0.991	(1,7,11,31)	1024	1.043*
97	(1,10)	101	1.020*	(1,6,22)	509	1.069*	(1,22,33,47)	1261	1.149*
98	(1,27)	98	1.000	(1,9,23)	491	1.042*	(1,9,15,43)	1176	1.101*
99	(1,10)	101	1.010*	(1,8,29)	465	1.008*	(1,5,16,29)	1114	1.063*
100	(1,9)	82	0.906	(1,13,21)	475	1.012*	(1,7,19,29)	1092	1.045*

ρ	$n = 5$			$n = 6$		
	d	D	D/D_c	**d**	D	D/D_c
2	(1,1,1,1,1)	5	1.284*	(1,1,1,1,1,1)	6	1.375**
3	(1,1,1,1,1)	5	0.929	(1,1,1,1,1,1)	6	0.981
4	(1,1,1,1,1)	5	0.738	(1,1,1,1,1,1)	6	0.772
5	(1,1,1,2,2)	11	0.915	(1,1,1,2,2,2)	15	1.013*
6	(1,1,1,1,1)	5	0.533	(1,1,1,1,1,1)	6	0.550
7	(1,1,2,2,3)	19	0.919	(1,1,2,2,3,3)	28	1.046*
8	(1,1,1,3,3)	20	0.847	(1,1,1,3,3,3)	24	0.866
9	(1,1,2,2,4)	26	0.879	(1,1,2,2,4,4)	42	1.039*
10	(1,1,1,3,3)	21	0.726	(1,1,1,3,3,3)	30	0.804
11	(1,2,3,4,5)	55	1.089*	(1,1,2,3,4,5)	56	1.015*
12	(1,1,1,5,5)	20	0.613	(1,1,1,5,5,5)	24	0.618
13	(1,2,3,4,5)	55	0.953	(1,2,3,4,5,6)	91	1.125*
14	(1,1,3,3,5)	45	0.812	(1,1,3,3,5,5)	70	0.928
15	(1,1,2,4,7)	71	0.966	(1,1,2,4,4,7)	87	0.977
16	(1,1,3,5,7)	80	0.973	(1,1,3,3,5,7)	94	0.962
17	(1,2,3,4,8)	94	1.005*	(1,2,3,4,6,8)	124	1.050*
18	(1,1,5,5,7)	45	0.664	(1,1,5,5,7,7)	54	0.661
19	(1,2,3,5,9)	118	1.030*	(1,2,3,5,7,8)	152	1.060*
20	(1,1,3,7,9)	84	0.834	(1,1,3,3,7,9)	120	0.902
21	(1,1,5,8,10)	134	1.013*	(1,2,4,5,8,10)	210	1.146*
22	(1,3,5,7,9)	165	1.083*	(1,1,3,5,7,9)	166	0.980
23	(1,2,4,7,10)	165	1.046*	(1,2,3,6,7,11)	220	1.088*
24	(1,1,5,7,11)	80	0.704	(1,1,5,7,11,11)	96	0.693
25	(1,2,3,7,11)	184	1.033*	(1,1,3,6,8,12)	230	1.037*
26	(1,3,5,7,9)	165	0.948	(1,3,5,7,9,11)	286	1.120*
27	(1,2,4,7,13)	206	1.028*	(1,2,4,5,7,13)	255	1.024*
28	(1,3,5,9,11)	180	0.933	(1,3,5,9,11,13)	280	1.041*
29	(1,2,4,7,14)	254	1.078*	(1,2,5,8,9,12)	319	1.080*
30	(1,1,7,11,13)	125	0.736	(1,1,7,7,11,13)	150	0.720
31	(1,2,4,8,15)	310	1.129*	(1,2,4,7,10,15)	363	1.089*
32	(1,3,5,7,13)	253	0.994	(1,3,5,7,9,15)	352	1.045*
33	(1,2,4,8,16)	341	1.126*	(1,1,5,7,10,14)	372	1.047*
34	(1,3,5,7,15)	277	0.991	(1,3,5,7,9,15)	390	1.045*

ρ	$n=5$			$n=6$		
	d	D	D/D_c	**d**	D	D/D_c
35	(1,2,4,9,16)	358	1.101*	(1,2,4,8,11,17)	430	1.072*
36	(1,5,7,11,13)	180	0.763	(1,5,7,11,13,17)	216	0.742
37	(1,2,5,9,17)	400	1.113*	(1,6,8,10,11,14)	518	1.123*
38	(1,3,5,7,17)	352	1.022*	(1,3,5,7,11,17)	494	1.072*
39	(1,2,4,10,17)	410	1.080*	(1,2,4,8,11,17)	495	1.051*
40	(1,3,7,9,19)	336	0.958	(1,3,7,9,11,17)	480	1.013*
41	(1,4,10,16,18)	492	1.137*	(1,2,6,9,13,18)	615	1.123*
42	(1,5,11,13,17)	245	0.787	(1,5,11,13,17,19)	294	0.761
43	(1,2,6,10,20)	486	1.088*	(1,2,7,11,15,20)	631	1.093*
44	(1,5,7,9,19)	517	1.101*	(1,3,5,7,9,21)	606	1.051*
45	(1,2,4,11,19)	477	1.039*	(1,2,8,11,13,17)	648	1.067*
46	(1,3,5,11,21)	493	1.038*	(1,3,5,11,17,21)	662	1.059*
47	(1,2,6,10,22)	551	1.079*	(1,2,4,11,17,22)	746	1.104*
48	(1,5,7,11,23)	320	0.808	(1,5,7,11,13,23)	384	0.778
49	(1,2,5,12,20)	560	1.052*	(1,2,6,10,13,22)	794	1.100*
50	(1,3,11,13,19)	525	1.002*	(1,3,7,11,21,23)	750	1.051*
51	(1,2,5,13,22)	623	1.074*	(1,2,4,10,16,22)	822	1.083*
52	(1,3,5,11,23)	628	1.062*	(1,3,5,7,11,25)	824	1.067*
53	(1,3,5,12,23)	697	1.102*	(1,4,6,13,21,24)	949	1.127*
54	(1,5,11,17,19)	405	0.828	(1,5,7,13,17,19)	486	0.794
55	(1,2,6,16,26)	702	1.074*	(1,2,6,13,21,24)	952	1.094*
56	(1,3,5,11,23)	685	1.045*	(1,5,9,11,13,25)	1022	1.117*
57	(1,2,5,13,23)	719	1.056*	(1,2,4,8,13,28)	990	1.083*
58	(1,3,5,17,27)	709	1.034*	(1,3,5,7,13,27)	982	1.063*
59	(1,2,17,23,28)	810	1.090*	(1,6,8,10,21,26)	1142	1.130*
60	(1,7,11,13,29)	500	0.845	(1,7,11,13,19,23)	600	0.808
61	(1,3,9,20,27)	854	1.090*	(1,2,8,13,17,27)	1215	1.134*
62	(1,15,23,27,29)	837	1.065*	(1,3,7,15,27,29)	1118	1.073*
63	(1,4,10,22,29)	812	1.036*	(1,2,4,8,16,31)	1302	1.142*
64	(1,3,5,15,27)	916	1.086*	(1,3,5,7,17,31)	1222	1.092*
65	(1,3,11,18,23)	909	1.069*	(1,2,8,12,16,31)	1430	1.167*
66	(1,17,25,29,31)	605	0.861	(1,5,7,13,19,23)	726	0.821
67	(1,3,13,20,28)	1019	1.105*	(1,2,12,16,20,25)	1430	1.137*

ρ	$n=5$			$n=6$		
	d	D	D/D_c	**d**	D	D/D_c
68	(1,3,9,19,33)	941	1.049*	(1,5,7,9,19,31)	1318	1.079*
69	(1,4,13,19,29)	972	1.054*	(1,2,4,8,17,32)	1398	1.097*
70	(1,3,19,27,33)	1021	1.068*	(1,3,17,19,27,31)	1270	1.034*
71	(1,5,14,17,25)	1136	1.113*	(1,2,12,18,22,26)	1633	1.158*
72	(1,5,7,17,35)	720	0.877	(1,5,7,11,13,35)	864	0.833
73	(1,3,9,14,33)	1197	1.118*	(1,4,9,15,17,35)	1660	1.141*
74	(1,3,7,21,33)	1104	1.062*	(1,11,23,27,29,31)	1702	1.142*
75	(1,4,11,13,34)	1142	1.069*	(1,2,11,16,23,31)	1737	1.141*
76	(1,3,11,23,29)	1140	1.056*	(1,13,23,27,29,31)	1862	1.169*
77	(1,5,12,27,30)	1267	1.102*	(1,2,8,18,30,34)	1935	1.178*
78	(1,5,7,19,37)	845	0.891	(1,17,23,25,29,35)	1014	0.844
79	(1,3,22,29,37)	1347	1.113*	(1,2,10,15,19,38)	1900	1.143*
80	(1,3,9,17,39)	1300	1.083*	(1,3,11,17,21,37)	1750	1.085*
81	(1,5,11,14,32)	1307	1.075*	(1,2,8,19,31,35)	1917	1.124*
82	(1,3,7,25,37)	1324	1.071*	(1,5,7,9,19,37)	1886	1.104*
83	(1,3,9,25,39)	1446	1.109*	(1,4,14,22,24,30)	2064	1.143*
84	(1,5,11,13,41)	980	0.904	(1,5,11,13,23,41)	1176	0.854
85	(1,3,8,26,39)	1434	1.083*	(1,4,7,9,23,41)	2088	1.127*
86	(1,3,9,17,41)	1420	1.068*	(1,7,9,11,23,37)	2150	1.133*
87	(1,4,10,23,35)	1467	1.075*	(1,2,14,22,28,32)	2151	1.122*
88	(1,3,19,27,37)	1488	1.073*	(1,3,7,17,29,41)	2240	1.134*
89	(1,5,12,32,35)	1666	1.125*	(1,2,14,22,30,41)	2326	1.145*
90	(1,11,17,23,31)	1125	0.917	(1,7,13,19,29,31)	1350	0.864
91	(1,3,10,18,43)	1710	1.120*	(1,4,15,27,34,36)	2352	1.130*
92	(1,3,21,33,43)	1588	1.070*	(1,3,5,21,35,45)	2392	1.130*
93	(1,4,16,23,29)	1643	1.079*	(1,2,13,17,23,44)	2391	1.119*
94	(1,3,9,25,39)	1669	1.078*	(1,3,5,11,21,45)	2404	1.112*
95	(1,3,8,21,36)	1811	1.114*	(1,3,9,23,37,42)	2548	1.135*
96	(1,7,13,17,47)	1280	0.929	(1,5,7,13,23,47)	1536	0.874
97	(1,5,17,37,45)	1836	1.103*	(1,4,15,29,36,38)	2686	1.145*
98	(1,3,9,29,45)	1756	1.070*	(1,3,5,13,27,47)	2646	1.127*
99	(1,4,13,23,34)	1827	1.082*	(1,4,7,19,28,41)	2703	1.130*
100	(1,3,29,37,47)	1844	1.079*	(1,3,7,17,29,41)	2720	1.124*

ρ	$n = 7$			$n = 8$		
	d	D	D/D_c	**d**	D	D/D_c
2	(1,1,1,1,1,1,1)	7	1.461**	(1,1,1,1,1,1,1,1)	8	1.542**
3	(1,1,1,1,1,1,1)	7	1.032*	(1,1,1,1,1,1,1,1)	8	1.082*
4	(1,1,1,1,1,1,1)	7	0.806	(1,1,1,1,1,1,1,1)	8	0.841
5	(1,1,1,1,2,2,2)	16	1.007*	(1,1,1,1,2,2,2,2)	20	1.094*
6	(1,1,1,1,1,1,1)	7	0.570	(1,1,1,1,1,1,1,1)	8	0.590
7	(1,1,1,2,2,3,3)	29	1.016*	(1,1,1,2,2,2,3,3)	33	1.047*
8	(1,1,1,1,3,3,3)	28	0.890	(1,1,1,1,3,3,3,3)	32	0.917
9	(1,1,1,2,2,4,4)	43	0.997	(1,1,1,2,2,2,4,4)	47	1.003*
10	(1,1,1,1,3,3,3)	31	0.774	(1,1,1,1,3,3,3,3)	40	0.843
11	(1,1,2,3,3,4,5)	65	1.032*	(1,1,2,2,3,4,4,5)	76	1.070*
12	(1,1,1,1,5,5,5)	28	0.629	(1,1,1,1,5,5,5,5)	32	0.643
13	(1,1,2,3,4,5,6)	92	1.064*	(1,1,2,3,4,5,5,6)	104	1.081*
14	(1,1,1,3,3,5,5)	71	0.877	(1,1,1,3,3,3,5,5)	80	0.889
15	(1,1,2,2,4,4,7)	91	0.936	(1,1,2,2,4,4,7,7)	140	1.107*
16	(1,1,3,3,5,5,7)	112	0.983	(1,1,3,3,5,5,7,7)	128	1.000
17	(1,1,2,4,5,7,8)	146	1.065*	(1,2,3,4,5,6,7,8)	204	1.197*
18	(1,1,1,5,5,7,7)	63	0.666	(1,1,1,5,5,5,7,7)	72	0.677
19	(1,2,3,4,5,8,9)	179	1.072*	(1,1,3,5,6,7,8,9)	209	1.099*
20	(1,1,3,3,7,7,9)	124	0.854	(1,1,3,3,7,7,9,9)	160	0.920
21	(1,1,2,4,5,8,10)	211	1.069*	(1,1,2,4,5,8,8,10)	239	1.077*
22	(1,1,3,5,5,7,9)	191	0.977	(1,1,3,3,5,7,7,9)	224	1.001*
23	(1,2,3,5,6,9,10)	256	1.089*	(1,1,2,4,6,8,8,11)	302	1.118*
24	(1,1,5,5,7,7,11)	112	0.694	(1,1,5,5,7,7,11,11)	128	0.701
25	(1,2,3,4,7,8,12)	287	1.073*	(1,1,2,4,6,8,9,12)	347	1.114*
26	(1,1,3,5,7,9,11)	287	1.038*	(1,1,3,5,7,7,9,11)	320	1.034*
27	(1,2,4,5,7,10,13)	331	1.079*	(1,1,2,4,7,8,10,13)	395	1.111*
28	(1,1,3,5,9,11,13)	284	0.969	(1,1,3,5,5,9,11,13)	320	0.969
29	(1,2,3,5,7,11,14)	400	1.116*	(1,2,3,4,8,9,10,14)	455	1.120*
30	(1,1,7,7,11,11,13)	175	0.717	(1,1,7,7,11,11,13,13)	200	0.721
31	(1,2,3,6,7,11,15)	445	1.111*	(1,2,3,4,8,9,13,15)	539	1.150*
32	(1,3,5,7,9,11,13)	448	1.085*	(1,3,5,7,9,11,13,15)	512	1.091*
33	(1,2,4,5,8,10,16)	466	1.078*	(1,2,4,5,7,8,14,16)	560	1.110*
34	(1,3,5,7,9,11,13)	455	1.038*	(1,3,5,7,9,11,13,15)	680	1.192*

ρ	$n = 7$			$n = 8$		
	d	D	D/D_c	d	D	D/D_c
35	(1,2,3,4,9,11,17)	521	1.084*	(1,2,3,6,8,12,13,17)	659	1.144*
36	(1,1,5,7,11,13,17)	252	0.736	(1,1,5,7,11,11,13,17)	288	0.738
37	(1,2,4,5,10,12,18)	607	1.115*	(1,2,3,6,10,12,14,18)	740	1.155*
38	(1,3,5,7,9,11,17)	575	1.061*	(1,3,5,7,9,11,13,15)	680	1.081*
39	(1,2,4,7,10,14,19)	675	1.124*	(1,2,4,5,7,10,16,19)	785	1.136*
40	(1,3,7,9,11,13,17)	496	0.943	(1,3,7,9,11,13,17,19)	640	1.003*
41	(1,2,3,7,11,15,20)	735	1.124*	(1,2,4,7,10,14,16,19)	900	1.164*
42	(1,1,5,11,13,17,19)	343	0.752	(1,1,5,11,11,13,17,19)	392	0.752
43	(1,2,4,8,11,16,21)	903	1.196*	(1,2,4,8,11,15,18,20)	979	1.164*
44	(1,3,5,7,9,13,21)	764	1.079*	(1,3,5,7,9,13,17,19)	896	1.092*
45	(1,2,4,8,11,16,22)	819	1.095*	(1,2,4,7,8,14,17,19)	980	1.120*
46	(1,3,5,7,11,17,19)	855	1.098*	(1,3,5,7,9,13,17,19)	984	1.100*
47	(1,2,4,7,13,16,21)	936	1.128*	(1,2,4,8,13,15,18,22)	1162	1.174*
48	(1,5,7,11,13,17,19)	448	0.767	(1,5,7,11,13,17,19,23)	512	0.765
49	(1,2,3,8,12,18,24)	1023	1.138*	(1,2,4,8,13,15,18,24)	1288	1.191*
50	(1,3,7,9,11,21,23)	775	0.974	(1,3,7,9,11,13,21,23)	1000	1.031*
51	(1,2,5,8,11,19,23)	1054	1.116*	(1,2,4,8,13,16,19,25)	1496	1.240*
52	(1,3,5,7,11,17,25)	1031	1.086*	(1,3,5,7,9,11,19,25)	1272	1.124*
53	(1,2,4,8,15,18,24)	1183	1.144*	(1,2,4,8,11,15,20,25)	1435	1.174*
54	(1,5,7,11,13,23,25)	567	0.780	(1,5,7,11,13,17,19,23)	648	0.776
55	(1,2,3,8,14,19,26)	1231	1.131*	(1,2,4,7,14,17,23,26)	1540	1.177*
56	(1,3,5,9,15,25,27)	1136	1.070*	(1,3,5,9,11,13,19,25)	1280	1.057*
57	(1,2,4,7,14,22,26)	1258	1.109*	(1,2,4,5,11,14,22,28)	1586	1.158*
58	(1,5,7,9,13,23,25)	1479	1.184*	(1,3,5,7,9,17,21,27)	1508	1.112*
59	(1,2,5,8,15,22,26)	1452	1.156*	(1,2,5,9,15,18,21,28)	1755	1.182*
60	(1,7,11,13,17,19,23)	700	0.791	(1,7,11,13,17,19,23,29)	800	0.786
61	(1,2,4,8,15,20,29)	1551	1.162*	(1,2,5,8,11,15,25,29)	1855	1.180*
62	(1,3,5,7,13,19,29)	1455	1.109*	(1,3,5,7,11,13,25,27)	1704	1.115*
63	(1,2,4,8,16,23,29)	1602	1.148*	(1,2,4,8,11,16,25,31)	1955	1.178*
64	(1,3,5,9,17,19,29)	1559	1.118*	(1,3,5,7,11,17,23,31)	1856	1.132*
65	(1,2,4,8,16,21,31)	1743	1.166*	(1,2,4,8,11,16,23,32)	2015	1.164*
66	(1,5,7,13,19,23,25)	847	0.802	(1,5,7,13,17,19,23,29)	968	0.796
67	(1,2,4,10,18,23,30)	1791	1.152*	(1,2,6,9,14,19,26,30)	2162	1.174*

ρ	$n=7$			$n=8$		
	d	D	D/D_c	d	D	D/D_c
68	(1,3,15,21,27,29,31)	1703	1.109*	(1,3,5,7,13,19,27,31)	2112	1.145*
69	(1,2,4,8,17,22,32)	1822	1.133*	(1,2,4,7,16,22,26,34)	2219	1.159*
70	(1,3,17,19,27,31,33)	1519	1.022*	(1,3,9,11,13,17,23,29)	1960	1.076*
71	(1,20,23,26,30,32,34)	2130	1.195*	(1,2,4,8,13,18,28,34)	2455	1.189*
72	(1,5,7,11,13,23,35)	1008	0.812	(1,5,7,11,13,19,23,35)	1152	0.805
73	(1,2,12,16,22,25,30)	2078	1.153*	(1,2,4,9,15,20,27,34)	2563	1.186*
74	(1,3,5,7,19,25,35)	2007	1.120*	(1,3,5,7,13,19,31,33)	2516	1.161*
75	(1,2,4,14,22,29,34)	2128	1.140*	(1,2,4,7,14,19,29,37)	2628	1.173*
76	(1,3,5,7,17,25,35)	2071	1.112*	(1,3,5,13,17,23,29,37)	2616	1.156*
77	(1,2,4,9,19,25,36)	2262	1.149*	(1,2,9,13,19,24,27,30)	2821	1.187*
78	(1,5,7,11,17,19,37)	1183	0.822	(1,5,7,11,17,19,31,37)	1352	0.813
79	(1,2,13,19,24,28,35)	2404	1.159*	(1,2,4,8,15,20,31,38)	2914	1.180*
80	(1,7,9,11,13,17,37)	1984	1.041*	(1,3,7,9,17,19,31,37)	2560	1.094*
81	(1,2,5,11,20,29,37)	2491	1.154*	(1,2,4,8,17,20,31,37)	3017	1.175*
82	(1,3,7,19,21,31,37)	2359	1.112*	(1,5,7,9,19,23,37,39)	2952	1.149*
83	(1,2,8,18,22,33,37)	2702	1.177*	(1,2,6,14,16,25,34,37)†	3193	1.183*
84	(1,5,11,13,17,23,37)	1372	0.830	(1,5,11,13,17,23,25,41)	1568	0.820
85	(1,2,14,19,23,26,32)	2726	1.159*	(1,2,4,8,16,21,32,42)	3570	1.225*
86	(1,3,5,11,21,35,39)	2583	1.117*	(1,3,5,11,21,23,35,41)	3144	1.138*
87	(1,5,7,16,25,35,38)	3103	1.212*	(1,2,4,11,20,28,34,41)	3431	1.177*
88	(1,3,7,13,21,37,39)	2703	1.120*	(1,3,5,7,13,21,31,43)	3328	1.147*
89	(1,2,6,11,23,27,42)	3062	1.181*	(1,3,5,15,22,26,33,39)‡	3641	1.188*
90	(1,7,11,17,23,29,31)	1575	0.839	(1,7,11,17,19,23,31,37)	1800	0.827
91	(1,2,6,10,23,36,41)	3031	1.152*	(1,2,4,8,17,20,33,44)	3727	1.179*
92	(1,3,5,11,25,31,43)	2975	1.131*	(1,3,5,7,13,25,31,45)	3680	1.160*
93	(1,2,8,20,25,37,41)	3154	1.154*	(1,2,4,13,20,28,38,43)	3836	1.174*
94	(1,3,11,17,23,41,45)	3071	1.128*	(1,3,5,7,13,27,29,45)	3744	1.149*
95	(1,2,12,16,22,41,46)	3324	1.163*	(1,2,9,12,16,22,41,46)	4058	1.185*
96	(1,5,7,11,29,31,43)	1792	0.846	(1,5,7,11,19,29,31,43)	2048	0.834
97	(1,2,10,14,18,35,40)	3450	1.164*	(1,2,4,10,18,25,38,46)‡	4263	1.192*
98	(1,3,5,13,17,23,47)	3215	1.114*	(1,3,9,11,23,37,39,43)	3920	1.133*
99	(1,2,7,25,37,41,47)	3535	1.158*	(1,2,4,14,23,32,40,47)	4319	1.179*
100	(1,7,9,11,13,23,49)	3100	1.075*	(1,3,7,23,29,37,39,49)	4000	1.125*

ρ	$n = 9$		
	d	D	D/D_c
2	(1,1,1,1,1,1,1,1,1)	9	1.620**
3	(1,1,1,1,1,1,1,1,1)	9	1.130*
4	(1,1,1,1,1,1,1,1,1)	9	0.875
5	(1,1,1,1,1,2,2,2,2)	21	1.096*
6	(1,1,1,1,1,1,1,1,1)	9	0.610
7	(1,1,1,2,2,2,3,3,3)	42	1.149*
8	(1,1,1,1,1,3,3,3,3)	36	0.945
9	(1,1,1,2,2,2,4,4,4)	63	1.126*
10	(1,1,1,1,1,3,3,3,3)	41	0.827
11	(1,1,1,2,3,3,4,5,5)	89	1.119*
12	(1,1,1,1,1,5,5,5,5)	36	0.659
13	(1,1,2,2,3,4,5,5,6)	120	1.121*
14	(1,1,1,3,3,3,5,5,5)	105	0.981
15	(1,1,1,2,2,4,4,7,7)	141	1.070*
16	(1,1,1,3,3,5,5,7,7)	144	1.021*
17	(1,1,2,3,4,5,6,7,8)	205	1.154*
18	(1,1,1,5,5,5,7,7,7)	81	0.689
19	(1,2,3,4,5,6,7,8,9)	285	1.232*
20	(1,1,1,3,3,7,7,9,9)	164	0.893
21	(1,1,2,2,4,5,8,8,10)	255	1.067*
22	(1,1,1,3,3,5,7,9,9)	257	1.027*
23	(1,2,3,4,5,6,8,10,11)	349	1.151*
24	(1,1,1,5,5,7,7,11,11)	144	0.712
25	(1,1,3,4,6,7,9,11,12)	387	1.125*
26	(1,1,3,3,5,7,9,9,11)	377	1.073*
27	(1,2,4,5,7,8,10,11,13)	549	1.252*
28	(1,1,3,3,5,5,9,11,13)	420	1.060*
29	(1,2,3,4,5,7,8,13,14)	524	1.148*
30	(1,1,1,7,7,11,11,13,13)	225	0.730
31	(1,2,3,4,7,8,12,13,14)	624	1.180*
32	(1,1,3,5,7,9,11,13,15)	576	1.102*
33	(1,1,2,4,5,8,10,13,16)	636	1.127*
34	(1,1,3,5,7,9,11,13,15)	681	1.136*

ρ	$n = 9$		
	d	D	D/D_c
35	(1,1,4,6,8,9,11,13,16)	735	1.150*
36	(1,1,5,7,11,11,13,13,17)	324	0.745
37	(1,2,4,5,7,8,10,16,18)	839	1.169*
38	(1,3,5,7,9,11,13,15,17)	969	1.227*
39	(1,2,4,5,7,10,14,16,17)	936	1.179*
40	(1,1,3,7,9,11,13,17,19)	656	0.965
41	(1,2,4,5,8,11,14,17,20)	1020	1.177*
42	(1,1,5,5,11,11,13,17,19)	441	0.757
43	(1,2,4,7,8,11,13,16,21)	1121	1.183*
44	(1,1,3,5,7,9,13,19,21)	1028	1.110*
45	(1,2,4,7,8,11,13,16,22)	1164	1.157*
46	(1,3,5,7,9,11,13,19,21)	1161	1.133*
47	(1,2,4,7,10,13,15,18,23)	1352	1.200*
48	(1,1,5,7,11,13,17,19,23)	576	0.769
49	(1,2,5,8,9,13,17,20,23)	1421	1.185*
50	(1,3,7,9,11,13,17,19,21)	1025	0.989
51	(1,2,4,5,8,13,16,19,25)	1521	1.184*
52	(1,3,5,7,9,11,17,23,25)	1508	1.158*
53	(1,2,4,9,12,15,18,20,25)	1655	1.193*
54	(1,5,7,11,13,17,19,23,25)	729	0.779
55	(1,2,3,7,12,13,17,21,27)	1815	1.209*
56	(1,3,5,9,11,13,17,23,27)	1680	1.145*
57	(1,2,4,7,8,14,16,25,28)	1995	1.228*
58	(1,3,5,7,9,13,17,25,27)	1889	1.177*
59	(1,2,5,9,12,18,20,24,28)	2040	1.204*
60	(1,1,7,11,13,17,19,23,29)	900	0.788
61	(1,2,4,8,15,18,22,27,29)	2171	1.206*
62	(1,3,5,7,11,13,15,27,29)	2068	1.160*
63	(1,2,4,8,11,13,16,25,31)	2217	1.184*
64	(1,3,5,7,11,17,19,27,31)	2304	1.191*
65	(1,2,4,7,8,16,18,28,32)	2483	1.219*
66	(1,5,7,13,17,19,23,25,29)	1089	0.796
67	(1,2,5,9,15,18,21,25,32)	2573	1.208*

ρ	$n = 9$		
	d	D	D/D_c
68	(1,3,5,7,11,13,19,29,33)	2481	1.171*
69	(1,2,4,7,8,17,20,29,34)	2604	1.184*
70	(1,3,9,11,13,17,19,23,29)	2009	1.027*
71	(1,2,4,8,11,18,24,27,32)†	2859	1.209*
72	(1,5,7,11,13,17,19,29,31)	1296	0.804
73	(1,2,4,8,9,16,18,32,36)†	3066	1.222*
74	(1,3,7,9,11,21,25,27,33)	3108	1.215*
75	(1,2,4,7,8,17,23,26,37)	2997	1.179*
76	(1,3,5,7,11,15,17,31,37)	2980	1.162*
77	(1,2,3,13,20,24,29,32,38)	3277	1.205*
78	(1,5,7,11,17,19,23,31,37)	1521	0.811
79	(1,2,4,7,13,21,24,29,39)‡	3500	1.217*
80	(1,3,7,9,11,17,19,31,37)	2624	1.042*
81	(1,2,4,5,10,20,23,32,40)	3618	1.210*
82	(1,3,5,7,9,19,25,31,39)	3452	1.169*
83	(1,2,3,7,14,20,26,31,41)‡	3872	1.225*
84	(1,5,11,13,17,23,25,37,41)	1764	0.818
85	(1,2,3,13,22,26,31,36,42)†	3884	1.201*
86	(1,3,5,7,13,19,23,33,41)	3913	1.193*
87	(1,2,4,7,10,17,22,35,43)	3975	1.190*
88	(1,3,5,7,13,19,25,35,39)	3961	1.176*
89	(1,2,4,8,16,23,29,34,43)‡	4484	1.239*
90	(1,7,11,13,17,23,29,31,43)	2025	0.824
91	(1,2,4,8,15,24,29,34,45)†	4467	1.212*
92	(1,3,5,7,11,19,25,35,45)	4388	1.190*
93	(1,2,5,13,23,29,32,40,44)	4563	1.202*
94	(1,3,7,11,13,19,25,39,41)	4460	1.177*
95	(1,2,4,8,16,23,31,34,44)†	4917	1.224*
96	(1,5,7,11,13,17,23,37,47)	2304	0.830
97	(1,2,3,9,18,24,31,36,47)‡	5003	1.212*
98	(1,3,5,17,19,29,31,37,41)	4900	1.189*
99	(1,2,4,8,16,23,32,37,49)	5211	1.215*
100	(1,3,7,19,23,29,37,39,49)	4100	1.068*

ρ	$n = 10$		
	d	D	D/D_c
2	$(1,1,1,1,1,1,1,1,1,1)$	10	1.695**
3	$(1,1,1,1,1,1,1,1,1,1)$	10	1.176*
4	$(1,1,1,1,1,1,1,1,1,1)$	10	0.908
5	$(1,1,1,1,1,2,2,2,2,2)$	25	1.175*
6	$(1,1,1,1,1,1,1,1,1,1)$	10	0.630
7	$(1,1,1,1,2,2,2,3,3,3)$	43	1.138*
8	$(1,1,1,1,1,3,3,3,3,3)$	40	0.973
9	$(1,1,1,1,2,2,2,4,4,4)$	64	1.107*
10	$(1,1,1,1,1,3,3,3,3,3)$	50	0.890
11	$(1,1,2,2,3,3,4,4,5,5)$	110	1.212*
12	$(1,1,1,1,1,5,5,5,5,5)$	40	0.676
13	$(1,1,2,2,3,3,4,5,6,6)$	137	1.164*
14	$(1,1,1,1,3,3,3,5,5,5)$	106	0.957
15	$(1,1,1,2,2,4,4,4,7,7)$	157	1.095*
16	$(1,1,1,3,3,3,5,5,7,7)$	160	1.043*
17	$(1,1,2,2,4,4,5,7,7,8)$	222	1.164*
18	$(1,1,1,1,5,5,5,7,7,7)$	90	0.704
19	$(1,1,2,3,4,5,6,7,8,9)$	286	1.195*
20	$(1,1,1,3,3,3,7,7,9,9)$	200	0.954
21	$(1,1,2,2,4,5,5,8,8,10)$	304	1.126*
22	$(1,1,3,3,5,5,7,7,9,9)$	330	1.125*
23	$(1,1,2,3,5,6,7,9,10,11)$	392	1.178*
24	$(1,1,1,5,5,7,7,11,11,11)$	160	0.724
25	$(1,2,3,4,6,7,8,9,11,12)$	525	1.265*
26	$(1,1,3,3,5,5,7,9,11,11)$	442	1.120*
27	$(1,1,2,4,5,7,8,10,11,13)$	550	1.208*
28	$(1,1,3,3,5,5,9,11,11,13)$	424	1.026*
29	$(1,2,3,4,6,7,8,11,12,13)$	613	1.196*
30	$(1,1,1,7,7,7,11,11,13,13)$	250	0.741
31	$(1,2,3,4,6,7,8,12,14,15)$	744	1.240*
32	$(1,1,3,5,5,7,9,11,13,15)$	640	1.118*
33	$(1,2,4,5,7,8,10,13,14,16)$	880	1.275*
34	$(1,1,3,5,7,9,9,11,13,15)$	738	1.137*

ρ	$n = 10$		
	d	D	D/D_c
35	(1,2,3,4,6,8,11,12,13,17)	853	1.191*
36	(1,1,5,5,7,7,11,13,17,17)	360	0.754
37	(1,1,3,5,7,9,11,13,15,18)	975	1.211*
38	(1,1,3,5,7,9,11,13,15,17)	970	1.179*
39	(1,2,4,5,7,8,10,14,16,19)	1021	1.182*
40	(1,1,3,7,7,9,11,13,17,19)	800	1.023*
41	(1,2,4,5,8,9,10,16,18,20)	1271	1.261*
42	(1,1,5,5,11,11,13,13,17,19)	490	0.766
43	(1,2,3,4,8,9,11,16,18,21)	1276	1.210*
44	(1,3,5,7,9,13,15,17,19,21)	1210	1.154*
45	(1,2,4,7,8,11,13,16,19,22)	1305	1.175*
46	(1,1,3,5,7,11,13,17,19,21)	1362	1.177*
47	(1,2,3,4,8,11,13,17,18,23)	1526	1.221*
48	(1,1,5,7,11,11,13,17,19,23)	610	0.700
49	(1,2,3,5,9,12,13,19,20,23)	1621	1.213*
50	(1,3,7,9,11,13,17,19,21,23)	1250	1.046*
51	(1,2,4,5,10,11,13,19,22,25)	1696	1.196*
52	(1,3,5,7,9,11,15,17,19,23)	1690	1.174*
53	(1,2,4,7,9,12,15,20,23,25)	1924	1.231*
54	(1,1,5,7,11,13,17,19,23,25)	810	0.785
55	(1,2,3,7,8,12,14,18,24,27)	2021	1.220*
56	(1,3,5,9,11,13,15,17,19,27)	1696	1.100*
57	(1,2,4,5,10,13,16,20,22,28)	2106	1.206*
58	(1,1,9,11,13,15,17,17,21,23)	2146	1.199*
59	(1,2,4,7,11,15,17,20,23,29)†	2356	1.237*
60	(1,1,7,11,13,13,17,19,23,29)	1000	0.794
61	(1,2,4,7,11,16,17,22,25,30)†	2501	1.237*
62	(1,3,5,7,9,11,15,19,27,29)	2354	1.182*
63	(1,2,4,5,11,13,17,22,25,31)	2520	1.206*
64	(1,3,5,7,9,11,13,19,29,31)	2536	1.193*
65	(1,2,4,7,8,16,18,21,27,32)†	2812	1.238*
66	(1,5,7,13,17,19,23,25,29,31)	1210	0.801
67	(1,2,4,8,11,17,22,24,27,31)‡	2936	1.231*

ρ	$n = 10$		
	d	D	D/D_c
68	(1,3,5,7,9,11,13,19,31,33)	2760	1.178*
69	(1,2,4,7,11,16,19,25,28,31)	3040	1.220*
70	(1,3,9,11,13,17,19,23,27,29)	2450	1.081*
71	(1,2,5,10,14,17,21,25,28,34)‡	3408	1.259*
72	(1,5,7,11,13,17,19,23,25,35)	1440	0.808
73	(1,2,4,9,12,17,22,26,32,33)‡	3425	1.231*
74	(1,3,5,7,11,19,21,29,33,35)	3338	1.201*
75	(1,2,4,7,14,16,22,28,31,34)	3550	1.223*
76	(1,3,5,7,13,17,21,27,31,33)	3496	1.200*
77	(1,2,4,8,13,16,17,27,32,38)†	3804	1.237*
78	(1,5,7,11,17,19,23,25,29,35)	1690	0.815
79	(1,2,3,9,14,20,24,28,33,39)‡	3950	1.231*
80	(1,3,7,9,11,13,17,19,31,37)	3200	1.096*
81	(1,2,4,8,13,19,22,29,34,37)†	4086	1.225*
82	(1,3,7,11,13,23,25,29,31,37)	4264	1.237*
83	(1,2,3,7,14,19,25,30,35,38)‡	4316	1.231*
84	(1,5,11,13,17,19,23,25,29,37)	1960	0.821
85	(1,2,4,14,21,24,29,32,37,41)†	4499	1.230*
86	(1,3,5,7,13,19,23,25,33,39)	4378	1.201*
87	(1,2,4,8,13,22,25,31,38,41)†	4567	1.214*
88	(1,3,5,7,9,17,23,29,37,41)	4552	1.200*
89	(1,2,3,6,15,23,27,34,36,43)‡	4808	1.220*
90	(1,7,11,13,17,19,23,31,37,41)	2250	0.827
91	(1,2,4,9,16,19,27,33,38,41)‡	5150	1.238*
92	(1,3,5,7,13,17,19,29,37,45)	4930	1.200*
93	(1,2,4,8,16,23,29,32,35,46)†	5580	1.264*
94	(1,3,5,7,11,19,25,31,41,43)†	5314	1.222*
95	(1,2,6,11,14,18,28,33,37,41)‡	5605	1.243*
96	(1,5,7,11,13,17,23,29,31,43)†	2560	0.832
97	(1,2,3,4,15,24,30,36,40,47)‡	5548	1.213*
98	(1,3,15,23,27,31,37,39,41,47)	5194	1.163*
99	(1,2,4,7,14,23,28,32,38,49)†	5980	1.237*
100	(1,3,7,11,13,17,19,21,41,47)	5000	1.121*

ρ	$n = 11$		
	d	D	D/D_c
2	(1,1,1,1,1,1,1,1,1,1,1)	11	1.766**
3	(1,1,1,1,1,1,1,1,1,1,1)	11	1.222*
4	(1,1,1,1,1,1,1,1,1,1,1)	11	0.941
5	(1,1,1,1,1,1,2,2,2,2,2)	26	1.180*
6	(1,1,1,1,1,1,1,1,1,1,1)	11	0.651
7	(1,1,1,1,2,2,2,2,3,3,3)	47	1.169*
8	(1,1,1,1,1,1,3,3,3,3,3)	44	1.002*
9	(1,1,1,1,2,2,2,2,4,4,4)	68	1.119*
10	(1,1,1,1,1,1,3,3,3,3,3)	51	0.880
11	(1,1,1,2,2,3,3,4,4,5,5)	111	1.191*
12	(1,1,1,1,1,1,5,5,5,5,5)	44	0.693
13	(1,1,2,2,3,3,4,4,5,5,6)	146	1.174*
14	(1,1,1,1,3,3,3,3,5,5,5)	115	0.974
15	(1,1,1,2,2,2,4,4,4,7,7)	161	1.082*
16	(1,1,1,3,3,3,5,5,5,7,7)	176	1.067*
17	(1,1,2,2,3,4,5,6,6,7,8)	245	1.191*
18	(1,1,1,1,5,5,5,5,7,7,7)	99	0.719
19	(1,1,2,3,4,4,5,6,7,8,9)	302	1.195*
20	(1,1,1,3,3,3,7,7,7,9,9)	204	0.938
21	(1,1,1,2,4,5,5,8,8,10,10)	344	1.165*
22	(1,1,1,3,3,5,5,7,7,9,9)	331	1.095*
23	(1,2,3,4,5,6,7,8,9,10,11)	506	1.301**
24	(1,1,1,5,5,5,7,7,7,11,11)	176	0.738
25	(1,1,2,3,4,6,7,8,9,11,12)	526	1.229*
26	(1,1,3,3,5,5,7,7,9,9,11)	451	1.098*
27	(1,1,2,4,5,7,8,8,10,11,13)	585	1.209*
28	(1,1,3,3,5,5,9,9,11,11,13)	460	1.037*
29	(1,2,3,4,5,6,7,10,11,13,14)	700	1.239*
30	(1,1,1,7,7,7,11,11,11,13,13)	275	0.753
31	(1,2,3,4,5,6,7,8,12,14,15)	769	1.222*
32	(1,1,3,5,5,7,7,9,11,13,15)	704	1.136*
33	(1,1,2,4,5,7,8,10,13,14,16)	881	1.236*
34	(1,1,3,3,5,7,9,11,13,15,15)	811	1.154*

ρ	$n = 11$		
	d	D	D/D_c
35	(1,1,2,4,6,8,9,11,13,16,16)	925	1.201*
36	(1,1,5,5,7,7,11,11,13,13,17)	396	0.766
37	(1,2,3,4,5,8,9,11,15,16,18)	1082	1.234*
38	(1,1,3,5,7,9,9,11,13,15,17)	1043	1.183*
39	(1,1,2,4,5,7,10,11,14,17,19)	1142	1.209*
40	(1,1,3,3,7,9,11,13,17,19,19)	816	0.999
41	(1,2,3,5,9,10,12,16,17,18,20)	1323	1.243*
42	(1,1,5,5,11,11,13,13,17,17,19)	539	0.776
43	(1,2,4,5,7,8,9,10,18,19,21)	1433	1.239*
44	(1,1,3,5,7,9,13,15,17,19,21)	1324	1.167*
45	(1,2,4,7,8,11,13,14,16,17,19)	1449	1.196*
46	(1,3,5,7,9,11,13,15,17,19,21)	1771	1.296*
47	(1,2,3,5,6,10,12,14,18,21,22)†	1734	1.257*
48	(1,1,5,7,11,11,13,13,17,19,23)	704	0.786
49	(1,2,3,5,9,10,12,16,18,20,24)†	1814	1.238*
50	(1,1,3,7,9,11,13,17,19,21,23)	1275	1.019*
51	(1,2,4,5,7,10,11,13,19,22,25)	1955	1.239*
52	(1,3,5,7,9,11,15,17,19,21,23)	1804	1.170*
53	(1,2,3,4,8,11,14,16,20,21,26)†	2166	1.260*
54	(1,1,5,7,11,13,13,17,19,23,25)	891	0.794
55	(1,2,3,7,8,12,14,18,23,24,27)	2230	1.236*
56	(1,3,5,9,11,13,15,17,19,23,25)	1840	1.104*
57	(1,2,4,5,7,10,13,14,20,26,28)	2405	1.243*
58	(1,3,5,7,9,11,13,17,23,25,27)	2419	1.227*
59	(1,2,4,7,9,10,14,18,22,24,29)†	2601	1.252*
60	(1,1,7,11,11,13,13,17,19,23,29)	1100	0.802
61	(1,2,3,6,11,12,15,19,22,24,29)†	2792	1.259*
62	(1,3,5,7,9,11,13,15,23,27,29)	2651	1.209*
63	(1,2,4,5,8,10,16,19,22,25,31)	2873	1.240*
64	(1,3,5,7,9,11,13,17,19,29,31)	2816	1.210*
65	(1,2,3,4,8,12,17,18,24,28,31)†	3133	1.259*
66	(1,1,5,7,13,17,19,23,25,29,31)	1331	0.809
67	(1,2,3,6,10,12,17,21,25,26,33)‡	3376	1.271*

ρ	$n = 11$		
	d	D	D/D_c
68	(1,3,5,7,9,11,13,19,21,29,33)	3179	1.217*
69	(1,2,4,5,11,13,19,22,25,29,32)†	3398	1.241*
70	(1,3,9,11,13,17,19,23,27,29,31)	2499	1.051*
71	(1,2,3,7,9,13,17,21,25,31,35)‡	3702	1.263*
72	(1,5,7,11,13,17,19,23,25,29,31)	1584	0.815
73	(1,2,3,7,11,15,17,23,27,31,36)‡	3880	1.260*
74	(1,3,5,7,9,13,15,21,29,31,33)	3708	1.217*
75	(1,2,4,7,8,13,19,22,29,32,37)	3950	1.241*
76	(1,3,5,7,9,13,17,23,27,29,35)	3916	1.221*
77	(1,2,3,6,12,16,20,24,29,31,38)†	4276	1.260*
78	(1,5,7,11,17,19,23,25,29,31,35)	1859	0.821
79	(1,2,3,4,10,15,20,23,31,32,39)‡	4425	1.253*
80	(1,3,7,9,11,13,17,19,23,37,39)	3264	1.064*
81	(1,2,4,7,11,16,20,23,29,35,37)†	1685	1.200*
82	(1,3,5,7,9,13,19,23,31,37,39)	4547	1.228*
83	(1,2,3,4,10,16,20,25,32,34,40)‡	4828	1.251*
84	(1,5,11,13,17,19,23,25,29,31,37)	2156	0.827
85	(1,2,3,8,14,19,21,28,32,37,41)‡	5125	1.261*
86	(1,3,5,7,11,15,17,19,27,39,41)†	4955	1.227*
87	(1,2,4,11,16,19,22,26,28,34,40)†	5273	1.253*
88	(1,3,5,7,9,13,17,23,29,41,43)	5195	1.230*
89	(1,2,3,4,8,16,18,24,32,35,44)‡	5364	1.238*
90	(1,7,11,13,17,19,23,29,31,37,41)	2475	0.832
91	(1,2,3,5,11,17,23,27,33,37,45)‡	5765	1.257*
92	(1,7,9,11,13,15,19,25,29,41,43)†	5819	1.251*
93	(1,2,4,7,8,16,23,25,35,38,44)†	5975	1.255*
94	(1,3,5,7,9,13,23,25,37,43,45)†	5859	1.231*
95	(1,2,3,4,11,17,24,29,34,39,47)‡	6237	1.258*
96	(1,5,7,11,13,17,23,25,29,37,47)†	2816	0.837
97	(1,2,3,4,5,16,24,29,36,38,48)‡	6122	1.223*
98	(1,3,5,11,13,19,23,27,41,45,47)†	5635	1.162*
99	(1,2,4,5,13,17,25,31,34,41,49)†	6795	1.264*
100	(1,3,7,11,13,17,19,21,27,41,47)†	5100	1.085*

ρ	$n = 12$		
	d	D	D/D_c
2	(1,1,1,1,1,1,1,1,1,1,1,1)	12	1.835**
3	(1,1,1,1,1,1,1,1,1,1,1,1)	12	1.265*
4	(1,1,1,1,1,1,1,1,1,1,1,1)	12	0.972
5	(1,1,1,1,1,1,2,2,2,2,2,2)	30	1.253*
6	(1,1,1,1,1,1,1,1,1,1,1,1)	12	0.670
7	(1,1,1,1,2,2,2,2,3,3,3,3)	56	1.257*
8	(1,1,1,1,1,1,3,3,3,3,3,3)	48	1.030*
9	(1,1,1,1,2,2,2,2,4,4,4,4)	84	1.223*
10	(1,1,1,1,1,1,3,3,3,3,3,3)	60	0.938
11	(1,1,1,2,2,3,3,3,4,4,5,5)	120	1.216*
12	(1,1,1,1,1,1,5,5,5,5,5,5)	48	0.710
13	(1,1,2,2,3,3,4,4,5,5,6,6)	182	1.285*
14	(1,1,1,1,3,3,3,3,5,5,5,5)	140	1.053*
15	(1,1,1,2,2,2,4,4,4,7,7,7)	210	1.211*
16	(1,1,1,3,3,3,5,5,5,7,7,7)	192	1.091*
17	(1,1,2,2,3,4,4,5,6,7,8,8)	289	1.266*
18	(1,1,1,1,5,5,5,5,7,7,7,7)	108	0.735
19	(1,1,2,2,3,5,5,6,6,7,9,9)	337	1.235*
20	(1,1,1,3,3,3,7,7,7,9,9,9)	240	0.994
21	(1,1,2,2,4,4,5,5,8,8,10,10)	420	1.258*
22	(1,1,1,3,3,5,5,5,7,7,9,9)	356	1.110*
23	(1,1,2,3,4,5,6,7,8,9,10,11)	507	1.271*
24	(1,1,1,5,5,5,7,7,7,11,11,11)	192	0.752
25	(1,1,2,3,4,6,7,7,8,9,11,12)	550	1.227*
26	(1,1,3,3,5,5,7,7,9,9,11,11)	572	1.207*
27	(1,1,2,4,4,5,7,8,10,10,11,13)	639	1.232*
28	(1,1,3,3,5,5,9,9,11,11,13,13)	560	1.116*
29	(1,2,3,4,5,6,7,8,10,12,13,14)	770	1.267*
30	(1,1,1,7,7,7,11,11,11,13,13,13)	300	0.767
31	(1,2,3,4,5,6,7,9,10,13,14,15)	859	1.259*
32	(1,1,3,3,5,7,7,9,11,13,15,15)	768	1.156*
33	(1,1,2,4,5,7,8,10,10,13,14,16)	945	1.247*
34	(1,1,3,3,5,5,7,9,11,13,15,15)	908	1.189*

ρ	$n = 12$		
	d	D	D/D_c
35	(1,2,3,4,6,8,9,11,12,13,16,17)	1190	1.325**
36	(1,1,5,5,7,7,11,11,13,13,17,17)	432	0.778
37	(1,2,3,4,5,6,8,10,11,15,17,18)	1193	1.261*
38	(1,1,3,5,7,7,9,11,11,13,15,17)	1140	1.203*
39	(1,2,4,5,7,8,10,11,14,16,17,19)	1482	1.340**
40	(1,1,3,3,7,7,9,11,13,17,19,19)	960	1.053*
41	(1,2,3,4,5,8,9,10,14,16,18,20)†	1476	1.277*
42	(1,1,5,5,11,11,13,13,17,17,19,19)	588	0.788
43	(1,2,3,4,6,7,11,12,14,16,19,21)†	1591	1.269*
44	(1,1,3,5,7,9,13,13,15,17,19,21)	1424	1.176*
45	(1,2,4,7,8,11,13,14,16,17,19,22)	1890	1.327**
46	(1,1,3,5,7,9,11,13,15,17,19,21)	1772	1.259*
47	(1,2,3,8,9,12,13,14,17,18,19,23)†	1881	1.272*
48	(1,1,5,7,11,11,13,13,17,19,23,23)	768	0.797
49	(1,2,3,4,6,10,11,12,17,19,22,24)†	2015	1.267*
50	(1,1,3,7,9,11,13,13,17,19,21,23)	1500	1.073*
51	(1,2,4,5,7,8,13,14,16,19,22,25)	2166	1.266*
52	(1,3,5,7,9,11,15,17,19,21,23,25)	2028	1.204*
53	(1,2,4,5,8,10,11,13,17,20,24,26)†	2417	1.291*
54	(1,1,5,7,11,13,17,17,19,19,23,25)	972	0.805
55	(1,2,3,4,6,9,13,14,19,21,24,26)†	2509	1.272*
56	(1,3,5,9,11,13,15,17,19,23,25,27)	2240	1.182*
57	(1,2,4,5,7,8,14,16,17,22,25,28)	2793	1.299*
58	(1,3,5,7,9,11,13,15,17,19,25,27)	2684	1.253*
59	(1,2,4,7,9,12,15,18,20,23,26,28)†	2964	1.296*
60	(1,1,7,11,11,13,13,17,19,23,23,29)	1200	0.812
61	(1,2,3,4,8,9,15,16,21,22,27,29)‡	3105	1.287*
62	(1,3,5,7,9,11,13,15,19,25,27,29)	3020	1.250*
63	(1,2,4,5,8,10,16,17,20,23,29,31)	3486	1.324**
64	(1,3,5,7,9,11,13,15,17,21,29,31)	3072	1.225*
65	(1,2,4,7,8,9,14,16,18,28,29,32)†	3640	1.314**
66	(1,1,5,7,13,17,19,23,25,25,29,31)	1452	0.819
67	(1,2,3,5,8,12,16,18,22,25,31,32)‡	3709	1.290*

ρ	$n = 12$		
	d	D	D/D_c
68	(1,3,5,7,9,11,13,15,19,25,29,31)	3468	1.231*
69	(1,2,4,7,8,11,16,19,22,25,31,32)†	3861	1.282*
70	(1,3,9,11,13,17,19,23,27,29,31,33)	2940	1.104*
71	(1,2,3,4,9,10,15,17,20,28,31,35)‡	4095	1.286*
72	(1,5,7,11,13,17,19,23,25,29,31,35)	1728	0.825
73	(1,2,3,4,8,12,18,19,25,27,32,35)‡	4380	1.296*
74	(1,3,5,7,11,13,19,23,27,29,31,33)	4440	1.289*
75	(1,2,4,7,8,13,16,19,23,29,32,34)†	4395	1.267*
76	(1,3,5,7,11,13,17,23,27,29,31,37)	4332	1.242*
77	(1,2,3,5,9,13,19,20,26,30,34,37)‡	4777	1.289*
78	(1,5,7,11,17,19,23,25,29,31,35,37)	2028	0.830
79	(1,2,3,4,6,12,16,21,24,29,32,39)‡	4829	1.266*
80	(1,3,7,9,11,13,17,19,23,29,37,39)	3840	1.116*
81	(1,2,4,7,8,14,19,23,25,28,31,40)†	5157	1.279*
82	(1,3,5,7,9,11,17,19,23,27,37,39)†	5084	1.255*
83	(1,2,3,4,5,11,16,22,23,31,35,40)‡	5231	1.259*
84	(1,5,11,13,17,19,23,25,29,31,37,41)	2352	0.835
85	(1,2,3,8,9,14,18,21,28,33,37,41)‡	5742	1.291*
86	(1,3,5,7,9,11,19,23,29,31,39,41)†	5508	1.251*
87	(1,2,4,5,8,16,19,22,28,32,34,43)†	5931	1.284*
88	(1,3,5,7,9,13,17,23,29,31,41,43)†	5696	1.245*
89	(1,2,3,4,5,10,16,22,24,33,37,44)‡	5865	1.251*
90	(1,7,11,13,17,19,23,29,31,37,41,43)	2700	0.840
91	(1,2,3,4,9,15,17,23,27,34,37,45)‡	6349	1.275*
92	(1,3,5,7,9,17,25,27,29,35,41,45)†	6348	1.262*
93	(1,2,4,5,13,16,23,26,32,34,40,46)‡	6684	1.283*
94	(1,3,5,9,11,13,19,23,25,41,43,45)†	6588	1.261*
95	(1,2,3,4,12,18,24,28,32,37,41,47)‡	6844	1.273*
96	(1,5,7,11,13,17,19,23,29,31,43,47)†	3072	0.845
97	(1,2,3,4,5,7,17,23,27,36,37,48)‡	6618	1.228*
98	(1,3,5,11,13,19,23,27,31,41,45,47)†	6860	1.238*
99	(1,2,4,7,13,19,26,29,35,41,43,46)‡	7533	1.286*
100	(1,3,7,9,11,13,17,19,23,29,47,49)†	6000	1.137*

ρ	$n = 13$		
	d	D	D/D_c
2	(1,1,1,1,1,1,1,1,1,1,1,1,1)	13	1.902**
3	(1,1,1,1,1,1,1,1,1,1,1,1,1)	13	1.308**
4	(1,1,1,1,1,1,1,1,1,1,1,1,1)	13	1.003*
5	(1,1,1,1,1,1,1,2,2,2,2,2,2)	31	1.260*
6	(1,1,1,1,1,1,1,1,1,1,1,1,1)	13	0.690
7	(1,1,1,1,1,2,2,2,2,3,3,3,3)	57	1.253*
8	(1,1,1,1,1,1,1,3,3,3,3,3,3)	52	1.058*
9	(1,1,1,1,1,2,2,2,2,4,4,4,4)	85	1.213*
10	(1,1,1,1,1,1,1,3,3,3,3,3,3)	61	0.932
11	(1,1,1,2,2,2,3,3,4,4,4,5,5)	131	1.251*
12	(1,1,1,1,1,1,1,5,5,5,5,5,5)	52	0.728
13	(1,1,1,2,2,3,3,4,4,5,5,6,6)	183	1.268*
14	(1,1,1,1,1,3,3,3,3,5,5,5,5)	141	1.039*
15	(1,1,1,1,2,2,2,4,4,4,7,7,7)	211	1.193*
16	(1,1,1,1,3,3,3,5,5,5,7,7,7)	208	1.116*
17	(1,1,2,2,3,3,4,4,5,6,7,8,8)	298	1.263*
18	(1,1,1,1,1,5,5,5,5,7,7,7,7)	117	0.751
19	(1,1,2,2,3,4,4,5,6,7,8,8,9)	370	1.270*
20	(1,1,1,1,3,3,3,7,7,7,9,9,9)	244	0.983
21	(1,1,1,2,2,4,4,5,5,8,8,10,10)	421	1.235*
22	(1,1,1,3,3,3,5,5,7,7,7,9,9)	389	1.137*
23	(1,1,2,3,4,5,6,7,8,9,10,11)	531	1.275*
24	(1,1,1,1,5,5,5,7,7,7,11,11,11)	208	0.767
25	(1,1,2,2,3,4,6,7,8,8,9,12,12)	617	1.273*
26	(1,1,1,3,3,5,5,7,7,9,9,11,11)	573	1.183*
27	(1,1,2,2,4,5,7,7,8,10,11,11,13)	684	1.248*
28	(1,1,1,3,3,5,5,9,9,11,11,13,13)	564	1.096*
29	(1,1,2,3,4,6,7,8,9,11,12,13,14)	827	1.285*
30	(1,1,1,1,7,7,7,11,11,11,13,13,13)	325	0.781
31	(1,2,3,4,5,6,7,8,9,12,13,14,15)	951	1.296*
32	(1,1,3,3,5,5,7,7,9,11,13,13,15)	832	1.177*
33	(1,1,2,2,4,5,8,8,10,13,13,14,16)	1008	1.259*
34	(1,1,3,3,5,5,7,9,9,11,13,13,15)	965	1.198*

ρ	$n = 13$		
	d	D	D/D_c
35	(1,1,2,3,4,6,8,9,11,12,13,16,17)	1191	1.296*
36	(1,1,1,5,5,7,7,11,11,13,13,17,17)	468	0.792
37	(1,2,3,4,5,6,8,9,12,13,14,16,18)†	1323	1.298*
38	(1,1,3,3,5,5,7,9,11,13,15,15,17)	1229	1.220*
39	(1,1,2,4,5,7,8,10,11,14,16,17,19)	1483	1.309**
40	(1,1,3,3,7,7,9,9,11,13,17,17,19)	976	1.037*
41	(1,2,3,4,5,8,9,10,11,16,17,18,20)†	1597	1.297*
42	(1,1,1,5,5,11,11,13,13,17,17,19,19)	637	0.801
43	(1,2,3,4,5,7,10,11,13,15,16,19,21)†	1755	1.301**
44	(1,1,3,5,5,7,7,9,13,15,17,19,21)	1556	1.199*
45	(1,1,2,4,7,8,11,13,14,16,17,19,22)	1899	1.298*
46	(1,1,3,5,7,9,11,11,13,15,17,19,21)	1877	1.264*
47	(1,2,3,4,6,9,11,14,15,16,19,21,23)†	2054	1.297*
48	(1,1,5,5,7,7,11,13,17,17,19,19,23)	832	0.809
49	(1,2,3,4,6,8,11,13,16,17,18,22,23)†	2245	1.304**
50	(1,1,3,7,9,11,13,13,17,19,19,21,23)	1525	1.055*
51	(1,2,4,5,7,8,10,13,14,16,19,22,25)	2350	1.286*
52	(1,1,3,5,7,9,11,15,17,19,21,23,25)	2197	1.222*
53	(1,2,3,4,7,8,13,14,16,19,22,24,25)†	2654	1.319**
54	(1,1,5,7,7,11,13,13,17,17,19,23,25)	1053	0.817
55	(1,2,3,4,6,8,12,13,17,18,21,23,27)†	2740	1.295*
56	(1,1,3,5,9,11,13,15,17,19,23,25,27)	2256	1.156*
57	(1,2,4,5,7,8,11,13,14,17,23,26,28)†	2923	1.295*
58	(1,3,5,7,9,11,13,15,17,19,21,23,25)	2925	1.274*
59	(1,2,3,6,7,11,12,15,16,20,24,25,29)‡	3209	1.314**
60	(1,1,7,7,11,11,13,13,17,19,23,23,29)	1300	0.823
61	(1,2,3,6,10,11,14,15,18,22,23,27,30)‡	3456	1.322**
62	(1,3,5,7,9,11,13,15,17,19,21,27,29)	3341	1.281*
63	(1,2,4,5,8,10,11,16,17,20,23,29,31)†	3607	1.311**
64	(1,3,5,7,9,11,13,15,17,21,23,29,31)	3328	1.241*
65	(1,2,3,4,9,11,16,17,19,23,24,29,31)†	3815	1.310**
66	(1,1,5,7,13,17,17,19,23,25,29,29,31)	1573	0.829
67	(1,2,3,4,8,11,13,18,19,23,27,28,33)‡	4070	1.316**

ρ	$n = 13$		
	d	D	D/D_c
68	(1,3,5,7,9,11,13,15,19,21,27,31,33)	3757	1.247*
69	(1,2,4,7,8,11,14,16,19,25,28,31,32)†	4255	1.309**
70	(1,1,3,9,11,13,17,19,23,27,29,31,33)	2989	1.083*
71	(1,2,3,4,6,11,14,17,22,23,27,29,35)‡	4446	1.304**
72	(1,1,5,7,11,13,17,19,23,25,29,31,35)†	1872	0.835
73	(1,2,3,4,5,12,13,19,21,26,27,30,36)‡	4594	1.292*
74	(1,3,5,7,9,11,13,17,19,21,29,33,35)†	4644	1.282*
75	(1,2,4,7,8,11,16,17,23,26,29,32,37)†	4981	1.312**
76	(1,3,5,7,9,11,13,17,23,27,29,31,37)†	4693	1.258*
77	(1,2,4,5,9,12,17,20,23,30,31,32,38)‡	5254	1.315**
78	(1,1,5,7,11,17,19,23,25,29,31,35,37)†	2197	0.840
79	(1,2,3,4,5,8,14,18,20,27,30,33,39)‡	5233	1.282*
80	(1,3,7,9,11,13,17,19,21,23,29,37,39)	3904	1.094*
81	(1,2,4,5,8,14,16,19,25,28,31,35,38)†	5785	1.317**
82	(1,3,5,7,9,11,13,21,23,29,33,35,39)†	5700	1.292*
83	(1,2,3,4,5,6,13,19,20,27,31,34,41)‡	5548	1.261*
84	(1,1,5,11,13,17,19,23,25,29,31,37,41)†	2548	0.845
85	(1,2,3,4,11,13,19,21,26,31,36,37,42)‡	6300	1.314**
86	(1,3,5,7,9,11,15,17,23,25,35,39,41)†	6213	1.291*
87	(1,2,4,5,8,14,16,22,25,31,32,37,43)†	6547	1.311**
88	(1,3,5,7,9,13,15,17,23,25,29,41,43)†	6224	1.265*
89	(1,2,3,4,5,6,13,20,21,28,34,36,44)‡	6224	1.252*
90	(1,1,7,11,13,17,19,23,29,31,37,41,43)†	2925	0.849
91	(1,2,3,4,5,10,15,22,23,31,34,40,45)‡	6973	1.298*
92	(1,3,5,9,11,13,17,19,21,33,39,43,45)†	6877	1.276*
93	(1,2,4,7,8,14,19,23,29,32,35,38,46)‡	7402	1.311**
94	(1,3,5,7,9,17,19,21,27,35,39,43,45)†	7356	1.294*
95	(1,2,3,4,6,12,16,24,26,31,37,39,47)‡	7658	1.308**
96	(1,5,7,11,13,17,19,23,29,31,37,41,43)†	3328	0.854
97	(1,2,3,4,5,6,8,17,24,26,37,39,48)‡	6890	1.217*
98	(1,3,5,9,11,13,19,23,27,31,41,45,47)†	6909	1.207*
99	(1,2,4,7,8,13,16,26,28,32,35,46,49)‡	8326	1.312**
100	(1,3,7,9,11,13,17,19,23,29,37,47,49)†	6100	1.113*

ρ	$n = 14$		
	d	D	D/D_c
2	(1,1,1,1,1,1,1,1,1,1,1,1,1,1)	14	1.966**
3	(1,1,1,1,1,1,1,1,1,1,1,1,1,1)	14	1.349**
4	(1,1,1,1,1,1,1,1,1,1,1,1,1,1)	14	1.033*
5	(1,1,1,1,1,1,1,2,2,2,2,2,2,2)	35	1.327**
6	(1,1,1,1,1,1,1,1,1,1,1,1,1,1)	14	0.709
7	(1,1,1,1,1,2,2,2,2,2,3,3,3,3)	61	1.282*
8	(1,1,1,1,1,1,1,3,3,3,3,3,3,3)	56	1.085*
9	(1,1,1,1,1,2,2,2,2,2,4,4,4,4)	89	1.226*
10	(1,1,1,1,1,1,1,3,3,3,3,3,3,3)	70	0.986
11	(1,1,1,1,2,2,3,3,3,4,4,5,5,5)	144	1.295*
12	(1,1,1,1,1,1,1,5,5,5,5,5,5,5)	56	0.745
13	(1,1,1,2,2,3,3,4,4,5,5,5,6,6)	195	1.290*
14	(1,1,1,1,1,3,3,3,3,3,5,5,5,5)	150	1.056*
15	(1,1,1,1,2,2,2,4,4,4,4,4,7,7,7)	227	1.219*
16	(1,1,1,1,3,3,3,3,5,5,5,7,7,7)	224	1.140*
17	(1,1,2,2,3,3,4,4,5,6,6,7,8,8)	328	1.304**
18	(1,1,1,1,1,5,5,5,5,5,7,7,7,7)	126	0.767
19	(1,1,2,2,3,3,4,5,5,6,7,8,9,9)	403	1.304**
20	(1,1,1,1,3,3,3,3,7,7,7,9,9,9)	280	1.036*
21	(1,1,1,2,2,4,4,5,5,8,8,8,10,10)	449	1.254*
22	(1,1,1,1,3,3,3,5,5,7,7,9,9,9)	422	1.164*
23	(1,1,2,2,3,4,5,6,7,7,8,9,11,11)	577	1.307**
24	(1,1,1,1,5,5,5,7,7,7,11,11,11,11)	224	0.783
25	(1,1,2,3,4,4,6,6,7,8,9,9,11,12)	651	1.284*
26	(1,1,1,3,3,5,5,7,7,7,9,9,11,11)	606	1.195*
27	(1,1,2,2,4,4,5,7,7,8,10,11,13,13)	755	1.288*
28	(1,1,1,3,3,5,5,5,9,9,11,11,13,13)	600	1.110*
29	(1,2,3,4,5,6,7,8,9,10,11,12,13,14)	1015	1.397**
30	(1,1,1,1,7,7,7,7,11,11,11,13,13,13)	350	0.795
31	(1,1,2,3,4,6,7,8,9,11,12,13,14,15)	1023	1.319**
32	(1,1,3,3,5,5,7,7,9,9,11,13,13,15)	896	1.198*
33	(1,1,2,2,4,5,7,8,10,10,13,14,14,16)	1094	1.287*
34	(1,1,3,3,5,5,7,7,9,9,11,13,15,15)	1070	1.238*

ρ	$n = 14$		
	d	D	D/D_c
35	(1,1,2,3,4,6,6,8,9,11,12,13,16,17)	1227	1.290*
36	(1,1,1,5,5,7,7,11,11,11,13,13,17,17)	504	0.806
37	(1,2,3,4,5,6,7,8,9,10,14,16,17,18)†	1439	1.327**
38	(1,1,3,3,5,5,7,9,11,11,13,15,15,17)	1350	1.254*
39	(1,1,2,4,5,7,8,10,11,14,14,16,17,19)	1571	1.320**
40	(1,1,3,3,7,7,9,9,11,11,13,17,17,19)	1120	1.089*
41	(1,2,3,4,5,6,8,10,11,13,14,15,19,20)†	1727	1.321**
42	(1,1,1,5,5,11,11,11,13,13,17,17,19,19)	686	0.814
43	(1,2,3,4,5,6,8,10,11,12,16,19,20,21)†	1978	1.353**
44	(1,1,3,3,5,7,9,13,13,15,17,19,21,21)	1688	1.224*
45	(1,1,2,4,7,8,11,13,14,16,17,19,19,22)	2043	1.318**
46	(1,1,3,3,5,7,9,11,13,15,17,19,21,21)	1998	1.277*
47	(1,2,3,4,6,7,10,11,12,15,18,19,20,23)†	2280	1.338**
48	(1,1,5,5,7,7,11,11,13,13,17,19,23,23)	896	0.822
49	(1,2,3,4,5,9,10,11,15,16,17,19,22,24)†	2450	1.334**
50	(1,1,3,7,7,9,9,11,13,13,17,19,21,23)	1750	1.106*
51	(1,2,4,5,7,8,10,11,13,14,20,22,23,25)†	2606	1.326**
52	(1,1,3,5,7,9,11,11,15,17,19,21,23,25)	2366	1.240*
53	(1,2,3,4,8,9,10,14,15,16,20,21,22,26)†	2846	1.337**
54	(1,1,5,5,7,11,11,13,17,17,19,19,23,25)	1134	0.829
55	(1,2,3,6,7,8,12,13,16,17,18,21,23,27)†	3000	1.326**
56	(1,1,3,5,9,11,13,15,17,17,19,23,25,27)	2400	1.166*
57	(1,2,4,5,7,8,10,11,14,17,20,23,26,28)†	3222	1.329**
58	(1,3,5,7,9,11,13,15,17,19,21,23,25,27)	3654	1.393**
59	(1,2,3,6,7,8,11,12,15,16,20,25,27,29)‡	3504	1.343**
60	(1,1,7,7,11,11,13,13,17,19,19,23,23,29)	1400	0.835
61	(1,2,3,4,9,11,15,17,18,22,23,25,28,30)‡	3782	1.352**
62	(1,1,3,5,7,11,13,15,17,19,23,25,27,29)	3700	1.317**
63	(1,2,4,5,8,10,11,16,17,20,23,25,29,31)†	4041	1.357**
64	(1,3,5,7,9,11,13,15,17,19,21,23,27,31)†	3584	1.259*
65	(1,2,3,7,8,12,14,16,18,23,27,28,29,32)†	4189	1.342**
66	(1,1,5,7,13,17,17,19,23,25,25,29,29,31)	1694	0.841
67	(1,2,3,4,6,9,11,16,17,21,24,28,29,31)‡	4398	1.337**

ρ	$n = 14$		
	d	D	D/D_c
68	(1,3,5,7,9,11,13,15,19,21,23,25,27,33)†	4046	1.264*
69	(1,2,4,5,7,8,11,16,19,20,22,25,32,34)†	4572	1.326**
70	(1,1,3,9,11,13,17,17,19,23,27,29,31,33)	3430	1.133*
71	(1,2,3,4,5,8,11,17,18,19,27,29,31,35)‡	4828	1.327**
72	(1,1,5,7,11,11,13,17,19,23,25,29,31,35)†	2016	0.846
73	(1,2,3,4,5,6,10,16,18,19,27,29,32,36)‡	4957	1.310**
74	(1,3,5,7,9,11,13,15,17,21,27,31,33,35)†	5140	1.318**
75	(1,2,4,7,8,11,14,16,17,19,28,32,34,37)†	5375	1.331**
76	(1,3,5,7,9,11,13,15,17,23,27,29,31,37)†	5054	1.275*
77	(1,2,3,4,9,10,15,17,18,23,29,31,34,38)‡	5741	1.342**
78	(1,1,5,7,11,17,19,23,25,29,29,31,35,37)†	2366	0.851
79	(1,2,3,4,5,6,10,16,20,21,28,29,37,39)‡	5674	1.303**
80	(1,3,7,9,11,13,17,19,21,23,27,31,37,39)†	4480	1.144*
81	(1,2,4,7,10,13,14,16,22,23,28,32,34,40)†	6317	1.343**
82	(1,3,5,7,9,11,13,17,19,23,25,35,37,39)†	6262	1.322**
83	(1,2,3,4,5,6,8,18,19,20,30,31,37,41)‡	5974	1.277*
84	(1,1,5,11,11,13,17,19,23,25,29,31,37,41)†	2744	0.856
85	(1,2,3,4,8,14,16,21,23,27,32,33,38,42)‡	6901	1.342**
86	(1,3,5,7,9,11,13,17,19,21,31,35,39,41)†	6782	1.316**
87	(1,2,4,7,8,13,14,16,23,25,28,34,40,43)†	7271	1.348**
88	(1,3,5,7,9,13,17,19,25,27,31,37,41,43)†	6752	1.286*
89	(1,2,3,4,5,6,7,15,22,23,31,33,39,43)‡	6625	1.260*
90	(1,1,7,11,13,17,19,23,23,29,31,37,41,43)†	3150	0.860
91	(1,2,3,4,5,8,16,17,23,27,32,34,38,45)‡	7534	1.316**
92	(1,3,5,7,9,11,13,19,21,25,31,41,43,45)†	7406	1.292*
93	(1,2,4,5,8,10,16,20,23,29,32,35,40,46)‡	8201	1.346**
94	(1,3,5,7,9,11,17,23,27,29,31,39,41,45)†	8078	1.323**
95	(1,2,3,4,6,8,14,21,23,26,36,37,42,47)‡	8125	1.314**
96	(1,5,7,11,13,17,19,23,25,29,31,41,43,47)†	3584	0.864
97	(1,2,3,4,5,6,7,10,20,24,28,37,41,48)‡	7300	1.221*
98	(1,3,5,9,11,13,15,19,23,27,31,41,45,47)†	7350	1.214*
99	(1,2,4,5,8,10,17,20,26,29,34,40,41,47)‡	9062	1.335**
100	(1,3,7,9,11,13,17,19,23,29,31,37,47,49)†	7000	1.163*

ρ	$n = 15$		
	d	D	D/D_c
2	(1,1,1,1,1,1,1,1,1,1,1,1,1,1,1)	15	2.028**
3	(1,1,1,1,1,1,1,1,1,1,1,1,1,1,1)	15	1.389**
4	(1,1,1,1,1,1,1,1,1,1,1,1,1,1,1)	15	1.062*
5	(1,1,1,1,1,1,1,1,2,2,2,2,2,2,2)	36	1.336**
6	(1,1,1,1,1,1,1,1,1,1,1,1,1,1,1)	15	0.727
7	(1,1,1,1,1,2,2,2,2,2,3,3,3,3,3)	70	1.361**
8	(1,1,1,1,1,1,1,1,3,3,3,3,3,3,3)	60	1.112*
9	(1,1,1,1,1,2,2,2,2,2,4,4,4,4,4)	105	1.318**
10	(1,1,1,1,1,1,1,1,3,3,3,3,3,3,3)	71	0.982
11	(1,1,1,2,2,2,3,3,3,4,4,4,5,5,5)	165	1.370**
12	(1,1,1,1,1,1,1,1,5,5,5,5,5,5,5)	60	0.762
13	(1,1,1,2,2,2,3,3,4,4,5,5,5,6,6)	211	1.326**
14	(1,1,1,1,1,3,3,3,3,3,5,5,5,5,5)	175	1.127*
15	(1,1,1,1,2,2,2,2,4,4,4,4,7,7,7)	231	1.214*
16	(1,1,1,1,3,3,3,3,5,5,5,5,7,7,7)	240	1.165*
17	(1,1,1,2,2,3,4,4,5,5,6,7,7,8,8)	350	1.329**
18	(1,1,1,1,1,5,5,5,5,5,7,7,7,7,7)	135	0.783
19	(1,1,2,2,3,3,4,5,5,6,7,7,8,8,9)	437	1.339**
20	(1,1,1,1,3,3,3,3,7,7,7,7,9,9,9)	284	1.029*
21	(1,1,1,2,2,2,4,4,5,5,8,8,8,10,10)	465	1.258*
22	(1,1,1,3,3,3,5,5,5,7,7,7,9,9,9)	495	1.243*
23	(1,1,2,2,3,4,5,6,6,7,8,9,10,10,11)	623	1.338**
24	(1,1,1,1,5,5,5,5,7,7,7,7,11,11,11)	240	0.798
25	(1,1,2,3,4,4,6,6,7,8,9,9,11,11,12)	750	1.358**
26	(1,1,1,3,3,3,5,5,7,7,9,9,9,11,11)	663	1.231*
27	(1,1,2,2,4,4,5,5,7,8,8,10,11,13,13)	810	1.313**
28	(1,1,1,3,3,3,5,5,5,9,9,11,11,13,13)	700	1.180*
29	(1,1,2,3,4,5,6,7,8,9,10,11,12,13,14)	1016	1.376**
30	(1,1,1,1,7,7,7,7,11,11,11,11,13,13,13)	375	0.810
31	(1,2,3,4,5,6,7,8,9,10,11,12,13,14,15)	1240	1.428**
32	(1,1,3,3,5,5,7,7,9,9,11,11,13,13,15)	960	1.220*
33	(1,1,2,2,4,4,5,7,8,8,10,13,14,16,16)	1221	1.337**
34	(1,1,1,3,3,5,5,7,9,11,11,13,13,15,15)	1151	1.262*

ρ	$n = 15$		
	d	D	D/D_c
35	(1,1,2,3,4,4,6,6,8,9,11,12,16,16,17)	1330	1.321**
36	(1,1,1,5,5,7,7,11,11,11,13,13,13,17,17)	540	0.820
37	(1,2,3,4,5,6,7,8,10,11,13,14,15,17,18)†	1591	1.371**
38	(1,1,3,3,5,5,7,7,9,11,11,13,15,17,17)	1463	1.283*
39	(1,1,2,2,4,5,7,8,10,11,14,14,16,17,19)	1623	1.319**
40	(1,1,3,3,7,7,9,9,11,11,13,13,17,17,19)	1136	1.078*
41	(1,2,3,4,5,6,7,9,11,12,13,14,17,19,20)†	1901	1.362**
42	(1,1,1,5,5,5,11,11,11,13,13,17,17,19,19)	735	0.828
43	(1,2,3,4,5,6,7,11,12,13,14,15,19,20,21)†	2079	1.363**
44	(1,1,3,5,5,7,7,9,9,13,15,17,19,19,21)	1815	1.246*
45	(1,1,2,2,4,7,8,8,11,13,14,16,17,19,22)	2079	1.306**
46	(1,1,3,3,5,7,7,9,11,13,15,17,19,19,21)	2151	1.301**
47	(1,2,3,4,5,6,8,12,13,14,15,16,18,22,23)†	2457	1.363**
48	(1,1,5,5,7,7,11,11,13,13,17,17,19,19,23)	960	0.836
49	(1,2,3,4,5,6,8,11,12,13,15,19,20,22,24)†	2635	1.358**
50	(1,1,3,3,7,7,9,11,11,13,17,19,21,23,23)	1775	1.094*
51	(1,2,4,5,7,8,10,11,13,14,16,19,20,22,23)†	2775	1.342**
52	(1,1,3,5,7,9,9,11,15,17,19,21,23,23,25)	2535	1.260*
53	(1,2,3,4,6,7,10,11,12,15,16,20,23,24,25)†	3091	1.367**
54	(1,1,5,5,7,7,11,11,13,17,17,19,19,23,25)	1215	0.842
55	(1,2,3,4,6,7,8,12,13,16,17,21,23,24,26)†	3246	1.353**
56	(1,1,3,5,9,9,11,13,15,17,17,19,23,25,27)	2800	1.236*
57	(1,2,4,5,7,8,10,11,13,14,16,22,25,26,28)†	3459	1.351**
58	(1,1,3,5,7,9,11,13,15,17,19,21,23,25,27)	3655	1.366**
59	(1,2,3,4,6,9,10,14,15,17,21,22,25,26,29)‡	3827	1.376**
60	(1,1,7,7,11,11,13,13,17,17,19,19,23,23,29)	1500	0.848
61	(1,2,3,4,6,10,11,15,17,18,22,23,25,27,30)‡	4087	1.378**
62	(1,3,5,7,9,11,13,15,17,19,21,23,25,27,29)†	4495	1.424**
63	(1,2,4,5,8,10,11,13,16,17,20,23,25,29,31)†	4185	1.354**
64	(1,3,5,7,9,11,13,15,17,19,21,23,25,27,29)†	3840	1.278**
65	(1,2,3,4,6,11,12,14,16,19,22,23,24,28,32)†	4525	1.367**
66	(1,1,5,7,13,17,17,19,23,25,25,29,29,31,31)	1815	0.853
67	(1,2,3,4,5,8,10,14,16,18,20,25,27,31,32)‡	4734	1.359**

ρ	$n = 15$		
	d	D	D/D_c
68	(1,3,5,7,9,11,13,15,19,21,23,25,27,29,31)†	4335	1.283*
69	(1,2,4,5,7,10,13,16,17,20,26,28,29,32,34)†	5013	1.361**
70	(1,1,3,9,9,11,13,17,17,19,23,27,29,31,33)	3479	1.118*
71	(1,2,3,4,5,6,10,12,18,19,20,27,28,32,35)‡	5179	1.347**
72	(1,1,1,5,7,11,11,13,17,19,23,25,29,31,35)†	2160	0.858
73	(1,2,3,4,5,6,9,13,18,20,21,26,28,35,36)‡	5450	1.346**
74	(1,3,5,7,9,11,13,15,17,19,21,25,31,33,35)†	5631	1.351**
75	(1,2,4,7,8,11,13,14,16,19,22,28,31,32,37)†	5775	1.351**
76	(1,3,5,7,9,11,13,15,17,21,25,27,33,35,37)†	5415	1.292*
77	(1,2,3,4,8,13,18,19,24,25,29,30,32,36,38)‡	6207	1.367**
78	(1,1,1,5,7,11,17,19,23,25,29,29,31,35,37)†	2535	0.863
79	(1,2,3,4,5,6,7,13,19,20,22,30,31,34,39)‡	6085	1.321**
80	(1,3,7,9,11,13,17,19,21,23,27,29,31,33,37)†	4544	1.129*
81	(1,2,4,5,7,0,10,10,20,22,23,31,34,37,40)†	6834	1.368**
82	(1,3,5,7,9,11,13,21,23,25,29,31,33,37,39)†	7052	1.374**
83	(1,2,3,4,5,6,7,10,18,21,22,29,31,40,41)‡	6521	1.306**
84	(1,1,1,5,11,13,17,19,23,25,29,31,37,37,41)†	2940	0.867
85	(1,2,3,4,6,8,12,16,21,23,28,29,32,39,42)‡	7406	1.361**
86	(1,3,5,7,9,11,13,21,23,25,29,31,35,37,41)†	7439	1.350**
87	(1,2,4,5,7,8,14,16,22,25,28,31,32,41,43)†	7803	1.367**
88	(1,3,5,7,9,13,15,17,19,21,23,29,35,39,41)†	7260	1.305**
89	(1,2,3,4,5,6,7,8,17,22,23,31,33,40,44)‡	7092	1.276*
90	(1,1,1,7,11,13,17,19,23,23,29,31,37,41,43)†	3375	0.871
91	(1,2,3,4,5,6,10,15,22,23,30,31,34,43,44)‡	7990	1.327*
92	(1,3,5,7,9,11,13,17,19,21,25,31,39,43,45)†	7935	1.309**
93	(1,2,4,5,8,13,16,20,22,23,29,32,35,41,46)‡	8835	1.367*
94	(1,3,5,7,9,11,13,19,25,27,29,31,39,41,45)†	8852	1.355**
95	(1,2,3,4,6,7,11,18,23,24,32,33,36,46,47)‡	8866	1.343*
96	(1,5,7,11,13,17,19,23,25,29,31,35,37,41,43)†	3840	0.875
97	(1,2,3,4,5,6,7,8,12,21,25,26,37,47,48)‡	7929	1.245*
98	(1,3,5,9,11,13,15,19,23,25,27,31,41,45,47)†	8575	1.283*
99	(1,2,4,5,10,13,14,20,23,28,31,35,40,43,47)‡	9936	1.368**
100	(1,3,7,9,11,13,17,19,23,29,31,37,43,47,49)†	7100	1.145*

Appendix C

Associative Memory Network Software Structure

The software developed to implement the B-spline and CMAC networks is now described. The network descriptions given in this book have shown their basic structures to be very similar, and this has been exploited so that the same software data structures and functional interfaces can be used for both networks. The data structures can be simply initialised from a data file which specifies the network's parameters, and the trained network can be saved to a similar file, allowing the network to be shared between different applications.

C.1 DATA STRUCTURES

The organisation of the software data structures reflects the physical implementation, and is designed in a hierarchical fashion which allows its functionality to be extended with a minimal amount of effort. The basic data structure is shown in Figure C.1, and this is used to implement the CMAC and the B-spline network. Each substructure in this figure has a number of parameters associated with it, as well as pointers to other substructures, and some of these will now be described. The parameter associated with the AMN substructure is simply the network type (CMAC, B-spline, fuzzy), and there are three pointers to other substructures: Dimension, Update and Overlay. The Dimension substructure is of the form:

```
struct Dimension {
    inputDimension;
    outputDimension;
    numberOfOverlays;
    Axis pointerVector; }
```

and contains the basic network structure parameters. Also included is an array of pointers to the parameters associated with each input axis, which are the basis function shape, its width and the knot vector.

The Overlay substructure includes the following data:

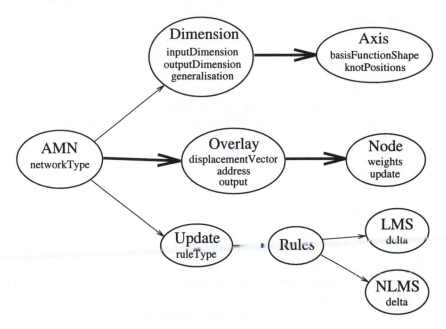

Figure C.1   The software data structure which stores the network's parameters. Arrows signify pointers whereas bold arrows represent pointer vectors.

```
struct Overlay {
    displacementVector;
    activeAddress;
    activeOutput;
    Node pointerVector; }
```

where a unique displacement vector is associated with each overlay, and for each input one and only one basis function is non-zero. Therefore only the address and the output of one basis function needs to be calculated and these are stored for use in the network output and updating routines. Finally, the Node substructure is composed of the weight (vector) and the update parameter, which remembers how many times the weight has been updated, and the appropriate local Node vector associated with each overlay is stored in the Overlay substructure. Thus each Overlay substructure has *local* memory requirements and a similar computational cost, and is therefore suitable for implementation on a Single Instruction Multiple Data (SIMD) parallel computer.

The Update substructure is also designed so that new learning rules can be easily incorporated into the software routines. Thus the data structure has been abstracted to a level which represents several networks, and was constructed so that the functionality was easily extendable.

C.2   INTERFACE FUNCTIONS

A set of interface functions has also been developed which allows the user to re-call information and manipulate the data structure. These routines control the user's access to the data, without limiting the range of operations which can be performed. They are used to hide the software complexity, so that a programmer can concentrate on the application without worrying about developing standard AMN code.

The data structure and the interface function declarations are contained in a header file "amn.h", which must be included at the top of any C file. The programmer can then use any of the following routines:

- initFile(), which allocates and sets the data structure by reading parameters from a network initialisation file. Two examples of these files are contained in the next section.
- output(), which calculates the network output for a given input.
- learn(), which adjusts the weight vector using any one of a range of different LMS learning rules.
- dumpNetwork(), which saves the modified network to a file.
- deleteNetwork(), which deallocates the memory used by the data structure.

Other functions allow the whole weight vector to be set to zero, and the output of a particular basis function to be accessed and updated. All of the parameters which are necessary to implement these functions (type of learning rule, basis function shape, etc.) are contained in the network initialisation file and two examples are now given.

C.2.1   Initialisation Files

A typical B-spline network initialisation file is given by:

```
network_type:            bspline
input_dimension:         2
output_dimension:        1
output_scaling:          yes
update_rule:             LMS   0.5
dead_zone:               0.0
field_shape:             constant    linear
number_of_interior_knots: 1  3
input_knots:        1 -1.5  1 0.0   1 1.5
                    2 -3.0  1 -1.0  1 1.0  1 3.0  2 5.0
weight_setting:     weights  1 0.0  10 3.4  5 5.3  8 2.1
                    3 1.1 12 0.5  7 2.5  6 4.3  9 1.7  6 3.9
```

and a CMAC network initialisation file is similar to:

```
network_type:            cmac
input_dimension:         2
output_dimension:        1
number_of_layers:        5
output_scaling:          yes
update_rule:             SALMS  0.5  30.0  0.01
dead_zone:               0.0
displacement_vector:     manual  1  2
field_shape:             linear    linear
number_of_interior_knots: 10  20
input_knots:             even  -1.5   1.5
                         even  -3.0   5.0
weight_setting:          no
```

The network_type parameter specifies the network structure and determines which keywords are given in the initialisation file, and this is the *only* time that a distinction is made between the two networks.

C.3   SAMPLE C CODE

To illustrate the potential of these software library routines, the C code necessary to store the data pair {0.0, 1.0} in an AMN is shown below. A pointer to the network data structure AMN is declared and is set using the parameters in the network initialisation file "Bspline.cfg". The response of the network is calculated by calling the output() function and the network is trained as the learn() function is called. The network is saved to a file using the dumpNetwork() routine, and the data structure is deallocated using the deleteNetwork() function.

```c
#include "header/amn.h"

main()
{
    double x[1], y[1], yd[1];
    AMN *amn;                            /* declare network */

    amn = initFile("Bspline.cfg");       /* initialise network */

    x[0] = 0.0;   yd[0] = 1.0;

    output(x, y, amn);          /* get network output */
```

```
    learn(yd, y, amn);          /* train network */

    dumpNetwork("Bspline1.cfg", amn);  /* save network to file */
    deleteNetwork(amn);               /* deallocate network */

    return 0;
}
```

The software complexity has been completely hidden from the user, while a complete set of functions has been made available. In addition, a high-level, graphical user interface has been developed which means that the developer does not even have to write any C code [An *et al.*, 1994a].

Appendix D

Fuzzy Intersection

The following work compares using the *min* and the *product* operator to generalise the logical intersection for an n-dimensional input space, where $n \geq 2$. The analysis covers normalised triangular and trapezoidal fuzzy sets, and can be extended to incorporate B-splines of order > 2. The defuzzified output is calculated using the *product* operator to implement implication and centre of area defuzzification. It shows that the *min* operator gives a larger confidence (relative to the *product* operator) to the set centres further away. Similarly, it is also proven that the *min* operator gives a relatively smaller confidence to the set centres close by.

Without loss of generality, the input space is assumed to be an n-dimensional unit hypercube. There exist 2^n, n-dimensional fuzzy sets, each centred at a unique corner of the input space and their supports are equal to the input space (this was the arrangement considered in Section 10.4.1 for $n = 2$). Then a *membership tree* may be constructed, as shown in Figure D.1, where the normalised cost of going

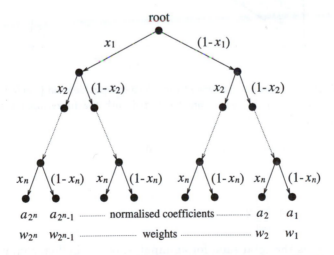

Figure D.1 The fuzzy input membership tree: the normalised membership of each fuzzy input is given by the normalised cost of going from the root to the appropriate leaf.

from the root to a leaf gives the membership of the appropriate multi-dimensional fuzzy input set for a particular input. By relabelling the axes (if necessary), it may also be assumed that:

$$x_1 \leq \cdots \leq x_n \leq 0.5 \tag{D.1}$$

This gives the membership of one of the univariate fuzzy sets for each input coordinate. The memberships of the other univariate fuzzy sets (because they are normalised) are given by $(1-x_i)$ $(i = 1, \ldots, n)$. In addition it may also be assumed that $x_1 > 0$, or else we may consider only the right-hand subtree in Figure D.1, where the root is translated down one level and $n \leftarrow n - 1$ (implicitly assuming that the new n exceeds 1). Hence Equation D.1 becomes:

$$0 < x_1 \leq \cdots \leq x_n \leq 0.5 \tag{D.2}$$

Additionally, the case when $x_1 = \cdots = x_n = 0.5$ is considered later, so it may be assumed that $x_1 \in (0, 0.5)$ and so $(1 - x_1) \in (0.5, 1)$. This arrangement is shown in Figure D.2 for $n = 2$.

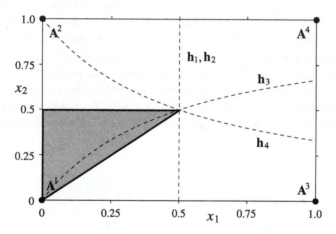

Figure D.2   A two-dimensional example of the relevant input domain (shaded area) and the hypercurves (dashed lines), h_i, which are associated with the fuzzy input sets, $\mathbf{A}^i$, whose centres are shown as bold circles.

It can easily be shown, by induction, that:

$$\sum_i \mu_{A^i}^*(\mathbf{x}) \equiv 1 \tag{D.3}$$

$$\sum_i \mu_{A^i}^M(\mathbf{x}) \equiv 1 + \sum_{j=1}^{n-1} 2^{n-j} x_j \tag{D.4}$$

where $\sum_i \mu_{A_i}^*(\mathbf{x})$ is the total sum, for an input $\mathbf{x}$, of all the fuzzy input sets, using the *product* operator to represent the logical AND, whilst $\sum_i \mu_{A_i}^M(\mathbf{x})$ is the total sum using *min* to represent the logical AND.

Let a_i^* and a_i^M be the coefficients of the weight w_i when the *product* and *min* operators are used to generalise the logical AND, respectively. Then:

$$a_i^* = \prod_{j=1}^{n} t_{ij}$$

where t_{ij} is either x_l or $(1 - x_l)$ for each $j, l = 1, \ldots, n$: the appropriate values can be obtained from the membership tree in Figure D.1. In addition, it may be assumed that the t_{ij} are ordered such that $t_{i1} \leq \cdots \leq t_{in}$. So by Equation D.4:

$$a_i^M = \frac{t_{i1}}{1 + \sum_{j=1}^{n-1} 2^{n-j} x_j}$$

Hence, $a_i^* > a_i^M$ if and only if:

$$\prod_{j=2}^{n} t_{ij} > \frac{1}{1 + \sum_{j=1}^{n-1} 2^{n-j} x_j} \tag{D.5}$$

Similarly, $a_i^* < a_i^M$ if and only if:

$$\prod_{j=2}^{n} t_{ij} < \frac{1}{1 + \sum_{j=1}^{n-1} 2^{n-j} x_j} \tag{D.6}$$

So for each coefficient, there exists an $(n-1)$-dimensional hypercurve partitioning the domain given by D.2 into regions where $a_i^M > a_i^*$ or vice versa. This hypercurve is given by:

$$\prod_{j=2}^{n} t_{ij} = \frac{1}{1 + \sum_{j=1}^{n-1} 2^{n-j} x_j}$$

For most of the coefficients, there exist at least two inputs, lying in the domain described by D, such that one satisfies Equation D.5 and the other Equation D.6. This is illustrated in Figure D.2, where the hypercurve corresponding to the third coefficient is drawn, $n = 2$. In order to proceed with the analysis, it is shown that Equation D.5 is always true for the first two coefficients (these correspond to the fuzzy sets closest to x). Similarly, it is also proved that Equation D.6 is always true for the last coefficient (corresponding to the fuzzy set furthest away from x). Now consider the coefficients associated with weight w_{2^n}:

$$a_{2^n}^* = \prod_{j=1}^{n} x_j, \quad a_{2^n}^M = \frac{x_1}{1 + \sum_{j=1}^{n-1} 2^{n-j} x_j}$$

so D.6 holds when:

$$\prod_{j=2}^{n} x_j - \frac{1}{1 + \sum_{j=1}^{n-1} 2^{n-j} x_j} < 0 \tag{D.7}$$

by D.2, the first term on the left-hand side is bounded above by 2^{1-n}. The second term can be bounded below by $(1 + (2^n - 2)x_{n-1})^{-1}$ which in turn exceeds 2^{1-n} (by D.2). Hence D.7 is always true, so for the fuzzy set furthest from **x**:

$$a_{2^n}^* < a_{2^n}^M \qquad \forall \mathbf{x}$$

Consider the coefficients associated with the weight w_1:

$$a_1^* = \prod_{j=1}^{n}(1 - x_j), \qquad a_1^M = \frac{1 - x_n}{1 + \sum_{j=1}^{n-1} 2^{n-j} x_j}$$

so Equation D.6 holds if and only if:

$$\prod_{j=1}^{n-1} x_j - \frac{1}{1 + \sum_{j=1}^{n-1} 2^{n-j} x_j} > 0 \tag{D.8}$$

This will be proved by induction, so first consider $n = 2$. Then the LHS of D.8 becomes $1 - x_1 - (1 + 2x_1)^{-1}$, now this is greater than zero (for $x_1 \in (0, 0.5)$) if and only if $x_1(1 - 2x_1) > 0$. Clearly this is true for $x_1 \in (0, 0.5)$, so D.8 holds for $n = 2$.

Now assume D.8 holds true for an $(n-1)$-dimensional input space and consider the LHS of D.8 for an n-dimensional input.

$$\prod_{j=1}^{n-1}(1 - x_j) - \frac{1}{1 + \sum_{j=1}^{n-1} 2^{n-j} x_j}$$

This is positive when:

$$\prod_{j=1}^{n-1}(1 - x_j)\left(1 + 2\sum_{j=1}^{n-2} 2^{n-(j+1)} x_j + 2x_{n-1}\right) - 1 > 0$$

Expanding the left-hand side gives:

$$\left(\left(1 + \sum_{j=1}^{n-2} 2^{n-(j+1)} x_j\right)\prod_{j=1}^{n-2}(1 - x_j) - 1\right) -$$

$$x_{n-1}\prod_{j=1}^{n-2}(1 - x_j)\left(1 + \sum_{j=1}^{n-2} 2^{n-(j+1)} x_j\right) +$$

$$\prod_{j=1}^{n-1}(1 - x_j)\left(\sum_{j=1}^{n-2} 2^{n-(j+1)} x_j + 2x_{n-1}\right)$$

By assumption the first term is greater than zero, and so combining the second and third terms gives:

$$\prod_{j=1}^{n-2}\left((1 - x_{n-1})\left(\sum_{j=1}^{n-2} 2^{n-(j+1)} x_j + 2x_{n-1}\right) - x_{n-1}\left(1 + \sum_{j=1}^{n-2} 2^{n-(j+1)} x_j\right)\right)$$

By definition, $\prod_{j=1}^{n-2}(1-x_j)$ is greater than zero, thus the above expression is greater than or equal to zero when:

$$\sum_{j=1}^{n-2} 2^{n-(j+1)} x_j + x_{n-1} - x_{n-1} \sum_{j=1}^{n-1} 2^{n-j} x_j \geq 0$$

which is true if and only if:

$$\sum_{j=1}^{n-1} 2^{n-(j+1)} x_j (1 - 2x_{n-1}) \geq 0$$

But, by D.2, $x_i \in (0, 0.5]$ $\forall i$, so the above expression is true and the proof by induction is complete. So, for the nearest fuzzy set:

$$a_1^* > a_1^M$$

The same conclusion can be drawn about the coefficients of w_2 as they are given by:

$$a_2^* = x_m \prod_{j=1}^{n-1}(1-x_j) \qquad a_2^M = \frac{x_n}{1 + \sum_{j=1}^{n-1} 2^{n-j} x_j}$$

So D.6 holds when D.8 is satisfied. However, it has just been shown that D.8 is true for all **x** satisfying D.2, so:

$$a_2^* > a_2^M \qquad \forall \mathbf{x}$$

It now only remains to consider the case when $x_1 = \cdots = x_n = 0.5$. It can easily be shown that:

$$a_i^* = a_i^M = 2^{-n} \qquad \forall i = 1, \ldots, 2^n$$

and so the normalised *min* and *product* operators are equivalent at the centre of the input space.

Appendix E

Weight to Rule Confidence Vector Map

In this appendix, *symmetric* B-splines are used to define the output fuzzy sets, the product operator is used to implement implication and intersection, B-splines are used to define the input fuzzy sets and a modified centre of area defuzzification strategy is employed. It it shown that when a weight is transformed into a rule confidence vector, the mapping is invertible (for B-splines of order 2 and greater) and that the centre of area defuzzification strategy employed is simply to multiply the rule confidence by the *centre* of the output fuzzy sets.

Without loss of generality, the B-spline knot set can be assumed to be defined on the integers and the desired value w can be assumed to lie in the first interval (i.e. $w \in [0,1)$). By induction it will be proved that the desired value to rule confidence mapping is invertible. Initially consider the case when B-splines of order 2 are used to define the output fuzzy sets. Then:

$$
\begin{aligned}
y_{2,1}^c &= 0.0 \\
N_{2,1}(w) &= (1-w) \\
y_{2,2}^c &= 1.0 \\
N_{2,2}(w) &= w
\end{aligned}
\tag{E.1}
$$

where $N_{k,j}$ is the jth B-spline of order k and $y_{k,j}^c$ is its centre. It can be clearly seen that:

$$
w = \sum_{j=1}^{2} y_{2,j}^c N_{2,j}(w)
$$

Assume that the relationship:

$$
w = \sum_{j=1}^{k-1} y_{k-1,j}^c N_{k-1,j}(w)
\tag{E.2}
$$

holds for B-splines of order $(k-1)$. For B-splines of order k, the value of:

$$
\sum_{j=1}^{k} y_{k,j}^c N_{k,j}(w)
\tag{E.3}
$$

must be found. Now $N_{k,j}$ is related to $N_{k-1,j}$ through the recurrence relation:

$$N_{k,j}(w) = \frac{1}{k-1}\left((w-(j-k))N_{k-1,j-1}(w) + (j-w)N_{k-1,j}(w)\right)$$

Hence Expression E.3 becomes:

$$\frac{1}{k-1}\sum_{j=2}^{k-1} y_{k,j}^c \left((w-(j-k))N_{k-1,j-1}(w) + (j-w)N_{k-1,j}(w)\right) +$$

$$\frac{1}{k-1}\left(y_{k,1}^c(1-w)N_{k-1,1}(w) + y_{k,k}^c w N_{k-1,k-1}(w)\right)$$

Similarly the centres of the B-splines of order k are offset from the centres of the B-splines of order $k-1$ by 0.5 and so the above expression becomes:

$$\frac{1}{k-1}\sum_{j=2}^{k-1}\left((y_{k-1,j-1}^c+0.5)(w-(j-k))N_{k-1,j-1}(w)+\right.$$

$$\left.(y_{k-1,j}^c-0.5)(j-w)N_{k-1,j}(w)\right) +$$

$$\frac{1}{k-1}\left((y_{k-1,1}^c-0.5)(1-w)N_{k-1,1}(w) + (y_{k-1,k-1}^c+0.5)w N_{k-1,k-1}(w)\right)$$

which can be rearranged to give:

$$\sum_{j=1}^{k-1}\left(y_{k-1,j}^c + \frac{w}{k-1} + \frac{k-1-2j}{2(k-1)}\right)N_{k-1,j}(w)$$

Using Equation E.2 and the normalising property of the B-splines, the above expression becomes:

$$w + \frac{w}{k-1} + \frac{1}{2} - \frac{1}{k-1}\left(\sum_{j=1}^{k-1} j N_{k-1,j}(w)\right)$$

Since the desired value w is $\in [0,1)$, $y_{k,j}^c = j - \frac{k}{2}$ and this expression becomes:

$$w + \frac{w}{k-1} + \frac{1}{2} - \frac{1}{k-1}\left(\sum_{j=1}^{k-1}\left(y_{k-1,j}^c + \frac{k-1}{2}\right)N_{k-1,j}(w)\right)$$

and hence using Equation E.2, this becomes:

$$w + \frac{w}{k-1} + \frac{1}{2} - \frac{w}{k-1} - \frac{1}{2}$$

which reduces to w, and therefore by induction it has been shown that for B-splines of order ≥ 2, defined on equispaced, simple knots:

$$w = \sum_{j=1}^{k} y_{k,j}^c N_{k,j}(w)$$

Similarly, it can also be shown that the above relationship holds for univariate B-splines of order 2 which are defined on simple knots (not necessarily equispaced). In this case, the centre of the B-spline is simply defined to be the knot λ_i when the support of the basis function is the interval $(\lambda_{i-1}, \lambda_{i+1})$.

Index